Bad Bug Book

U.S. Food & Drug Administration
Center for Food Safety & Applied Nutrition

Foodborne Pathogenic Microorganisms and Natural Toxins Handbook

International Medical Publishing

The Bad Bug Book has been created from the website
of the same name at the **www.fda.gov**.
Please refer to the Food & Drug Administration website
for the latest updates to the Bad Bug Book.

Trade names used in this book are the property
of their manufacturers.

First Printing.

The Bad Bug Book
US Food & Drug Administration
Center for Food Safety & Applied Nutition
Copyright 2004 (c) International Medical Publishing, Inc.

Published by:
International Medical Publishing
PO Box 479
McLean, Virginia 22101

www.MedicalPublishing.com
orders 800-530-4146

Inquiries:
contact@medicalpublishing.com

ISBN 1-58808-266-0
Retail $21.95

Table of Contents

PATHOGENIC BACTERIA

DATE DUE

BRODART, CO.

Cat. No. 23-221-003

Salmonella spp.

1. Name of Organism: *Salmonella* spp.

Salmonella is a rod-shaped, motile bacterium -- nonmotile exceptions *S. gallinarum* and *S. pullorum*--, nonsporeforming and Gram-negative. There is a widespread occurrence in animals, especially in poultry and swine. Environmental sources of the organism include water, soil, insects, factory surfaces, kitchen surfaces, animal feces, raw meats, raw poultry, and raw seafoods, to name only a few.

2. Nature of Acute Disease: *S. typhi* and the paratyphoid bacteria are normally caused sep ticemic and produce typhoid or typhoid-like fever in humans. Other forms of salmonellosis generally produce milder symptoms.

3. Nature of Disease: Acute symptoms -- Nausea, vomiting, abdominal cramps,minal diarrhea, fever, and headache. Chronic consequences -- arthritic symptoms may follow 3-4 weeks after onset of acute symptoms.

Onset time -- 6-48 hours.

Infective dose -- As few as 15-20 cells; depends upon age and health of host, and strain differences among the members of the genus.

Duration of symptoms -- Acute symptoms may last for 1 to 2 days or may be prolonged, again depending on host factors, ingested dose, and strain characteristics.

Cause of disease -- Penetration and passage of Salmonella organisms from gut lumen into epithelium of small intestine where inflammation occurs; there is evidence that an enterotoxin may be produced, perhaps within the enterocyte.

4. Diagnosis of Human Illness: Serological identification of culture isolated from stool.

5. Associated Foods: Raw meats, poultry, eggs, milk and dairy products, fish, shrimp, frog legs, yeast, coconut, sauces and salad dressing, cake mixes, cream-filled desserts and toppings, dried gelatin, peanut butter, cocoa, and chocolate.

Various Salmonella species have long been isolated from the outside of egg shells. The present situation with *S. enteritidis* is complicated by the presence of the organism inside the egg, in the yolk. This and other information strongly suggest vertical transmission, i.e., deposition of the organism in the yolk by an infected layer hen prior to shell deposition. Foods other than eggs have also caused outbreaks of *S. enteritidis* disease.

6. Relative Frequency of Disease: It is estimated that from 2 to 4 million cases of salmonellosis occur in the U.S. annually.

The incidence of salmonellosis appears to be rising both in the U.S. and in other industrialized nations. *S. enteritidis* isolations from humans have shown a dramatic rise in the past decade, particularly in the northeast United States (6-fold or more), and the increase in human infections is spreading south and west, with sporadic outbreaks in other regions.

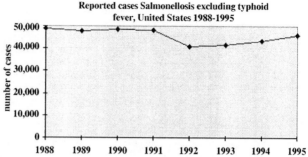

Reported cases Salmonellosis excluding typhoid fever, United States 1988-1995

Summary of Notifiable Diseases, United States MMWR 44(53): 1996 October 25

7. Complications:
S. typhi and *S. paratyphi* A, B, and C produce typhoid and typhoid-like fever in humans. Various organs may be infected, leading to lesions. The fatality rate of typhoid fever is 10% compared to less than 1% for most forms of salmonellosis. *S. dublin* has a 15% mortality rate when septicemic in the elderly, and *S. enteritidis* is demonstrating approximately a 3.6% mortality rate in hospital/nursing home outbreaks, with the elderly being particularly affected.

Salmonella septicemia has been associated with subsequent infection of virtually every organ system.

Postenteritis reactive arthritis and *Reiter's syndrome* have also been reported to occur generally after 3 weeks. Reactive arthritis may occur with a frequency of about 2% of culture-proven cases. Septic arthritis, subsequent or coincident with septicemia, also occurs and can be difficult to treat.

8. Target Populations:
All age groups are susceptible, but symptoms are most severe in the elderly, infants, and the infirm. AIDS patients suffer salmonellosis frequently (estimated 20-fold more than general population) and suffer from recurrent episodes.

9. Foods Analysis:
Methods have been developed for many foods having prior history of Salmonella contamination. Although conventional culture methods require 5 days for presumptive results, several rapid methods are available which require only 2 days.

10. Selected Outbreaks:
In 1985, a salmonellosis outbreak involving 16,000 confirmed cases in 6 states was caused by low fat and whole milk from one Chicago dairy. This was the largest outbreak of foodborne salmonellosis in the U.S. FDA inspectors discovered that the pasteuriza-

tion equipment had been modified to facilitate the running off of raw milk, resulting in the pasteurized milk being contaminated with raw milk under certain conditions. The dairy has subsequently disconnected the cross-linking line. Persons on antibiotic therapy were more apt to be affected in this outbreak.

In August and September, 1985, *S. enteritidis* was isolated from employees and patrons of three restaurants of a chain in Maryland. The outbreak in one restaurant had at least 71 illnesses resulting in 17 hospitalizations. Scrambled eggs from a breakfast bar were epidemiologically implicated in this outbreak and in possibly one other of the three restaurants. The plasmid profiles of isolates from patients all three restaurants matched.
The Centers for Disease Control (CDC) has recorded more than 120 outbreaks of S. enteritidis to date, many occurring in restaurants, and some in nursing homes, hospitals and prisons.

In 1984, 186 cases of salmonellosis (*S. enteritidis*) were reported on 29 flights to the United States on a single international airline. An estimated 2,747 passengers were affected overall. No specific food item was implicated, but food ordered from the first class menu was strongly associated with disease.

S. enteritidis outbreaks continue to occur in the U.S. (Table 1). The CDC estimates that 75% of those outbreaks are associated with the consumption of raw or inadequately cooked Grade A whole shell eggs. The U.S. Department of Agriculture published Regulations on February 16, 1990, in the Federal Register establishing a mandatory testing program for egg-producing breeder flocks and commercial flocks implicated in causing human illnesses. This testing should lead to a reduction in cases of gastroenteritis caused by the consumption of Grade A whole shell eggs.

Salmonellosis associated with a Thanksgiving Dinner in Nevada in 1995 is reported in MMWR 45(46):1996 Nov 22.

MMWR 45(34):1996 Aug 30 reports on several outbreaks of *Salmonella enteritidis* infection associated with the consumption of raw shell eggs in the United States from 1994 to 1995.

A report of an outbreak of *Salmonella* Serotype Typhimurium infection associated with the consumption of raw ground beef may be found in MMWR 44(49):1995 Dec 15.

MMWR 44(42):1995 Oct 27 reports on an outbreak of Salmonellosis associated with beef jerky in New Mexico in 1995.

The report on the outbreak of Salmonella from commercially prepared ice cream is found in MMWR 43(40):1994 Oct 14.

An outbreak of *S. enteritidis* in homemade ice cream is reported in this MMWR 43(36):1994 Sep 16.

A series of *S. enteritidis* outbreaks in California are summarized in the following MMWR 42(41):1993 Oct 22.

For information on an outbreak of Salmonella Serotype Tennessee in Powdered Milk Products and Infant Formula -- see this MMWR 42(26):1993 Jul 09.

Summaries of Salmonella outbreaks associated with Grade A eggs are reported in MMWR 37(32):1988 Aug 19 and MMWR 39(50):1990 Dec 21.

For more information on recent outbreaks see the Morbidity and Mortality Weekly Reports from CDC.

11. Education:

The CDC provides an informational brochure on preventing Salmonella enteritidis infection.

Safe Egg Handling (FDA Consumer Sep - Oct 1998)

12. Other Resources:

A Loci index for genome *Salmonella enteritidis* is available from GenBank.

Clostridium botulinum

1. Name of organism: *Clostridium botulinum*

Clostridium botulinum is an anaerobic, Gram-positive, spore-forming rod that produces a potent neurotoxin. The spores are heat-resistant and can survive in foods that are incorrectly or minimally processed. Seven types (A, B, C, D, E, F and G) of botulism are recognized, based on the antigenic specificity of the toxin produced by each strain. Types A, B, E and F cause human botulism. Types C and D cause most cases of botulism in animals. Animals most commonly affected are wild fowl and poultry, cattle, horses and some species of fish. Although type G has been isolated from soil in Argentina, no outbreaks involving it have been recognized.

Foodborne botulism (as distinct from wound botulism and infant botulism) is a severe type of food poisoning caused by the ingestion of foods containing the potent neurotoxin formed during growth of the organism. The toxin is heat labile and can be destroyed if heated at 80oC for 10 minutes or longer. The incidence of the disease is low, but the disease is of considerable concern because of its high mortality rate if not treated immediately and properly. Most of the 10 to 30 outbreaks that are reported annually in the United States are associated with inadequately processed, home-canned foods, but occasionally commercially produced foods have been involved in outbreaks. Sausages, meat products, canned vegetables and seafood products have been the most frequent vehicles for human botulism.

The organism and its spores are widely distributed in nature. They occur in both cultivated and forest soils, bottom sediments of streams, lakes, and coastal waters, and in the intestinal tracts of fish and mammals, and in the gills and viscera of crabs and other shellfish.

2. Name of the Disease: Four types of botulism are recognized: foodborne, infant, wound, and a form of botulism whose classification is as yet undetermined. Certain foods have been reported as sources of spores in cases of infant botulism and the undetermined category; wound botulism is not related to foods.

Foodborne botulism is the name of the disease (actually a foodborne intoxication) caused by the consumption of foods containing the neurotoxin produced by *C. botulinum*.

Infant botulism, first recognized in 1976, affects infants under 12 months of age. This type of botulism is caused by the ingestion of *C. botulinum* spores which colonize and produce toxin in the intestinal tract of infants (intestinal toxemia botulism). Of the various potential environmental sources such as soil, cistern water, dust and foods, honey is the one dietary reservoir of *C. botulinum* spores thus far definitively linked to infant botulism by both laboratory and epidemiologic studies. The number of confirmed infant botulism cases has increased significantly as a result of greater awareness by health officials since its recognition in 1976. It is now internationally recognized, with cases being reported in more countries.

Wound botulism is the rarest form of botulism. The illness results when *C. botulinum* by itself or with other microorganisms infects a wound and produces toxins which reach other parts of the body via the blood stream. Foods are not involved in this type of botulism.

Undetermined category of botulism involves adult cases in which a specific food or wound source cannot be identified. It has been suggested that some cases of botulism assigned to this category might result from intestinal colonization in adults, with in vivo production of toxin. Reports in the medical literature suggest the existence of a form of botulism similar to infant botulism, but occurring in adults. In these cases, the patients had surgical alterations of the gastrointestinal tract and/or antibiotic therapy. It is proposed that these procedures may have altered the normal gut flora and allowed *C. botulinum* to colonize the intestinal tract.

3. Nature of the Disease:

Infective dose -- a very small amount (a few nanograms) of toxin can cause illness.

Onset of symptoms in foodborne botulism is usually 18 to 36 hours after ingestion of the food containing the toxin, although cases have varied from 4 hours to 8 days. Early signs of intoxication consist of marked lassitude, weakness and vertigo, usually followed by double vision and progressive difficulty in speaking and swallowing. Difficulty in breathing, weakness of other muscles, abdominal distention, and constipation may also be common symptoms.

Clinical symptoms of infant botulism consist of constipation that occurs after a period of normal development. This is followed by poor feeding, lethargy, weakness, pooled oral secretions, and wail or altered cry. Loss of head control is striking. Recommended treatment is primarily supportive care. Antimicrobial therapy is not recommended. Infant botulism is diagnosed by demonstrating botulinal toxins and the organism in the infants' stools.

4. Diagnosis of Human Illness:

Although botulism can be diagnosed by clinical symptoms alone, differentiation from other diseases may be difficult. The most direct and effective way to confirm the clinical diagnosis of botulism in the laboratory is to demonstrate the presence of toxin in the serum or feces of the patient or in the food which the patient consumed. Currently, the most sensitive and widely used method for detecting toxin is the mouse neutralization test. This test takes 48 hours. Culturing of specimens takes 5-7 days.

5. Associated Foods:

The types of foods involved in botulism vary according to food preservation and eating habits in different regions. Any food that is conducive to outgrowth and toxin production, that when processed allows spore survival, and is not subsequently heated before consumption can be associated with botulism. Almost any type of food that is not very acidic (pH above 4.6) can support growth and toxin production by *C. botulinum*. Botulinal toxin has been demonstrated in a considerable variety of foods, such as canned corn, peppers, green beans, soups, beets, asparagus, mushrooms, ripe olives, spinach, tuna fish, chicken and chicken livers and liver pate, and luncheon meats, ham, sausage, stuffed eggplant, lobster, and smoked and salted fish.

6. Frequency: The incidence of the disease is low, but the mortality rate is high if not treated immediately and properly. There are generally between 10 to 30 outbreaks a year in the United States. Some cases of botulism may go undiagnosed because symptoms are transient or mild, or misdiagnosed as Guillain-Barre syndrome.

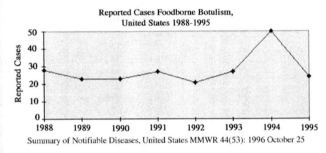

Reported Cases Foodborne Botulism, United States 1988-1995

Summary of Notifiable Diseases, United States MMWR 44(53): 1996 October 25

7. The Usual Course of Disease and Complications: Botulinum toxin causes flaccid paralysis by blocking motor nerve terminals at the myoneural junction. The flaccid paralysis progresses symmetrically downward, usually starting with the eyes and face, to the throat, chest and extremities. When the diaphragm and chest muscles become fully involved, respiration is inhibited and death from asphyxia results. Recommended treatment for foodborne botulism includes early administration of botulinal antitoxin (available from CDC) and intensive supportive care (including mechanical breathing assistance).

8. Target Populations: All people are believed to be susceptible to the foodborne intoxication.

9. Food Analysis: Since botulism is foodborne and results from ingestion of thet toxin of C. botulinum, determination of the source of an outbreak is based on detection and identification of toxin in the food involved. The most widely accepted method is the injection of extracts of the food into passively immunized mice (mouse neutralization test). The test takes 48 hours. This analysis is followed by culturing all suspect food in an enrichment medium for the detection and isolation of the causative organism. This test takes 7 days.

10. Selected Outbreaks: Two separate outbreaks of botulism have occurred involving commercially canned salmon. Restaurant foods such as sauteed onions, chopped bottled garlic, potato salad made from baked potatoes and baked potatoes themselves have been responsible for a number of outbreaks. Also, smoked fish, both hot and cold-smoke (e.g., Kapchunka) have caused outbreaks of type E botulism. In October and November, 1987, 8 cases of type E botulism occurred, 2 in New York City and 6 in Israel. All 8 patients had consumed Kapchunka, an uneviscerated, dry-salted, air-dried, whole whitefish. The product was made in New York City and some of it was transported by individuals to Israel. All 8 patients with botulism developed symptoms within 36 hours of consuming the Kapchunka. One female died, 2 required breathing assistance, 3 were treated therapeutically with antitoxin, and 3 recovered spontaneously. The Kapchunka involved in this outbreak contained high levels of type E botulinal toxin despite salt levels that exceeded those sufficient to inhibit C. botulinum type E outgrowth. One possible explanation was that the fish contained low salt levels when air-dried at room temperature, became toxic, and then were re-brined. Regulations were published to prohibit the processing, distribution and sale of Kapchunka and Kapchunka-type products in the United States.

A bottled chopped garlic-in-oil mix was responsible for three cases of botulism in Kingston, N.Y. Two men and a woman were hospitalized with botulism after consuming a chopped garlic-in-oil mix that had been used in a spread for garlic bread. The bottled chopped garlic relied solely on refrigeration to ensure safety and did not contain any additional antibotulinal additives or barriers. The FDA has ordered companies to stop making the product and to withdraw from the market any garlic-in-oil mix which does not include microbial inhibitors or acidifying agents and does not require refrigeration for safety.

Since botulism is a life-threatening disease, FDA always initiates a Class I recall. January 1992

An incident of foodborne botulism in Oklahoma is reported in MMWR 44(11):1995 Mar 24.

A botulism type B outbreak in Italy associated with eggplant in oil is reported in MMWR 44(2):1995 Jan 20.

The botulism outbreak associated with salted fish mentioned above is reported in greater detail in MMWR 36(49):1987 Dec 18.

For more information on recent outbreaks see the Morbidity and Mortality Weekly Reports from CDC.

11. Education: The December 1995 issue of "FDA Consumer" has an article titled Botulism Toxin: a Poison That Can Heal which discusses Botulism toxin with an emphasis on its medical uses.

12. Other Resources: FDA Warns Against Consuming Certain Italian Mascarpone Cream Cheese Because of Potential Serious Botulism Risk (Sept. 9, 1996)

A Loci index for genome Clostridium botulinum is available from GenBank.

Staphylococcus aureus

1. Name of Organism: *Staphylococcus aureus*

S. aureus is a spherical bacterium (coccus) which on microscopic examination appears in pairs, short chains, or bunched, grape-like clusters. These organisms are Gram-positive. Some strains are capable of producing a highly heat-stable protein toxin that causes illness in humans.

2. Name of Acute Disease: Staphylococcal food poisoning (staphyloenterotoxicosis; staphy loenterotoxemia) is the name of the condition caused by the enterotoxins which some strains of *S. aureus* produce.

3. Nature of the Disease: The onset of symptoms in staphylococcal food poisoning is usually rapid and in many cases acute, depending on individual susceptibility to the toxin, the amount of contaminated food eaten, the amount of toxin in the food ingested, and the general health of the victim. The most common symptoms are nausea, vomiting, retching, abdominal cramping, and prostration. Some individuals may not always demonstrate all the symptoms associated with the illness. In more severe cases, headache, muscle cramping, and transient changes in blood pressure and pulse rate may occur. Recovery generally takes two days, However, it us not unusual for complete recovery to take three days and sometimes longer in severe cases.

Infective dose--a toxin dose of less than 1.0 microgram in contaminated food will produce symptoms of staphylococcal intoxication. This toxin level is reached when *S. aureus* populations exceed 100000 per gram.

4. Diagnosis of Human Illness: In the diagnosis of staphylococcal foodborne illness, proper inter views with the victims and gathering and analyzing epidemiologic data are essential. Incriminated foods should be collected and examined for staphylococci. The presence of relatively large numbers of enterotoxigenic staphylococci is good circumstantial evidence that the food contains toxin. The most conclusive test is the linking of an illness with a specific food or in cases where multiple vehicles exist, the detection of the toxin in the food sample(s). In cases where the food may have been treated to kill the staphylococci, as in pasteurization or heating, direct microscopic observation of the food may be an aid in the diagnosis. A number of serological methods for determining the enterotoxigenicity of *S. aureus* isolated from foods as well as methods for the separation and detection of toxins in foods have been developed and used successfully to aid in the diagnosis of the illness. Phage typing may also be useful when viable staphylococci can be isolated from the incriminated food, from victims, and from suspected carrier such as food handlers.

5. Foods Incriminated: Foods that are frequently incriminated in staphylococcal food poisoning include meat and meat products; poultry and egg products;

salads such as egg, tuna, chicken, potato, and macaroni; bakery products such as cream-filled pastries, cream pies, and chocolate eclairs; sandwich fillings; and milk and dairy products. Foods that require considerable handling during preparation and that are kept at slightly elevated temperatures after preparation are frequently involved in staphylococcal food poisoning.

Staphylococci exist in air, dust, sewage, water, milk, and food or on food equipment, environmental surfaces, humans, and animals. Humans and animals are the primary reservoirs. Staphylococci are present in the nasal passages and throats and on the hair and skin of 50 percent or more of healthy individuals. This incidence is even higher for those who associate with or who come in contact with sick individuals and hospital environments. Although food handlers are usually the main source of food contamination in food poisoning outbreaks, equipment and environmental surfaces can also be sources of contamination with *S. aureus*. Human intoxication is caused by ingesting enterotoxins produced in food by some strains of *S. aureus*, usually because the food has not been kept hot enough (60°C, 140°F, or above) or cold enough (7.2°C, 45°F, or below).

6. Frequency of Illness:

The true incidence of staphylococcal food poisoning is unknown for a number of reasons, including poor responses from victims during interviews with health officials; misdiagnosis of the illness, which may be symptomatically similar to other types of food poisoning (such as vomiting caused by *Bacillus cereus* toxin); inadequate collection of samples for laboratory analyses; and improper laboratory examination. Of the bacterial pathogens causing foodborne illnesses in the U.S. (127 outbreaks, 7,082 cases recorded in 1983), 14 outbreaks involving 1,257 cases were caused by S. aureus. These outbreaks were followed by 11 outbreaks (1,153 cases) in 1984, 14 outbreaks (421 cases) in 1985, 7 outbreaks (250 cases) in 1986 and one reported outbreak (100 cases) in 1987.

7. Complications:

Death from staphylococcal food poisoning is very rare, although such cases have occurred among the elderly, infants, and severely debilitated persons.

8. Target Population:

All people are believed to be susceptible to this type of bacterial intoxication; however, intensity of symptoms may vary.

9. Analysis of Foods:

For detecting trace amounts of staphylococcal enterotoxin in foods incriminated in food poisoning, the toxin must be separated from food constituents and concentrated before identification by specific precipitation with antiserum (antienterotoxin) as follows. Two principles are used for the purpose: (1) the selective adsorption of the enterotoxin from an extract of the food onto ion exchange resins and (2) the use of physical and chemical procedures for the selective removal of food constituents from the extract, leaving the enterotoxin(s) in solution. The use of these techniques and concentration of the resulting products (as much as possible) has made it possible to detect small amounts of enterotoxin in food.

There are developed rapid methods based on monoclonal antibodies (e.g., ELISA, Reverse Passive Latex Agglutination), which are being evaluated for their efficacy in the detection of enterotoxins in food. These rapid methods can detect approximately 1.0 nanogram of toxin/g of food.

10. Typical Outbreak: 1,364 children became ill out of a total of 5,824 who had eaten lunch served at 16 elementary schools in Texas. The lunches were prepared in a central kitchen and transported to the schools by truck. Epidemiological studies revealed that 95% of the children who became ill had eaten a chicken salad. The afternoon of the day preceding the lunch, frozen chickens were boiled for 3 hours. After cooking, the chickens were deboned, cooled to room temperature with a fan, ground into small pieces, placed into 12-inch-deep aluminum pans and stored overnight in a walk-in refrigerator at 42-45°F.

The following morning, the remaining ingredients of the salad were added and the mixture was blended with an electric mixer. The food was placed in thermal containers and transported to the various schools at 9:30 AM to 10:30 AM, where it was kept at room temperature until served between 11:30 AM and noon. Bacteriological examination of the chicken salad revealed the presence of large numbers of *S. aureus*.

Contamination of the chicken probably occurred when it was deboned. The chicken was not cooled rapidly enough because it was stored in 12-inch-deep layers. Growth of the staphylococcus probably occurred also during the period when the food was kept in the warm classrooms. Prevention of this incident would have entailed screening the individuals who deboned the chicken for carriers of the staphylococcus, more rapid cooling of the chicken, and adequate refrigeration of the salad from the time of preparation to its consumption.

11. Atypical Outbreaks: In 1989, multiple staphylococcal foodborne diseases were associated with the consumption of canned mushrooms. (CDC Morbidity and Mortality Weekly Report, June 23, 1989, Vol. 38, #24.)

Starkville, Mississippi. On February 13, 22 people became ill with gastroenteritis several hours after eating at a university cafeteria. Symptoms included nausea, vomiting, diarrhea, and abdominal cramps. Nine people were hospitalized. Canned mushrooms served with omelets and hamburgers were associated with illness. No deficiencies in food handling were found. Staphylococcal enterotoxin type A was identified in a sample of implicated mushrooms from the omelet bar and in unopened cans from the same lot.

Queens, New York, On February 28, 48 people became ill a median of 3 hours after eating lunch in a hospital employee cafeteria. One person was hospitalized. Canned mushrooms served at the salad bar were epidemiologically implicated. Two unopened cans of mushrooms from the same lot as the implicated can contained staphylococcal enterotoxin A.

McKeesport, Pennsylvania. On April 17, 12 people became ill with gastroenteritis a median of 2 hours after eating lunch or dinner at a restaurant. Two people were hospitalized. Canned mushrooms, consumed on pizza or with a parmigiana sauce, were associated with illness. No deficiencies were found in food preparation or storage. Staphylococcal enterotoxin was found in samples of remaining mushrooms and in unopened cans from the same lot.

Philipsburg, Pennsylvania. On April 22, 20 people developed illness several hours after eating food from a take-out pizzeria. Four people were hospitalized. Only pizza served with canned mushrooms was associated with illness. Staphylococcal enterotoxin was found in a sample of mushrooms from the pizzeria and in unopened cans with the same lot number.

For more information on recent outbreaks see the Morbidity and Mortality Weekly Reports from CDC.

12. Other Resources: A Loci index for genome *Staphylococcus aureus* is available from GenBank.

Campylobacter jejuni

1. Name of Organism: *Campylobacter jejuni* (formerly known as *Campylobacter fetus* subsp. jejuni)

Campylobacter jejuni is a Gram-negative slender, curved, and motile rod. It is a microaerophilic organism, which means it has a requirement for reduced levels of oxygen. It is relatively fragile, and sensitive to environmental stresses (e.g., 21% oxygen, drying, heating, disinfectants, acidic conditions). Because of its microaerophilic characteristics the organism requires 3 to 5% oxygen and 2 to 10% carbon dioxide for optimal growth conditions. This bacterium is now recognized as an important enteric pathogen. Before 1972, when methods were developed for its isolation from feces, it was believed to be primarily an animal pathogen causing abortion and enteritis in sheep and cattle. Surveys have shown that *C. jejuni* is the leading cause of bacterial diarrheal illness in the United States. It causes more disease than *Shigella* spp. and *Salmonella* spp. combined.

Although *C. jejuni* is not carried by healthy individuals in the United States or Europe, it is often isolated from healthy cattle, chickens, birds and even flies. It is sometimes present in non-chlorinated water sources such as streams and ponds.

Because the pathogenic mechanisms of *C. jejuni* are still being studied, it is difficult to differentiate pathogenic from nonpathogenic strains. However, it appears that many of the chicken isolates are pathogens.

2. Name of Disease: Campylobacteriosis is the name of the illness caused by *C. jejuni*. It is also often known as campylobacter enteritis or gastroenteritis.

3. Major Symptoms: *C. jejuni* infection causes diarrhea, which may be watery or sticky and can contain blood (usually occult) and fecal leukocytes (white cells). Other symptoms often present are fever, abdominal pain, nausea, headache and muscle pain. The illness usually occurs 2-5 days after ingestion of the contaminated food or water. Illness generally lasts 7-10 days, but relapses are not uncommon (about 25% of cases). Most infections are self-limiting and are not treated with antibiotics. However, treatment with erythromycin does reduce the length of time that infected individuals shed the bacteria in their feces.

The infective dose of *C. jejuni* is considered to be small. Human feeding studies suggest that about 400-500 bacteria may cause illness in some individuals, while in others, greater numbers are required. A conducted volunteer human feeding study suggests that host susceptibility also dictates infectious dose to some degree. The pathogenic mechanisms of *C. jejuni* are still not completely understood, but it does produce a heat-labile toxin that may cause diarrhea. *C. jejuni* may also be an invasive organism.

4. Isolation Procedures: *C. jejuni* is usually present in high numbers in the diarrheal stools of individuals, but isolation requires special antibiotic-containing media and a special microaerophilic atmosphere (5% oxygen). However, most clinical laboratories are equipped to isolate Campylobacter spp. if requested.

5. Associated Foods: *C. jejuni* frequently contaminates raw chicken. Surveys show that 20 to 100% of retail chickens are contaminated. This is not overly surprising since many healthy chickens carry these bacteria in their intestinal tracts. Raw milk is also a source of infections. The bacteria are often carried by healthy cattle and by flies on farms. Non-chlorinated water may also be a source of infections. However, properly cooking chicken, pasteurizing milk, and chlorinating drinking water will kill the bacteria.

6. Frequency of the Disease: *C. jejuni* is the leading cause of bacterial diarrhea in the U.S. There are probably numbers of cases in excess of the estimated cases of salmonellosis (2- to 4,000,000/year).

7. Complications: Complications are relatively rare, but infections have been associated with reactive arthritis, hemolytic uremic syndrome, and following septicemia, infections of nearly any organ. The estimated case/fatality ratio for all *C. jejuni* infections is 0.1, meaning one death per 1,000 cases. Fatalities are rare in healthy individuals and usually occur in cancer patients or in the otherwise debilitated. Only 20 reported cases of septic abortion induced by *C. jejuni* have been recorded in the literature.

Meningitis, recurrent colitis, acute cholecystitis and Guillain-Barre syndrome are very rare complications.

8. Target Populations: Although anyone can have a *C. jejuni* infection, children under 5 years and young adults (15-29) are more frequently afflicted than other age groups. Reactive arthritis, a rare complication of these infections, is strongly associated with people who have the human lymphocyte antigen B27 (HLA-B27).

9. Recovery from Foods: Isolation of *C. jejuni* from food is difficult because the bacteria are usually present in very low numbers (unlike the case of diarrheal stools in which 10/6 bacteria/gram is not unusual). The methods require an enrichment broth containing antibiotics, special antibiotic-containing plates and a microaerophilic atmosphere generally a microaerophilic atmosphere with 5% oxygen and an elevated concentration of carbon dioxide (10%). Isolation can take several days to a week.

10. Selected Outbreaks: Usually outbreaks are small (less than 50 people), but in Bennington, VT a large outbreak involving about 2,000 people occurred while the town was temporarily using an non-chlorinated water source as a water supply. Several small outbreaks have been reported among children who were taken on a class trip to a dairy and given raw milk to drink. An outbreak was also associated with consumption of raw clams. However, a survey showed that about 50% of infections are associated with either eating inadequately cooked or recontaminated chicken meat or handling chickens. It is the leading bacterial cause of sporadic (non-clustered cases) diarrheal disease in the U.S.

In April, 1986, an elementary school child was cultured for bacterial pathogens (due to bloody diarrhea), and *C. jejuni* was isolated. Food consumption/gastrointestinal illness questionnaires were administered to other students and faculty at the school. In all, 32 of 172 students reported symptoms of diarrhea (100%), cramps (80%), nausea (51%), fever (29%), vomiting (26%), and bloody stools (14%). The food questionnaire clearly implicated milk as the common source, and a dose/response was evident (those drinking more milk were more likely to be ill). Investigation of the dairy supplying the milk showed that they vat pasteurized the milk at 135°F for 25 minutes rather than the required 145°F for 30 minutes. The dairy processed surplus raw milk for the school, and this milk had a high somatic cell count. Cows from the herd supplying the dairy had *C. jejuni* in their feces. This outbreak points out the variation in symptoms which may occur with campylobacteriosis and the absolute need to adhere to pasteurization time/temperature standards.

Although other *Campylobacter* spp. have been implicated in human gastroenteritis (e.g. *C. laridis*, *C. hyointestinalis*), it is believed that 99% of the cases are caused by C. jejuni.

Information regarding an outbreak of Campylobacter in New Zealand is found in this MMWR 40(7):1991 Feb 22.

For more information on recent outbreaks see the Morbidity and Mortality Weekly Reports from CDC.

11. Education: The Food Safety Inspection Service of the U.S. Department of Agriculture has produced a background document on *Campylobacter*.

12. Other Resources: A Loci index for genome *Campylobacter jejuni* is available from GenBank.

Yersinia enterocolitica

1. Name of Organism: *Yersinia enterocolitica* (and *Yersinia pseudotuberculosis*)

Y. enterocolitica, a small rod-shaped, Gram-negative bacterium, is often isolated from clinical specimens such as wounds, feces, sputum and mesenteric lymph nodes. However, it is not part of the normal human flora. *Y. pseudotuberculosis* has been isolated from the diseased appendix of humans.

Both organisms have often been isolated from such animals as pigs, birds, beavers, cats, and dogs. Only *Y. enterocolitica* has been detected in environmental and food sources, such as ponds, lakes, meats, ice cream, and milk. Most isolates have been found not to be pathogenic.

2. Name of Disease: Yersiniosis

There are 3 pathogenic species in the genus Yersinia, but only *Y. enterocolitica* and *Y. pseudotuberculosis* cause gastroenteritis. To date, no foodborne outbreaks caused by *Y. pseudotuberculosis* have been reported in the United States, but human infections transmitted via contaminated water and foods have been reported in Japan. *Y. pestis*, the causative agent of "the plague," is genetically very similar to *Y. pseudotuberculosis* but infects humans by routes other than food.

3. Nature of Disease: Yersiniosis is frequently characterized by such symptoms as gastroenteritis with diarrhea and/or vomiting; however, fever and abdominal pain are the hallmark symptoms. *Yersinia* infections mimic appendicitis and mesenteric lymphadenitis, but the bacteria may also cause infections of other sites such as wounds, joints and the urinary tract.

4. Infective dose: Unknown.

Illness onset is usually between 24 and 48 hours after ingestion, which (with food or drink as vehicle) is the usual route of infection.

5. Diagnosis of Human Illness: Diagnosis of yersiniosis begins with isolation of the organism from the human host's feces, blood, or vomit, and sometimes at the time of appendectomy. Confirmation occurs with the isolation, as well as biochemical and serological identification, of *Y. enterocolitica* from both the human host and the ingested foodstuff. Diarrhea is reported to occur in about 80% of cases; abdominal pain and fever are the most reliable symptoms.

Because of the difficulties in isolating *yersiniae* from feces, several countries rely on serology. Acute and convalescent patient sera are titered against the suspect serotype of *Yersinia spp.*

Yersiniosis has been misdiagnosed as Crohn's disease (regional enteritis) as well as appendicitis.

6. Associated Foods: Strains of *Y. enterocolitica* can be found in meats (pork, beef, lamb, etc.), oysters, fish, and raw milk. The exact cause of the food contamination is unknown. However, the prevalence of this organism in the soil and water and in animals such as beavers, pigs, and squirrels, offers ample opportunities for it to enter our food supply. Poor sanitation and improper sterilization techniques by food handlers, including improper storage, cannot be overlooked as contributing to contamination.

7. Frequency of the Disease: Yersiniosis does not occur frequently. It is rare unless a breakdown occurs in food processing techniques. CDC estimates that about 17,000 cases occur annually in the USA. Yersiniosis is a far more common disease in Northern Europe, Scandinavia, and Japan.

8. Complications: The major "complication" is the performance of unnecessary appendectomies, since one of the main symptoms of infections is abdominal pain of the lower right quadrant.

Both *Y. enterocolitica* and *Y. pseudotuberculosis* have been associated with reactive arthritis, which may occur even in the absence of obvious symptoms. The frequency of such postenteritis arthritic conditions is about 2-3%.

Another complication is bacteremia (entrance of organisms into the blood stream), in which case the possibility of a disseminating disease may occur. This is rare, however, and fatalities are also extremely rare.

9. Target Populations: The most susceptible populations for the main disease and possible complications are the very young, the debilitated, the very old and persons undergoing immunosuppressive therapy. Those most susceptible to postenteritis arthritis are individuals with the antigen HLA-B27 (or related antigens such as B7).

10. Food Analysis: The isolation method is relatively easy to perform, but in some instances, cold enrichment may be required. *Y. enterocolitica* can be presumptively identified in 36-48 hours. However, confirmation may take 14-21 days or more. Determination of pathogenicity is more complex. The genes encoding for invasion of mammalian cells are located on the chromosome while a 40-50 MDal plasmid encodes most of the other virulence associated phenotypes. The 40-50 MDal plasmid is present in almost all the pathogenic *Yersinia* species, and the plasmids appear to be homologous.

10. Selected Outbreaks: 1976. A chocolate milk outbreak in Oneida County, N.Y. involving school children (first reported yersiniosis incident in the United States in which a food vehicle was identified). A research laboratory was set up by FDA to investigate and study *Y. enterocolitica* and *Y. pseudotuberculosis* in the human food supply.

Dec. 1981 - Feb. 1982. *Y. enterocolitica* enteritis in King County, Washington caused by ingestion of tofu, a soybean curd. FDA investigators and researchers determined the source of the infection to be an non-chlorinated water supply. Manufacturing was halted until uncontaminated product was produced.

June 11 to July 21, 1982. *Y. enterocolitica* outbreak in Arkansas, Tennessee, and Mississippi associated with the consumption of pasteurized milk. FDA personnel participated in the investigation, and presumptively identified the infection source to be externally contaminated milk containers.

A report of Yersinia enterocolitica incidents associated with raw chitterlings may be found in MMWR 39(45):1990 Nov 16

For more information on recent outbreaks see the Morbidity and Mortality Weekly Reports from CDC.

12. Other Resources: A Loci index for genome *Yersinia enterocolitica* and Loci index for genome *Yersinia pseudotuberculosis* are available from GenBank.

Listeria monocytogenes

1. Name of Organism: *Listeria monocytogenes*

This is a Gram-positive bacterium, motile by means of flagella. Some studies suggest that 1-10% of humans may be intestinal carriers of *L. monocytogenes*. It has been found in at least 37 mammalian species, both domestic and feral, as well as at least 17 species of birds and possibly some species of fish and shellfish. It can be isolated from soil, silage, and other environmental sources. *L. monocytogenes* is quite hardy and resists the deleterious effects of freezing, drying, and heat remarkably well for a bacterium that does not form spores. Most *L. monocytogenes* are pathogenic to some degree.

2. Name of Acute Disease:

Listeriosis is the name of the general group of disorders caused by *L. monocytogenes*.

3. Nature of Disease:

Listeriosis is clinically defined when the organism is isolated from blood, cerebrospinal fluid, or an otherwise normally sterile site (e.g. placenta, fetus).

The manifestations of listeriosis include septicemia, meningitis (or meningoencephalitis), encephalitis, and intrauterine or cervical infections in pregnant women, which may result in spontaneous abortion (2nd/3rd trimester) or stillbirth. The onset of the afore-mentioned disorders is usually preceded by influenza-like symptoms including persistent fever. It was reported that gastrointestinal symptoms such as nausea, vomiting, and diarrhea may precede more serious forms of listeriosis or may be the only symptoms expressed. Gastrointestinal symptoms were epidemiologically associated with use of antacids or cimetidine. The onset time to serious forms of listeriosis is unknown but may range from a few days to three weeks. The onset time to gastrointestinal symptoms is unknown but is probably greater than 12 hours.

The infective dose of *L. monocytogenes* is unknown but is believed to vary with the strain and susceptibility of the victim. From cases contracted through raw or supposedly pasteurized milk, it is safe to assume that in susceptible persons, fewer than 1,000 total organisms may cause disease. *L. monocytogenes* may invade the gastrointestinal epithelium. Once the bacterium enters the host's monocytes, macrophages, or polymorphonuclear leukocytes, it is bloodborne (septicemic) and can grow. Its presence intracellularly in phagocytic cells also permits access to the brain and probably transplacental migration to the fetus in pregnant women. The pathogenesis of *L. monocytogenes* centers on its ability to survive and multiply in phagocytic host cells.

4. Diagnosis of Human Illness:

Listeriosis can only be positively diagnosed by culturing the organism from blood, cerebrospinal fluid, or stool (although the latter is difficult and of limited value).

5. Associated Foods: *L. monocytogenes* has been associated with such foods as raw milk, supposedly pasteurized fluid milk, cheeses (particularly soft-ripened varieties), ice cream, raw vegetables, fermented raw-meat sausages, raw and cooked poultry, raw meats (all types), and raw and smoked fish. Its ability to grow at temperatures as low as 3oC permits multiplication in refrigerated foods.

6. Frequency of the Disease: The 1987 incidence data prospectively collected by CDC suggests that there are at least 1600 cases of listeriosis with 415 deaths per year in the U.S. The vast majority of cases are sporadic, making epidemiological links to food very difficult.

7. Complications: Most healthy persons probably show no symptoms. The "complications" are the usual clinical expressions of the disease.

When listeric meningitis occurs, the overall mortality may be as high as 70%; from septicemia 50%, from perinatal/neonatal infections greater than 80%. In infections during pregnancy, the mother usually survives. Successful treatment with parenteral penicillin or ampicillin has been reported. Trimethoprim-sulfamethoxazole has been shown effective in patients allergic to penicillin.

8. Target Populations: The main target populations for listeriosis are:

- pregnant women/fetus - perinatal and neonatal infections;
- persons immunocompromised by corticosteroids, anticancer drugs, graft suppression therapy, AIDS;
- cancer patients - leukemic patients particularly;
- less frequently reported - diabetic, cirrhotic, asthmatic, and ulcerative colitis patients;
- the elderly;
- normal people--some reports suggest that normal, healthy people are at risk, although antacids or cimetidine may predispose. A listerosis outbreak in Switzerland involving cheese suggested that healthy uncompromised individuals could develop the disease, particularly if the foodstuff was heavily contaminated with the organism.

9. Food Analysis: The methods for analysis of food are complex and time consuming. The present FDA method, revised in September, 1990, requires 24 and 48 hours of enrichment, followed by a variety of other tests. Total time to identification is from 5 to 7 days, but the announcement of specific nonradiolabled DNA probes should soon allow a simpler and faster confirmation of suspect isolates.

Recombinant DNA technology may even permit 2-3 day positive analysis in the future. Currently, FDA is collaborating in adapting its methodology to quantitate very low numbers of the organisms in foods.

10. Selected Outbreaks: Outbreaks include the California episode in 1985, which was due to Mexican-style cheese and led to numerous stillbirths. As a result of this episode, FDA has been monitoring domestic and imported cheeses and has taken numerous actions to remove these products from the market when *L. monocytogenes* is found.

There have been other clustered cases, such as in Philadelphia, PA, in 1987. Specific food linkages were only made epidemiologically in this cluster.

CDC has established an epidemiological link between consumption of raw hot dogs or undercooked chicken and approximately 20% of the sporadic cases under prospective study.

For more information on recent outbreaks see the Morbidity and Mortality Weekly Reports from CDC.

11. Education: The FDA health alert for hispanic pregnant women concerns the risk of listeriosis from soft cheeses.The CDC provides similar information in spanish.

The Food Safety and Inspection Service of the U.S. Department of Agriculture has jointly produced with the FDA a background document on *Listeria* and Listeriosis.

The CDC produces an information brochure on preventing Listeriosis.

12. Other Resources: A Loci index for genome *Listeria monocytogenes* is available from GenBank.

Vibrio cholerae Serogroup O1

1. Name of Organism: *Vibrio cholerae* Serogroup O1

This bacterium is responsible for Asiatic or epidemic cholera. No major outbreaks of this disease have occurred in the United States since 1911. However, sporadic cases occurred between 1973 and 1991, suggesting the possible reintroduction of the organism into the U.S. marine and estuarine environment. The cases between 1973 and 1991 were associated with the consumption of raw shellfish or of shellfish either improperly cooked or recontaminated after proper cooking. Environmental studies have demonstrated that strains of this organism may be found in the temperate estuarine and marine coastal areas surrounding the United States.

In 1991 outbreaks of cholera in Peru quickly grew to epidemic proportions and spread to other South American and Central American countries, including Mexico. Over 340,000 cases and 3,600 deaths have been reported in the Western Hemisphere since January 1991. However, only 24 cases of cholera have been reported in the United States. The U.S. cases were brought into the country by travelers returning from South America, or were associated with illegally smuggled, temperature-abused crustaceans.

2. Name of the Acute Disease:

Cholera is the name of the infection caused by *V. cholerae*.

3. Nature of the Disease:

Symptoms of Asiatic cholera may vary from a mild, watery diarrhea to an acute diarrhea, with characteristic rice water stools. Onset of the illness is generally sudden, with incubation periods varying from 6 hours to 5 days. Abdominal cramps, nausea, vomiting, dehydration, and shock; after severe fluid and electrolyte loss, death may occur. Illness is caused by the ingestion of viable bacteria, which attach to the small intestine and produce cholera toxin. The production of cholera toxin by the attached bacteria results in the watery diarrhea associated with this illness.

Infective dose -- Human volunteer feeding studies utilizing healthy individuals have demonstrated that approximately one million organisms must be ingested to cause illness. Antacid consumption markedly lowers the infective dose.

4. Diagnosis of Human Illness:

Cholera can be confirmed only by the isolation of the causative organism from the diarrheic stools of infected individuals.

5. Foods in which it Occurs:

Cholera is generally a disease spread by poor sanitation, resulting in contaminated water supplies. This is clearly the main mechanism for the spread of cholera in poor communities in South America. The excellent sanitation facilities in the U.S. are responsible for the near eradication of epidemic cholera. Sporadic cases occur when shellfish harvested from fecally polluted coastal waters are consumed raw. Cholera may also be transmitted by shellfish harvested from nonpolluted waters since *V. cholerae* O1 is part of the autochthonous microbiota of these waters.

6. Frequency of Disease

Fewer than 80 proven cases of cholera have been reported in the U.S. since 1973. Most of these cases were detected only after epidemiological investigation. Probably more sporadic cases have occurred, but have gone undiagnosed or unreported.

7. The Usual Course of Disease and Some Complications:

Individuals infected with cholera require rehydration either intravenously or orally with a solution containing sodium chloride, sodium bicarbonate, potassium chloride, and dextrose (glucose). The illness is generally self-limiting. Antibiotics such as tetracycline have been demonstrated to shorten the course of the illness. Death occurs from dehydration and loss of essential electrolytes. Medical treatment to prevent dehydration prevents all complications.

8. Target Populations:

All people are believed to be susceptible to infection, but individuals with damaged or undeveloped immunity, reduced gastric acidity, or malnutrition may suffer more severe forms of the illness.

9. Analysis of Foods:

V. cholerae serogroup O1 may be recovered from foods by methods similar to those used for recovering the organism from the feces of infected individuals. Pathogenic and non- pathogenic forms of the organism exist, so all food isolates must be tested for the production of cholera enterotoxin.

10. Selected Outbreaks:

An incident of cholera in Indiana from imported food is reported in MMWR 44(20):1995 May 20 .

See MMWR 44(11):1995 Mar 24 for an updated report on *Vibrio cholerae* O1 in the Western Hemisphere 1991-1994 and on *V. cholerae* O139 in Asia, 1994.

Surveillance for cholera in Cochabamba Department, Bolivia is discussed in in this MMWR 42(33):1993 Aug 27.

The cholera outbreak in Burundi and Zimbabwe is detailed in the following MMWR 42(21):1993 Jun 04.

MMWR 40(49):1991 Dec 13 reports on a cholera outbreak associated with imported coconut milk.

A report of a cholera incident in New York is found in MMWR 40(30):1991 Aug 01.

Similar incidents in New Jersey and Florida are reported in MMWR 40(17):1991 May 03.

A case of importation of cholera from Peru to the United States is detailed in MMWR 40(15):1991 Apr 19.

The cholera outbreak in Peru is reported on in MMWR:40(6):1991 Feb 15, and the update of the South American endemic is in MMWR 40(13):1991 Apr 5.

For more information on recent outbreaks see the Morbidity and Mortality Weekly Reports from CDC.

11. Education: The CDC has a brochure on the prevention of cholera in English, Spanish and Portuguese

12. Other Resources A Loci index for genome *Vibrio cholerae* is available from GenBank.

Vibrio cholerae Serogroup Non-O1

1. Name of Organism: *Vibrio cholerae* Serogroup Non-O1

This bacterium infects only humans and other primates. It is related to *V. cholerae* Serogroup O1, the organism that causes Asiatic or epidemic cholera, but causes a disease less severe than cholera. Both pathogenic and nonpathogenic strains of the organism are normal inhabitants of marine and estuarine environments of the United States. This organism has been referred to as non-cholera vibrio (NCV) and nonagglutinable vibrio (NAG) in the past.

2. Name of Acute Disease: Non-O1 *V. cholerae* gastroenteritis is the name associated with this illness.

3. Nature of the Disease: Diarrhea, abdominal cramps, and fever are the predominant symptoms associated with this illness, with vomiting and nausea occurring in approximately 25% of infected individuals. Approximately 25% of infected individuals will have blood and mucus in their stools. Diarrhea may, in some cases, be quite severe, lasting 6-7 days. Diarrhea will usually occur within 48 hours following ingestion of the organism. It is unknown how the organism causes the illness, although an enterotoxin is suspected as well as an invasive mechanism. Disease is caused when the organism attaches itself to the small intestine of infected individuals and perhaps subsequently invades.

Infective dose -- It is suspected that large numbers (more than one million) of the organism must be ingested to cause illness.

4. Diagnosis of Human Illness: Diagnosis of a *V. cholerae* non-O1 infection is made by culturing the organism from an individual's diarrheic stool.

5. Foods in which it Occurs: Shellfish harvested from U.S. coastal waters frequently contain *V. cholerae* serogroup non-O1. Consumption of raw, improperly cooked or cooked, recontaminated shellfish may lead to infection.

6. Relative Frequency of Disease: No major outbreaks of diarrhea have been attributed to this organism. Sporadic cases occur frequently mainly along the coasts of the U.S., and are usually associated with the consumption of raw oysters during the warmer months.

7. The Usual Course of Disease and Some Complications: Diarrhea resulting from ingestion of the organism usually lasts 7 days and is self-limiting. Antibiotics such as tetracycline shorten the severity and duration of the illness. Septicemia (bacteria gaining entry into the blood stream and multiplying therein) can occur. This complication is associated with individuals with cirrhosis of the liver, or who are immunosuppressed, but this is relatively rare. FDA has warned individuals with liver disease to refrain from consuming raw or improperly cooked shellfish.

8. Target Populations: All individuals who consume raw shellfish are susceptible to diarrhea caused by this organism. Cirrhotic or immunosuppressed individuals may develop severe complications such as septicemia.

9. Analysis of Foods: Methods used to isolate this organism from foods are similar to those used with diarrheic stools. Because many food isolates are nonpathogenic, pathogenicity of all food isolates must be demonstrated. All virulence mechanisms of this group have not been elucidated; therefore, pathogenicity testing must be performed in suitable animal models.

10. Selected Outbreaks: Sporadic cases continue to occur all year, increasing in frequency during the warmer months.

An update report from CDC on *Vibrio cholerae* O139 in Asia may be found in MMWR 44(11):1995 Mar 24.

See MMWR 42(26):1993 Jul 09 for a report on the new O139 Non-O1 *Vibrio cholerae* (Bengal).

For more information on recent outbreaks see the Morbidity and Mortality Weekly Reports from CDC.

Vibrio parahaemolyticus

1. Name of Organism: *Vibrio parahaemolyticus* (and other marine *Vibrio* spp.**)

This bacterium is frequently isolated from the estuarine and marine environment of the United States. Both pathogenic and non-pathogenic forms of the organism can be isolated from marine and estuarine environments and from fish and shellfish dwelling in these environments.

2. Name of Acute Disease: *V. parahaemolyticus*-associated gastroenteritis is the name of the infection caused by this organism.

3. Nature of the Disease: Diarrhea, abdominal cramps, nausea, vomiting, headache, fever, and chills may be associated with infections caused by this organism. The illness is usually mild or moderate, although some cases may require hospitalization. The median duration of the illness is 2.5 days. The incubation period is 4-96 hours after the ingestion of the organism, with a mean of 15 hours. Disease is caused when the organism attaches itself to an individuals' small intestine and excretes an as yet unidentified toxin.

Infective dose -- A total dose of greater than one million organisms may cause disease; this is markedly lowered by antacids (or presumably by food with buffering capability).

4. Diagnosis of Human Illness: Diagnosis of gastroenteritis caused by this organism is made by culturing the organism from the diarrheic stools of an individual.

5. Associated Foods: Infections with this organism have been associated with the consumption of raw, improperly cooked, or cooked, recontaminated fish and shellfish. A correlation exists between the probability of infection and warmer months of the year. Improper refrigeration of seafoods contaminated with this organism will allow its proliferation, which increases the possibility of infection.

6. Relative Frequency of Disease: Major outbreaks have occurred in the U.S. during the warmer months of the year. Sporadic cases occur frequently along all coasts of the U.S.

7. The Usual Course the Disease: Diarrhea caused by this organism is usually self-limiting, with few cases requiring hospitalization and/or antibiotic treatment.

8. Target populations: All individuals who consume raw or improperly cooked fish and shellfish are susceptible to infection by this organism.

9. Analysis of Foods: Methods used to isolate this organism from foods are similar to those used with diarrheic stools. Because many food isolates are nonpathogenic, pathogenicity of all food isolates must be demonstrated. Although the demonstration of the Kanagawa hemolysin was long considered indicative of pathogenicity, this is now uncertain.

10. Selected Outbreaks: Sporadic outbreaks of gastroenteritis caused by this organism have occurred in the U.S. and cases are more common during the warmer months. It is very common in Japan, where large outbreaks occur with regularity.

**OTHER MARINE VIBRIOS IMPLICATED IN FOODBORNE DISEASE:

Several other marine vibrios have been implicated in human disease. Some may cause wound or ear infections, and others, gastroenteritis. The amount of evidence for certain of these organisms as being causative of human gastroenteritis is small. Nonetheless, several have been isolated from human feces from diarrhea patients from which no other pathogens could be isolated. Methods for recovery of these organisms from foods are similar to those used for recovery of *V. parahaemolyticus*. The species implicated in human disease include:

Vibrio alginolyticus	*Vibrio furnissii*
Vibrio carchariae	*Vibrio hollisae*
Vibrio cincinnatiensis	*Vibrio metschnikovii*
Vibrio damsela	*Vibrio mimicus*
Vibrio fluvialis	

For more information on recent outbreaks see the Morbidity and Mortality Weekly Reports from CDC.

11. Other Resources: A Loci index for genome Vibrio parahaemolyticus is available from GenBank.

Vibrio vulnificus

1. Name of Organism: *Vibrio vulnificus*

This bacterium infects only humans and other primates. It has been isolated from a wide range of environmental sources, including water, sediment, plankton, and shellfish (oysters, clams, and crabs) and a variety of locations, including the Gulf of Mexico, the Atlantic Coast as far north as Cape Cod, and the entire U.S. west coast. Cases of illness have also been associated with brackish lakes in New Mexico and Oklahoma.

2. Name of the Acute Disease: This organism causes wound infections, gastroenteritis, or a syndrome known as "primary septicemia."

3. Nature of the Disease: Wound infections result either from contaminating an open wound with sea water harboring the organism, or by lacerating part of the body on coral, fish, etc., followed by contamination with the organism. The ingestion of *V. vulnificus* by healthy individuals can result in gastroenteritis. The "primary septicemia" form of the disease follows consumption of raw seafood containing the organism by individuals with underlying chronic disease, particularly liver disease (see below). In these individuals, the microorganism enters the blood stream, resulting in septic shock, rapidly followed by death in many cases (about 50%). Over 70% of infected individuals have distinctive bulbous skin lesions.

Infective dose -- The infective dose for gastrointestinal symptoms in healthy individuals is unknown but for predisposed persons, septicemia can presumably occur with doses of less than 100 total organisms.

4. Diagnosis of Human Illness: The culturing of the organism from wounds, diarrheic stools, or blood is diagnostic of this illness.

5. Associated Foods: This organism has been isolated from oysters, clams, and crabs. Consumption of these products raw or recontaminated may result in illness.

6. Relative Frequency of Disease: No major outbreaks of illness have been attributed to this organism. Sporadic cases occur frequently, becoming more prevalent during the warmer months.

In a survey of cases of *V. vulnificus* infections in Florida from 1981 to 1987, Klontz et al. (Annals of Internal Medicine 109:318-23;1988) reported that 38 cases of primary septicemia (ingestion), 17 wound infections, and 7 cases gastroenteritis were associated with the organism. Mortality from infection varied from 55% for primary septicemia cases, to 24% with wound infections, to no deaths associated with gastroenteritis. Raw oyster consumption was a common feature of primary septicemia and gastroenteritis, and liver disease was a feature of primary septicemia.

7. The Usual Course of Disease and Some Complications: In healthy individuals, gastroenteritis usually occurs within 16 hours of ingesting the organism. Ingestion of the organism by individuals with some type of chronic underlying disease [such as diabetes, cirrhosis, leukemia , lung carcinoma, acquired immune deficiency syndrome (AIDS), AIDS- related complex (ARC), or asthma requiring the use of steroids] may cause the "primary septicemia" form of illness. The mortality rate for individuals with this form of the disease is over 50%.

8. Target Populations: All individuals who consume foods contaminated with this organism are susceptible to gastroenteritis. Individuals with diabetes, cirrhosis, or leukemia, or those who take immunosuppressive drugs or steroids are particularly susceptible to primary septicemia. These individuals should be strongly advised not to consume raw or inadequately cooked seafood, as should AIDS/ARC patients.

9. Analysis of Foods: Methods used to isolate this organism from foods are similar to those used with diarrheic stools. To date, all food isolates of this organism have been pathogenic in animal models.

FDA has a genetic probe for *V. vulnificus*; its target is a cytotoxin gene which appears not to correlate with the organism's virulence.

10. Selected Outbreaks: Sporadic cases continue to occur all year, increasing in frequency during the warmer months.

MMWR 45(28):1996 Jul 26 reports on three incidents of *V. vulnificus* infection in Los Angeles, California.

A multi-year summary of *V. vulnificus* incidents associated with the consumption of raw oysters is reported in MMWR 42(21):1993 Jun 04

For more information on recent outbreaks see the Morbidity and Mortality Weekly Reports from CDC.

11. Education: More information for consumers of raw shellfish is available in the FDA brochure *If You Eat Raw Oysters, You Need to Know...*

Clostridium perfringens

1. Name of Organism: *Clostridium perfringens*

Clostridium perfringens is an anaerobic, Gram-positive, sporeforming rod (anaerobic means unable to grow in the presence of free oxygen). It is widely distributed in the environment and frequently occurs in the intestines of humans and many domestic and feral animals. Spores of the organism persist in soil, sediments, and areas subject to human or animal fecal pollution.

2. Name of Acute Disease: Perfringens food poisoning is the term used to describe the common foodborne illness caused by *C. perfringens*. A more serious but rare illness is also caused by ingesting food contaminated with Type C strains. The latter illness is known as enteritis necroticans or pig-bel disease.

3. Nature of Disease: The common form of perfringens poisoning is characterized by intense abdominal cramps and diarrhea which begin 8-22 hours after consumption of foods containing large numbers of those *C. perfringens* bacteria capable of producing the food poisoning toxin. The illness is usually over within 24 hours but less severe symptoms may persist in some individuals for 1 or 2 weeks. A few deaths have been reported as a result of dehydration and other complications.

Necrotic enteritis (pig-bel) caused by *C. perfringens* is often fatal. This disease also begins as a result of ingesting large numbers of the causative bacteria in contaminated foods. Deaths from necrotic enteritis (pig-bel syndrome) are caused by infection and necrosis of the intestines and from resulting septicemia. This disease is very rare in the U.S.

Infective dose--The symptoms are caused by ingestion of large numbers (greater than 10 to the 8th) vegetative cells. Toxin production in the digestive tract (or in test tubes) is associated with sporulation. This disease is a food infection; only one episode has ever implied the possibility of intoxication (i.e., disease from preformed toxin).

4. Diagnosis of Human Illness: Perfringens poisoning is diagnosed by its symptoms and the typical delayed onset of illness. Diagnosis is confirmed by detecting the toxin in the feces of patients. Bacteriological confirmation can also be done by finding exceptionally large numbers of the causative bacteria in implicated foods or in the feces of patients.

5. Associated Foods and Food Handling: In most instances, the actual cause of poisoning by C. perfringens is temperature abuse of prepared foods. Small numbers of the organisms are often present after cooking and multiply to food poisoning levels during cool down and storage of prepared foods. Meats, meat products, and gravy are the foods most frequently implicated.

6. Frequency: Perfringens poisoning is one of the most commonly reported foodborne illnesses in the U.S. There were 1,162 cases in 1981, in 28 separate outbreaks. At least 10-20 outbreaks have been reported annually in the U.S. for the past 2 decades. Typically, dozens or even hundreds of person are affected. It is probable that many outbreaks go unreported because the implicated foods or patient feces are not tested routinely for *C. perfringens* or its toxin. CDC estimates that about 10,000 actual cases occur annually in the U.S.

7. Usual Course of Disease and Complications:
The disease generally lasts 24 hours. In the elderly or infirm, symptoms may last 1-2 weeks. Complications and/or death only very rarely occur.

8. Target Populations:
Institutional feeding (such as school cafeterias, hospitals, nursing homes, prisons, etc.) where large quantities of food are prepared several hours before serving is the most common circumstance in which perfringens poisoning occurs. The young and elderly are the most frequent victims of perfringens poisoning. Except in the case of pig-bel syndrome, complications are few in persons under 30 years of age. Elderly persons are more likely to experience prolonged or severe symptoms.

9. Analysis of Food and Feces:
Standard bacteriological culturing procedures are used to detect the organism in implicated foods and in feces of patients. Serological assays are used for detecting enterotoxin in the feces of patients and for testing the ability of strains to produce toxin. The procedures take 1-3 days.

10. Selected Outbreaks:
Since December 1981, FDA has investigated 10 outbreaks in 5 states. In two instances, more than one outbreak occurred in the same feeding facility within a 3-week period. One such outbreak occurred on 19 March 1984, involving 77 prison inmates. Roast beef served as a luncheon meat was implicated as the food vehicle and *C. perfringens* was confirmed as the cause by examining stools of 24 patients. Most of the patients became ill 8-16 hours after the meal. Eight days later, on 27 March 1984, a second outbreak occurred involving many of the same persons. The food vehicle was ham. Inadequate refrigeration and insufficient reheating of the implicated foods caused the outbreaks. Most of the other outbreaks occurred in institutional feeding environments: a hospital, nursing home, labor camp, school cafeteria, and at a fire house luncheon.

In November, 1985, a large outbreak of *C. perfringens* gastroenteritis occurred among factory workers in Connecticut. Forty-four percent of the 1,362 employees were affected. Four main-course foods served at an employee banquet were associated with illness, but gravy was implicated by stratified analysis. The gravy had been prepared 12-24 hours before serving, had been improperly cooled, and was reheated shortly before serving. The longer the reheating period, the less likely the gravy was to cause illness.

A outbreak of *C. perfringens* in corned beef was reported in MMWR 43(8):1994 Mar 04.

For more information on recent outbreaks see the Morbidity and Mortality Weekly Reports from CDC.

Bacillus cereus and other Bacillus spp.

1. Name of Organism: *Bacillus cereus* and other *Bacillus* spp.

Bacillus cereus is a Gram-positive, facultatively aerobic spore-former whose cells are large rods and whose spores do not swell the sporangium. These and other characteristics, including biochemical features, are used to differentiate and confirm the presence *B. cereus*, although these characteristics are shared with *B. cereus* var. mycoides, *B. thuringiensis* and *B. anthracis*. Differentiation of these organisms depends upon determination of motility (most *B. cereus* are motile), presence of toxin crystals (*B. thuringiensis*), hemolytic activity (*B. cereus* and others are beta hemolytic whereas *B. anthracis* is usually nonhemolytic), and rhizoid growth which is characteristic of *B. cereus* var. mycoides.

2. Name of Illness: *B. cereus* food poisoning is the general description, although two recognized types of illness are caused by two distinct metabolites. The diarrheal type of illness is caused by a large molecular weight protein, while the vomiting (emetic) type of illness is believed to be caused by a low molecular weight, heat-stable peptide.

3. Nature of Illness: The symptoms of *B. cereus* diarrheal type food poisoning mimic those of *Clostridium perfringens* food poisoning. The onset of watery diarrhea, abdominal cramps, and pain occurs 6-15 hours after consumption of contaminated food. Nausea may accompany diarrhea, but vomiting (emesis) rarely occurs. Symptoms persist for 24 hours in most instances.

The emetic type of food poisoning is characterized by nausea and vomiting within 0.5 to 6 h after consumption of contaminated foods. Occasionally, abdominal cramps and/or diarrhea may also occur. Duration of symptoms is generally less than 24 h. The symptoms of this type of food poisoning parallel those caused by *Staphylococcus aureus* foodborne intoxication. Some strains of *B. subtilis* and *B. licheniformis* have been isolated from lamb and chicken incriminated in food poisoning episodes. These organisms demonstrate the production of a highly heat-stable toxin which may be similar to the vomiting type toxin produced by *B. cereus*.

The presence of large numbers of *B. cereus* (greater than 10^{6} organisms/g) in a food is indicative of active growth and proliferation of the organism and is consistent with a potential hazard to health.

4. Diagnosis of Human Illness: Confirmation of *B. cereus* as the etiologic agent in a foodborne outbreak requires either (1) isolation of strains of the same serotype from the suspect food and feces or vomitus of the patient, (2) isolation of large numbers of a *B. cereus* serotype known to cause foodborne illness from the suspect food or from the feces or vomitus of the patient, or (3) isolation of *B. cereus* from suspect foods and determining their enterotoxigenicity by serological (diarrheal toxin) or biological (diarrheal and emetic) tests. The rapid onset time to symptoms in the emetic form of disease, coupled with some food evidence, is often sufficient to diagnose this type of food poisoning.

5. Foods Incriminated: A wide variety of foods including meats, milk, vegetables, and fish have been associated with the diarrheal type food poisoning. The vomiting-type outbreaks have generally been associated with rice products; however, other starchy foods such as potato, pasta and cheese products have also been implicated. Food mixtures such as sauces, puddings, soups, casseroles, pastries, and salads have frequently been incriminated in food poisoning outbreaks.

6. Relative Frequency of Illness: In 1980, 9 outbreaks were reported to the Centers for Disease Control and included such foods as beef, turkey, and Mexican foods. In 1981, 8 outbreaks were reported which primarily involved rice and shellfish. Other outbreaks go unreported or are misdiagnosed because of symptomatic similarities to *Staphylococcus aureus* intoxication (*B. cereus* vomiting-type) or *C. perfringens* food poisoning (B. cereus diarrheal type).

7. Complications: Although no specific complications have been associated with the diarrheal and vomiting toxins produced by *B. cereus*, other clinical manifestations of *B. cereus* invasion or contamination have been observed. They include bovine mastitis, severe systemic and pyogenic infections, gangrene, septic meningitis, cellulitis, panophthalmitis, lung abscesses, infant death, and endocarditis.

8. Target Populations: All people are believed to be susceptible to *B. cereus* food poisoning.

9. Food Analysis: A variety of methods have been recommended for the recovery, enumeration and confirmation of *B. cereus* in foods. More recently, a serological method has been developed for detecting the putative enterotoxin of *B. cereus* (diarrheal type) isolates from suspect foods. Recent investigations suggest that the vomiting type toxin can be detected by animal models (cats, monkeys) or possibly by cell culture.

10. Selected Outbreaks: On September 22, 1985, the Maine Bureau of Health was notified of gastrointestinal illness among patrons of a Japanese restaurant. Because the customers were exhibiting symptoms of illness while still on the restaurant premises, and because uncertainty existed as to the etiology of the problem, the local health department, in concurrence with the restaurant owner, closed the restaurant at 7:30 p.m. that same day.

Eleven (31%) of the approximately 36 patrons reportedly served on the evening of September 22, were contacted in an effort to determine the etiology of the outbreak. Those 11 comprised the last three dining parties served on September 22. Despite extensive publicity, no additional cases were reported.

A case was defined as anyone who demonstrated vomiting or diarrhea within 6 hours of dining at the restaurant. All 11 individuals were interviewed for symptoms, time of onset of illness, illness duration, and foods ingested. All 11 reported nausea and vomiting; nine reported diarrhea; one reported headache; and one reported abdominal cramps. Onset of illness ranged from 30 minutes to 5 hours (mean 1 hour, 23 minutes) after eating at the restaurant. Duration of illness ranged from 5 hours to several days, except for two individuals still symptomatic with diarrhea 2 weeks after dining at the restaurant. Ten persons sought medical treatment at local emergency rooms on September 22; two ultimately required hospitalization for rehydration.

Analysis of the association of specific foods with illness was not instructive, since all persons consumed the same food items; chicken soup, fried shrimp, stir-fried rice, fried zucchini, onions, bean sprouts, cucumber, cabbage, and lettuce salad, ginger salad dressing, hibachi chicken and steak, and tea. Five persons ordered hibachi scallops, and one person ordered hibachi swordfish. However, most individuals sampled each other's entrees. One vomitus specimen and two stool specimens from the three separate individuals yielded an overgrowth of *B. cereus*, although an accurate bacterial count could not be made because an inadequate amount of the steak remained for laboratory analysis. No growth of *B. cereus* was reported from the fried rice, mixed fried vegetables, or hibachi chicken.

According to the owner, all meat was delivered 2-3 times a week from a local meat supplier and refrigerated until ordered by restaurant patrons. Appropriate-sized portions for a dining group were taken from the kitchen to the dining area and diced or sliced, then sauteed at the table directly in front of restaurant patrons. The meat was seasoned with soy sauce salt and white pepper, open containers of which had been used for at least 2 months by the restaurant. The hibachi steak was served immediately after cooking.

The fried rice served with the meal was customarily made from leftover boiled rice. It could not be established whether the boiled rice had been stored refrigerated or at room temperature.

Fresh, rapidly cooked meat, eaten immediately, seems an unlikely vehicle of *B. cereus* food poisoning. The laboratory finding of *B. cereus* in a foodstuff without quantitative cultures and without accompanying epidemiologic data is insufficient to establish its role in the outbreak. Although no viable *B. cereus* organisms were isolated from the fried rice eaten with the meal, it does not exclude this food as the common vehicle. Reheating during preparation may have eliminated the bacteria in the food without decreasing the activity of the heat-stable toxin. While the question of the specific vehicle remains incompletely resolved, the clinical and laboratory findings substantially support *B. cereus* as the cause of the outbreak.

Most episodes of food poisoning undoubtedly go unreported, and in most of those reported, the specific pathogens are never identified. Alert recognition of the clinical syndrome and appropriate laboratory work permitted identification of the role of *B. cereus* in this outbreak.

For a report on a *B. cereus* outbreak in northern Virginia see this MMWR 43(10):1994 Mar 18.

For more information on recent outbreaks see the Morbidity and Mortality Weekly Reports from CDC.

Enterotoxigenic Escherichia coli

1. Name of Organism: Enterotoxigenic *Escherichia coli* (ETEC)

Currently, there are four recognized classes of enterovirulent *E. coli* (collectively referred to as the EEC group) that cause gastroenteritis in humans. Among these are the enterotoxigenic (ETEC) strains. They comprise a relatively small proportion of the species and have been etiologically associated with diarrheal illness of all age groups from diverse global locations. The organism frequently causes diarrhea in infants in less developed countries and in visitors there from industrialized countries. The etiology of this cholera-like illness has been recognized for about 20 years.

2. Name of Acute Disease: Gastroenteritis is the common name of the illness caused by ETEC, although travelers' diarrhea is a frequent sobriquet.

3. Nature of Disease: The most frequent clinical syndrome of infection includes watery diarrhea, abdominal cramps, low-grade fever, nausea and malaise.

Infective dose--Volunteer feeding studies indicate that a relatively large dose (100 million to 10 billion bacteria) of enterotoxigenic *E. coli* is probably necessary to establish colonization of the small intestine, where these organisms proliferate and produce toxins which induce fluid secretion. With high infective dose, diarrhea can be induced within 24 hours. Infants may require fewer organisms for infection to be established.

4. Diagnosis of Human Illness: During the acute phase of infection, large numbers of enterotoxigenic cells are excreted in feces. These strains are differentiated from nontoxigenic *E. coli* present in the bowel by a variety of in vitro immunochemical, tissue culture, or gene probe tests designed to detect either the toxins or genes that encode for these toxins. The diagnosis can be completed in about 3 days.

5. Associated Foods: ETEC is not considered a serious foodborne disease hazard in countries having high sanitary standards and practices. Contamination of water with human sewage may lead to contamination of foods. Infected food handlers may also contaminate foods. These organisms are infrequently isolated from dairy products such as semi-soft cheeses.

6. Relative Frequency of Disease: Only four outbreaks in the U.S. have been documented, one resulting from consumption of water contaminated with human sewage, another from consumption of Mexican food prepared by an infected food handler. In two others, one in a hospital cafeteria and one aboard a cruise ship, food was the probable cause. The disease among travelers to foreign countries, however, is common.

7. Complications: The disease is usually self-limiting. In infants or debilitated elderly persons, appropriate electrolyte replacement therapy may be necessary.

8. Target Populations: Infants and travelers to underdeveloped countries are most at-risk of infection.

9. Analysis of Food: With the availability of a gene probe method, foods can be analyzed directly for the presence of enterotoxigenic *E. coli*, and the analysis can be completed in about 3 days. Alternative methods which involve enrichment and plating of samples for isolation of *E. coli* and their subsequent confirmation as toxigenic strains by conventional toxin assays may take at least 7 days.

10. Selected Outbreaks: In the last decade, four major common-source outbreaks of ETEC gastroenteritis occurred in the U.S. In late 1975 one-third of the passengers on two successive cruises of a Miami-based ship experienced diarrheal illness. A CDC investigation found ETEC to be the cause, presumably linkedto consumption of crabmeat cocktail. In early 1980, 415 persons eating at a Mexican restaurant experienced diarrhea. The source of the causative organism was an ill food handler. In 1981, 282 of 3,000 personnel at a Texas hospital acquired ETEC gastroenteritis after eating in the hospital cafeteria. No single food was identified by CDC.

Outbreaks of ETEC in Rhode Island and New Hampshire are reported in this MMWR 43(5):1994 Feb 11.

For more information on recent outbreaks see the Morbidity and Mortality Weekly Reports from CDC.

Enteropathogenic Escherichia coli

1. Name of Organism: Enteropathogenic *Escherichia coli* (EPEC)

Currently, there are four recognized classes of enterovirulent *E. coli* (collectively referred to as the EEC group) that cause gastroenteritis in humans. Among these are the enteropathogenic (EPEC) strains. EPEC are defined as *E. coli* belonging to serogroups epidemiologically implicated as pathogens but whose virulence mechanism is unrelated to the excretion of typical *E. coli* enterotoxins. *E. coli* are Gram-negative, rod-shaped bacteria belonging the family Enterobacteriaceae. Source(s) and prevalence of EPEC are controversial because foodborne outbreaks are sporadic. Humans, bovines, and swine can be infected, and the latter often serve as common experimental animal models. *E. coli* are present in the normal gut flora of these mammals. The proportion of pathogenic to nonpathogenic strains, although the subject of intense research, is unknown.

2. Name of Acute Disease: Infantile diarrhea is the name of the disease usually associated with EPEC.

3. Nature of Disease: EPEC cause either a watery or bloody diarrhea, the former associated with the attachment to, and physical alteration of, the integrity of the intestine. Bloody diarrhea is associated with attachment and an acute tissue-destructive process, perhaps caused by a toxin similar to that of *Shigella dysenteriae*, also called verotoxin. In most of these strains the shiga-like toxin is cell-associated rather than excreted.

Infective dose -- EPEC are highly infectious for infants and the dose is presumably very low. In the few documented cases of adult diseases, the dose is presumably similar to other colonizers (greater than 10^6 total dose).

4. Diagnosis of Human Illness: The distinction of EPEC from other groups of pathogenic *E. coli* isolated from patients' stools involves serological and cell culture assays. Serotyping, although useful, is not strict for EPEC.

5. Associated Foods: Common foods implicated in EPEC outbreaks are raw beef and chicken, although any food exposed to fecal contamination is strongly suspect.

6. Relative Frequency of Disease: Outbreaks of EPEC are sporadic. Incidence varies on a worldwide basis; countries with poor sanitation practices have the most frequent outbreaks.

7. Usual Course of Disease and Some Complications: Occasionally, diarrhea in infants is prolonged, leading to dehydration, electrolyte imbalance and death (50% mortality rates have been reported in third world countries).

8. Target Populations: EPEC outbreaks most often affect infants, especially those that are bottle fed, suggesting that contaminated water is often used to rehydrate infant formulae in underdeveloped countries.

9. Analysis of Foods: The isolation and identification of *E. coli* in foods follows standard enrichment and biochemical procedures. Serotyping of isolates to distinguish EPEC is laborious and requires high quality, specific antisera, and technical expertise. The total analysis may require from 7 to 14 days.

10. Selected Outbreaks: Sporadic outbreaks of EPEC diarrhea have occurred for half a century in infant nurseries, presumably derived from the hospital environment or contaminated infant formula. Common-source outbreaks of EPEC diarrhea involving healthy young adults were reported in the late 1960s. Presumably a large inoculum was ingested.

For more information on recent outbreaks see the Morbidity and Mortality Weekly Reports from CDC.

Escherichia coli O157:H7

1. Name of Organism: *Escherichia coli* O157:H7 (enterohemorrhagic *E. coli* or EHEC)

Currently, there are four recognized classes of enterovirulent *E. coli* (collectively referred to as the EEC group) that cause gastroenteritis in humans. Among these is the enterohemorrhagic (EHEC) strain designated *E. coli* O157:H7. *E. coli* is a normal inhabitant of the intestines of all animals, including humans. When aerobic culture methods are used, *E. coli* is the dominant species found in feces. Normally *E. coli* serves a useful function in the body by suppressing the growth of harmful bacterial species and by synthesizing appreciable amounts of vitamins. A minority of *E. coli* strains are capable of causing human illness by several different mechanisms. *E. coli* serotype O157:H7 is a rare variety of *E. coli* that produces large quantities of one or more related, potent toxins that cause severe damage to the lining of the intestine. These toxins [verotoxin (VT), shiga-like toxin] are closely related or identical to the toxin produced by *Shigella dysenteriae*.

2. Name of Acute Disease: Hemorrhagic colitis is the name of the acute disease caused by *E. coli* O157:H7.

3. Nature of Disease: The illness is characterized by severe cramping (abdominal pain) and diarrhea which is initially watery but becomes grossly bloody. Occasionally vomiting occurs. Fever is either low-grade or absent. The illness is usually self-limited and lasts for an average of 8 days. Some individuals exhibit watery diarrhea only.

Infective dose -- Unknown, but from a compilation of outbreak data, including the organism's ability to be passed person-to-person in the day-care setting and nursing homes, the dose may be similar to that of *Shigella* spp. (10 organisms).

4. Diagnosis of Illness: Hemorrhagic colitis is diagnosed by isolation of *E. coli* of **Human** serotype O157:H7 or other verotoxin-producing *E. coli* from diarrheal stools. Alternatively, the stools can be tested directly for the presence of verotoxin. Confirmation can be obtained by isolation of *E. coli* of the same serotype from the incriminated food.

5. Associated Foods: Undercooked or raw hamburger (ground beef) has been implicated in nearly all documented outbreaks and in other sporadic cases. Raw milk was the vehicle in a school outbreak in Canada. These are the only two demonstrated food causes of disease, but other meats may contain *E. coli* O157:H7.

6. Relative Frequency of Disease: Hemorrhagic colitis infections are not too common, but this is probably not reflective of the true frequency. In the Pacific Northwest, *E. coli* O157:H7 is thought to be second only to Salmonella as a cause of bacterial diarrhea. Because of the unmistakable symptoms of profuse, visible blood in severe cases, those victims probably seek medical attention, but less severe cases are probably more numerous.

Reported Cases of *E. coli* O157, United States 1994-1995

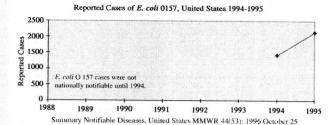

Summary Notifiable Diseases, United States MMWR 44(53): 1996 October 25

7. Usual Course of Disease and Some Complications:

Some victims, particularly the very young, have developed the hemolytic uremic syndrome (HUS), characterized by renal failure and hemolytic anemia. From 0 to 15% of hemorrhagic colitis victims may develop HUS. The disease can lead to permanent loss of kidney function.

In the elderly, HUS, plus two other symptoms, fever and neurologic symptoms, constitutes thrombotic thrombocytopenic purpura (TTP). This illness can have a mortality rate in the elderly as high as 50%.

8. Target Populations:

All people are believed to be susceptible to hemorrhagic colitis, but larger outbreaks have occurred in institutional settings.

9. Analysis of Foods:

E. coli O157:H7 will form colonies on agar media that are selective for *E. coli*. However, the high temperature growth procedure normally performed to eliminate background organisms before plating cannot be used because of the inability of these organisms to grow at temperatures of 44.0 - 45.5°C that support the growth of most *E. coli*. The use of DNA probes to detect genes encoding for the production of verotoxins (VT1 and VT2) is the most sensitive method devised.

10. Selected Outbreaks:

Three outbreaks occurred in 1982. Two of them, one in Michigan and one in Oregon, involved hamburgers from a national fast-food chain. The third occurred in a home for the aged in Ottawa, Ontario; club sandwiches were implicated, and 19 people died. More recently, several outbreaks in nursing homes and a day-care center have been investigated. Two large outbreaks occurred in 1984, one in 1985, three in 1986. Larger outbreaks have occurred in the Northwest U.S. and Canada.

In October-November, 1986, an outbreak of hemorrhagic colitis caused by *E. coli* O157:H7 occurred in Walla Walla, WA. Thirty-seven people, aged 11 months to 78 years developed diarrhea caused by the organism. All isolates from patients (14) had a unique plasmid profile and produced Shiga-like toxin II. In addition to diarrhea, 36 persons reported grossly bloody stools and 36 of the 37 reported abdominal cramps. Seventeen patients were hospitalized. One patient developed HUS (4 years old) and three developed TTP (70, 78, and 78 years old). Two patients with TTP died. Ground beef was the implicated food vehicle.

An excellent summary of nine *E. coli* O157:H7 outbreaks

appeared in the Annals of Internal Medicine, 1 November, 1988, pp. 705-712.

There was a recall of frozen hamburger underway (12 Aug 1997). For more information, see the USDA announcement and follow-up announcement (15 Aug 1997) on the U.S. Department of Agriculture web site concerning the recall of Hudson frozen ground beef.

The Centers for Disease Control and Prevention have reported on the above outbreak in preliminary (MMWR 45(44):975, 1996 November 8) and in updated (MMWR 46(1):4-8, 1997 January 10) form.

The FDA has issued on 31 October 1996 a press release concerning an outbreak of *E. coli* O157:H7 associated with Odwalla brand apple juice products.

A non-food related outbreak of *E. coli* O157:H7 is reported in MMWR 45(21):1996 May 31. While, the source of the outbreak is thought to be waterborne, the article is linked to this chapter to provide updated reference information on enterohemorrhagic *E. coli.*

MMWR 45(12):1996 Mar 29 reports on an outbreak of O157:H7 that occured in Georgia and Tennessee in June of 1995.

A community outbreak of hemolytic uremic syndrome attributable to *Escherichia coli* O111:NM in southern Australia in 1995 is reported in MMWR 44(29):1995 Jul 28.

A report on enhanced detection of sporadic *E. coli* O157:H7 infections in New Jersey and on an E. coli O157:H7 outbreak at a summer camp are in MMWR 44(22): 1995 Jun 9.

An outbreak of *E. coli* O157:H7 in Washington and California associated with dry-cured salami is reported in MMWR 44(9):1995 Mar 10.

Information concerning an outbreak that occured because of home-cooked hamburger can be found in this MMWR 43(12):1994 Apr 01.

MMWR 43(10):1994 Mar 18 reports on laboratory screening for *E. coli* O157 in Connecticut.

The outbreak of EHEC in the western states of the US is reported in preliminary form in this MMWR 42(4):1993 Feb 5, and in updated form in this MMWR 42(14):1993 Apr 16.

An outbreak of *E. coli* O157 in 1990 in North Dakota is reported in the MMWR 40(16):1991 Apr 26.

The Centers for Disease Control and Prevention has reissued the 5

November 1982 MMWR report that was the first to describe the diarrheal illness of *E. coli* O157:H7. This reissue is a part of the commemoration of CDC's 50th anniversary.

For more information on recent outbreaks see the Morbidity and Mortality Weekly Reports from CDC.

11. Education:

USDA Urges Consumers To Use Food Thermometer When Cooking Ground Beef Patties (Aug 11 1998)

The CDC has an information brochure on preventing *Escherichia coli* O157:H7 infections.

12. Other Resources:

Dr. Feng of FDA/CFSAN has written a monograph on *E. coli* O157:H7 which appeared in the CDC journal Emerging Infectious Diseases Vol. 1 No. 2, April-June 1995.

Enteroinvasive Escherichia coli

1. Name of Organism: Enteroinvasive *Escherichia coli* or (EIEC)

Currently, there are four recognized classes of enterovirulent *E. coli* (collectively referred to as the EEC group) that cause gastroenteritis in humans. *E. coli* is part of the normal intestinal flora of humans and other primates. A minority of *E. coli* strains are capable of causing human illness by several different mechanisms. Among these are the enteroinvasive (EIEC) strains. It is unknown what foods may harbor these pathogenic enteroinvasive (EIEC) strains responsible for a form of bacillary dysentery.

2. Name of Disease: Enteroinvasive *E. coli* (EIEC) may produce an illness known as bacillary dysentery. The EIEC strains responsible for this syndrome are closely related to *Shigella* spp.

3. Nature of the Disease: Following the ingestion of EIEC, the organisms invade the epithelial cells of the intestine, resulting in a mild form of dysentery, often mistaken for dysentery caused by *Shigella* species. The illness is characterized by the appearance of blood and mucus in the stools of infected individuals.

Infective dose -- The infectious dose of EIEC is thought to be as few as 10 organisms (same as *Shigella*).

4. Diagnosis of Human Illness: The culturing of the organism from the stools of infected individuals and the demonstration of invasiveness of isolates in tissue culture or in a suitable animal model is necessary to diagnose dysentery caused by this organism.

More recently, genetic probes for the invasiveness genes of both EIEC and *Shigella* spp. have been developed.

5. Associated Foods: It is currently unknown what foods may harbor EIEC, but any food contaminated with human feces from an ill individual, either directly or via contaminated water, could cause disease in others. Outbreaks have been associated with hamburger meat and unpasteurized milk.

6. Relative Frequency of Disease: One major foodborne outbreak attributed to enteroinvasive *E. coli* in the U.S. occurred in 1973. It was due to the consumption of imported cheese from France. The disease caused by EIEC is uncommon, but it may be confused with shigellosis and its prevalence may be underestimated.

7. The Usual Course of Disease and Some Complications: Dysentery caused by EIEC usually occurs within 12 to 72 hours following the ingestion of contaminated food. The illness is characterized by abdominal cramps, diarrhea, vomiting, fever, chills, and a generalized malaise. Dysentery caused by this organism is generally self-limiting with no known complications. A common sequelus associated with infection, especially in pediatric cases, is hemolytic uremic syndrome (HUS).

8. Target Populations: All people are subject to infection by this organism.

9. Analysis of Foods: Foods are examined as are stool cultures. Detection of this organism in foods is extremely difficult because undetectable levels may cause illness. It is estimated that the ingestion of as few as 10 organisms may result in dysentery.

10. Selected Outbreaks: Several outbreaks in the U.S. have been attributed to this organism. One outbreak occurred in 1973 and was due to the consumption of imported cheese. More recently, a cruise ship outbreak was attributed to potato salad, and an outbreak occurred in a home for the mentally retarded where subsequent person-to-person transmission occurred.

Aeromonas hydrophila

1. Name of Organism: *Aeromonas hydrophila, Aeromonas caviae, Aeromonas sobria &* *(Aeromonas veronii?)*

Aeromonas hydrophila is a species of bacterium that is present in all freshwater environments and in brackish water. Some strains of *A. hydrophila* are capable of causing illness in fish and amphibians as well as in humans who may acquire infections through open wounds or by ingestion of a sufficient number of the organisms in food or water.

Not as much is known about the other *Aeromonas* spp., but they too are aquatic microorganisms and have been implicated in human disease.

2. Name of Acute Disease:

A. hydrophila may cause gastroenteritis in healthy individuals or septicemia in individuals with impaired immune systems or various malignancies.

A. caviae and *A. sobria* also may cause enteritis in anyone or septicemia in immunocompromised persons or those with malignancies.

3. Nature of Disease: At the present time, there is controversy as to whether *A. hydrophila* is a cause of human gastroenteritis. Although the organism possesses several attributes which could make it pathogenic for humans, volunteer human feeding studies, even with enormous numbers of cells (i.e. 10^{11}), have failed to elicit human illness. Its presence in the stools of individuals with diarrhea, in the absence of other known enteric pathogens, suggests that it has some role in disease.

Likewise, *A. caviae* and *A. sobria* are considered by many as "putative pathogens," associated with diarrheal disease, but as of yet they are unproven causative agents.

Two distinct types of gastroenteritis have been associated with *A. hydrophila*: a cholera-like illness with a watery (rice and water) diarrhea and a dysenteric illness characterized by loose stools containing blood and mucus. The infectious dose of this organism is unknown, but SCUBA divers who have ingested small amounts of water have become ill, and *A. hydrophila* has isolated from their stools.

A general infection in which the organisms spread throughout the body has been observed in individuals with underlying illness (septicemia).

4. Diagnosis of Human Illness:

A. hydrophila can be cultured from stools or from blood by plating the organisms on an agar medium containing sheep blood and the antibiotic ampicillin. Ampicillin prevents the growth of most competing microorganisms. The species identification is confirmed by a series of biochemical tests. The ability of the organism to produce the enterotoxins believed to cause the gastrointestinal symptoms can be confirmed by tissue culture assays.

5. Associated Foods: *A. hydrophila* has frequently been found in fish and shellfish. It has also been found in market samples of red meats (beef, pork, lamb) and poultry. Since little is known about the virulence mechanisms of *A. hydrophila*, it is presumed that not all strains are pathogenic, given the ubiquity of the organism.

6. Relative Frequency of Disease: The relative frequency of *A. hydrophila* disease in the U.S. is unknown since efforts to ascertain its true incidence have only recently been attempted. Most cases have been sporadic rather than associated with large outbreaks, but increased reports have been noted from several clinical centers.

7. Usual Course of Disease and Some Complications: On rare occasions the dysentery-like syndrome is severe and may last for several weeks.

A. hydrophila may spread throughout the body and cause a general infection in persons with impaired immune systems. Those at risk are individuals suffering from leukemia, carcinoma, and cirrhosis and those treated with immunosuppressive drugs or who are undergoing cancer chemotherapy.

8. Target Populations: All people are believed to be susceptible to gastroenteritis, although it is most frequently observed in very young children. People with impaired immune systems or underlying malignancy are susceptible to the more severe infections.

9. Analysis of Foods: *A. hydrophila* can be recovered from most foods by direct plating onto a solid medium containing starch as the sole carbohydrate source and ampicillin to retard the growth of most competing microorganisms.

10. Selected Outbreaks: Most cases have been sporadic, rather than associated with large outbreaks.

For more information on recent outbreaks see the Morbidity and Mortality Weekly Reports from CDC.

Plesiomonas shigelloides

1. Name of Organism: *Plesiomonas shigelloides*

This is a Gram-negative, rod-shaped bacterium which has been isolated from freshwater, freshwater fish, and shellfish and from many types of animals including cattle, goats, swine, cats, dogs, monkeys, vultures, snakes, and toads.

Most human *P. shigelloides* infections are suspected to be water-borne. The organism may be present in unsanitary water which has been used as drinking water, recreational water, or water used to rinse foods that are consumed without cooking or heating. The ingested *P. shigelloides* organism does not always cause illness in the host animal but may reside temporarily as a transient, nonin-fectious member of the intestinal flora. It has been isolated from the stools of patients with diarrhea, but is also sometimes isolated from healthy individuals (0.2-3.2% of population).

It cannot yet be considered a definite cause of human disease, although its association with human diarrhea and the virulence factors it demonstrates make it a prime candidate.

2. Name of Acute Disease:
Gastroenteritis is the disease with which *P. shigelloides* has been implicated.

3. Nature of Disease: *P. shigelloides* gastroenteritis is usually a mild self-limiting dis-ease with fever, chills, abdominal pain, nausea, diarrhea, or vomit-ing; symptoms may begin 20-24 hours after consumption of con-taminated food or water; diarrhea is watery, non-mucoid, and non-bloody; in severe cases, diarrhea may be greenish-yellow, foamy, and blood tinged; duration of illness in healthy people may be 1-7 days.

The infectious dose is presumed to be quite high, at least greater than one million organisms.

4. Diagnosis of Human Illness:
The pathogenesis of *P. shigelloides* infection is not known. The organism is suspected of being toxigenic and invasive. Its signifi-cance as an enteric (intestinal) pathogen is presumed because of its predominant isolation from stools of patients with diarrhea. It is identified by common bacteriological analysis, serotyping, and antibiotic sensitivity testing.

5. Associated Foods: Most *P. shigelloides* infections occur in the summer months and correlate with environmental contamination of freshwater (rivers, streams, ponds, etc.). The usual route of transmission of the organism in sporadic or epidemic cases is by ingestion of contam-inated water or raw shellfish.

6. Frequency of Disease:
Most *P. shigelloides* strains associated with human gastrointestinal disease have been from stools of diarrheic patients living in tropi-cal and subtropical areas. Such infections are rarely reported in

the U.S. or Europe because of the self-limiting nature of the disease.

7. Usual Course of Disease and Some Complications:

P. shigelloides infection may cause diarrhea of 1-2 days duration in healthy adults. However, there may be high fever and chills and protracted dysenteric symptoms in infants and children under 15 years of age. Extra- intestinal complications (septicemia and death) may occur in people who are immunocompromised or seriously ill with cancer, blood disorders, or hepatobiliary disease.

8. Target Populations:

All people may be susceptible to infection. Infants, children and chronically ill people are more likely to experience protracted illness and complications.

9. Food Analysis:

P. shigelloides may be recovered from food and water by methods similar to those used for stool analysis. The keys to recovery in all cases are selective agars which enhance the survival and growth of these bacteria over the growth of the background microflora. Identification following recovery may be completed in 12-24 hours.

10. Selected Outbreaks:

Gastrointestinal illness in healthy people caused by *P. shigelloides* infection may be so mild that they do not seek medical treatment. Its rate of occurrence in the U.S. is unknown. It may be included in the group of diarrheal diseases "of unknown etiology" which are treated with and respond to broad spectrum antibiotics.

Most cases reported in the United States involve individuals with preexisting health problems such as cancer, sickle cell anemia, immunoincompetence, the aged, and the very young, who develop complications.

A case cluster occurred in North Carolina in November, 1980, following an oyster roast. Thirty-six out of 150 people who had eaten roasted oysters experienced nausea, chills, fever, vomiting, diarrhea, and abdominal pain beginning 2 days after the roast. The average duration of these symptoms was 2 days. *P. shigelloides* was recovered from oyster samples and patient stools.

A non-food related outbreak of *P. shigelloides* is reported in MMWR 38(36):1989 Sep 15.

For more information on recent outbreaks see the Morbidity and Mortality Weekly Reports from CDC.

Shigella spp.

1. Name of Organism: *Shigella* spp. (*Shigella sonnei, S. boydii, S. flexneri,* and *S. dysenteriae*)

Shigella are Gram-negative, nonmotile, nonsporeforming rod-shaped bacteria. The illness caused by *Shigella* (shigellosis) accounts for less than 10% of the reported outbreaks of foodborne illness in this country. *Shigella* rarely occurs in animals; principally a disease of humans except other primates such as monkeys and chimpanzees. The organism is frequently found in water polluted with human feces.

2. Name of Disease: *Shigellosis* (bacillary dysentery).

3. Nature of Disease: Symptoms -- Abdominal pain; cramps; diarrhea; fever; vomiting; blood, pus, or mucus in stools; tenesmus.

Onset time -- 12 to 50 hours.

Infective dose -- As few as 10 cells depending on age and condition of host. The *Shigella* spp. are highly infectious agents that are transmitted by the fecal-oral route.

The disease is caused when virulent *Shigella* organisms attach to, and penetrate, epithelial cells of the intestinal mucosa. After invasion, they multiply intracellularly, and spread to contiguous epitheleal cells resulting in tissue destruction. Some strains produce enterotoxin and Shiga toxin (very much like the verotoxin of *E. coli* O157:H7).

4. Diagnosis of Human Illness: Serological identification of culture isolated from stool.

5. Associated Foods: Salads (potato, tuna, shrimp, macaroni, and chicken), raw vegetables, milk and dairy products, and poultry. Contamination of these foods is usually through the fecal-oral route. Fecally contaminated water and unsanitary handling by food handlers are the most common causes of contamination.

6. Relative Frequency of Disease: An estimated 300,000 cases of shigellosis occur annually in the U.S. The number attributable to food is unknown, but given the low infectious dose, it is probably substantial.

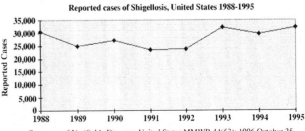

Reported cases of Shigellosis, United States 1988-1995

Summary of Notifiable Diseases, United States MMWR 44(53): 1996 October 25

7. Complications: Infections are associated with mucosal ulceration, rectal bleeding, drastic dehydration; fatality may be as high as 10-15% with some strains. Reiter's disease, reactive arthritis, and hemolytic uremic syndrome are possible sequelae that have been reported in the aftermath of shigellosis.

8. Target Populations: Infants, the elderly, and the infirm are susceptible to the severest symptoms of disease, but all humans are susceptible to some degree. Shigellosis is a very common malady suffered by individuals with acquired immune deficiency syndrome (AIDS) and AIDS-related complex, as well as non-AIDS homosexual men.

9. Food Analysis: Organisms are difficult to demonstrate in foods because methods are not developed or are insensitive. A genetic probe to the virulence plasmid has been developed by FDA and is currently under field test. However, the isolation procedures are still poor.

10. Selected Outbreaks: In 1985, a huge outbreak of foodborne shigellosis occurred in Midland-Odessa, Texas, involving perhaps as many as 5,000 persons. The implicated food was chopped, bagged lettuce, prepared in a central location for a Mexican restaurant chain. FDA research subsequently showed that S. sonnei, the isolate from the lettuce, could survive in chopped lettuce under refrigeration, and the lettuce remained fresh and appeared to be quite edible.

In 1985-1986, several outbreaks of shigellosis occurred on college campuses, usually associated with fresh vegetables from the salad bar. Usually an ill food service worker was shown to be the cause.

In 1987, several very large outbreaks of shigellosis (S. sonnei) occurred involving thousands of persons, but no specific food vector could be proven.

In 1988, numerous individuals contracted shigellosis from food consumed aboard Northwest Airlines flights; food on these flights had been prepared in one central commisary. No specific food item was implicated, but various sandwiches were suspected.

**NOTE - Although all *Shigella* spp. have been implicated in foodborne outbreaks at some time, S. sonnei is clearly the leading cause of shigellosis from food. The other species are more closely associated with contaminated water. One in particular, *S. flexneri*, is now thought to be in large part sexually transmitted.

For information on the outbreak of *Shigella* on a cruise ship, see MMWR 43(35):1994 Sep 09.

MMWR 40(25):1991 Jun 28 reports on a *Shigella dysenteriae* Type 1 outbreak in Guatemala, 1991.

For more information on recent outbreaks see the Morbidity and Mortality Weekly Reports from CDC.

Miscellaneous enterics

1. Name of Organism: Miscellaneous enterics, Gram-negative genera including: *Klebsiella, Enterobacter, Proteus, Citrobacter, Aerobacter, Providencia, Serratia*

These rod-shaped enteric (intestinal) bacteria have been suspected of causing acute and chronic gastrointestinal disease. The organisms may be recovered from natural environments such as forests and freshwater as well as from farm produce (vegetables) where they reside as normal microflora. They may be recovered from the stools of healthy individuals with no disease symptoms. The relative proportion of pathogenic to nonpathogenic strains is unknown.

2. Name of Acute Disease: Gastroenteritis is name of the disease occasionally and sporadically caused by these genera.

3. Nature of Disease: Acute gastroenteritis is characterized by two or more of the symptoms of vomiting, nausea, fever, chills, abdominal pain, and watery (dehydrating) diarrhea occurring 12-24 hours after ingestion of contaminated food or water. Chronic diarrheal disease is characterized by dysenteric symptoms: foul-smelling, mucus-containing, diarrheic stool with flatulence and abdominal distention. The chronic disease may continue for months and require antibiotic treatment.

Infectious dose--unknown. Both the acute and chronic forms of the disease are suspected to result from the elaboration of enterotoxins. These organisms may become transiently virulent by gaining mobilizeable genetic elements from other pathogens. For example, pathogenic *Citrobacter freundii* which elaborated a toxin identical to *E. coli* heat-stable toxin was isolated from the stools of ill children.

4. Diagnosis of Human Illness: Recovery and identification methods for these organisms from food, water or diarrheal specimens are based upon the efficacy of selective media and results of microbiological and biochemical assays. The ability to produce enterotoxin(s) may be determined by cell culture assay and animal bioassays, serological methods, or genetic probes.

5. Associated Foods: These bacteria have been recovered from dairy products, raw shellfish, and fresh raw vegetables. The organisms occur in soils used for crop production and shellfish harvesting waters and, therefore, may pose a health hazard.

6. Relative Frequency of Disease: Acute gastrointestinal illness may occur more frequently in undeveloped areas of the world. The chronic illness is common in malnourished children living in unsanitary conditions in tropical countries.

7. Usual Course of Disease and Some Complications: Healthy individuals recover quickly and without treatment from the acute form of gastrointestinal disease. Malnourished children (1-4 years) and infants who endure chronic diarrhea soon develop structural and functional abnormalities of their intestinal tracts resulting in loss of ability to absorb nutrients. Death is not uncommon in these children and results indirectly from the chronic toxigenic effects which produce the malabsorption and malnutrition.

8. Target Populations: All people may be susceptible to pathogenic forms of these bacteria. Protracted illness is more commonly experienced by the very young.

9. Food Analysis: These strains are recovered by standard selective and differential isolation procedures for enteric bacteria. Biochemical and in vitro assays may be used to determine species and pathogenic potential. Not being usually thought of as human pathogens, they may easily be overlooked by the clinical microbiology laboratory.

10. Selected Outbreaks: Intestinal infections with these species in the U.S. have usually taken the form of sporadic cases of somewhat doubtful etiology.

Citrobacter freundii was suspected by CDC of causing an outbreak of diarrheal disease in Washington, DC. Imported Camembert cheese was incriminated.

For more information on recent outbreaks see the Morbidity and Mortality Weekly Reports from CDC.

Streptococcus spp.

1. Name of Organism: *Streptococcus* spp.

The genus Streptococcus is comprised of Gram-positive, microaerophilic cocci (round), which are not motile and occur in chains or pairs. The genus is defined by a combination of antigenic, hemolytic, and physiological characteristics into Groups A, B, C, D, F, and G. Groups A and D can be transmitted to humans via food.

Group A: one species with 40 antigenic types (*S. pyogenes*).

Group D: five species (*S. faecalis, S. faecium, S. durans, S. avium,* and *S. bovis*).

2. Name of Acute Disease:
Group A: Cause septic sore throat and scarlet fever as well as other pyogenic and septicemic infections.

Group D: May produce a clinical syndrome similar to staphylococcal intoxication.

3. Nature of Illness/ Disease:
Group A: Sore and red throat, pain on swallowing, tonsilitis, high fever, headache, nausea, vomiting, malaise, rhinorrhea; occasionally a rash occurs, onset 1-3 days; the infectious dose is probably quite low (less than 1,000 organisms).

Group D: Diarrhea, abdominal cramps, nausea, vomiting, fever, chills, dizziness in 2-36 hours. Following ingestion of suspect food, the infectious dose is probably high (greater than 107 organisms).

4. Diagnosis of Human Disease:
Group A: Culturing of nasal and throat swabs, pus, sputum, blood, suspect food, environmental samples.

Group D: Culturing of stool samples, blood, and suspect food.

5. Associated Foods:
Group A: Food sources include milk, ice cream, eggs, steamed lobster, ground ham, potato salad, egg salad, custard, rice pudding, and shrimp salad. In almost all cases, the foodstuffs were allowed to stand at room temperature for several hours between preparation and consumption. Entrance into the food is the result of poor hygiene, ill food handlers, or the use of unpasteurized milk.

Group D: Food sources include sausage, evaporated milk, cheese, meat croquettes, meat pie, pudding, raw milk, and pasteurized milk. Entrance into the food chain is due to underprocessing and/or poor and unsanitary food preparation.

6. Relative Frequency of Infection:
Group A infections are low and may occur in any season, whereas Group D infections are variable.

7. Usual Course of Disease and Complications:

Group A: Streptococcal sore throat is very common, especially in children. Usually it is successfully treated with antibiotics. Complications are rare and the fatality rate is low.

Group D: Diarrheal illness is poorly characterized, but is acute and self-limiting.

8. Target Population:

All individuals are susceptible. No age or race susceptibilities have been found.

9. Analysis of Foods:

Suspect food is examined microbiologically by selective enumeration techniques which can take up to 7 days. Group specificities are determined by Lancefield group-specific antisera.

10. Selected Outbreaks:

Group A: Outbreaks of septic sore throat and scarlet fever were numerous before the advent of milk pasteurization. Salad bars have been suggested as possible sources of infection. Most current outbreaks have involved complex foods (i.e., salads) which were infected by a food handler with septic sore throat. One ill food handler may subsequently infect hundreds of individuals.

Group D: Outbreaks are not common and are usually the result of preparing, storing, or handling food in an unsanitary manner.

For more information on recent outbreaks see the Morbidity and Mortality Weekly Reports from CDC.

PARASITIC PROTOZOA and WORMS

Giardia lamblia

1. Name of Organism: *Giardia lamblia*

Giardia lamblia (intestinalis) is a single celled animal, i.e., a protozoa, that moves with the aid of five flagella. In Europe, it is sometimes referred to as Lamblia intestinalis.

2. Disease Name: Giardiasis is the most frequent cause of non-bacterial diarrhea in North America.

3. Nature of the Disease: Organisms that appear identical to those that cause human illness have been isolated from domestic animals (dogs and cats) and wild animals (beavers and bears). A related but morphologically distinct organism infects rodents, although rodents may be infected with human isolates in the laboratory. Human giardiasis may involve diarrhea within 1 week of ingestion of the cyst, which is the environmental survival form and infective stage of the organism. Normally illness lasts for 1 to 2 weeks, but there are cases of chronic infections lasting months to years. Chronic cases, both those with defined immune deficiencies and those without, are difficult to treat. The disease mechanism is unknown, with some investigators reporting that the organism produces a toxin while others are unable to confirm its existence. The organism has been demonstrated inside host cells in the duodenum, but most investigators think this is such an infrequent occurrence that it is not responsible for disease symptoms. Mechanical obstruction of the absorptive surface of the intestine has been proposed as a possible pathogenic mechanism, as has a synergistic relationship with some of the intestinal flora. Giardia can be excysted, cultured and encysted in vitro; new isolates have bacterial, fungal, and viral symbionts. Classically the disease was diagnosed by demonstration of the organism in stained fecal smears. Several strains of *G. lamblia* have been isolated and described through analysis of their proteins and DNA; type of strain, however, is not consistently associated with disease severity. Different individuals show various degrees of symptoms when infected with the same strain, and the symptoms of an individual may vary during the course of the disease.

Infectious Dose -- Ingestion of one or more cysts may cause disease, as contrasted to most bacterial illnesses where hundreds to thousands of organisms must be consumed to produce illness.

4. Diagnosis of Human Illness: *Giardia lamblia* is frequently diagnosed by visualizing the organism, either the trophozoite (active reproducing form) or the cyst (the resting stage that is resistant to adverse environmental conditions) in stained preparations or unstained wet mounts with the aid of a microscope. A commercial fluorescent antibody kit is available to stain the organism. Organisms may be concentrated by sedimentation or flotation; however, these procedures reduce the number of recognizable organisms in the sample. An enzyme linked immunosorbant assay (ELISA) that detects excretory secretory products of the organism is also available. So far, the increased sensitivity of indirect serological detection has not been consistently demonstrated.

5. Associated Foods: Giardiasis is most frequently associated with the consumption of contaminated water. Five outbreaks have been traced to food contamination by infected or infested food handlers, and the possibility of infections from contaminated vegetables that are eaten raw cannot be excluded. Cool moist conditions favor the survival of the organism.

6. Relative Frequency of Disease: Giardiasis is more prevalent in children than in adults, possibly because many individuals seem to have a lasting immunity after infection. This organism is implicated in 25% of the cases of gastrointestinal disease and may be present asymptomatically. The overall incidence of infection in the United States is estimated at 2% of the population. This disease afflicts many homosexual men, both HIV-positive and HIV-negative individuals. This is presumed to be due to sexual transmission. The disease is also common in child day care centers, especially those in which diapering is done.

7. Complications: About 40% of those who are diagnosed with giardiasis demonstrate disaccharide intolerance during detectable infection and up to 6 months after the infection can no longer be detected. Lactose (i.e., milk sugar) intolerance is most frequently observed. Some individuals (less than 4%) remain symptomatic more than 2 weeks; chronic infections lead to a malabsorption syndrome and severe weight loss. Chronic cases of giardiasis in immunodeficient and normal individuals are frequently refractile to drug treatment. Flagyl is normally quite effective in terminating infections. In some immune deficient individuals, giardiasis may contribute to a shortening of the life span.

8. Target Populations: Giardiasis occurs throughout the population, although the prevalence is higher in children than adults. Chronic symptomatic giardiasis is more common in adults than children.

9. Food Analysis: Food is analyzed by thorough surface cleaning of the suspected food and sedimentation of the organisms from the cleaning water. Feeding to specific pathogen-free animals has been used to detect the organism in large outbreaks associated with municipal water systems. The precise sensitivity of these methods has not been determined, so that negative results are questionable. Seven days may be required to detect an experimental infection.

10. Selected Outbreaks: Major outbreaks are associated with contaminated water systems that do not use sand filtration or have a defect in the filtration system. The largest reported foodborne outbreak involved 24 of 36 persons who consumed macaroni salad at a picnic.

For more information on recent outbreaks see the Morbidity and Mortality Weekly Reports from CDC.

11. FDA Regulations or Activity: FDA is actively developing and improving methods of recovering parasitic protozoa and helminth eggs from foods. Current recovery methods are published in the FDA's *Bacteriological Analytical Manual.*

Entamoeba histolytica

1. Name of Organism: *Entamoeba histolytica*

This is a single celled parasitic animal, i.e., a protozoa, that infects predominantly humans and other primates. Diverse mammals such as dogs and cats can become infected but usually do not shed cysts (the environmental survival form of the organism) with their feces, thus do not contribute significantly to transmission. The active (trophozoite) stage exists only in the host and in fresh feces; cysts survive outside the host in water and soils and on foods, especially under moist conditions on the latter. When swallowed they cause infections by excysting (to the trophozoite stage) in the digestive tract.

2. Name of Acute Disease: Amebiasis (or amoebiasis) is the name of the infection caused by *E. histolytica*.

3. Nature of the Acute Disease: Infections that sometimes last for years may be accompanied by 1) no symptoms, 2) vague gastrointestinal distress, 3) dysentery (with blood and mucus). Most infections occur in the digestive tract but other tissues may be invaded. Complications include 4) ulcerative and abscess pain and, rarely, 5) intestinal blockage. Onset time is highly variable. It is theorized that the absence of symptoms or their intensity varies with such factors as 1) strain of amoeba, 2) immune health of the host, and 3) associated bacteria and, perhaps, viruses. The amoeba's enzymes help it to penetrate and digest human tissues; it secretes toxic substances.

Infectious Dose--Theoretically, the ingestion of one viable cyst can cause an infection.

4. Diagnosis of Human Illness: Human cases are diagnosed by finding cysts shed with the stool; various flotation or sedimentation procedures have been developed to recover the cysts from fecal matter; stains (including fluorescent antibody) help to visualize the isolated cysts for microscopic examination. Since cysts are not shed constantly, a minimum of 3 stools should be examined. In heavy infections, the motile form (the trophozoite) can be seen in fresh feces. Serological tests exist for long-term infections. It is important to distinguish the *E. histolytica* cyst from the cysts of nonpathogenic intestinal protozoa by its appearance.

5. Transmission: Amebiasis is transmitted by fecal contamination of drinking water and foods, but also by direct contact with dirty hands or objects as well as by sexual contact.

6. Frequency of Infections: The infection is "not uncommon" in the tropics and arctics, but also in crowded situations of poor hygiene in temperate-zone urban environments. It is also frequently diagnosed among homosexual men.

7. Usual Course of the Disease and Some Complications: In the majority of cases, amoebas remain in the gastrointestinal tract of the hosts. Severe ulceration of the gastrointestinal mucosal surfaces occurs in less than 16% of cases. In fewer cases, the parasite invades the soft tissues, most commonly the liver. Only rarely are masses formed (amoebomas) that lead to intestinal obstruction. Fatalities are infrequent.

8. Target Populations: All people are believed to be susceptible to infection, but individuals with a damaged or undeveloped immunity may suffer more severe forms of the disease. AIDS/ARC patients are very vulnerable.

9. Analysis of Foods: *E. histolytica* cysts may be recovered from contaminated food by methods similar to those used for recovering *Giardia lamblia* cysts from feces. Filtration is probably the most practical method for recovery from drinking water and liquid foods. *E. histolytica* cysts must be distinguished from cysts of other parasitic (but non-pathogenic) protozoa and from cysts of free-living protozoa. Recovery procedures are not very accurate; cysts are easily lost or damaged beyond recognition, which leads to many falsely negative results in recovery tests. (See the FDA Bacteriological AnalyticalManual.)

10. Selected Outbreaks: The most dramatic incident in the USA was the Chicago World's Fair outbreak in 1933 caused by contaminated drinking water; defective plumbing permitted sewage to contaminate the drinking water. There were 1,000 cases (with 58 deaths). In recent times, food handlers are suspected of causing many scattered infections, but there has been no single large outbreak.

For more information on recent outbreaks see the Morbidity and Mortality Weekly Reports from CDC.

Cryptosporidium parvum

1. Name of Organism: *Cryptosporidium parvum*

Cryptosporidium parvum, a single-celled animal, i.e., a protozoa, is an obligate intracellular parasite. It has been given additional species names when isolated from different hosts. It is currently thought that the form infecting humans is the same species that causes disease in young calves. The forms that infect avian hosts and those that infect mice are not thought capable of infecting humans. *Cryptosporidium* sp. infects many herd animals (cows, goats, sheep among domesticated animals, and deer and elk among wild animals). The infective stage of the organism, the oocyst is 3 um in diameter or about half the size of a red blood cell. The sporocysts are resistant to most chemical disinfectants, but are susceptible to drying and the ultraviolet portion of sunlight. Some strains appear to be adapted to certain hosts but cross-strain infectivity occurs and may or may not be associated with illness. The species or strain infecting the respiratory system is not currently distinguished from the form infecting the intestines.

2. Disease Name: Intestinal, tracheal, or pulmonary cryptosporidiosis.

3. Nature of Acute Disease: Intestinal cryptosporidiosis is characterized by severe watery diarrhea but may, alternatively, be asymptomatic. Pulmonary and tracheal cryptosporidiosis in humans is associated with coughing and frequently a low-grade fever; these symptoms are often accompanied by severe intestinal distress.

Infectious dose--Less than 10 organisms and, presumably, one organism can initiate an infection. The mechanism of disease is not known; however, the intracellular stages of the parasite can cause severe tissue alteration.

4. Diagnosis of Human Illness: Oocysts are shed in the infected individual's feces. Sugar flotation is used to concentrate the organisms and acid fast staining is used to identify them. A commercial kit is available that uses fluorescent antibody to stain the organisms isolated from feces. Diagnosis has also been made by staining the trophozoites in intestinal and biopsy specimens. Pulmonary and tracheal cryptosporidiosis are diagnosed by biopsy and staining.

5. Food Occurence: *Cryptosporidium* sp. could occur, theoretically, on any food touched by a contaminated food handler. Incidence is higher in child day care centers that serve food. Fertilizing salad vegetables with manure is another possible source of human infection. Large outbreaks are associated with contaminated water supplies.

6. Relative Frequency the Disease: Direct human surveys indicate a prevalence of about 2% of the **of** population in North America. Serological surveys indicate that 80% of the population has had cryptosporidiosis. The extent of illness associated with reactive sera is not known.

7. Usual Course of the Disease and Complications: Intestinal cryptosporidiosis is self-limiting in most healthy individuals, with watery diarrhea lasting 2-4 days. In some outbreaks at day care centers, diarrhea has lasted 1 to 4 weeks. To date, there is no known effective drug for the treatment of cryptosporidiosis. Immunodeficient individuals, especially AIDS patients, may have the disease for life, with the severe watery diarrhea contributing to death. Invasion of the pulmonary system may also be fatal.

8. Target Populations: In animals, the young show the most severe symptoms. For the most part, pulmonary infections are confined to those who are immunodeficient. However, an infant with a presumably normal immune system had tracheal cryptosporidiosis (although a concurrent viremia may have accounted for lowered resistance). Child day care centers, with a large susceptible population, frequently report outbreaks.

9. Analysis of Foods: The 7th edition of FDA's Bacteriological Analytical Manual will contain a method for the examination of vegetables for *Cryptosporidium* sp.

10. Selected Outbreaks: Since 1984, cryptosporidiosis has been associated with outbreaks of diarrheal illness in child day care centers throughout the United States and Canada. During 1987 a waterborne outbreak in Georgia produced illness in an estimated 13,000 individuals, and exposure to contaminated drinking water was the major distinction between those that were ill and those that were not. This was the first report of disease transmission by a municipal water system that was in compliance with all state and federal standards for

An outbreak of cryptosporidiosis associated with the consumption of apple cider is reported in MMWR 46(1):1997 Jan 10.

MMWR 45(36):1996 Sep 13 reports on an outbreak of cryptosporidiosis associated with the consumption of home-made chicken salad in Minnesota.

A non-food outbreak of cryptosporidiosis in a day-camp is reported in MMWR 45(21):1995 May 31. This report is linked to this chapter to provide reference information.

MMWR 39(20):1990 May 25 reports on a non-food related outbreak of cryptosporidiosis, but contains useful information on *Cryptosporidium* sp.

For more information on recent outbreaks see the Morbidity and Mortality Weekly Reports from CDC.

11. FDA Regulations or Activity: FDA is developing and improving methods for the recovery of cysts of parasitic protozoa from fresh vegetables. Current recovery methods are published in the Bacteriological Analytical Manual.

12. Education: The CDC has information on Cryptosporidium.

13. Other Resources: From GenBank there is a Loci index for genome *Cryptosporidium parvum.*

Anisakis simplex and related worms

1. Name of Organism: *Anisakis simplex* and related worms

Anisakis simplex (herring worm), *Pseudoterranova (Phocanema, Terranova) decipiens* (cod or seal worm), *Contracaecum* spp., and *Hysterothylacium (Thynnascaris)* spp. are anisakid nematodes (roundworms) that have been implicated in human infections caused by the consumption of raw or undercooked seafood. To date, only *A. simplex* and *P. decipiens* are reported from human cases in North America.

2. Name of Acute Disease: Anisakiasis is generally used when referring to the acute disease in humans. Some purists utilize generic names (e.g., contracaeciasis) in referring to the disease, but the majority consider that the name derived from the family is specific enough. The range of clinical features is not dependent on species of anisakid parasite in cases reported to date.

3. Nature of the Acute Disease: In North America, anisakiasis is most frequently diagnosed when the affected individual feels a tingling or tickling sensation in the throat and coughs up or manually extracts a nematode. In more severe cases there is acute abdominal pain, much like acute appendicitis accompanied by a nauseous feeling. Symptoms occur from as little as an hour to about 2 weeks after consumption of raw or undercooked seafood. One nematode is the usual number recovered from a patient. With their anterior ends, these larval nematodes from fish or shellfish usually burrow into the wall of the digestive tract to the level of the muscularis mucosae (occasionally they penetrate the intestinal wall completely and are found in the body cavity). They produce a substance that attracts eosinophils and other host white blood cells to the area. The infiltrating host cells form a granuloma in the tissues surrounding the penetrated worm. In the digestive tract lumen, the worm can detach and reattach to other sites on the wall. Anisakids rarely reach full maturity in humans and usually are eliminated spontaneously from the digestive tract lumen within 3 weeks of infection. Penetrated worms that die in the tissues are eventually removed by the host's phagocytic cells.

4. Diagnosis of Human Illness: In cases where the patient vomits or coughs up the worm, the disease may be diagnosed by morphological examination of the nematode. (*Ascaris lumbricoides*, the large roundworm of humans, is a terrestrial relative of anisakines and sometimes these larvae also crawl up into the throat and nasal passages.) Other cases may require a fiber optic device that allows the attending physician to examine the inside of the stomach and the first part of the small intestine. These devices are equipped with a mechanical forceps that can be used to remove the worm. Other cases are diagnosed upon finding a granulomatous lesion with a worm on laparotomy. A specific radioallergosorbent test has been developed for anasakiasis, but is not yet commercially marketed.

FDA 69

5. Associated Foods: Seafoods are the principal sources of human infections with these larval worms. The adults of *A. simplex* are found in the stomachs of whales and dolphins. Fertilized eggs from the female parasite pass out of the host with the host's feces. In seawater, the eggs embryonate, developing into larvae that hatch in sea water. These larvae are infective to copepods (minute crustaceans related to shrimp) and other small invertebrates. The larvae grow in the invertebrate and become infective for the next host, a fish or larger invertebrate host such as a squid. The larvae may penetrate through the digestive tract into the muscle of the second host. Some evidence exists that the nematode larvae move from the viscera to the flesh if the fish hosts are not gutted promptly after catching. The life cycles of all the other anisakid genera implicated in human infections are similar. These parasites are known to occur frequently in the flesh of cod, haddock, fluke, pacific salmon, herring, flounder, and monkfish.

6. Relative Frequency of the Disease: Fewer than 10 cases are diagnosed in the U.S. annually. However, it is suspected that many other cases go undetected. The disease is transmitted by raw, undercooked or insufficiently frozen fish and shellfish, and its incidence is expected to increase with the increasing popularity of sushi and sashimi bars.

7. Usual Disease Course and Complications: Severe cases of anisakiasis are extremely painful and require surgical intervention. Physical removal of the nematode(s) from the lesion is the only known method of reducing the pain and eliminating the cause (other than waiting for the worms to die). The symptoms apparently persist after the worm dies since some lesions are found upon surgical removal that contain only nematode remnants. Stenosis (a narrowing and stiffening) of the pyloric sphincter was reported in a case in which exploratory laparotomy had revealed a worm that was not removed.

8. Target Populations: The target population consists of consumers of raw or underprocessed seafood.

9. Analysis of Foods: Candling or examining fish on a light table is used by commercial processors to reduce the number of nematodes in certain whiteflesh fish that are known to be infected frequently. This method is not totally effective, nor is it very adequate to remove even the majority of nematodes from fish with pigmented flesh.

10. Selected Outbreaks: This disease is known primarily from individual cases. Japan has the greatest number of reported cases because of the large volume of raw fish consumed there.

A recent letter to the editor of the New England Journal of Medicine (319:1128-29, 1988) stated that approximately 50 cases of anisakiasis have been documented in the United States, to date. Three cases in the San Francisco Bay area involved ingestion of sushi or undercooked fish. The letter also points out that anasakiasis is easily misdiagnosed as acute appendicitis, Crohn's disease, gastric ulcer , or gastrointestinal cancer.

For more information on recent outbreaks see the Morbidity and Mortality Weekly Reports from CDC.

11. FDA Activity and Regulations: FDA recommends that all fish and shellfish intended for raw (or semiraw such as marinated or partly cooked) consumption be blast frozen to -35°C (-31°F) or below for 15 hours, or be regularly frozen to -20°C (-4°F) or below for 7 days.

Diphyllobothrium spp.

1. Name of Organism: *Diphyllobothrium* spp.

Diphyllobothrium latum and other members of the genus are broad fish tapeworms reported from humans. They are parasitic flatworms.

2. Name of the Acute Disease: Diphyllobothriasis is the name of the disease caused by broad fish tapeworm infections.

3. Nature of the Acute Disease: Diphyllobothriasis is characterized by abdominal distention, flatulence, intermittent abdominal cramping, and diarrhea with onset about 10 days after consumption of raw or insufficiently cooked fish. The larva that infects people, a "plerocercoid," is frequently encountered in the viscera of freshwater and marine fishes. *D. latum* is sometimes encountered in the flesh of freshwater fish or fish that are anadromous (migrating from salt water to fresh water for breeding). Bears and humans are the final or definitive hosts for this parasite. *D. latum* is a broad, long tapeworm, often growing to lengths between 1 and 2 meters (3-7 feet) and potentially capable of attaining 10 meters (32 feet); the closely related *D. pacificum* normally matures in seals or other marine mammals and reaches only about half the length of *D. latum*. Treatment consists of administration of the drug, niclosamide, which is available to physicians through the Centers for Disease Control's Parasitic Disease Drug Service.

4. Diagnosis of Human Illness: The disease is diagnosed by finding operculate eggs (eggs with a lid) in the patient's feces on microscopical examination. These eggs may be concentrated by sedimentation but not by flotation. They are difficult to distinguish from the eggs of *Nanophyetus* spp..

5. Associated Foods: The larvae of these parasites are sometimes found in the flesh of fish.

6. Relative Frequency of Disease: Diphyllobothriasis is rare in the United States, although it was formerly common around the Great Lakes and known as "Jewish or Scandinavian housewife's disease" because the preparers of gefillte fish or fish balls tended to taste these dishes before they were fully cooked. The parasite is now supposedly absent from Great Lakes fish. Recently, cases have been reported from the West Coast.

7. Usual Course of the Disease and Complications: In persons that are genetically susceptible, usually persons of Scandinavian heritage, a severe anemia may develop as the result of infection with broad fish tapeworms. The anemia results from the tapeworm's great requirement for and absorption of Vitamin B12.

8. Target Populations: Consumers of raw and underprocessed fish are the target population for diphyllobothriasis.

9. Analysis of Foods: Foods are not routinely analyzed for larvae of *D. latum*, but microscopic inspection of thin slices of fish, or digestion, can be used to detect this parasite in fish flesh.

10. Selected Outbreaks: An outbreak involving four Los Angeles physicians occurred in 1980. These physicians all consumed sushi (a raw fish dish) made of tuna, red snapper, and salmon. Others who did not consume the sushi made with salmon did not contract diphyllobothriasis. At the time of this outbreak there was also a general increase in requests for niclosamide from CDC; interviews of 39 patients indicated that 32 recalled consuming salmon prior to their illness.

For more information on recent outbreaks see the Morbidity and Mortality Weekly Reports from CDC.

11. FDA Activity and Regulations: FDA is determining whether the freezing recommendations (see chapter 25) for raw or semiraw seafood with anisakid nematodes will also prevent infections with the broad fish tapeworms.

Nanophyetus spp.

1. Name of Organism: *Nanophyetus* spp.

Nanophyetus salmincola or *N. schikhobalowi* are the names, respectively, of the North American and Russian troglotrematoid trematodes (or flukes). These are parasitic flatworms.

2. Name of the Acute Disease: Nanophyetiasis is the name of the human disease caused by these flukes. At least one newspaper referred to the disease as "fish flu." *N. salmincola* is responsible for the transmission of *Neorickettsia helminthoeca*, which causes an illness in dogs that may be serious or even fatal.

3. Nature of the Acute Disease: Knowledge of nanophyetiasis is limited. The first reported cases are characterized by an increase of bowel movements or diarrhea, usually accompanied by increased numbers of circulating eosinophils, abdominal discomfort and nausea. A few patients reported weight loss and fatigue, and some were asymptomatic. The rickettsia, though fatal to 80% of untreated dogs, is not known to infect humans.

4. Diagnosis of Human Infections: Detection of operculate eggs of the characteristic size and shape in the feces is indicative of nanophyetiasis. The eggs are difficult to distinguish from those of *Diphyllobothrium latum*.

5. Relative Frequency of the Disease: There have been no reported outbreaks of nanophyetiasis in North America; the only scientific reports are of 20 individual cases referred to in one Oregon clinic. A report in the popular press indicates that the frequency is significantly higher. It is significant that two cases occurred in New Orleans well outside the endemic area. In Russia's endemic area the infection rate is reported to be greater than 90% and the size of the endemic area is growing.

6. Associated Foods: Nanophyetiasis is transmitted by the larval stage (metacercaria) of a worm that encysts in the flesh of freshwater fishes. In anadromous fish, the parasite's cysts can survive the period spent at sea. Although the metacercaria encysts in many species of fish, North American cases were all associated with salmonids. Raw, underprocessed, and smoked salmon and steelhead were implicated in the cases to date.

7. Usual Course of the Disease and Treatment: Mebendazole was ineffective as a treatment; patients kept shedding eggs, and symptoms gradually decreased over 2 months or more. Treatment with two doses of bithionol or three doses of niclosamide resulted in the resolution of symptoms and disappearance of eggs in the feces. These drugs are available in the U.S. from the Centers for Disease Control's Parasitic Drug Service.

8. Target Population: Consumers of raw or underprocessed freshwater or anadromous fish, especially salmonids.

9. Analysis of Foods: There are no tested methods for detection of *Nanophyetus* spp. in fishes. Candling with the aid of a dissecting microscope, or pepsin HCl digestion should detect heavily infected fish.

10. Selected Outbreaks: None

For more information on recent outbreaks see the Morbidity and Mortality Weekly Reports from CDC.

11. FDA Activity and Regulations: FDA has no specific regulation or activity regarding these trematodes. As pathogens, however, they should not be live in fish consumed raw or semiraw.

Eustrongylides sp.

1. Name of Organism: *Eustrongylides* sp.

Larval Eustrongylides sp. are large, bright red roundworms (nemotodes), 25-150 mm long, 2 mm in diameter. They occur in freshwater fish, brackish water fish and in marine fish. The larvae normally mature in wading birds such as herons, egrets, and flamingos.

2. Nature of the Acute Disease: If the larvae are consumed in undercooked or raw fish, they can attach to the wall of the digestive tract. In the five cases for which clinical symptoms have been reported, the penetration into the gut wall was accompanied by severe pain. The nematodes can perforate the gut wall and probably other organs. Removal of the nematodes by surgical resection or fiber optic devices with forceps is possible if the nematodes penetrate accessible areas of the gut.

3. Infective Dose: One live larva can cause an infection.

4. Diagnosis of Human Illness: In three of the five reported cases, the worms were diagnosed by surgical resection of the intestine. In one case, there was no clinical data and in one other, the patient was treated medically and recovered in 4 days.

5. Associated Foods: Fish from fresh, brackish or salt water.

6. Relative Frequency of Disease: The disease is extremely rare; there have been only five cases reported in the U.S.

7. Complications: Septicemia, which is due to the perforated digestive tract.

8. Target Populations: Those consuming whole minnows are at greatest risk. One case was reported from the consumption of sashimi.

9. Food Analysis: These large worms may be seen without magnification in the flesh of fish and are normally very active after death of the fish.

10. Selected Outbreaks: There have been no major outbreaks.

For more information on recent outbreaks see the Morbidity and Mortality Weekly Reports from CDC.

11. FDA Regulation or Activity: FDA has no specific regulation or activity regarding these worms; however, as pathogens, no live *Eustrongylides* sp. should be present in fish consumed raw or semiraw.

Acanthamoeba spp., Naegleria fowleri and other amobae

1. Name of Organisms: *Acanthamoeba* spp., *Naegleria fowleri* and other amobae

Members of the two genera named above are the principal examples of protozoa commonly referred to as pathogenic free-living amoebae.

2. Disease Name: Primary amoebic meningoencephalitis (PAM), *Naegleria fowleri* and granulomatious amoebic encephalitis (GAE), acanthamoebic keratitis or acanthamoebic uveitis.

These organisms are ubiquitous in the environment, in soil, water, and air. Infections in humans are rare and are acquired through water entering the nasal passages (usually during swimming) and by inhalation. They are discussed here because the FDA receives inquiries about them.

3. Nature of the Acute Disease: PAM occurs in persons who are generally healthy prior to infection. Central nervous system involvement arises from organisms that penetrate the nasal passages and enter the brain through the cribriform plate. The organisms can multiply in the tissues of the central nervous system and may be isolated from spinal fluid. In untreated cases death occurs within 1 week of the onset of symptoms. Amphotercin B is effective in the treatment of PAM. At least four patients have recovered when treated with Amphotercin B alone or in combination with micronazole administered both intravenously and intrathecally or intraventrically.

GAE occurs in persons who are immunodeficient in some way; the organisms cause a granulomatous encephalitis that leads to death in several weeks to a year after the appearance of symptoms. The primary infection site is thought to be the lungs, and the organisms in the brain are generally associated with blood vessels, suggesting vascular dissemination. Treatment with sulfamethazine may be effective in controling the amobae.

Prior to 1985 amoebae had been reported isolated from diseased eyes only rarely; cases were associated with trauma to the eye. In 1985-1986, 24 eye cases were reported to CDC and most of these occurred in wearers of contact lenses. It has been demonstrated that many of these infections resulted from the use of home-made saline solutions with the contact lenses. Some of the lenses had been heat treated and others had been chemically disinfected. The failure of the heat treatment was attributed to faulty equipment, since the amoebae are killed by 65°C (149°F) for 30 minutes. The failure of the chemical disinfection resulted from insufficient treatment or rinsing the lenses in contaminated saline after disinfection. The following agents have been used to successfully eliminate the amoebic infection in the eye: ketoconazole , microconazole, and propamidine isothionate; however, penetrating keratoplasty has been necessary to restore useful vision.

4. Diagnosis of Human Illness: PAM is diagnosed by the presence of amoebae in the spinal fluid. GAE is diagnosed by biopsy of the lesion. Ocular amoebic keratitis may be diagnosed by culturing corneal scrapings on nonnutrient agar overlaid with viable *Escherichia coli*; amoebae from PAM and GAE may be cultured by the same method. Clinical diagnosis by experienced practitioners is based on the characteristic stromal infiltrate.

5. Transmission: Transmission is through water based fluids or the air.

6. Frequency of Infections: PAM and GAE are rare in occurrence; fewer than 100 cases have been reported in the United States in the 25 years since these diseases were recognized.

7. Complications: PAM and GAE both lead to death in most cases. Eye infections may lead to blindness.

8. Target Populations: Immunodeficients, especially those infected with HIV, may be at risk for atypical infections. PAM, GAE, and eye infections have occurred in otherwise healthy individuals.

9. Food Analysis: Foods are not analyzed for these amoebae since foods are not implicated in the infection of individuals.

10. Selected Outbreaks: These diseases are known only from isolated cases. For more information on recent outbreaks see the Morbidity and Mortality Weekly Reports from CDC.

11. FDA Activity and Regulations: Since infection is not known to be by way of the digestive tract, the FDA has no regulations concerning these organisms. Eye infections are indirectly regulated by FDA's Center for Medical Devices and Radiological Health; FDA's Center for Drug Evaluation and Research regulates heat sterilization units and saline solutions for ophthalmological use. FDA has published a paper documenting the presence of amoebae in eye wash stations, and warning about the potential danger of such contamination.

Ascaris lumbricoides and Trichuris trichiura

1. Name of Organisms: *Ascaris lumbricoides* and *Trichuris trichiura*

Humans worldwide are infected with *Ascaris lumbricoides* and *Trichuris trichiura*; the eggs of these roundworms (nematode) are "sticky" and may be carried to the mouth by hands, other body parts, fomites (inanimate objects), or foods.

2. Name of Acute Disease: Ascariasis and trichuriasis are the scientific names of these infections. Ascariasis is also known commonly as the "large roundworm" infection and trichuriasis as "whip worm" infection.

3. Nature of the Acute Disease: Infection with one or a few *Ascaris* sp. may be inapparent unless noticed when passed in the feces, or, on occasion, crawling up into the throat and trying to exit through the mouth or nose. Infection with numerous worms may result in a pneumonitis during the migratory phase when larvae that have hatched from the ingested eggs in the lumen of the small intestine penetrate into the tissues and by way of the lymph and blood systems reach the lungs. In the lungs, the larvae break out of the pulmonary capillaries into the air sacs, ascend into the throat and descend to the small intestine again where they grow, becoming as large as 31 X 4 cm. Molting (ecdysis) occurs at various points along this path and, typically for roundworms, the male and female adults in the intestine are 5th-stage nematodes. Vague digestive tract discomfort sometimes accompanies the intestinal infection, but in small children with more than a few worms there may be intestinal blockage because of the worms' large size. Not all larval or adult worms stay on the path that is optimal for their development; those that wander may locate in diverse sites throughout the body and cause complications. Chemotherapy with anthelmintics is particularly likely to cause the adult worms in the intestinal lumen to wander; a not unusual escape route for them is into the bile duct which they may occlude. The larvae of ascarid species that mature in hosts other than humans may hatch in the human intestine and are especially prone to wander; they may penetrate into tissues and locate in various organ systems of the human body, perhaps eliciting a fever and diverse complications.

Trichuris sp. larvae do not migrate after hatching but molt and mature in the intestine. Adults are not as large as *A. lumbricoides*. Symptoms range from inapparent through vague digestive tract distress to emaciation with dry skin and diarrhea (usually mucoid). Toxic or allergic symptoms may also occur.

4. Diagnosis of Human Illness: Both infections are diagnosed by finding the typical eggs in the patient's feces; on occasion the larval or adult worms are found in the feces or, especially for *Ascaris* sp., in the throat, mouth, or nose.

5. Associated Foods: The eggs of these worms are found in insufficiently treated sewage-fertilizer and in soils where they embryonate (i.e., larvae develop in fertilized eggs). The eggs may contaminate crops grown in soil or fertilized with sewage that has received nonlethal treatment; humans are infected when such produce is consumed raw. Infected foodhandlers may contaminate a wide variety of foods.

6. Relative Frequency of Disease: These infections are cosmopolitan, but ascariasis is more common in North America and trichuriasis in Europe. Relative infection rates on other continents are not available.

7. Usual Course of Disease and Complications: Both infections may self-cure after the larvae have matured into adults or may require anthelmintic treatment. In severe cases, surgical removal may be necessary. Allergic symptoms (especially but not exclusively of the asthmatic sort) are common in long-lasting infections or upon reinfection in ascariasis.

8. Target Populations: Particularly consumers of uncooked vegetables and fruits grown in or near soil fertilized with sewage.

9. Analysis of Foods: Eggs of *Ascaris* spp. have been detected on fresh vegetables (cabbage) sampled by FDA. Methods for the detection of *Ascaris* spp. and *Trichuris* spp. eggs are detailed in the FDA's Bacteriological Analytical Manual.

10. Selected Outbreaks: Although no major outbreaks have occurred, there are many individual cases. The occurrence of large numbers of eggs in domestic municipal sewage implies that the infection rate, especially with *A. lumbricoides*, is high in the U.S.

For more information on recent outbreaks see the Morbidity and Mortality Weekly Reports from CDC.

11. FDA Activity and Regulations: Ascarids and trichurids are considered pathogens and foods eaten without further cooking should not be contaminated with viable embryonated eggs of either genus.

VIRUSES

Hepatitis A Virus

1. Name of Organism: Hepatitis A Virus

Hepatitis A virus (HAV) is classified with the enterovirus group of the Picornaviridae family. HAV has a single molecule of RNA surrounded by a small (27 nm diameter) protein capsid and a buoyant density in CsCl of 1.33 g/ml. Many other picornaviruses cause human disease, including polioviruses, coxsackieviruses, echoviruses, and rhinoviruses (cold viruses).

2. Name of Acute Disease: The term hepatitis A (HA) or type A viral hepatitis has replaced all previous designations: infectious hepatitis, epidemic hepatitis, epidemic jaundice, catarrhal jaundice, infectious icterus, Botkins disease, and MS-1 hepatitis.

3. Nature of Disease: Hepatitis A is usually a mild illness characterized by sudden onset of fever, malaise, nausea, anorexia, and abdominal discomfort, followed in several days by jaundice. The infectious dose is unknown but presumably is 10-100 virus particles.

4. Diagnosis of Human Illness: Hepatitis A is diagnosed by finding IgM-class anti-HAV in serum collected during the acute or early convalescent phase of disease. Commercial kits are available.

5. Associated Foods: HAV is excreted in feces of infected people and can produce clinical disease when susceptible individuals consume contaminated water or foods. Cold cuts and sandwiches, fruits and fruit juices, milk and milk products, vegetables, salads, shellfish, and iced drinks are commonly implicated in outbreaks. Water, shellfish, and salads are the most frequent sources. Contamination of foods by infected workers in food processing plants and restaurants is common.

6. Frequency of Disease: Hepatitis A has a worldwide distribution occurring in both epidemic and sporadic fashions. About 22,700 cases of hepatitis A representing 38% of all hepatitis cases (5-year average from all routes of transmission) are reported annually in the U.S. In 1988 an estimated 7.3% cases were foodborne or waterborne. HAV is primarily transmitted by person-to-person contact through fecal contamination, but common-source epidemics from contaminated food and water also occur. Poor sanitation and crowding facilitate transmission. Outbreaks of HA are common in institutions, crowded house projects, and prisons and in military forces in adverse situations. In developing countries, the incidence of disease in adults is relatively low because of exposure to the virus in childhood. Most individuals 18 and older demonstrate an immunity that provides lifelong protection against reinfection. In the U.S., the percentage of adults with immunity increases with age (10% for those 18-19 years of age to 65% for those over 50). The increased number of susceptible individuals allows common source epidemics to evolve rapidly.

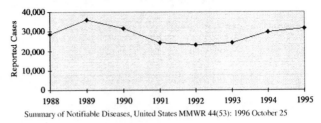

Reported cases of Hepatitis A, United States 1988-1995

Summary of Notifiable Diseases, United States MMWR 44(53): 1996 October 25

7. Usual Course of Disease: The incubation period for hepatitis A, which varies from 10 to 50 days (mean 30 days), is dependent upon the number of infectious particles consumed. Infection with very few particles results in longer incubation periods. The period of communicability extends from early in the incubation period to about a week after the development of jaundice. The greatest danger of spreading the disease to others occurs during the middle of the incubation period, well before the first presentation of symptoms. Many infections with HAV do not result in clinical disease, especially in children. When disease does occur, it is usually mild and recovery is complete in 1-2 weeks. Occasionaly, the symptoms are severe and convalescence can take several months. Patients suffer from feeling chronically tired during convalescence, and their inability to work can cause financial loss. Less than 0.4% of the reported cases in the U.S. are fatal. These rare deaths usually occur in the elderly.

8. Target Population: All people who ingest the virus and are immunologically unprotected are susceptible to infection. Disease however, is more common in adults than in children.

9. Analysis of Foods: The virus has not been isolated from any food associated with an outbreak. Because of the long incubation period, the suspected food is often no longer available for analysis. No satisfactory method is presently available for routine analysis of food, but sensitive molecular methods used to detect HAV in water and clinical specimens, should prove useful to detect virus in foods. Among those, the PCR amplification method seems particularly promising.

10. Selected Outbreaks: Hepatitis A is endemic throughout much of the world. Major national epidemics occurred in 1954, 1961 and 1971. Although no major epidemic occurred in the 1980s, the incidence of hepatitis A in the U.S. increased 58% from 1983 to 1989. Foods have been implicated in over 30 outbreaks since 1983. The most recent ones and the suspected contaminated foods include:

- 1987 - Louisville, Kentucky. Suspected source: imported lettuce.
- 1988 - Alaska. Ice-slush beverage prepared in a local market. - North Carolina. Iced tea prepared in a restaurant. - Florida. Raw oysters harvested from nonapproved bed.
- 1989 - Washington. Unidentified food in a restaurant chain.
- 1990 - North Georgia. Frozen strawberries. - Montana. Frozen strawberries. - Baltimore. Shellfish.

A summary of foodborne Hepatitis A outbreaks in Missouri, Wisconsin, and Alaska is found in MMWR 42(27):1993 Jul 16.

MMWR 39(14):1990 Apr 13 summarizes foodborne outbreaks of Hepatitis A in Alaska, Florida, North Carolina, Washington.

For more information on recent outbreaks see the Morbidity and Mortality Weekly Reports from CDC.

Hepatitis E Virus

1. Name of Organism: Hepatitis E Virus

Hepatitis E Virus (HEV) has a particle diameter of 32-34 nm, a buoyant density of 1.29 g/ml in KTar/Gly gradient, and is very labile. Serologically related smaller (27-30 nm) particles are often found in feces of patients with Hepatitis E and are presumed to represent degraded viral particles. HEV has a single-stranded polyadenylated RNA genome of approximately 8 kb. Based on its physicochemical properties it is presumed to be a calici-like virus.

2. Name of Acute Disease: The disease caused by HEV is called hepatitis E, or enterically transmitted non-A non-B hepatitis (ET-NANBH). Other names include fecal-oral non-A non-B hepatitis,and A-like non-A non-B hepatitis.

Note: This disease should not be confused with hepatitis C, also called parenterally transmitted non-A non-B hepatitis (PT-NANBH), or B-like non-A non-B hepatitis, which is a common cause of hepatitis in the U.S.

3. Nature of Disease: Hepatitis caused by HEV is clinically indistinguishable from hepatitis A disease. Symptoms include malaise, anorexia, abdominal pain, arthralgia, and fever. The infective dose is not known.

4. Diagnosis of Human Illness: Diagnosis of HEV is based on the epidemiological characteristics of the outbreak and by exclusion of hepatitis A and B viruses by serological tests. Confirmation requires identification of the 27-34 nm virus-like particles by immune electron microscopy in feces of acutely ill patients.

5. Associated Foods: HEV is transmitted by the fecal-oral route. Waterborne and person-to-person spread have been documented. The potential exists for foodborne transmission.

6. Frequency of Disease: Hepatitis E occurs in both epidemic and sporadic-endemic forms, usually associated with contaminated drinking water. Major waterborne epidemics have occurred in Asia and North and East Africa. To date no U.S. outbreaks have been reported.

7. Usual Course of Disease and Some Complications: The incubation period for hepatitis E varies from 2 to 9 weeks. The disease usually is mild and resolves in 2 weeks, leaving no sequelae. The fatality rate is 0.1-1% except in pregnant women. This group is reported to have a fatality rate approaching 20%.

8. Target Populations: The disease is most often seen in young to middle aged adults (15-40 years old). Pregnant women appear to be exceptionally susceptible to severe disease, and excessive mortality has been reported in this group.

9. Analysis of Foods: HEV has not been isolated from foods. No method is currently available for routine analysis of foods.

10. History of Recent Outbreaks: Major waterborne epidemics have occurred in India (1955 and 1975-1976), USSR (1955-1956), Nepal (1973), Burma (1976-1977), Algeria (1980-1981), Ivory Coast (1983-1984), in refugee camps in Eastern Suddan and Somalia (1985-6), and most recently in Borneo (1987). The first outbreaks reported in the American continents occurred in Mexico in late 1986. To date, no outbreak has occurred in the U.S., but imported cases were identified in Los Angeles in 1987. There is no evidence for immunity against this agent in the American population. Thus, unless other factors (such as poor sanitation or prevalence of other enteric pathogens) are important, the potential for spread to the U.S. is great. Good sanitation and personal hygiene are the best preventive measures.

For more information on recent outbreaks see the Morbidity and Mortality Weekly Reports from CDC.

Rotavirus

1. Name of Organism: Rotavirus

Rotaviruses are classified with the Reoviridae family. They have a genome consisting of 11 double-stranded RNA segments surrounded by a distinctive two-layered protein capsid. Particles are 70 nm in diameter and have a buoyant density of 1.36 g/ml in CsCl. Six serological groups have been identified, three of which (groups A, B, and C) infect humans.

2. Name of Acute Disease: Rotaviruses cause acute gastroenteritis. Infantile diarrhea, winter diarrhea, acute nonbacterial infectious gastroenteritis, and acute viral gastroenteritis are names applied to the infection caused by the most common and widespread group A *rotavirus.*

3. Nature of Disease: Rotavirus gastroenteritis is a self-limiting, mild to severe disease characterized by vomiting, watery diarrhea, and low-grade fever. The infective dose is presumed to be 10-100 infectious viral particles. Because a person with rotavirus diarrhea often excretes large numbers of virus (108-1010 infectious particles/ml of feces), infection doses can be readily acquired through contaminated hands, objects, or utensils. Asymptomatic rotavirus excretion has been well documented and may play a role in perpetuating endemic disease.

4. Diagnosis of Human Illness: Specific diagnosis of the disease is made by identification of the virus in the patient's stool. Enzyme immunoassay (EIA) is the test most widely used to screen clinical specimens, and several commercial kits are available for group A rotavirus. Electron microscopy (EM) and polyacrylamide gel electrophoresis (PAGE) are used in some laboratories in addition or as an alternative to EIA. A reverse transcription-polymerase chain reaction (RT-PCR) has been developed to detect and identify all three groups of human rotaviruses.

5. Associated Foods: Rotaviruses are transmitted by the fecal-oral route. Person-to-person spread through contaminated hands is probably the most important means by which rotaviruses are transmitted in close communities such as pediatric and geriatric wards, day care centers and family homes. Infected food handlers may contaminate foods that require handling and no further cooking, such as salads, fruits, and hors d'oeuvres. Rotaviruses are quite stable in the environment and have been found in estuary samples at levels as high as 1-5 infectious particles/gal. Sanitary measures adequate for bacteria and parasites seem to be ineffective in endemic control of rotavirus, as similar incidence of rotavirus infection is observed in countries with both high and low health standards.

6. Frequency of Disease: Group A rotavirus is endemic worldwide. It is the leading cause of severe diarrhea among infants and children, and accounts for about half of the cases requiring hospitalization. Over 3 million cases of rotavirus gastroenteritis occur annually in the U.S. In temperate areas, it occurs primarily in the winter, but in the tropics it occurs throughout the year. The number attributable to food contamination is unknown.

Group B rotavirus, also called adult diarrhea rotavirus or ADRV, has caused major epidemics of severe diarrhea affecting thousands of persons of all ages in China.

Group C rotavirus has been associated with rare and sporadic cases of diarrhea in children in many countries. However, the first outbreaks were reported from Japan and England.

7. Usual Course of Disease: The incubation period ranges from 1-3 days. Symptoms often start with vomiting followed by 4-8 days of diarrhea. Temporary lactose intolerance may occur. Recovery is usually complete. However, severe diarrhea without fluid and electrolyte replacement may result in severe diarrhea and death. Childhood mortality caused by rotavirus is relatively low in the U.S., with an estimated 100 cases/year, but reaches almost 1 million cases/year worldwide. Association with other enteric pathogens may play a role in the severity of the disease.

8. Target Populations: Humans of all ages are susceptible to rotavirus infection. Children 6 months to 2 years of age, premature infants, the elderly, and the immunocompromised are particularly prone to more severe symptoms caused by infection with group A *rotavirus*.

9. Analysis of Foods: The virus has not been isolated from any food associated with an outbreak, and no satisfactory method is available for routine analysis of food. However, it should be possible to apply procedures that have been used to detect the virus in water and in clinical specimens, such as enzyme immunoassays, gene probing, and PCR amplification to food analysis.

10. Selected Outbreaks: Outbreaks of group A rotavirus diarrhea are common among hospitalized infants, young children attending day care centers, and elder persons in nursing homes. Among adults, multiple foods served in banquets were implicated in 2 outbreaks. An outbreak due to contaminated municipal water occurred in Colorado, 1981.

Several large outbreaks of group B rotavirus involving millions of persons as a result of sewage contamination of drinking water supplies have occurred in China since 1982. Although to date outbreaks caused by group B rotavirus have been confined to mainland China, seroepidemiological surveys have indicated lack of immunity to this group of virus in the U.S.

The newly recognized group C rotavirus has been implicated in rare and isolated cases of gastroenteritis. However, it was associated with three outbreaks among school children: one in Japan, 1989, and two in England, 1990.

For a discussion of rotavirus surveillance in the US, see MMWR 40(5)1991 Feb 8.

For more information on recent outbreaks see the Morbidity and Mortality Weekly Reports from CDC.

11. Other Resources: From GenBank there is a Loci index for genome Rotavirus sp.

The Norwalk virus family

1. Name of Organism: The Norwalk virus family

Norwalk virus is the prototype of a family of unclassified small round structured viruses (SRSVs) which may be related to the caliciviruses. They contain a positive strand RNA genome of 7.5 kb and a single structural protein of about 60 kDa. The 27-32 nm viral particles have a buoyant density of 1.39-1.40 g/ml in CsCl. The family consists of several serologically distinct groups of viruses that have been named after the places where the outbreaks occurred. In the U.S., the Norwalk and Montgomery County agents are serologically related but distinct from the Hawaii and Snow Mountain agents. The Taunton, Moorcroft, Barnett, and Amulree agents were identified in the U.K., and the Sapporo and Otofuke agents in Japan. Their serological relationships remain to be determined.

2. Name of Acute Disease: Common names of the illness caused by the Norwalk and Norwalk-like viruses are viral gastroenteritis, acute nonbacterial gastroenteritis, food poisoning, and food infection.

3. Nature of Disease: The disease is self-limiting, mild, and characterized by nausea, vomiting, diarrhea, and abdominal pain. Headache and low-grade fever may occur. The infectious dose is unknown but presumed to be low.

4. Diagnosis of Human Illness: Specific diagnosis of the disease can only be made by a few laboratories possessing reagents from human volunteer studies. Identification of the virus can be made on early stool specimens using immune electron microscopy and various immunoassays. Confirmation often requires demonstration of seroconversion, the presence of specific IgM antibody, or a four-fold rise in antibody titer to Norwalk virus on paired acute-convalescent sera.

5. Associated Foods: Norwalk gastroenteritis is transmitted by the fecal-oral route via contaminated water and foods. Secondary person-to-person transmission has been documented. Water is the most common source of outbreaks and may include water from municipal supplies, well, recreational lakes, swiming pools, and water stored aboard cruise ships.

Shellfish and salad ingredients are the foods most often implicated in Norwalk outbreaks. Ingestion of raw or insufficiently steamed clams and oysters poses a high risk for infection with Norwalk virus. Foods other than shellfish are contaminated by ill food handlers.

6. Frequency of Disease: Only the common cold is reported more frequently than viral gastroenteritis as a cause of illness in the U.S. Although viral gastroenteritis is caused by a number of viruses, it is estimated that Norwalk viruses are responsible for about 1/3 of the cases not involving the 6-to-24-month age group. In developing countries the percentage of individuals who have developed immunity is very high at an early age. In the U.S. the percentage increases gradually with age, reaching 50% in the population over 18 years of age. Immunity, however, is not permanent and reinfection can occur.

7. Usual Course of Disease and Some Complications: A mild and brief illness usually develops 24-48 h after contaminated food or water is consumed and lasts for 24-60 hours. Severe illness or hospitalization is very rare.

8. Target Populations: All individuals who ingest the virus and who have not (within 24 months) had an infection with the same or related strain, are susceptible to infection and can develop the symptoms of gastroenteritis. Disease is more frequent in adults and older children than in the very young.

9. Analysis of Foods: The virus has been identified in clams and oysters by radioimmunoassay. The genome of Norwalk virus has been cloned and development of gene probes and PCR amplification techniques to detect the virus in clinical specimens and possibly in food are under way.

10. Selected Outbreaks: Foodborne outbreaks of gastroenteritis caused by Norwalk virus are often related to consumption of raw shellfish. Frequent and widespread outbreaks, reaching epidemic proportions, occurred in Australia (1978) and in the state of New York (1982) among consumers of raw clams and oysters. From 1983 to 1987, ten well documented outbreaks caused by Norwalk virus were reported in the U.S., involving a variety of foods: fruits, salads, eggs, clams, and bakery items.

Preliminary evidence suggests that large outbreaks of gastroenteritis which occurred in Pennsylvania and Delaware in September, 1987, were caused by Norwalk virus. The source of both outbreaks was traced to ice made with water from a contaminated well. In Pennsylvania, the ice was consumed at a football game, and in Delaware, at a cocktail party. Norwalk virus is also suspected to have caused an outbreak aboard a cruise ship in Hawaii in 1990. Fresh fruits were the probable vehicle of contamination.

Snow Mountain virus was implicated in an outbreak in a retirement community in California (1988) which resulted in two deaths. Illness was associated with consumption of shrimp probably contaminated by food handlers.

For outbreaks of Norwalk virus see MMWR 42(49):1993 Dec 17 and this MMWR 43(24):1994 Jun 24 as well.

The multistate outbreak of viral gastroenteritis associated with consumption of oysters from Apalachicola Bay, Florida, December 1994-January 1995 is reported in MMWR 44(2):1995 Jan 20.

For more information on recent outbreaks see the Morbidity and Mortality Weekly Reports from CDC.

Other gastroenteritis viruses

1. Name of Organism: Other viruses associated with gastroenteritis

Although the rotavirus and the Norwalk family of viruses are the leading causes of viral gastroenteritis, a number of other viruses have been implicated in outbreaks, including astroviruses, caliciviruses, enteric adenoviruses and parvovirus. Astroviruses, caliciviruses, and the Norwalk family of viruses possess well-defined surface structures and are sometimes identified as "small round structured viruses" or SRSVs. Viruses with smooth edge and no discernible surface structure are designated "featureless viruses" or "small round viruses" (SRVs). These agents resemble enterovirus or parvovirus, and may be related to them.

Astroviruses are unclassified viruses which contain a single positive strand of RNA of about 7.5 kb surrounded by a protein capsid of 28-30 nm diameter. A five or six pointed star shape can be observed on the particles under the electron microscope. Mature virions contain two major coat proteins of about 33 kDa each and have a buoyant density in CsCl of 1.38 - 1.40 g/ml. At least five human serotypes have been identified in England. The Marin County agent found in the U.S. is serologically related to astrovirus type 5.

Caliciviruses are classified in the family Caliciviridae. They contain a single strand of RNA surrounded by a protein capsid of 31-40 nm diameter. Mature virions have cup-shaped indentations which give them a 'Star of David' appearance in the electron microscope. The particle contain a single major coat protein of 60 kDa and have a buoyant density in CsCl of 1.36 - 1.39 g/ml. Four serotypes have been identified in England.

Enteric adenoviruses represent serotypes 40 and 41 of the family Adenoviridae. These viruses contain a double-stranded DNA surrounded by a distinctive protein capsid of about 70 nm diameter. Mature virions have a buoyant density in CsCl of about 1.345 g/ml.

Parvoviruses belong to the family Parvoviridae, the only group of animal viruses to contain linear single-stranded DNA. The DNA genome is surrounded by a protein capsid of about 22 nm diameter. The buoyant density of the particle in CsCl is 1.39-1.42 g/ml. The Ditchling, Wollan, Paramatta, and cockle agents are candidate parvoviruses associated with human gastroenteritis.

2. Name of Acute Disease: Common names of the illness caused by these viruses are acute nonbacterial infectious gastroenteritis and viral gastroenteritis.

3. Nature of Disease: Viral gastroenteritis is usually a mild illness characterized by nausea, vomiting, diarrhea, malaise, abdominal pain, headache, and fever. The infectious dose is not known but is presumed to be low.

4. Diagnosis of Human Illness: Specific diagnosis of the disease can be made by some laboratories possessing appropriate reagents. Identification of the virus present in early acute stool samples is made by immune electron microscopy and various enzyme immunoassays. Confirmation often requires demonstration of seroconversion to the agent by serological tests on acute and convalescent serum pairs.

5. Associated Foods: Viral gastroenteritis is transmitted by the fecal-oral route via person-to-person contact or ingestion of contaminated foods and water. Ill food handlers may contaminate foods that are not further cooked before consumption. Enteric adenovirus may also be transmitted by the respiratory route. Shellfish have been implicated in illness caused by a parvo-like virus.

6. Frequency of Disease: Astroviruses cause sporadic gastroenteritis in children under 4 years of age and account for about 4% of the cases hospitalized for diarrhea. Most American and British children over 10 years of age have antibodies to the virus.

Caliciviruses infect children between 6 and 24 months of age and account for about 3% of hospital admissions for diarrhea. By 6 years of age, more than 90% of all children have developed immunity to the illness.

The enteric adenovirus causes 5-20% of the gastroenteritis in young children, and is the second most common cause of gastroenteritis in this age group. By 4 years of age, 85% of all children have developed immunity to the disease. Parvo-like viruses have been implicated in a number of shellfish-associated outbreaks, but the frequency of disease is unknown.

7. Usual Course of Disease and Some Complications: A mild, self limiting illness usually develops 10 to 70 hours after contaminated food or water is consumed and lasts for 2 to 9 days. The clinical features are milder but otherwise indistinguishable from rotavirus gastroenteritis. Co-infections with other enteric agents may result in more severe illness lasting a longer period of time.

8. Target Population: The target populations for astro and caliciviruses are young children and the elderly. Only young children seem to develop illness caused by the enteric adenoviruses. Infection with these viruses is widespread and seems to result in development of immunity. Parvoviruses infect all age groups and probably do not ilicit a permanent immunity.

9. Analysis of Foods: Only a parvovirus-like agent (cockle) has been isolated from seafood associated with an outbreak. Although foods are not routinely analyzed for these viruses, it may be possible to apply current immunological procedures to detect viruses in clinical specimens. Gene probes and PCR detection methods are currently being developed.

10. Selected Outbreaks: Outbreaks of astrovirus and calicivirus occur mainly in child care settings and nursing homes. In the past decade, 7 outbreaks of calicivirus and 4 of astrovirus have been reported from England

and Japan. In California, an outbreak caused by an astrovirus, the Marin County agent, occurred among elderly patients in a convalescent hospital. No typical calicivirus has been implicated in outbreaks in the U.S. However, if Norwalk and Norwalk-like viruses prove to be caliciviruses, they would account for most food and waterborne outbreaks of gastroenteritis in this country.

Outbreaks of adenovirus have been reported in England and Japan, all involving children in hospitals or day care centers.

The small featureless, parvo-like viruses caused outbreaks of gastroenteritis in primary and secondary schools in England (Ditchling and Wollan) and Australia (Paramatta). The cockle agent caused a large community-wide outbreak in England (1977) associated with consumption of contaminated seafood. Parvo-like viruses were also implicated in several outbreaks which occurred in the States of New York and Louisiana in 1982-1983.

For more information on recent outbreaks see the Morbidity and Mortality Weekly Reports from CDC.

NATURAL TOXINS

Ciguatera

1. Name of Toxin: Ciguatera

2. Name of Disease: Ciguatera Fish Poisoning

Ciguatera is a form of human poisoning caused by the consumption of subtropical and tropical marine finfish which have accumulated naturally occurring toxins through their diet. The toxins are known to originate from several dinoflagellate (algae) species that are common to ciguatera endemic regions in the lower latitudes.

3. Nature of Disease: Manifestations of ciguatera in humans usually involves a combination of gastrointestinal, neurological, and cardiovascular disorders. Symptoms defined within these general categories vary with the geographic origin of toxic fish.

4. Normal Course of Disease: Initial signs of poisoning occur within six hours after consumption of toxic fish and include perioral numbness and tingling (paresthesia), which may spread to the extremities, nausea, vomiting, and diarrhea. Neurological signs include intensified paresthesia, arthralgia, myalgia, headache, temperature sensory reversal and acute sensitivity to temperature extremes, vertigo, and muscular weakness to the point of prostration. Cardiovascular signs include arrhythmia, bradycardia or tachycardia, and reduced blood pressure. Ciguatera poisoning is usually self-limiting, and signs of poisoning often subside within several days from onset. However, in severe cases the neurological symptoms are known to persist from weeks to months. In a few isolated cases neurological symptoms have persisted for several years, and in other cases recovered patients have experienced recurrence of neurological symptoms months to years after recovery. Such relapses are most often associated with changes in dietary habits or with consumption of alcohol. There is a low incidence of death resulting from respiratory and cardiovascular failure.

5. Diagnosis of Human Illness: Clinical testing procedures are not presently available for the diagnosis of ciguatera in humans. Diagnosis is based entirely on symptomology and recent dietary history. An enzyme immunoassay (EIA) designed to detect toxic fish in field situations is under evaluation by the Association of Official Analytical Chemists (AOAC) and may provide some measure of protection to the public in the future.

6. Associated Foods: Marine finfish most commonly implicated in ciguatera fish poisoning include the groupers, barracudas, snappers, jacks, mackerel, and triggerfish. Many other species of warm-water fishes harbor ciguatera toxins. The occurrence of toxic fish is sporadic, and not all fish of a given species or from a given locality will be toxic.

7. Relative Frequency of Disease: The relative frequency of ciguatera fish poisoning in the United States is not known. The disease has only recently become known to the general medical community, and there is a concern that incidence is largely under-reported because of the generally non-fatal nature and short duration of the disease.

8. Target Population: All humans are believed to be susceptible to ciguatera toxins. Populations in tropical/subtropical regions are most likely to be affected because of the frequency of exposure to toxic fishes. However, the increasing per capita consumption of fishery products coupled with an increase in interregional transportation of seafood products has expanded the geographic range of human poisonings.

9. Analysis of Foods: The ciguatera toxins can be recovered from toxic fish through tedious extraction and purification procedures. The mouse bioassay is a generally accepted method of establishing toxicity of suspect fish. A much simplified EIA method intended to supplant the mouse bioassay for identifying ciguatera toxins is under evaluation.

10. Selected Outbreaks: Isolated cases of ciguatera fish poisoning have occurred along the eastern coast of the United States from south Florida to Vermont. Hawaii, the U.S. Virgin Islands, and Puerto Rico experience sporadic cases with some regularity. A major outbreak of ciguatera occurred in Puerto Rico between April and June 1981 in which 49 persons were afflicted and two fatalities occurred. This outbreak prompted government officials of the Commonwealth of Puerto Rico to ban the sale of barracuda, amberjack, and blackjack.

In February-March of 1987 a large common-source outbreak of ciguatera occurred among Canadian vacationers returning from a Caribbean resort. Of 147 tourists, 61 ate a fish casserole shortly before departure, resulting in 57 identified cases of ciguatera.

In May of 1988 several hundred pounds of fish (primarily hogfish) from the Dry Tortuga Bank were responsible for over 100 human poisonings in Palm Beach County, Florida. The fish were sold to a seafood distributor after the fishermen (sport spearfishermen) themselves were first afflicted but dismissed their illness as seasickness and hangover. The poisonings resulted in a statewide warning against eating hogfish, grouper, red snapper, amberjack, and barracuda caught at the Dry Tortuga Bank.

For a report on Ciguatera poisoning in Florida, see this MMWR 42(21):1993 Jun 04.

For more information on recent outbreaks see the Morbidity and Mortality Weekly Reports from CDC.

Various shellfish-associated toxins

1. Name of Toxins: Various Shellfish-Associated

Shellfish poisoning is caused by a group of toxins elaborated by planktonic algae (dinoflagellates, in most cases) upon which the shellfish feed. The toxins are accumulated and sometimes metabolized by the shellfish. The 20 toxins responsible for paralytic shellfish poisonings (PSP) are all derivatives of saxitoxin. Diarrheic shellfish poisoning (DSP) is presumably caused by a group of high molecular weight polyethers, including okadaic acid, the dinophysis toxins, the pectenotoxins, and yessotoxin. Neurotoxic shellfish poisoning (NSP) is the result of exposure to a group of polyethers called brevetoxins. Amnesic shellfish poisoning (ASP) is caused by the unusual amino acid, domoic acid, as the contaminant of shellfish.

2. Name of the Acute Diseases: Shellfish Poisoning:

Paralytic Shellfish Poisoning (PSP), Diarrheic Shellfish Poisoning (DSP), Neurotoxic Shellfish Poisoning (NSP), Amnesic Shellfish Poisoning (ASP).

3. Nature of the Diseases: Ingestion of contaminated shellfish results in a wide variety of symptoms, depending upon the toxins(s) present, their concentrations in the shellfish and the amount of contaminated shellfish consumed. In the case of PSP, the effects are predominantly neurological and include tingling, burning, numbness, drowsiness, incoherent speech, and respiratory paralysis. Less well characterized are the symptoms associated with DSP, NSP, and ASP. DSP is primarily observed as a generally mild gastrointestinal disorder, i.e., nausea, vomiting, diarrhea, and abdominal pain accompanied by chills, headache, and fever. Both gastrointestinal and neurological symptoms characterize NSP, including tingling and numbness of lips, tongue, and throat, muscular aches, dizziness, reversal of the sensations of hot and cold, diarrhea, and vomiting. ASP is characterized by gastrointestinal disorders (vomiting, diarrhea, abdominal pain) and neurological problems (confusion, memory loss, disorientation, seizure, coma).

4. Normal Course of the Disease: PSP: Symptoms of the disease develop fairly rapidly, within 0.5 to 2 hours after ingestion of the shellfish, depending on the amount of toxin consumed. In severe cases respiratory paralysis is common, and death may occur if respiratory support is not provided. When such support is applied within 12 hours of exposure, recovery usually is complete, with no lasting side effects. In unusual cases, because of the weak hypotensive action of the toxin, death may occur from cardiovascular collapse despite respiratory support.

NSP: Onset of this disease occurs within a few minutes to a few hours; duration is fairly short, from a few hours to several days.

Recovery is complete with few after effects; no fatalities have been reported.

DSP: Onset of the disease, depending on the dose of toxin ingested, may be as little as 30 minutes to 2 to 3 hours, with symptoms of the illness lasting as long as 2 to 3 days. Recovery is complete with no after effects; the disease is generally not life threatening.

ASP: The toxicosis is characterized by the onset of gastrointestinal symptoms within 24 hours; neurological symptoms occur within 48 hours. The toxicosis is particularly serious in elderly patients, and includes symptoms reminiscent of Alzheimer's disease. All fatalities to date have involved elderly patients.

5. Diagnosis of Human Illnesses: Diagnosis of shellfish poisoning is based entirely on observed symptomatology and recent dietary history.

6. Associated Foods: All shellfish (filter-feeding molluscs) are potentially toxic. However, PSP is generally associated with mussels, clams, cockles, and scallops; NSP with shellfish harvested along the Florida coast and the Gulf of Mexico; DSP with mussels, oysters, and scallops, and ASP with mussels.

7. Relative Frequency Disease: Good statistical data on the occurrence and severity of shellfish of poisoning are largely unavailable, which undoubtedly reflects the inability to measure the true incidence of the disease. Cases are frequently misdiagnosed and, in general, infrequently reported. Of these toxicoses, the most serious from a public health perspective appears to be PSP. The extreme potency of the PSP toxins has, in the past, resulted in an unusually high mortality rate.

8. Target Populations: All humans are susceptible to shellfish poisoning. Elderly people are apparently predisposed to the severe neurological effects of the ASP toxin. A disproportionate number of PSP cases occur among tourists or others who are not native to the location where the toxic shellfish are harvested. This may be due to disregard for either official quarantines or traditions of safe consumption, both of which tend to protect the local population.

9. Analysis of Foods: The mouse bioassay has historically been the most universally applied technique for examining shellfish (especially for PSP); other bioassay procedures have been developed but not generally applied. Unfortunately, the dose-survival times for the DSP toxins in the mouse assay fluctuate considerably and fatty acids interfere with the assay, giving false-positive results; consequently, a suckling mouse assay that has been developed and used for control of DSP measures fluid accumulation after injection of the shellfish extract. In recent years considerable effort has been applied to development of chemical assays to replace these bioassays. As a result a good high performance liquid chromatography (HPLC) procedure has been developed to identify individual PSP toxins (detection limit for saxitoxin = 20 fg/100 g of meats; 0.2 ppm), an excellent HPLC procedure (detection limit for okadaic acid = 400 ng/g; 0.4 ppm), a commercially available immunoassay (detection limit for okadaic acid = 1 fg/100 g of meats; 0.01 ppm) for DSP and a totally satisfactory HPLC procedure for ASP (detection limit for domoic acid = 750 ng/g; 0.75 ppm).

10. Selected Outbreaks: PSP is associated with relatively few outbreaks, most likely because of the strong control programs in the United States that prevent human exposure to toxic shellfish. That PSP can be a serious public health problem, however, was demonstrated in Guatemala, where an outbreak of 187 cases with 26 deaths, recorded in 1987, resulted from ingestion of a clam soup. The outbreak led to the establishment of a control program over shellfish harvested in Guatemala.

ASP first came to the attention of public health authorities in 1987 when 156 cases of acute intoxication occurred as a result of ingestion of cultured blue mussels (Mytilus edulis) harvested off Prince Edward Island, in eastern Canada; 22 individuals were hospitalized and three elderly patients eventually died.

The occurrence of DSP in Europe is sporadic, continuous and presumably widespread (anecdotal). DSP poisoning has not been confirmed in U.S. seafood, but the organisms that produce DSP are present in U.S. waters. An outbreak of DSP was recently confirmed in Eastern Canada. Outbreaks of NSP are sporadic and continuous along the Gulf coast of Florida and were recently reported in North Carolina and Texas.

For more information on recent outbreaks see the Morbidity and Mortality Weekly Reports from CDC.

Scombrotoxin

1. Name of Toxin: Scombrotoxin

2. Name of Acute Disease: Scombroid Poisoning (also called Histamine Poisoning)

Scombroid poisoning is caused by the ingestion of foods that contain high levels of histamine and possibly other vasoactive amines and compounds. Histamine and other amines are formed by the growth of certain bacteria and the subsequent action of their decarboxylase enzymes on histidine and other amino acids in food, either during the production of a product such as Swiss cheese or by spoilage of foods such as fishery products, particularly tuna or mahi mahi. However, any food that contains the appropriate amino acids and is subjected to certain bacterial contamination and growth may lead to scombroid poisoning when ingested.

3. Nature of Disease: Initial symptoms may include a tingling or burning sensation in the mouth, a rash on the upper body and a drop in blood pressure. Frequently, headaches and itching of the skin are encountered. The symptoms may progress to nausea, vomiting, and diarrhea and may require hospitalization, particularly in the case of elderly or impaired patients.

4. Normal Course of Disease: The onset of intoxication symptoms is rapid, ranging from immediate to 30 minutes. The duration of the illness is usually 3 hours, but may last several days.

5. Diagnosis of Human Illness: Diagnosis of the illness is usually based on the patient's symptoms, time of onset, and the effect of treatment with antihistamine medication. The suspected food must be analyzed within a few hours for elevated levels of histamine to confirm a diagnosis.

6. Associated Foods: Fishery products that have been implicated in scombroid poisoning include the tunas (e.g., skipjack and yellowfin), mahi mahi, bluefish, sardines, mackerel, amberjack, and abalone. Many other products also have caused the toxic effects. The primary cheese involved in intoxications has been Swiss cheese. The toxin forms in a food when certain bacteria are present and time and temperature permit their growth. Distribution of the toxin within an individual fish fillet or between cans in a case lot can be uneven, with some sections of a product causing illnesses and others not. Neither cooking, canning, or freezing reduces the toxic effect. Common sensory examination by the consumer cannot ensure the absence or presence of the toxin. Chemical testing is the only reliable test for evaluation of a product.

7. Relative Frequency of Disease: Scombroid poisoning remains one of the most common forms of fish poisoning in the United States. Even so, incidents of poisoning often go unreported because of the lack of required reporting, a lack of information by some medical personnel, and confusion with the symptoms of other illnesses. Difficulties with underreporting are a worldwide problem. In the United States from 1968 to 1980, 103 incidents of intoxication involving 827 people were reported. For the same period in Japan, where the quality of fish is a national priority, 42 incidents involving 4,122 people were recorded. Since 1978, 2 actions by FDA have reduced the frequency of intoxications caused by specific products. A defect action level for histamine in canned tuna resulted in increased industry quality control. Secondly, blocklisting of mahi mahi reduced the level of fish imported to the United States.

8. Target Population: All humans are susceptible to scombroid poisoning; however, the symptoms can be severe for the elderly and for those taking medications such as isoniazid. Because of the worldwide network for harvesting, processing, and distributing fishery products, the impact of the problem is not limited to specific geographical areas of the United States or consumption pattern. These foods are sold for use in homes, schools, hospitals, and restaurants as fresh, frozen, or processed products.

9. Analysis of Foods: An official method was developed at FDA to determine histamine, using a simple alcoholic extraction and quantitation by fluorescence spectroscopy. There are other untested procedures in the literature.

10. Selected Outbreaks: Several large outbreaks of scombroid poisoning have been reported. In 1970, some 40 children in a school lunch program became ill from imported canned tuna. In 1973, more than 200 consumers across the United States were affected by domestic canned tuna. In 1979-1980 more than 200 individuals became ill after consuming imported frozen mahi mahi. Symptoms varied with each incident. In the 1973 situation, of the interviewed patients, 86% experienced nausea, 55% diarrhea, 44% headaches and 32% rashes.

Other incidents of intoxication have resulted from the consumption of canned abalone-like products, canned anchovies, and fresh and frozen amberjack, bluefish sole, and scallops. In particular, shipments of unfrozen fish packed in refrigerated containers have posed a significant problem because of inadequate temperature control.

For more information on recent outbreaks see the Morbidity and Mortality Weekly Reports from CDC.

Tetrodotoxin

1. Name of Toxin: Tetrodotoxin (anhydrotetrodotoxin 4-epitetrodotoxin, tetrodonic acid)

2. Name of the Acute Disease: Pufferfish Poisoning, Tetradon Poisoning, Fugu Poisoning

3. Nature of the Disease: Fish poisoning by consumption of members of the order Tetraodontiformes is one of the most violent intoxications from marine species. The gonads, liver, intestines, and skin of pufferfish can contain levels of tetrodotoxin sufficient to produce rapid and violent death. The flesh of many pufferfish may not usually be dangerously toxic. Tetrodotoxin has also been isolated from widely differing animal species, including the California newt, parrotfish, frogs of the genus Atelopus, the blue-ringed octopus, starfish, angelfish, and xanthid crabs. The metabolic source of tetrodotoxin is uncertain. No algal source has been identified, and until recently tetrodotoxin was assumed to be a metabolic product of the host. However, recent reports of the production of tetrodotoxin/anhydrotetrodotoxin by several bacterial species, including strains of the family Vibrionaceae, Pseudomonas sp., and Photobacterium phosphoreum, point toward a bacterial origin of this family of toxins. These are relatively common marine bacteria that are often associated with marine animals. If confirmed, these findings may have some significance in toxicoses that have been more directly related to these bacterial species.

4. Normal Course of the Disease: The first symptom of intoxication is a slight numbness of the lips and tongue, appearing between 20 minutes to three hours after eating poisonous pufferfish. The next symptom is increasing paraesthesia in the face and extremities, which may be followed by sensations of lightness or floating. Headache, epigastric pain, nausea, diarrhea, and/or vomiting may occur. Occasionally, some reeling or difficulty in walking may occur. The second stage of the intoxication is increasing paralysis. Many victims are unable to move; even sitting may be difficult. There is increasing respiratory distress. Speech is affected, and the victim usually exhibits dyspnea, cyanosis, and hypotension. Paralysis increases and convulsions, mental impairment, and cardiac arrhythmia may occur. The victim, although completely paralyzed, may be conscious and in some cases completely lucid until shortly before death. Death usually occurs within 4 to 6 hours, with a known range of about 20 minutes to 8 hours.

5. Diagnosis of Human Illness: The diagnosis of pufferfish poisoning is based on the observed symptomology and recent dietary history.

6. Associated Foods: Poisonings from tetrodotoxin have been almost exclusively associated with the consumption of pufferfish from waters of the Indo-Pacific ocean regions. Several reported cases of poisonings, including fatalities, involved pufferfish from the Atlantic Ocean, Gulf of Mexico, and Gulf of California. There have been no confirmed cases of poisoning from the Atlantic pufferfish, Spheroides maculatus. However, in one study, extracts from fish of this species were highly toxic in mice. The trumpet shell Charonia sauliae has been implicated in food poisonings, and evidence suggests that it contains a tetrodotoxin derivative. There have been several reported poisonings from mislabelled pufferfish and at least one report of a fatal episode when an individual swallowed a California newt.

7. Relative Frequency of Disease: From 1974 through 1983 there were 646 reported cases of pufferfish poisoning in Japan, with 179 fatalities. Estimates as high as 200 cases per year with mortality approaching 50% have been reported. Only a few cases have been reported in the United States, and outbreaks in countries outside the Indo-Pacific area are rare.

8. Target Population: All humans are susceptible to tetrodotoxin poisoning. This toxicosis may be avoided by not consuming pufferfish or other animal species containing tetrodotoxin. Most other animal species known to contain tetrodotoxin are not usually consumed by humans. Poisoning from tetrodotoxin is of major public health concern primarily in Japan, where "fugu" is a traditional delicacy. It is prepared and sold in special restaurants where trained and licensed individuals carefully remove the viscera to reduce the danger of poisoning. Importation of pufferfish into the United States is not generally permitted, although special exceptions may be granted. There is potential for misidentification and/or mislabelling, particularly of prepared, frozen fish products.

9. Analysis of Foods: The mouse bioassay developed for paralytic shellfish poisoning (PSP) can be used to monitor tetrodotoxin in pufferfish and is the current method of choice. An HPLC method with post-column reaction with alkali and fluorescence has been developed to determine tetrodotoxin and its associated toxins. The alkali degradation products can be confirmed as their trimethylsilyl derivatives by gas chromatography/mass spectrometry. These chromatographic methods have not yet been validated.

10. Selected Outbreaks: Pufferfish poisoning is a continuing problem in Japan, affecting 30 - 100 persons/year. Most of these poisoning episodes occur from home preparation and consumption and not from commercial sources of the pufferfish. Three deaths were reported in Italy in 1977 following the consumption of frozen pufferfish imported from Taiwan and mislabelled as angler fish.

An incident of Fugu fish poisoning in the United States is reported in MMWR 45(19):1996 May 17.

For more information on recent outbreaks see the Morbidity and Mortality Weekly Reports from CDC.

Mushroom toxins

1. Name of Toxin(s): Amanitin, Gyromitrin, Orellanine, Muscarine, Ibotenic Acid, Muscimol, Psilocybin, Coprine

2. Name of Acute Disease: Mushroom Poisoning, Toadstool Poisoning

Mushroom poisoning is caused by the consumption of raw or cooked fruiting bodies (mushrooms, toadstools) of a number of species of higher fungi. The term toadstool (from the German Todesstuhl, death's stool) is commonly given to poisonous mushrooms, but for individuals who are not experts in mushroom identification there are generally no easily recognizable differences between poisonous and nonpoisonous species. Old wives' tales notwithstanding, there is no general rule of thumb for distinguishing edible mushrooms and poisonous toadstools. The toxins involved in mushroom poisoning are produced naturally by the fungi themselves, and each individual specimen of a toxic species should be considered equally poisonous. Most mushrooms that cause human poisoning cannot be made nontoxic by cooking, canning, freezing, or any other means of processing. Thus, the only way to avoid poisoning is to avoid consumption of the toxic species. Poisonings in the United States occur most commonly when hunters of wild mushrooms (especially novices) misidentify and consume a toxic species, when recent immigrants collect and consume a poisonous American species that closely resembles an edible wild mushroom from their native land, or when mushrooms that contain psychoactive compounds are intentionally consumed by persons who desire these effects.

3. Nature of Disease(s): Mushroom poisonings are generally acute and are manifested by a variety of symptoms and prognoses, depending on the amount and species consumed. Because the chemistry of many of the mushroom toxins (especially the less deadly ones) is still unknown and positive identification of the mushrooms is often difficult or impossible, mushroom poisonings are generally categorized by their physiological effects. There are four categories of mushroom toxins: protoplasmic poisons (poisons that result in generalized destruction of cells, followed by organ failure); neurotoxins (compounds that cause neurological symptoms such as profuse sweating, coma, convulsions, hallucinations, excitement, depression, spastic colon); gastrointestinal irritants (compounds that produce rapid, transient nausea, vomiting, abdominal cramping, and diarrhea); and disulfiram-like toxins. Mushrooms in this last category are generally nontoxic and produce no symptoms unless alcohol is consumed within 72 hours after eating them, in which case a short-lived acute toxic syndrome is produced.

4. Normal Course of Disease(s): The normal course of the disease varies with the dose and the mushroom species eaten. Each poisonous species contains one or more toxic compounds which are unique to few other species. Therefore, cases of mushroom poisonings generally do not resembles each other unless they are caused by the same or very closely

related mushroom species. Almost all mushroom poisonings may
be grouped in one of the categories outlined above.

PROTOPLASMIC POISONS

Amatoxins:
Several mushroom species, including the Death Cap or
Destroying Angel (*Amanita phalloides, A. virosa*), the Fool's
Mushroom (*A. verna*) and several of their relatives, along with the
Autumn Skullcap (*Galerina autumnalis*) and some of its relatives,
produce a family of cyclic octapeptides called amanitins.
Poisoning by the amanitins is characterized by a long latent period (range 6-48 hours, average 6-15 hours) during which the
patient shows no symptoms. Symptoms appear at the end of the
latent period in the form of sudden, severe seizures of abdominal
pain, persistent vomiting and watery diarrhea, extreme thirst, and
lack of urine production. If this early phase is survived, the
patient may appear to recover for a short time, but this period will
generally be followed by a rapid and severe loss of strength, prostration, and pain-caused restlessness. Death in 50-90% of the cases from progressive and irreversible liver, kidney, cardiac, and
skeletal muscle damage may follow within 48 hours (large dose),
but the disease more typically lasts 6 to 8 days in adults and 4 to
6 days in children. Two or three days after the onset of the later
phase, jaundice, cyanosis, and coldness of the skin occur. Death
usually follows a period of coma and occasionally convulsions. If
recovery occurs, it generally requires at least a month and is
accompanied by enlargement of the liver. Autopsy will usually
reveal fatty degeneration and necrosis of the liver andkidney.

Hydrazines:
Certain species of False Morel (*Gyromitra esculenta* and *G.
gigas*) contain the protoplasmic poison gyromitrin, a volatile
hydrazine derivative. Poisoning by this toxin superficially resembles Amanita poisoning but is less severe. There is generally a
latent period of 6 - 10 hours after ingestion during which no
symptoms are evident, followed by sudden onset of abdominal
discomfort (a feeling of fullness), severe headache, vomiting, and
sometimes diarrhea. The toxin affects primarily the liver, but there
are additional disturbances to blood cells and the central nervous
system. The mortality rate is relatively low (2-4%). Poisonings
with symptoms almost identical to those produced by Gyromitra
have also been reported after ingestion of the Early False Morel
(Verpa bohemica). The toxin is presumed to be related to
gyromitrin but has not yet been identified.

Orellanine:
The final type of protoplasmic poisoning is caused by the Sorrel
Webcap mushroom (*Cortinarius orellanus*) and some of its relatives. This mushroom produces orellanine, which causes a type of
poisoning characterized by an extremely long asymptomatic latent
period of 3 to 14 days. An intense, burning thirst (*polydipsia*) and
excessive urination (*polyuria*) are the first symptoms. This may be

followed by nausea, headache, muscular pains, chills, spasms, and loss of consciousness. In severe cases, severe renal tubular necrosis and kidney failure may result in death (15%) several weeks after the poisoning. Fatty degeneration of the liver and severe inflammatory changes in the intestine accompany the renal damage, and recovery in less severe cases may require several months.

NEUROTOXINS
Poisonings by mushrooms that cause neurological problems may be divided into three groups, based on the type of symptoms produced, and named for the substances responsible for these symptoms.

Muscarine Poisoning:
Ingestion of any number of Inocybe or Clitocybe species (e.g., *Inocybe geophylla, Clitocybe dealbata*) results in an illness characterized primarily by profuse sweating. This effect is caused by the presence in these mushrooms of high levels (3- 4%) of muscarine. Muscarine poisoning is characterized by increased salivation, perspiration, and lacrimation within 15 to 30 minutes after ingestion of the mushroom. With large doses, these symptoms may be followed by abdominal pain, severe nausea, diarrhea, blurred vision, and labored breathing. Intoxication generally subsides within 2 hours. Deaths are rare, but may result from cardiac or respiratory failure in severe cases.

Ibotenic acid/Muscimol Poisoning:
The Fly Agaric (*Amanita muscaria*) and Panthercap (*Amanita pantherina*) mushrooms both produce ibotenic acid and muscimol. Both substances produce the same effects, but muscimol is approximately 5 times more potent than ibotenic acid. Symptoms of poisoning generally occur within 1 - 2 hours after ingestion of the mushrooms. An initial abdominal discomfort may be present or absent, but the chief symptoms are drowsiness and dizziness (sometimes accompanied by sleep), followed by a period of hyperactivity, excitability, illusions, and delirium. Periods of drowsiness may alternate with periods of excitement, but symptoms generally fade within a few hours. Fatalities rarely occur in adults, but in children, accidental consumption of large quantities of these mushrooms may cause convulsions, coma, and other neurologic problems for up to 12 hours.

Psilocybin Poisoning:
A number of mushrooms belonging to the genera *Psilocybe, Panaeolus, Copelandia, Gymnopilus, Conocybe,* and *Pluteus,* when ingested, produce a syndrome similar to alcohol intoxication (sometimes accompanied by hallucinations). Several of these mushrooms (e.g., *Psilocybe cubensis, P. mexicana, Conocybe cyanopus*) are eaten for their psychotropic effects in religious ceremonies of certain native American tribes, a practice which dates to the pre- Columbian era. The toxic effects are caused by psilocin and psilocybin. Onset of symptoms is usually rapid and the effects generally subside within 2 hours. Poisonings by these mushrooms are rarely fatal in adults and may be distinguished

from ibotenic acid poisoning by the absence of drowsiness or coma. The most severe cases of psilocybin poisoning occur in small children, where large doses may cause the hallucinations accompanied by fever, convulsions, coma, and death. These mushrooms are generally small, brown, nondescript, and not particularly fleshy; they are seldom mistaken for food fungi by innocent hunters of wild mushrooms. Poisonings caused by intentional ingestion of these mushrooms by people with no legitimate religious justification must be handled with care, since the only cases likely to be seen by the physician are overdoses or intoxications caused by a combination of the mushroom and some added psychotropic substance (such as PCP).

GASTROINTESTINAL IRRITANTS

Numerous mushrooms, including the Green Gill (*Chlorophyllum molybdites*), Gray Pinkgill (*Entoloma lividum*), Tigertop (*Tricholoma pardinum*), Jack O'Lantern (*Omphalotus illudens*), Naked Brimcap (*Paxillus involutus*), Sickener (*Russula emetica*), Early False Morel (*Verpa bohemica*), Horse mushroom (*Agaricus arvensis*) and Pepper bolete (*Boletus piperatus*), contain toxins that can cause gastrointestinal distress, including but not limited to nausea, vomiting, diarrhea, and abdominal cramps. In many ways these symptoms are similar to those caused by the deadly protoplasmic poisons. The chief and diagnostic difference is that poisonings caused by these mushrooms have a rapid onset, rather than the delayed onset seen in protoplasmic poisonings. Some mushrooms (including the first five species mentioned above) may cause vomiting and/or diarrhea which lasts for several days. Fatalities caused by these mushrooms are relatively rare and are associated with dehydration and electrolyte imbalances caused by diarrhea and vomiting, especially in debilitated, very young, or very old patients. Replacement of fluids and other appropriate supportive therapy will prevent death in these cases. The chemistry of the toxins responsible for this type of poisoning is virtually unknown, but may be related to the presence in some mushrooms of unusual sugars, amino acids, peptides, resins, and other compounds.

DISULFIRAM-LIKE POISONING

The Inky Cap Mushroom (*Coprinus atramentarius*) is most commonly responsible for this poisoning, although a few other species have also been implicated. A complicating factor in this type of intoxication is that this species is generally considered edible (i.e., no illness results when eaten in the absence of alcoholic beverages). The mushroom produces an unusual amino acid, coprine, which is converted to cyclopropanone hydrate in the human body. This compound interferes with the breakdown of alcohol, and consumption of alcoholic beverages within 72 hours after eating it will cause headache, nausea and vomiting, flushing, and cardiovascular disturbances that last for 2 - 3 hours.

MISCELLANEOUS POISONINGS

Young fruiting bodies of the sulfur shelf fungus Laetiporus sulphureus are considered edible. However, ingestion of this shelf fungus has caused digestive upset and other symptoms in adults and visual hallucinations and ataxia in a child.

5. Diagnosis of Human Illness: A clinical testing procedure is currently available only for the most serious types of mushroom toxins, the amanitins. The commercially available method uses a 3H-radioimmunoassay (RIA) test kit and can detect sub-nanogram levels of toxin in urine and plasma. Unfortunately, it requires a 2-hour incubation period, and this is an excruciating delay in a type of poisoning which the clinician generally does not see until a day or two has passed. A 125I-based kit which overcomes this problem has recently been reported, but has not yet reached the clinic. A sensitive and rapid HPLC technique has been reported in the literature even more recently, but it has not yet seen clinical application. Since most clinical laboratories in this country do not use even the older RIA technique,diagnosis is based entirely on symptomology and recent dietary history. Despite the fact that cases of mushroom poisoning may be broken down into a relatively small number of categories based on symptomatology, positive botanical identification of the mushroom species consumed remains the only means of unequivocally determining the particular type of intoxication involved, and it is still vitally important to obtain such accurate identification as quickly as possible. Cases involving ingestion of more than one toxic species in which one set of symptoms masks or mimics another set are among many reasons for needing this information. Unfortunately, a number of factors (not discussed here) often make identification of the causative mushroom impossible. In such cases, diagnosis must be based on symptoms alone. In order to rule out other types of food poisoning and to conclude that the mushrooms eaten were the cause of the poisoning, it must be established that everyone who ate the suspect mushrooms became ill and that no one who did not eat the mushrooms became ill. Wild mushrooms eaten raw, cooked, or processed should always be regarded as prime suspects. After ruling out other sources of food poisoning and positively implicating mushrooms as the cause of the illness, diagnosis may proceed in two steps. The first step, outlined in Table 1, provides an early indication of the seriousness of the disease and its prognosis.

As described above, the protoplasmic poisons are the most likely to be fatal or to cause irreversible organ damage. In the case of poisoning by the deadly Amanitas, important laboratory indicators of liver (elevated LDH, SGOT, and bilirubin levels) and kidney (elevated uric acid, creatinine, and BUN levels) damage will be present. Unfortunately, in the absence of dietary history, these signs could be mistaken for symptoms of liver or kidney impairment as the result of other causes (e.g., viral hepatitis). It is important that this distinction be made as quickly as possible, because the delayed onset of symptoms will generally mean that the organ has already been damaged. The importance of rapid diagnosis is obvious: victims who are hospitalized and given aggressive supporttherapy almost immediately after ingestion have a mortality rate of only 10%, whereas those admitted 60 or more hours after ingestion have a 50-90% mortality rate. Table 2 provides more accurate diagnoses and appropriate therapeutic measures. A recent report indicates that amanitins are observable in urine well before the onset of any symptoms, but that laboratory tests for liver dysfunction do not appear until well after the organ has been damaged.

6. Associated Foods: Mushroom poisonings are almost always caused by ingestion of wild mushrooms that have been collected by nonspecialists (although specialists have also been poisoned). Most cases occur when toxic species are confused with edible species, and a useful question to ask of the victims or their mushroom-picking benefactors is the identity of the mushroom they thought they were picking. In the absence of a well- preserved specimen, the answer to this question could narrow the possible suspects considerably. Intoxication has also occurred when reliance was placed on some folk method of distinguishing poisonous and safe species. Outbreaks have occurred after ingestion of fresh, raw mushrooms, stir-fried mushrooms, home-canned mushrooms, mushrooms cooked in tomato sauce (which rendered the sauce itself toxic, even when no mushrooms were consumed), and mushrooms that were blanched and frozen at home. Cases of poisoning by home-canned and frozen mushrooms are especially insidious because a single outbreak may easily become a multiple outbreak when the preserved toadstools are carried to another location and consumed at another time.

Specific cases of mistaken mushroom identity appears frequently. The Early False Morel *Gyromitra esculenta* is easily confused with the true Morel *Morchella esculenta*, and poisonings have occurred after consumption of fresh or cooked *Gyromitra*. *Gyromitra* poisonings have also occurred after ingestion of commercially available "morels" contaminated with *G. esculenta*. The commercial sources for these fungi (which have not yet been successfully cultivated on a large scale) are field collection of wild morels by semiprofessionals. Cultivated commercial mushrooms of whatever species are almost never implicated in poisoning outbreaks unless there are associated problems such as improper canning (which lead to bacterial food poisoning). A short list of the mushrooms responsible for serious poisonings and the edible mushrooms with which they are confused is presented in Table 3. Producers of mild gastroenteritis are too numerous to list here, but include members of many of the most abundant genera, including *Agaricus, Boletus, Lactarius, Russula, Tricholoma, Coprinus, Pluteus*, and others. The Inky Cap Mushroom (*Coprinus atrimentarius*) is considered both edible and delicious, and only the unwary who consume alcohol after eating this mushroom need be concerned. Some other members of the genus Coprinus (*Shaggy Mane, C. comatus; Glistening Inky Cap, C. micaceus*, and others) and some of the larger members of the Lepiota family such as the Parasol Mushroom (*Leucocoprinus procera*) do not contain coprine and do not cause this effect. The potentially deadly Sorrel Webcap Mushroom (*Cortinarius orellanus*) is not easily distinguished from nonpoisonous webcaps belonging to the same distinctive genus, and all should be avoided. Most of the psychotropic mushrooms (*Inocybe* spp., *Conocybe* spp., *Paneolus* spp., *Pluteus* spp.) are in general appearance small, brown, and leathery (the so-called "Little Brown Mushrooms" or LBMs) and relatively unattractive from a culinary standpoint. The Sweat Mushroom (*Clitocybe dealbata*) and the Smoothcap Mushroom (*Psilocybe cubensis*) are small, white, and leathery. These small, unattractive mushrooms are distinctive, fairly unappetizing, and not easily confused with the fleshier fungi normally considered edible. Intoxications associated with them are less likely to be accidental, although both *C. dealbata* and *Paneolus foenisicii* have been found growing in the same fairy ring area as the edible (and choice) Fairy Ring Mushroom (*Marasmius oreades*) and the Honey Mushroom (*Armillariella mellea*), and have been consumed when the picker has not carefully examined every mushroom picked from the ring. Psychotropic mushrooms, which are larger and therefore more easily confused with edible mushrooms, include the Showy Flamecap or Big Laughing Mushroom (*Gymnopilus spectabilis*), which has been mistaken for Chanterelles (*Cantharellus* spp.) and for Gymnopilus ventricosus found growing on wood of conifers in western North America. The Fly Agaric (*Amanita muscaria*) and Panthercap (*Amanita pantherina*) mushrooms are large, fleshy, and colorful. Yellowish cap colors on some varieties of the Fly Agaric and the Panthercap are similar to the edible Caesar's Mushroom (*Amanita caesarea*), which is considered a delicacy in Italy. Another edible yellow capped mushroom occasionally confused with yellow A. muscaria and A. pantherina varieties are the Yellow Blusher (*Amanita flavorubens*). Orange to yellow-orange A. muscaria and A. pantherina may also be confused with the Blusher (*Amanita rubescens*) and the Honey Mushroom (*Armillariella mellea*). White to pale forms of A. muscaria may be confused with edible field mushrooms (*Agaricus* spp.). Young (button stage) specimens of A. muscaria have also been confused with puffballs.

7. Relative Frequency of Disease:

Accurate figures on the relative frequency of mushroom poisonings are difficult to obtain. For the 5-year period between 1976 and 1981, 16 outbreaks involving 44 cases were reported to the Centers for Disease Control in Atlanta (Rattanvilay et al. MMWR 31(21): 287-288, 1982). The number of unreported cases is, of course, unknown. Cases are sporadic and large outbreaks are rare. Poisonings tend to be grouped in the spring and fall when most mushroom species are at the height of their fruiting stage. While the actual incidence appears to be very low, the potential exists for grave problems. Poisonous mushrooms are not limited in distribution as are other poisonous organisms (such as dinoflagellates). Intoxications may occur at any time and place, with dangerous species occurring in habitats ranging from urban lawns to deep woods. As Americans become more adventurous in their mushroom collection and consumption, poisonings are likely to increase.

8. Target Population:

All humans are susceptible to mushroom toxins. The poisonous species are ubiquitous, and geographical restrictions on types of poisoning that may occur in one location do not exist (except for some of the hallucinogenic LBMs, which occur primarily in the

American southwest and southeast). Individual specimens of poisonous mushrooms are also characterized by individual variations in toxin content based on genetics, geographic location, and growing conditions. Intoxications may thus be more or less serious, depending not on the number of mushrooms consumed, but on the dose of toxin delivered. In addition, although most cases of poisoning by higher plants occur in children, toxic mushrooms are consumed most often by adults. Occasional accidental mushroom poisonings of children and pets have been reported, but adults are more likely to actively search for and consume wild mushrooms for culinary purposes. Children are more seriously affected by the normally nonlethal toxins than are adults and are more likely to suffer very serious consequences from ingestion of relatively smaller doses. Adults who consume mushrooms are also more likely to recall what was eaten and when, and are able to describe their symptoms more accurately than are children. Very old, very young, and debilitated persons of both sexes are more likely to become seriously ill from all types of mushroom poisoning, even those types which are generally considered to be mild.

Many idiosyncratic adverse reactions to mushrooms have been reported. Some mushrooms cause certain people to become violently ill, while not affecting others who consumed part of the same mushroom cap. Factors such as age, sex, and general health of the consumer do not seem to be reliable predictors of these reactions, and they have been attributed to allergic or hypersensitivity reactions and to inherited inability of the unfortunate victim to metabolize certain unusual fungal constituents (such as the uncommon sugar, trehalose). These reactions are probably not true poisonings as the general population does not seem to be affected.

9. Analysis of Foods for Toxins: The mushroom toxins can with difficulty be recovered from poisonous fungi, cooking water, stomach contents, serum, and urine. Procedures for extraction and quantitation are generally elaborate and time-consuming, and the patient will in most cases have recovered by the time an analysis is made on the basis of toxin chemistry. The exact chemical natures of most of the toxins that produce milder symptoms are unknown. Chromatographic techniques (TLC, GLC, HPLC) exist for the amanitins, orellanine, muscimol/ibotenic acid, psilocybin, muscarine, and the gyromitrins. The amanitins may also be determined by commercially available 3H-RIA kits. The most reliable means of diagnosing a mushroom poisoning remains botanical identification of the fungus that was eaten. An accurate pre-ingestion determination of species will also prevent accidental poisoning in 100% of cases. Accurate post-ingestion analyses for specific toxins when no botanical identification is possible may be essential only in cases of suspected poisoning by the deadly Amanitas, since prompt and aggressive therapy (including lavage, activated charcoal, and plasmapheresis) can greatly reduce the mortality rate.

10. Selected Outbreaks: Isolated cases of mushroom poisoning have occurred throughout the continental United States. The occurred in Oregon in

October,1988, and involved the intoxication of five people who consumed stir-fried Amanita phalloides. The poisonings were severe, and at this writing three of the five people had undergone liver transplants for treatment of amanitin-induced liver failure. Other recent cases have included the July, 1986, poisoning of a family in Philadelphia, by Chlorophyllum molybdites; the September, 1987, intoxication of seven men in Bucks County, PA, by spaghetti sauce which contained Jack O'Lantern mushroom (*Omphalotus illudens*); and of 14 teenage campers in Maryland by the same species (July, 1987). A report of a North Carolina outbreak of poisoning by False Morel (*Gyromitra* spp.) appeared in 1986. A 1985 report details a case of *Chlorophyllum molybdites* which occurred in Arkansas; a fatal poisoning case caused by an amanitin containing Lepiota was described in 1986. In 1981, two Berks County, PA, people were poisoned (one fatally) after ingesting *Amanita phalloides*, while in the same year, seven Laotian refugees living in California were poisoned by *Russula* spp. In separate 1981 incidents, several people from New York State were poisoned by Omphalotus illudens, *Amanita muscaria*, *Entoloma lividum*, and *Amanita virosa*. An outbreak of gastroen- teritis during a banquet for 482 people in Vancouver, British Columbia, was reported by the Vancouver Health Department in June, 1991. Seventy-seven of the guests reported symptoms con- sisting of early onset nausea (15-30 min), diarrhea (20 min-13 h), vomiting (20-60 min), cramps and bloated feeling. Other symp- toms included feeling warm, clamminess, numbness of the tongue and extreme thirst along with two cases of hive-like rash with onset of 3-7 days. Bacteriological tests were negative. This intoxi- cation merits special attention because it involved consumption of species normally considered not only edible but choice. The fungi involved were the morels *Morchella esculenta* and *M. elata (M. angusticeps)*, which were prepared in a marinade and consumed raw. The symptoms were severe but not life threatening. Scattered reports of intoxications by these species and M. conica have appeared in anecodotal reports for many years.

Numerous other cases exist; however, the cases that appear in the literature tend to be the serious poisonings such as those causing more severe gastrointestinal symptoms, psychotropic reactions, and severe organ damage (deadly *Amanita*). Mild intoxications are probably grossly underreported, because of the lack of severi- ty of symptoms and the unlikeliness of a hospital admission.

For more information on recent outbreaks see the Morbidity and Mortality Weekly Reports from CDC.

Aflatoxins

1. Name of Toxin: Aflatoxins

2. Name of Acute Disease: Aflatoxicosis

Aflatoxicosis is poisoning that results from ingestion of aflatoxins in contaminated food or feed. The aflatoxins are a group of structurally related toxic compounds produced by certain strains of the fungi *Aspergillus flavus* and *A. parasiticus*. Under favorable conditions of temperature and humidity, these fungi grow on certain foods and feeds, resulting in the production of aflatoxins. The most pronounced contamination has been encountered in tree nuts, peanuts, and other oilseeds, including corn and cottonseed. The major aflatoxins of concern are designated B1, B2, G1, and G2. These toxins are usually found together in various foods and feeds in various proportions; however, aflatoxin B1 is usually predominant and is the most toxic. When a commodity is analyzed by thin-layer chromatography, the aflatoxins separate into the individual components in the order given above; however, the first two fluoresce blue when viewed under ultraviolet light and the second two fluoresce green. Aflatoxin M a major metabolic product of aflatoxin B1 in animals and is usually excreted in the milk and urine of dairy cattle and other mammalian species that have consumed aflatoxin-contaminated food or feed.

3. Nature of Disease: Aflatoxins produce acute necrosis, cirrhosis, and carcinoma of the liver in a number of animal species; no animal species is resistant to the acute toxic effects of aflatoxins; hence it is logical to assume that humans may be similarly affected. A wide variation in LD50 values has been obtained in animal species tested with single doses of aflatoxins. For most species, the LD50 value ranges from 0.5 to 10 mg/kg body weight. Animal species respond differently in their susceptibility to the chronic and acute toxicity of aflatoxins. The toxicity can be influenced by environmental factors, exposure level, and duration of exposure, age, health, and nutritional status of diet. Aflatoxin B1 is a very potent carcinogen in many species, including nonhuman primates, birds, fish, and rodents. In each species, the liver is the primary target organ of acute injury. Metabolism plays a major role in determining the toxicity of aflatoxin B1; studies show that this aflatoxion requires metabolic activation to exert its carcinogenic effect, and these effects can be modified by induction or inhibition of the mixed function oxidase system.

4. Normal Course of Disease: In well-developed countries, aflatoxin contamination rarely occurs in foods at levels that cause acute aflatoxicosis in humans. In view of this, studies on human toxicity from ingestion of aflatoxins have focused on their carcinogenic potential. The relative susceptibility of humans to aflatoxins is not known, even though epidemiological studies in Africa and Southeast Asia, where there is a high incidence of hepatoma, have revealed an association between cancer incidence and the aflatoxin content of the diet.

These studies have not proved a cause-effect relationship, but the evidence suggests an association.

One of the most important accounts of aflatoxicosis in humans occurred in more than 150 villages in adjacent districts of two neighboring states in northwest India in the fall of 1974. According to one report of this outbreak, 397 persons were affected and 108 persons died. In this outbreak, contaminated corn was the major dietary constituent, and aflatoxin levels of 0.25 to 15 mg/kg were found. The daily aflatoxin B1 intake was estimated to have been at least 55 ug/kg body weight for an undetermined number of days. The patients experienced high fever, rapid progressive jaundice, edema of the limbs, pain, vomiting, and swollen livers. One investigator reported a peculiar and very notable feature of the outbreak: the appearance of signs of disease in one village population was preceded by a similar disease in domestic dogs, which was usually fatal. Histopathological examination of humans showed extensive bile duct proliferation and periportal fibrosis of the liver together with gastrointestinal hemorrhages. A 10-year follow-up of the Indian outbreak found the survivors fully recovered with no ill effects from the experience.

A second outbreak of aflatoxicosis was reported from Kenya in 1982. There were 20 hospital admissions with a 60% mortality; daily aflatoxin intake was estimated to be at least 38 ug/kg body weight for an undetermined number of days.

In a deliberate suicide attempt, a laboratory worker ingested 12 ug/kg body weight of aflatoxin B1 per day over a 2-day period and 6 months later, 11 ug/kg body weight per day over a 14-day period. Except for transient rash, nausea and headache, there were no ill effects; hence, these levels may serve as possible no-effect levels for aflatoxin B1 in humans. In a 14-year follow-up, a physical examination and blood chemistry, including tests for liver function, were normal.

5. Diagnosis of Human Illnesses:

Aflatoxicosis in humans has rarely been reported; however, such cases are not always recognized. Aflatoxicosis may be suspected when a disease outbreak exhibits the following characteristics:

• the cause is not readily identifiable
• the condition is not transmissible
• syndromes may be associated with certain batches of food
• treatment with antibiotics or other drugs has little effect
• the outbreak may be seasonal, i.e., weather conditions may affect mold growth.

The adverse effects of aflatoxins in animals (and presumably in humans) have been categorized in two general forms.

A. (Primary) Acute aflatoxicosis is produced when moderate to high levels of aflatoxins are consumed. Specific, acute episodes of disease ensue may include hemorrhage, acute liver damage, edema, alteration in digestion, absorption and/or metabolism of nutrients, and possibly death.

B. (Primary) Chronic aflatoxicosis results from ingestion of low to moderate levels of aflatoxins. The effects are usually subclinical and difficult to recognize. Some of the common symptoms are impaired food conversion and slower rates of growth with or without the production of an overt aflatoxin syndrome.

7. Associated Foods: In the United States, aflatoxins have been identified in corn and corn products, peanuts and peanut products, cottonseed, milk, and tree nuts such as Brazil nuts, pecans, pistachio nuts, and walnuts. Other grains and nuts are susceptible but less prone to contamination.

8. Relative Frequency Disease: The relative frequency of aflatoxicosis in humans in the United of States is not known. No outbreaks have been reported in humans. Sporadic cases have been reported in animals.

9. Target Populations: Although humans and animals are susceptible to the effects of acute aflatoxicosis, the chances of human exposure to acute levels of aflatoxin is remote in well-developed countries. In undeveloped countries, human susceptibility can vary with age, health, and level and duration of exposure.

9. Analysis of Foods: Many chemical procedures have been developed to identify and measure aflatoxins in various commodities. The basic steps include extraction, lipid removal, cleanup, separation and quantification. Depending on the nature of the commodity, methods can sometimes be simplified by omitting unnecessary steps. Chemical methods have been developed for peanuts, corn, cottonseed, various tree nuts, and animal feeds. Chemical methods for aflatoxin in milk and dairy products are far more sensitive than for the above commodities because the aflatoxin M animal metabolite is usually found at much lower levels (ppb and ppt). All collaboratively studied methods for aflatoxin analysis are described in Chapter 26 of the AOAC Official Methods of Analysis.

10. Outbreaks: Very little information is available on outbreaks of aflatoxicosis in humans because medical services are less developed in the areas of the world where high levels of contamination of aflatoxins occur in foods, and, therefore, many cases go unnoticed.

For more information on recent outbreaks see the Morbidity and Mortality Weekly Reports from CDC.

Pyrrolizidine Alkaloids

1. Name of Toxin: Pyrrolizidine Alkaloids

2. Name of Acute Disease: Pyrrolizidine Alkaloids Poisoning

Pyrrolizidine alkaloid intoxication is caused by consumption of plant material containing these alkaloids. The plants may be consumed as food, for medicinal purposes, or as contaminants of other agricultural crops. Cereal crops and forage crops are sometimes contaminated with pyrrolizidine-producing weeds, and the alkaloids find their way into flour and other foods, including milk from cows feeding on these plants. Many plants from the Boraginaceae, Compositae, and Leguminosae families contain well over 100 hepatotoxic pyrrolizidine alkaloids.

3. Normal Course of Disease: Most cases of pyrrolizidine alkaloid toxicity result in moderate to severe liver damage. Gastrointestinal symptoms are usually the first sign of intoxication, and consist predominantly of abdominal pain with vomiting and the development of ascites. Death may ensue from 2 weeks to more than 2 years after poisoning, but patients may recover almost completely if the alkaloid intake is discontinued and the liver damage has not been too severe.

4. Diagnosis of Human Illness: Evidence of toxicity may not become apparent until sometime after the alkaloid is ingested. The acute illness has been compared to the Budd-Chiari syndrome (thrombosis of hepatic veins, leading to liver enlargement, portal hypertension, and ascites). Early clinical signs include nausea and acute upper gastric pain, acute abdominal distension with prominent dilated veins on the abdominal wall, fever, and biochemical evidence of liver disfunction. Fever and jaundice may be present. In some cases the lungs are affected; pulmonary edema and pleural effusions have been observed. Lung damage may be prominent and has been fatal. Chronic illness from ingestion of small amounts of the alkaloids over a long period proceeds through fibrosis of the liver to cirrhosis, which is indistinguishable from cirrhosis of other etiology.

5. Associated Foods: The plants most frequently implicated in pyrrolizidine poisoning are members of the Borginaceae, Compositae, and Leguminosae families. Consumption of the alkaloid-containing plants as food, contaminants of food, or as medicinals has occurred.

6. Relative Frequency of Disease: Reports of acute poisoning in the United States among humans are relatively rare. Most result from the use of medicinal preparations as home remedies. However, intoxications of range animals sometimes occur in areas under drought stress, where plants containing alkaloids are common. Milk from dairy animals can become contaminated with the alkaloids, and alkaloids have been found in the honey collected by bees foraging on toxic plants. Mass human poisonings have occurred in other countries when cereal crops used to prepare food were contaminated with seeds containing pyrrolizidine alkaloid.

7. Target Population: All humans are believed to be susceptible to the hepatotoxic pyrrolizidine alkaloids. Home remedies and consumption of herbal teas in large quantities can be a risk factor and are the most likely causes of alkaloid poisonings in the United States.

8. Analysis in Foods: The pyrrolizidine alkaloids can be isolated from the suspect commodity by any of several standard alkaloid extraction procedures. The toxins are identified by thin layer chromatography. The pyrrolizidine ring is first oxidized to a pyrrole followed by spraying with Ehrlich reagent, which gives a characteristic purple spot. Gas-liquid chromatographic and mass spectral methods also are available for identifying the alkaloids.

9. Selected Outbreaks: There have been relatively few reports of human poisonings in the United States. Worldwide, however, a number of cases have been documented. Most of the intoxications in the USA involved the consumption of herbal preparations either as a tea or as a medicine. The first patient diagnosed in the USA was a female who had used a medicinal tea for 6 months while in Ecuador. She developed typical hepatic veno-occlusive disease, with voluminous ascites, centrilobular congestion of the liver, and increased portal vein pressure. Interestingly, the patient completely recovered within one year after ceasing to consume the tea. Another herbal tea poisoning occurred when Senecio longilobus was mistaken for a harmless plant (called "gordolobo yerba" by Mexican Americans) and used to make herbal cough medicine. Two infants were given this medication for several days. The 2-month-old boy was ill for 2 weeks before being admitted to the hospital and died 6 days later. His condition was first diagnosed as Reye's syndrome, but was changed when jaundice, ascites, and liver necrosis were observed. The second child, a 6-month-old female, had acute hepatocellular disease, ascites, portal hypertension, and a right pleural effusion. The patient improved with treatment; however, after 6 months, a liver biopsy revealed extensive hepatic fibrosis, progressing to cirrhosis over 6 months. Another case of hepatic veno-occlusive disease was described in a 47-year-old nonalcoholic woman who had consumed large quantities of comfrey (Symphytum species) tea and pills for more than one year. Liver damage was still present 20 months after the comfrey consumption ceased.

For more information on recent outbreaks see the Morbidity and Mortality Weekly Reports from CDC.

Phytohaemagglutinin

1. Name of the Toxin: Phytohaemagglutinin (Kidney Bean Lectin)

This compound, a lectin or hemagglutinin, has been used by immunologists for years to trigger DNA synthesis in T lymphocytes, and more recently, to activate latent human immunodeficiency virus type 1 (HIV-1, AIDS virus) from human peripheral lymphocytes. Besides inducing mitosis, lectins are known for their ability to agglutinate many mammalian red blood cell types, alter cell membrane transport systems, alter cell permeability to proteins, and generally interfere with cellular metabolism.

2. Name of the Acute Disease: Red Kidney Bean (Phaseolus vulgaris) Poisoning, Kinkoti Bean Poisoning, and possibly other names.

3. Nature of the Acute Disease: The onset time from consumption of raw or undercooked kidney beans to symptoms varies from between 1 to 3 hours. Onset is usually marked by extreme nausea, followed by vomiting, which may be very severe. Diarrhea develops somewhat later (from one to a few hours), and some persons report abdominal pain. Some persons have been hospitalized, but recovery is usually rapid (3 - 4 h after onset of symptoms) and spontaneous.

4. Diagnosis of Human Illness: Diagnosis is made on the basis of symptoms, food history, and the exclusion of other rapid onset food poisoning agents (e.g., Bacillus cereus, Staphylococcus aureus, arsenic, mercury, lead, and cyanide).

5. Foods in Which it Occurs: Phytohaemagglutinin, the presumed toxic agent, is found in many species of beans, but it is in highest concentration in red kidney beans (Phaseolus vulgaris). The unit of toxin measure is the hemagglutinating unit (hau). Raw kidney beans contain from 20,000 to 70,000 hau, while fully cooked beans contain from 200 to 400 hau. White kidney beans, another variety of Phaseolus vulgaris, contain about one-third the amount of toxin as the red variety; broad beans (Vicia faba) contain 5 to 10% the amount that red kidney beans contain.

The syndrome is usually caused by the ingestion of raw, soaked kidney beans, either alone or in salads or casseroles. As few as four or five raw beans can trigger symptoms. Several outbreaks have been associated with "slow cookers" or crock pots, or in casseroles which had not reached a high enough internal temperature to destroy the glycoprotein lectin. It has been shown that heating to 80°C may potentiate the toxicity five-fold, so that these beans are more toxic than if eaten raw. In studies of casseroles cooked in slow cookers, internal temperatures often did not exceed 75°C.

6. Frequency of the Disease: This syndrome has occurred in the United Kingdom with some regularity. Seven outbreaks occurred in the U.K. between 1976 and 1979 and were reviewed (Noah et al. 1980. Br. Med. J. 19

July, 236-7). Two more incidents were reported by Public Health Laboratory Services (PHLS), Colindale, U.K. in the summer of 1988. Reports of this syndrome in the United States are anecdotal and have not been formally published.

7. Usual Course of the Disease and Some Complications: The disease course is rapid. All symptoms usually resolve within several hours of onset. Vomiting is usually described as profuse, and the severity of symptoms is directly related to the dose of toxin (number of raw beans ingested). Hospitalization has occasionally resulted, and intravenous fluids may have to be administered. Although of short duration, the symptoms are extremely debilitating.

8. Target Populations: All persons, regardless of age or gender, appear to be equally susceptible; the severity is related only to the dose ingested. In the seven outbreaks mentioned above, the attack rate was 100%.

9. Analysis of Food: The difficulty in food analysis is that this syndrome is not well known in the medical community. Other possible causes must be eliminated, such as Bacillus cereus, staphylococcal food poisoning, or chemical toxicity. If beans are a component of the suspected meal, analysis is quite simple, and based on hemagglutination of red blood cells (hau).

10. Selected Outbreaks: As previously stated, no major outbreaks have occurred in the U.S. Outbreaks in the U.K. are far more common. The syndrome is probably sporadic, affecting small numbers of persons or individuals, and is easily misdiagnosed or never reported due to the short duration of symptoms. Differences in reporting between the U.S. and U.K. may be attributed to greater use of dried kidney beans in the U.K., or better physician awareness. The U.K. has established a reference laboratory for the quantitation of hemagglutinins from suspected foods.

For more information on recent outbreaks see the Morbidity and Mortality Weekly Reports from CDC.

11. Education: NOTE: The following procedure has been recommended by the PHLS to render kidney, and other, beans safe for consumption:

Soak in water for at least 5 hours.
Pour away the water.
Boil briskly in fresh water for at least 10 minutes.
Undercooked beans may be more toxic than raw beans.

Grayanotoxin

1. Name of Toxin: Grayanotoxin (formerly known as andromedotoxin, acetylandromedol, and rhodotoxin)

2. Name of Acute Disease: Honey Intoxication

Honey intoxication is caused by the consumption of honey produced from the nectar of rhododendrons. The grayanotoxins cause the intoxication. The specific grayanotoxins vary with the plant species. These compounds are diterpenes, polyhydroxylated cyclic hydrocarbons that do not contain nitrogen. Other names associated with the disease is rhododendron poisoning, mad honey intoxication or grayanotoxin poisoning.

3. Nature of Disease: The intoxication is rarely fatal and generally lasts for no more than 24 hours. Generally the disease induces dizziness, weakness, excessive perspiration, nausea, and vomiting shortly after the toxic honey is ingested. Other symptoms that can occur are low blood pressure or shock, bradyarrhythima (slowness of the heart beat associated with an irregularity in the heart rhythm), sinus bradycardia (a slow sinus rhythm, with a heart rate less than 60), nodal rhythm (pertaining to anode, particularly the atrioventricular node), Wolff-Parkinson-White syndrome (anomalous atrioventricular excitation) and complete atrioventricular block.

4. Normal Course of the Disease: The grayanotoxins bind to sodium channels in cell membranes. The binding unit is the group II receptor site, localized on a region of the sodium channel that is involved in the voltage-dependent activation and inactivation. These compounds prevent inactivation; thus, excitable cells (nerve and muscle) are maintained in a state of depolarization, during which entry of calcium into the cells may be facilitated. This action is similar to that exerted by the alkaloids of veratrum and aconite. All of the observed responses of skeletal and heart muscles, nerves, and the central nervous system are related to the membrane effects.

Because the intoxication is rarely fatal and recovery generally occurs within 24 hours, intervention may not be required. Severe low blood pressure usually responds to the administration of fluids and correction of bradycardia; therapy with vasopressors (agents that stimulate contraction of the muscular tissue of the capillaries and arteries) is only rarely required. Sinus bradycardia and conduction defects usually respond to atropine therapy; however, in at least one instance the use of a temporary pacemaker was required.

5. Diagnosis of Human Illness: In humans, symptoms of poisoning occur after a dose-dependent latent period of a few minutes to two or more hours and include salivation, vomiting, and both circumoral (around or near the mouth) and extremity paresthesia (abnormal sensations). Pronounced low blood pressure and sinus bradycardia develop. In severe intoxication, loss of coordination and progressive muscular weakness result. Extrasystoles (a premature contraction of the heart that is independent of the normal rhythm and arises in response to an impulse in some part of the heart other than the sinoatrial node; called also premature beat) and ventricular tachycardia (an abnormally rapid ventricular rhythm with aberrant ventricular excitation, usually in excess of 150 per minute) with both atrioventricular and intraventricular conduction disturbances also may occur. Convulsions are reported occasionally.

6. Associated Foods: Grayanotoxin poisoning most commonly results from the ingestion of grayanotoxin-contaminated honey, although it may result from the ingestion of the leaves, flowers, and nectar of rhododendrons. Not all rhododendrons produce grayanotoxins. Rhododendron ponticum grows extensively on the mountains of the eastern Black Sea area of Turkey. This species has been associated with honey poisoning since 401 BC. A number of toxin species are native to the United States. Of particular importance are the western azalea (*Rhododendron occidentale*) found from Oregon to southern California, the California rosebay (*Rhododendron macrophyllum*) found from British Columbia to central California, and *Rhododendron albiflorum* found from British Columbia to Oregon and in Colorado. In the eastern half of the United States grayanotoxin-contaminated honey may be derived from other members of the botanical family Ericaceae, to which rhododendrons belong. Mountain laurel (*Kalmia latifolia*) and sheep laurel (*Kalmia angustifolia*) are probably the most important sources of the toxin.

7. Relative Frequency of Disease: Grayanotoxin poisoning in humans is rare. However, cases of honey intoxication should be anticipated everywhere. Some may be ascribed to a increase consumption of imported honey. Others may result from the ingestion of unprocessed honey with the increased desire of natural foods in the American diet.

8. Target Population: All people are believed to be susceptible to honey intoxication. The increased desire of the American public for natural (unprocessed) foods, may result in more cases of grayanotoxin poisoning. Individuals who obtain honey from farmers who may have only a few hives are at increased risk. The pooling of massive quantities of honey during commercial processing generally dilutes any toxic substance.

9. Analysis in Foods: The grayanotoxins can be isolated from the suspect commodity by typical extraction procedures for naturally occurring terpenes. The toxins are identified by thin layer chromatography.

10. Selected Outbreaks: Several cases of grayanotoxin poisonings in humans have been documented in the 1980s. These reports come from Turkey and Austria. The Austrian case resulted from the consumption of honey that was brought back from a visit to Turkey. From 1984 to 1986, 16 patients were treated for honey intoxication in Turkey. The symptoms started approximately 1 h after 50 g of honey was consumed. In an average of 24 h, all of the patients recovered. The case in Austria resulted in cardiac arrhythmia, which required a temporal pacemaker to prevent further decrease in heart rate. After a few hours, pacemaker simulation was no longer needed. The Austrian case shows that with increased travel throughout the world, the risk of grayanotoxin poisoning is possible outside the areas of Ericaceae-dominated vegetation, namely, Turkey, Japan, Brazil, United States, Nepal, and British Columbia. In 1983 several British veterinarians reported a incident of grayanotoxin poisoning in goats. One of the four animals died. Post-mortem examination showed grayanotoxin in the rumen contents.

For more information on recent outbreaks see the Morbidity and Mortality Weekly Reports from CDC.

APPENDICES

Infective dose information

Most chapters include a statement on infectious dose. These numbers should be viewed with caution for any of the following reasons:
- Often they were extrapolated from epidemiologic investigations.
- They were obtained by human feeding studies on healthy, young adult volunteers.
- They are best estimates based on a limited data base from outbreaks.
- They are worst case estimates.
- Because of the following variables they cannot be directly used to assess risk:

Variables of the Parasite or Microorganism
- Variability of gene expression of multiple pathogenic mechanism(s)
- Potential for damage or stress of the microorganism.
- Interaction of organism with food menstruum and environment
- pH susceptibility of organism
- Immunologic "uniqueness" of the organism
- Interactions with other organisms

Variables of the Host
- Age
- General health
- Pregnancy
- Medications--OTC or prescription
- Metabolic disorders
- Alcoholism, cirrhosis, hemochromatosis
- Malignancy
- Amount of food consumed
- Gastric acidity variation: antacids, natural variation, achlorhydria
- Genetic disturbances
- Nutritional status
- Immune competence
- Surgical history
- Occupation

Because of the complexity of factors involved in making risk decisions, the multidisciplinary Health Hazard Evaluation Board judges each situation on all available facts.

Approximate onset time to symptoms	Predominant symptoms	Associated organism or toxin
UPPER GASTROINTESTINAL TRACT SYMPTOMS (NAUSEA, VOMITING) OCCUR FIRST OR PREDOMINATE		
Less than 1 h	Nausea, vomiting, unusual taste, burning of mouth	Metallic salts
1-2 h	Nausea, vomiting, cyanosis, headache, dizziness, dyspnea, trembling, weakness, loss of consciousness	Nitrites
1-6 h mean 2-4 h	Nausea, vomiting, retching, diarrhea, abdominal pain, prostration	*Staphylococcus aureus* and its enterotoxins
8-16 h (2-4 h emesis possible)	Vomiting, abdominal cramps, diarrhea, nausea	*Bacillus cereus*
6-24 h	Nausea, vomiting, diarrhea, thirst, dilation of pupils, collapse, coma	Amanita species mushrooms
SORE THROAT AND RESPIRATORY SYMPTOMS OCCUR		
12-72 h	Sore throat, fever, nausea, vomiting, rhinorrhea, sometimes a rash	*Streptococcus pyogenes*
2-5 days	Inflamed throat and nose, spreading grayish exudate, fever, chills, sore throat, malaise, difficulty in swallowing, edema of cervical lymph node	*Corynebacterium diphtheriae*
LOWER GASTROINTESTINAL TRACT SYMPTOMS (ABDOMINAL CRAMPS, DIARRHEA) OCCUR FIRST OR PREDOMINATE		
2-36 h, mean 6-12 h	Abdominal cramps, diarrhea, putrefactive diarrhea associated with *C. perfringens*, sometimes nausea and vomiting	*Clostridium perfringens, Bacillus cereus, Streptococcus faecalis, S. faecium*
12-74 h, mean 18-36 h	Abdominal cramps, diarrhea, vomiting, fever, chills, malaise, nausea, headache, possible. Sometimes bloody or mucoid diarrhea, cutaneous lesions associated with V. vulnificus. Yersinia enterocolitica mimics flu and acute appendicitis	*Salmonella* species (including *S. arizonae*), *Shigella*, enteropathogenic *Escherichia coli*, other *Enterobacteriacae*, *Vibrio parahaemolyticus, Yersinia enterocolitica, Aeromonas hydrophila, Plesiomonas shigelloides, Campylobacter jejuni, Vibrio cholerae* (O1 and non-O1) *V.vulnificus, V. fluvialis*
3-5 days	Diarrhea, fever, vomiting abdominal pain, respiratory symptoms	Enteric viruses

1-6 weeks	Mucoid diarrhea (fatty stools) abdominal pain, weight loss	*Giardia lamblia*
1 to several weeks	Abdominal pain, diarrhea, constipation, headache, drowsiness, ulcers, variable -- often asymptomatic	*Entamoeba histolytica*
3-6 months	Nervousness, insomnia, hunger pains, anorexia, weight loss, abdominal pain, sometimes gastroenteritis	*Taenia saginata, T. solium*

NEUROLOGICAL SYMPTOMS (VISUAL DISTURBANCES, VERTIGO, TINGLING, PARALYSIS) OCCUR

Less than 1 h	*** SEE GASTROINTESTINAL AND/OR NEUROLOGIC SYMPTOMS (Shellfish Toxins) (this Appendix)	Shellfish toxin
	Gastroenteritis, nervousness, blurred vision, chest pain, cyanosis, twitching, convulsions	Organic phosphate
	Excessive salivation, perspiration, gastroenteritis, irregular pulse, pupils constricted, asthmatic breathing	Muscaria-type mushrooms
	Tingling and numbness, dizziness, pallor, gastro-hemmorrhage, and desquamation of skin, fixed eyes, loss of reflexes, twitching, paralysis	Tetradon (tetrodotoxin) toxins
1-6 h	Tingling and numbness, gastroenteritis, dizziness, dry mouth, muscular aches, dilated pupils, blurred vision, paralysis	Ciguatera toxin
	Nausea, vomiting, tingling, dizziness, weakness, anorexia, weight loss, confusion	Chlorinated hydrocarbons
2 h to 6 days, usually 12-36 h	Vertigo, double or blurred vision, loss of reflex to light, difficulty in swallowing. speaking, and breathing, dry mouth, weakness, respiratory paralysis	*Clostridium botulinum* and its neurotoxins
More than 72 h	Numbness, weakness of legs, spastic paralysis, impairment of vision, blindness, coma	Organic mercury
	Gastroenteritis, leg pain, ungainly high-stepping gait, foot and wrist drop	*Triorthocresyl phosphate*

ALLERGIC SYMPTOMS (FACIAL FLUSHING, ITCHING) OCCUR

Less than 1 h	Headache, dizziness, nausea, vomiting, peppery taste, burning of throat, facial swelling and flushing, stomach pain, itching of skin	Histamine (scombroid)
	Numbness around mouth, tingling sensation, lushing, dizziness, headache, nausea	Monosodium glutamate
	Flushing, sensation of warmth, itching, abdominal pain, puffing of face and knees	Nicotinic acid

GENERALIZED INFECTION SYMPTOMS
(FEVER, CHILLS, MALAISE, PROSTRATION, ACHES, SWOLLEN LYMPH NODES) OCCUR

4-28 days, mean 9 days	Gastroenteritis, fever, edema about eyes, perspiration, muscular pain, chills, prostration, labored breathing	*Trichinella spiralis*

7-28 days, mean 14 days	Malaise, headache, fever, cough, nausea, vomiting, constipation, abdominal pain, chills, rose spots, bloody stools	*Salmonella typhi*
10-13 days	Fever, headache, myalgia, rash	*Toxoplasma gondii*
10-50 days, mean 25-30 days	Fever, malaise, lassitude, anorexia, nausea, abdominal pain, jaundice	Etiological agent not yet isolated -- probably viral
Varying periods (depends on specific illness)	Fever, chills, head- or joint ache, prostration, malaise, swollen lymph nodes, and other specific symptoms of disease in question	*Bacillus anthracis, Brucella melitensis, B. abortus, B. suis, Coxiella burnetii, Francisella tularensis, Listeria monocytogenes, Mycobacterium tuberculosis, Mycobacterium species, Pasteurella multocida, Streptobacillus moniliformis, Campylobacter jejuni, Leptospira species.*

GASTROINTESTINAL AND/OR NEUROLOGIC SYMPTOMS – (SHELLFISH TOXINS)

0.5 to 2 h	Tingling, burning, numbness, drowsiness, incoherent speech, respiratory paralysis	Paralytic Shellfish Poisoning (PSP) (saxitoxins)
2-5 min to 3-4 h	Reversal of hot and cold sensation, tingling; numbness of lips, tongue & throat; muscle aches, dizziness, diarrhea, vomiting	Neurotoxic Shellfish Poisoning (NSP) (brevetoxins)
30 min to 2-3 h	Nausea, vomiting, diarrhea, abdominal pain, chills, fever	Diarrheic Shellfish Poisoning (DSP) (dinophysis toxin, okadaic acid, pectenotoxin, yessotoxin)
24 h (gastrointestinal) to 48 h (neurologic)	Vomiting, diarrhea, abdominal pain, confusion, memory loss, disorientation, seizure, coma	Amnesic Shellfish Poisoning (ASP) (domoic acid)

Factors affecting the growth of microorganisms in foods

Food is a chemically complex matrix, and predicting whether, or how fast, microorganisms will grow in any given food is difficult. Most foods contain sufficient nutrients to support microbial growth. Several factors encourage, prevent, or limit the growth of microorganisms in foods, the most important are aw, pH, and temperature.

a_w: (Water Activity or Water Availability). Water molecules are loosely oriented in pure liquid water and can easily rearrange. When other substances (solutes) are added to water, water molecules orient themselves on the surface of the solute and the properties of the solution change dramatically. The microbial cell must compete with solute molecules for free water molecules. Except for Staphylococcus aureus, bacteria are rather poor competitors, whereas molds are excellent competitors.

aw varies very little with temperature over the range of temperatures that support microbial growth. A solution of pure water has an a_w of 1.00. The addition of solute decreases the aw to less than 1.00.

Water Activity of Various NaCl Solutions

Percent NaCl (w/v)	Molal	Water Activity (a_w)
0.9	0.15	0.995
1.7	0.30	0.99
3.5	0.61	0.98
7.0	1.20	0.96
10.0	1.77	0.94
13.0	2.31	0.92
16.0	2.83	0.90
22.0	3.81	0.86

The a_w of a solution may dramatically affect the ability of heat to kill a bacterium at a given temperature. For example, a population of Salmonella typhimurium is reduced tenfold in 0.18 minutes at 60°C if the aw of the suspending medium is 0.995. If the a_w is lowered to 0.94, 4.3 min are required at 60°C to cause the same tenfold reduction.

An a_w value stated for a bacterium is generally the minimum aw which supports growth. At the minimum aw, growth is usually minimal, increasing as the aw increases. At a_w values below the minimum for growth, bacteria do not necessarily die, although some proportion of the population does die. The bacteria may remain dormant, but infectious. Most importantly, a_w is only one factor, and the other factors (e.g., pH, temperature) of the food must be considered. It is the interplay between factors that ultimately determines if a bacterium will grow or not. The a_w of a food may not be a fixed value; it may change over time, or may vary considerably between similar foods from different sources.

pH: (hydrogen ion concentration, relative acidity or alkalinity). The pH range of a microorganism is defined by a minimum value (at the acidic end of the scale) and a maximum value (at the basic end of the scale). There is a pH optimum for each microorganism at which growth is maximal. Moving away from the pH optimum in either direction slows microbial growth.

A range of pH values is presented here, as the pH of foods, even those of a similar type, varies considerably. Shifts in pH of a food with time may reflect microbial activity, and foods that are poorly buffered (i.e., do not resist changes in pH), such as vegetables, may shift pH values considerably. For meats, the pH of muscle from a rested animal may differ from that of a fatigued animal.

A food may start with a pH which precludes bacterial growth, but as a result of the metabolism of other microbes (yeasts or molds), pH shifts may occur and permit bacterial growth. Temperature. Temperature values for microbial growth, like pH values, have a minimum and maximum range with an optimum temperature for maximal growth. The rate of growth at extremes of temperature determines the classification of an organism (e.g., psychrotroph, thermotroph). The optimum growth temperature determines its classification as a thermophile, mesophile, or psychrophile.

INTERPLAY OF FACTORS AFFECTING MICROBIAL GROWTH IN FOODS:

Although each of the major factors listed above plays an important role, the interplay between the factors ultimately determines whether a microorganism will grow in a given food. Often, the results of such interplay are unpredictable, as poorly understood synergism or antagonism may occur. Advantage is taken of this interplay with regard to preventing the outgrowth of *C. botulinum*. Food with a pH of 5.0 (within the range for *C. botulinum*) and an aw of 0.935 (above the minimum for *C. botulinum*) may not support the growth of this bacterium. Certain processed cheese spreads take advantage of this fact and are therefore shelf stable at room temperature even though each individual factor would permit the outgrowth of *C. botulinum.*

Therefore, predictions about whether or not a particular microorganism will grow in a food can, in general, only be made through experimentation. Also, many microorganisms do not need to multiply in food to cause disease.

Foodborne disease outbreaks in the United States, graphs for 1988-1992

Outbreaks 1988

FIGURE 1. Number of reported foodborne-disease outbreaks, by state-United States, * 1988

* Includes Guam, Puerto Rico, and U.S. Virgin Islands

Outbreaks 1989

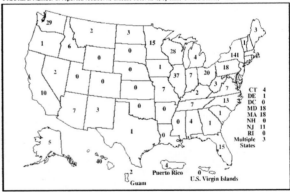

FIGURE 2. Number of reported foodborne-disease outbreaks, by state-United States, * 1989

* Includes Guam, Puerto Rico, and U.S. Virgin Islands

Outbreaks 1990

FIGURE 3. Number of reported foodborne-disease outbreaks, by state-United States, * 1990

* Includes Guam, Puerto Rico, and U.S. Virgin Islands

Outbreaks 1991

FIGURE 4. Number of reported foodborne-disease outbreaks, by state-United States, * 1991

* Includes Guam, Puerto Rico, and U.S. Virgin Islands

Outbreaks 1992

FIGURE 5. Number of reported foodborne-disease outbreaks, by state-United States, * 1992

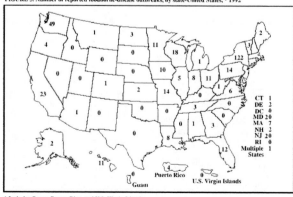

* Includes Guam, Puerto Rico, and U.S. Virgin Islands

133

Foodborne disease outbreak articles and databases of interest

"Impact of Changing Consumer Lifestyles on the Emergence and Reemergence of Foodborne Pathogens", Emerging Infectious Diseases 3(4)1997

Foodborne illness of microbial origin is the most serious food safety problem in the United States. The Centers for Disease Control and Prevention reports that 79% of outbreaks between 1987 and 1992 were bacterial; improper holding temperature and poor personal hygiene of food handlers contributed most to disease incidence. Some microbes have demonstrated resistance to standard methods of preparation and storage of foods. Nonetheless, food safety and public health officials attribute a rise in incidence of foodborne illness to changes in demographics and consumer lifestyles that affect the way food is prepared and stored. Food editors report that fewer than 50% of consumers are concerned about food safety. An American Meat Institute (1996) study details lifestyle changes affecting food behavior, including an increasing number of women in the workforce, limited commitment to food preparation, and a greater number of single heads of households. Consumers appear to be more interested in convenience and saving time than in proper food handling and preparation.

"Quantitative Risk Assessment: An Emerging Tool for Emerging Foodborne Pathogens", Emerging Infectious Diseases 3(4)1997

New challenges to the safety of the food supply require new strategies for evaluating and managing food safety risks. Changes in pathogens, food preparation, distribution, and consumption, and population immunity have the potential to adversely affect human health. Risk assessment offers a framework for predicting the impact of changes and trends on the provision of safe food. Risk assessment models facilitate the evaluation of active or passive changes in how foods are produced, processed, distributed, and consumed.

"Outbreak Investigations: A Perspective", Emerging Infectious Diseases 4(1)1998

Outbreak investigations, an important and challenging component of epidemiology and public health, can help identify the source of ongoing outbreaks and prevent additional cases. Even when an outbreak is over, a thorough epidemiologic and environmental investigation often can increase our knowledge of a given disease and prevent future outbreaks. Finally, outbreak investigations provide epidemiologic training and foster cooperation between the clinical and public health communities.

National Center for Environmental Health, Diseases Transmitted through the Food Supply

The objective of Environmental Health Services (EHS) is to strengthen the role of state, local, and national environmental public health programs and professionals to better anticipate, identify, and respond to adverse environmental exposures and the consequences of these exposures to human health. Section 103(d) of the Americans with Disabilities Act of 1990, Public Law 101-336, requires Secretary the Department of Health and Human Services to:

1. Review all infectious and communicable diseases which may be transmitted through handling the food supply;
2. Publish a list of infectious and communicable diseases which are transmitted through handling the food supply;
3. Publish the methods by which such diseases are transmitted;
4. Widely disseminate such information regarding the list of diseases and their modes of transmissibility to the general public;
5. Additionally, update the list annually.

Food Safety and Inspection Service Pathogen Reduction/HACCP & HACCP Implementation

FSIS links to federal documents concerning "Hazard Analysis Critical Control Points" implementation.

Food Safety and Inspection Service Active Recall Information Center

This page contains summary data on active recall cases. When a recall is completed, it will be removed from this listing, but will be included in the Recall Case Archive.

Food Safety and Inspection Service Office of Public Health and Science Publications

The Office of Public Health and Science (OPHS) provides expert scientific analysis, advice, data, and recommendations on all matters involving public health and science that are of concern to FSIS.

Food Safety and Inspection Service Food Safety Publications

Disaster Assistance, Fact Sheets, Food Safety Features, Food Safety Focus (Background), Seasonal Features (Press Kits) from the Meat and Poultry Hotline, Consumer Information From USDA, some available as one-page reproducibles), Brochures, Graphics, For Children, News Feature Stories and Technical Information From FSIS.

MORBIDITY AND MORTALITY WEEKLY REPORTS

PATHOGENIC BACTERIA

MMWR 45(46):1996 Nov 22

Salmonellosis Associated with a Thanksgiving Dinner--Nevada, 1995

On November 28, 1995, the county coroner's office notified the Clark County Health District in Las Vegas, Nevada, about a death suspected to have resulted from a foodborne disease. This report summarizes the investigation of the outbreak of gastroenteritis among persons who attended a Thanksgiving dinner. The investigation documented Salmonella serotype Enteritidis (SE) infection associated with eating improperly prepared turkey and stuffing containing eggs and emphasizes the need to use a meat thermometer to ensure complete cooking of turkey and stuffing.

During November 25-28, 1995, all six persons who attended a Thanksgiving dinner at a private home on November 23 and a seventh person who on November 25 ate food remaining from the dinner had onset of abdominal cramps, vomiting, and diarrhea. Two persons were hospitalized because of dehydration; a third person was found comatose at home and died from severe dehydration and sepsis. Stool cultures obtained from three persons, including the decedent, yielded SE phage type 13a. Turkey and stuff ing were the only foods eaten by all seven ill persons. No leftover food was available for culture.

The Clark County Health District interviewed the ill persons (including the cook) to obtain details about the preparation and cooking of the turkey and stuffing. On November 22, a 13-pound frozen turkey was thawed for 6 hours in a sink filled with cold water. After thawing, the packet of giblets (heart, liver, and gizzard) was removed, and the turkey was stored in a refrigerator overnight. However, on November 23, parts of the turkey were noted to be frozen. The turkey was filled with a stuffing made from bread, the giblets, and three raw eggs, and then placed for 1 hour in an oven set at 350 F (177 C). The setting was lowered to 300 F (149 C) while the turkey cooked for an estimated additional 4 hours. The turkey was removed from the oven when the exterior had browned. A meat thermometer was not used. The stuffing was removed immediately and was served with the turkey. After the outbreak, health officials tested the oven set at 300 F (149 C) and found the temperature to be 350 F (176 C).

Reported by: O Ravenholt, MD, CA Schmutz, LC Empey, DJ Maxson, PL Klouse, AJ Bryant, Clark County Health District, Las Vegas; R Todd, DrPH, State Epidemiologist, Nevada State Health Div. Foodborne and Diarrheal Diseases Br, Div of Bacterial and Mycotic Diseases, National Center for Infectious Diseases, CDC.

Editorial Note: Salmonellosis is frequently associated with eating undercooked eggs and poultry. Undercooked eggs are a particularly common source of SE infections. During 1988--1992, among foodborne disease outbreaks of salmonellosis reported to CDC in which a single food item was implicated, consumption of turkey and eggs accounted for 4% and 14% of cases, respectively. In addition, eggs or foods containing eggs as a principal ingredient caused 64% of the SE outbreaks (2).

Factors probably associated with the outbreak described in this report included inadequate thawing, use of raw eggs in the stuffing, and undercooking; in addition, the browned color of the turkey may have caused the cook to believe that the turkey and stuffing were thoroughly cooked. Although the original source of the Salmonella is unknown, the raw eggs used in the stuffing probably contained SE, and these eggs probably were incompletely cooked; undercooking may occur more commonly in turkeys that contain stuffing (J. Carpenter, Ph.D., University of Georgia, personal communication, 1996).

Each year, an estimated 45 million turkeys are eaten in the United States at Thanksgiving (J. DeYoung, National Turkey Federation, personnel communication, 1996). Salmonella infection may result from eating improperly cooked turkey and stuffing (3,4). This risk for infection can be reduced by cooking stuffing outside the turkey. Guidelines prepared by the U.S. Department of Agriculture (USDA) for persons who choose to cook stuffing inside the turkey recommend preparing the stuffing immediately before it is placed inside the turkey, stuffing the turkey loosely, inserting a meat thermometer into the center of the stuffing, and ensuring that the thermometer attains a temperature of at least 165 F (74 C). Additional recommendations for safely preparing and cooking a turkey include thawing the turkey completely before cooking, cooking in an oven set no lower than 325 F (163 C), and using a meat thermometer to ensure that the innermost part of the thigh attains a temperature of 180 F (82 C). Although the set temperature and cooking time can be used as guides to determine whether food is completely cooked, inaccuracies in the actual temperature and incomplete thawing before cooking can lead to undercooking. Use of a meat thermometer provides a more accurate determination of thorough cooking. Further advice on cooking turkeys and stuffing is available from USDA's Meat and Poultry Hotline, telephone (800) 535-4555.

References
1. Cohen ML, Tauxe RV. Drug-resistant Salmonella in the United States: an epidemiologic perspective. Science 1986;234:964-9.
2. Bean NH, Goulding JS, Loa C, Angulo FJ. Surveillance for foodborne-disease outbreaks--United States, 1988-1992. In: CDC surveillance summaries (October). MMWR 1996;45(no. SS-5).
3. CDC. Foodborne nosocomial outbreak of Salmonella reading--Connecticut. MMWR 1991;40:804-6.
4. CDC. Restaurant outbreak of salmonellosis due to undercooked turkey--Washington. MMWR 1978;27:514,519.

MMWR 45(34):1996 Aug 30

Outbreaks of *Salmonella* Serotype Enteritidis Infection Associated with Consumption of Raw Shell Eggs--United States, 1994-1995

Salmonella serotype Enteritidis (SE) accounts for an increasing proportion of all Salmonella serotypes reported to CDC's National Salmonella Surveillance System. During 1976-1994, the proportion of reported Salmonella isolates that were SE increased from 5% to 26% (Figure 1 [not present]). During 1985-1995, state and territorial health departments reported 582 SE outbreaks, which accounted for 24,058 cases of illness, 2290 hospitalizations, and 70 deaths. This report describes four SE outbreaks during 1994-1995 associated with consumption of raw shell eggs (i.e., unpasteurized eggs) and underscores that outbreaks of egg-associated SE infections remain a public health problem.

WASHINGTON, D.C.
In August 1994, a total of 56 persons who ate at a Washington, D.C., hotel had onset of diarrhea; 20 persons were hospitalized. Salmonella group D was isolated from stools of the 29 patrons who submitted specimens; 27 of the 29 isolates further typed were identified as SE.

An investigation by the District of Columbia Commission of Public Health (DCCPH) involved 41 ill patrons and 23 well patrons who had eaten brunch at the hotel on August 28. A case was defined as onset of diarrheal illness in a person who ate brunch at the hotel on August 28. All 39 patrons who had eaten hollandaise sauce became ill, compared with two (8%) of 25 persons who had not eaten the sauce (odds ratio [OR]=undefined; lower 95% confidence limit=52; p=<0.01).

Cultures of the sauce served on August 28 yielded SE. Of the 11 isolates tested (10 obtained from ill persons and one from the sauce), all were phage type 8. Cultures of pooled whole raw shell eggs, egg whites, and raw shell eggs from the same shipment as the implicated eggs did not yield Salmonella.

The hollandaise sauce was prepared on August 28 by hand-cracking and pooling the egg yolks from 36 extra-large grade A raw shell eggs. Lemon juice, melted butter, salt, and pepper were added to the egg yolk mixture while heating over a hot water bath. After preparation, the sauce was held in a hot water bath at an estimated temperature of 100 °F - 120 F (38 C-49 C) for 9 hours while being served. Traceback of the implicated eggs by DCCPH, the Maryland Veterinary Service, and the U.S. Department of Agriculture's (USDA's) Animal and Plant Health Inspection Service (APHIS) identified two flocks in Pennsylvania as possible sources for the eggs.

INDIANA

In June 1995, approximately 70 residents and staff of a nursing home in Indiana had onset of diarrhea and abdominal cramps. Stool cultures from symptomatic residents and staff yielded 39 confirmed cases of SE. The one isolate tested was phage type 13A. Three residents died from complications of SE infection.

An investigation by the Indiana State Department of Health and the Vanderburgh County Health Department involved seven of the initial 18 case-patients and 13 well residents. A case was defined as diarrheal illness in a nursing home resident with onset on June 9. Six (86%) of the seven patients had eaten baked eggs for breakfast on June 7, compared with three (23%) of 13 well persons (OR=16.5; 95% CI=1.3-1009; p=0.02).

The baked eggs were prepared by hand-cracking and pooling 180 medium grade A raw shell eggs, mixing the eggs with a hand whisk, and baking them in a single 8-inch deep pan at 400 F (204 C) for 45 minutes-1 hour. The eggs were then placed on a steam table where an internal temperature was obtained and reported in a chart log. Although recorded internal temperatures of eggs prepared during June ranged from 180 F-200 F (82 C-93 C), inadequate cooking may have contributed to the outbreak because the eggs were not stirred while being baked, and the internal temperature was obtained from only one place in the pan. The eggs were served within 30 minutes after cooking.

At the time of the investigation, none of the prepared eggs or raw shell eggs from the same shipment were available for testing. APHIS traced the implicated eggs to a distributor who received eggs from at least 35 different flocks.

GREENPORT, NEW YORK

On June 24, 1995, a total of 76 persons attended a catered wedding reception. Following the reception, attendees contacted the local health department to report onset of a gastrointestinal illness. Salmonella group D was isolated from stools of the 13 persons who submitted specimens; 11 of the 13 isolates further typed were identified as SE.

An investigation by the Suffolk County Health Department involved the 28 ill attendees and the 12 well attendees that were contacted. A case was defined as onset of diarrheal illness in an attendee of the reception. Twenty-six (93%) of 28 persons who had eaten Caesar salad became ill, compared with two (17%) of 12 persons who had not eaten the salad, (OR=52; 95% CI=6.2-849; p=<0.01).

The Caesar salad dressing was prepared with 18 raw shell eggs, olive oil, lemon juice, anchovies, Romano cheese, and Worcestershire sauce at 11:30 a.m. on June 24. The mixture was held unrefrigerated at the catering establishment for 2 hours, then placed in an unrefrigerated van until delivered and served at the reception at 6 p.m.

A traceback by the New York State Department of Agriculture and Markets (NYSDAM) identified the source of the eggs as a producer/distributor in Pennsylvania who received and commingled eggs from at least five flocks.

NEW YORK CITY

On July 23, 1995, three persons who lived in the same household drank a home-made beverage known as "jamaican malt" and had onset of diarrhea vomiting, and abdominal cramps; two were hospitalized. The mean period from consumption to onset of illness was 7.6 hours (range: 5.5-10.5 hours). Stool cultures from all three persons yielded SE.

The beverage was prepared at home the evening of July 22 by mixing beer, refrigerated raw shell eggs, sweetened condensed milk, oatmeal, and ice. Two patients drank the beverage immediately after preparation, and the third drank it 5 hours later. The beverage had been refrigerated after preparation. Cultures of the leftover beverage, raw eggs from the same carton used to prepare the drink, and leftover egg shells from the eggs used to prepare the drink all yielded SE. Isolates from the one patient tested and all three food samples were phage type 13A.

Traceback of the implicated eggs by NYSDAM identified a single flock in Pennsylvania. At the recommendation of the Pennsylvania Department of Health, eggs from the implicated flock were diverted to a pasteurization plant.

Reported by: M Levy, MD, M Fletcher, PhD, M Moody, MS, Bur of Epidemiology and Disease Control, District of Columbia Commission of Public Health. D Cory, W Corbitt, MS, C Borowiecki, D Gries, J Heidingsfelder, MD, Vanderburgh County Health Dept, Evansville; A Oglesby, MPH, J Butwin, MSN, D Ewert, MPH, D Bixler, MD, B Barrett, K Laurie, E Muniz, MD, G Steele, DrPH, State Epidemiologist, Indiana State Dept of Health. A Baldonti, MD, Albert Einstein College of Medicine, New York City; B Williamson, Suffolk County Health Dept, Hauppauge; M Layton, MD, Bur of Communicable Disease Control, L Kornstein, PhD, Bur of Laboratories, E Griffin, Bur of Environmental Investigation, New York City Health Dept; M Cambridge, N Fogg, J Guzewich, Bur of Community Sanitation and Food Protection, T Root, Wadsworth Center for Laboratories and Research, D Morse, MD, State Epidemiologist, New York State Dept of Health; J Wagoner, New York State Dept of Agriculture and Markets. M Deasey, Div of Epidemiology, Pennsylvania Dept of Health; K Miller, Pennsylvania Dept of Agriculture. Animal and Plant Health Inspection Service, Food Safety and Inspection Service, US Dept of Agriculture. Food and Drug Administration. Foodborne and Diarrheal Diseases Br, Div of Bacterial and Mycotic Diseases, National Center for Infectious Diseases, CDC.

Editorial Note: During 1976-1994, rates of isolation of SE increased in the United States from 0.5 to 3.9 per 100,000 population (Figure 2 [not present]). Two important factors probably contributed to the increase in 1994:

1. the effect of an outbreak of SE infections associated with a nationally distributed ice cream product (1) and
2. the expansion of the SE epidemic into California.

During 1990-1994, the SE isolation rate for the North-east region decreased from 8.9 to 7.0 per 100,000 population; the rate increased approximately threefold for the Pacific region (Figure 2 [not present]). This increase was primarily associated with reports from California, where the percentage of Salmonella isolates that were SE increased from 11% in 1990 to 38% in 1994. In 1994, 24% of all SE isolates in the United States were from California. In the United States, both sporadic and outbreak-associated cases of SE infection frequently have been associated with consumption of raw or undercooked shell eggs (2-4).

The findings in this report illustrate that outbreaks of egg-associated SE infections remain a public health problem in commercial food-service establishments, institutional facilities, and private homes throughout the United States. However, the risk for SE infection in humans can be reduced through public health prevention efforts (see box [addendum at end of document]) (5).

In 1994, no reported deaths resulted from SE outbreaks in the United States; however, in 1995, five deaths were associated with SE outbreaks, including the three in Indiana described in this report. One possible explanation for the lack of deaths in 1994 is that no nursing home outbreaks were reported that year; four of the five reported deaths in 1995 occurred among nursing home residents. During 1985-1991, a total of 59 SE outbreaks occurred in hospitals or nursing homes, accounting for only 12% of all outbreak-associated cases but 90% of all deaths. The case-fatality rate in these institutions was 70 times higher than in outbreaks in other settings (4). This underscores the importance of using pasteurized egg products for all recipes requiring pooled, raw, or undercooked shell eggs for the institutionalized elderly and other high-risk populations.

In 1990, USDA initiated a mandatory program to test for SE in breeder flocks that produce egg-laying chickens. In addition, USDA traced the eggs implicated in human foodborne SE outbreaks back to the farm of origin and, when feasible, conducted serologic and microbiologic assessments of the farm. If SE was detected at the source farm, the eggs were diverted to pasteurization. Funding for this program was discontinued effective October 1, 1995. As a result, these efforts are conducted by the Food and Drug Administration, which has regulatory authority for shell eggs in interstate commerce.

Further control of SE will require limiting the spread of SE on farms. In 1992, USDA, in collaboration with the industry, academia, and the Pennsylvania Department of Agriculture (PDA), initiated a flock-based intervention program, the Pennsylvania Pilot Project (6), which evolved in 1994 into the current Pennsylvania Egg Quality Assurance Program (PEQAP). USDA provided oversight for PEQAP until June 30, 1996, when the program was transferred to PDA and the industry. This prevention program uses many of the on-farm microbiologic testing and control procedures developed in the pilot project to reduce SE contamination of eggs. The decrease in SE infections in the

Northeast may reflect the collaborative prevention efforts in that region; similar efforts may be necessary to control the problem elsewhere in the country.

References

1. Hennessy TW, Hedberg CW, Slutsker L, et al. A national outbreak of Salmonella enteritidis infections from ice cream. N Engl J Med 1996; 334:1281-6.
2. St Louis ME, Morse DL, Potter ME, et al. The emergence of grade A eggs as a major source of Salmonella enteritidis infections: new implications for the control of salmonellosis. JAMA 1988; 259:2103-7.
3. Hedberg CW, David MJ, White KE, MacDonald KL, Osterholm MT. Role o egg consumption in sporadic Salmonella enteritidis and Salmonella typhimurium infections in Minnesota. J Infect Dis 1993;167:107-11.
4. Mishu B, Koehler J, Lee LA, et al. Outbreaks of Salmonella enteritidis infections in the United States, 1985-1991. J Infect Dis 1994;169:547-52.
5. Food and Drug Administration. Food code: 1995 recommendations of th United States Public Health Service. Washington, DC: US Department of Health and Human Services, Public Health Service, Food and Drug Administration, 1995.
6. Animal and Plant Inspection Service, US Department of Agriculture. Salmonella Enteritidis Pilot Project progress report. Washington, DC: US Department of Agriculture, 1995.

Recommendations for Preventing Salmonella Serotype Enteritidis Infections Associated with Eggs
* Consumption of raw or undercooked eggs should be avoided, especially by immunocompromised or other debilitated persons.
* In hospitals, nursing homes, and commercial kitchens, pooled eggs or raw or undercooked eggs should be substituted with pasteurized egg products.
* Eggs should be cooked at >= 145 F (63 C) for >= 15 seconds (until the white is completely set and the yolk begins to thicken) and eaten promptly after cooking.
* Hands, cooking utensils, and food-preparation surfaces should be washed with soap and water after contact with raw eggs.
* Eggs should be purchased refrigerated and stored refrigerated at <=41 F (5 C) at all times.
* Flock-based egg-quality-assurance programs that meet national standards and include microbiologic testing should be adopted by industry nationwide.

MMWR 44(49):1995 Dec 15

Outbreak of *Salmonella* Serotype Typhimurium Infection Associated with Eating Raw Ground Beef--Wisconsin, 1994

Despite previously publicized outbreaks of illness associated with and recommendations to avoid eating undercooked meat, some persons continue to eat undercooked or raw meat. This report summarizes the investigation of an outbreak of Salmonella serotype Typhimurium gastrointestinal illness in Wisconsin associated with eating contaminated raw ground beef during the 1994 winter holiday season.

On December 29, 1994, physicians in a group medical practice in Dodge County (1994 estimated population: 79,360), Wisconsin, reported to the Public Health Unit of the Dodge County Human Services and Health Department (DCHSHD) that during December 27-29 they had treated 17 patients with acute gastrointestinal illness characterized by diarrhea and abdominal cramps. At least 14 patients reported having eaten raw ground beef that was either plain or seasoned with onions and an herb mix during the 72 hours before illness onset. Stool samples for culture were obtained from 11 patients; Salmonella serotype Typhimurium that did not ferment tartrate was isolated from seven specimens. Based on these reports and findings, the DCHSHD issued a physician alert and press release that encouraged affected residents to report their illnesses and physicians to obtain stool cultures from case-patients. In addition, DCHSHD and the Bureau of Public Health, Wisconsin Division of Health (WDOH), initiated an investigation of this outbreak. A probable case of Salmonella infection was defined as diarrhea or abdominal cramps with onset during December 22, 1994-January 4, 1995, in a resident of or a visitor to Dodge County or any of the four contiguous counties. A confirmed case was defined as a stool culture positive for tartrate-negative Salmonella Typhimurium.

DCHSHD and WDOH identified 107 confirmed and 51 probable case-patients (Figure 1 [-not included]); of these, 17 (16%) were hospitalized. Predominant manifestations of illness included diarrhea (98%), abdominal cramps (88%), chills (77%), body aches (71%), fever (65%), nausea (60%), and bloody stools (43%). The ages of ill persons ranged from 2 years to 90 years; 62% were male.

To assess potential risk factors for illness, DCHSHD and WDOH conducted a case-control study including 40 case-patients who were randomly selected from the persons with a stool specimen culture positive for tartrate-negative Salmonella Typhimurium and 40 controls who were identified by random telephone digit dialing. The mean ages of cases and controls were similar (43 years for cases; 47 years for controls). Of 40 case-patients, 35 (88%) reported having eaten raw ground beef during December 22-January 4, compared with eight (20%) of 40 controls (odds ratio [OR]=28; 95% confidence interval [CI]=7-117). Among the 35 who ate raw ground beef, 34 (97%) had purchased the beef from one butcher shop, compared with three (37%) of the eight controls (OR=56; 95% CI=4-1881). Knowledge of previous reports of outbreaks related to eating raw or undercooked beef was less among ill persons than among controls (26 [65%] of 40 case-patients compared with 30 [75%] of 40 controls [OR=0.6; 95% CI=0.2-1.8]). However, 22 (85%) of the 26 case-patients who reported being aware of previous outbreaks associated with consumption of raw ground beef continued this behavior compared with seven (23%) of the 30 controls with knowledge of previous outbreaks (OR=18.1; 95% CI=4.0-92.0).

DCHSHD and WDOH obtained from case-patients six leftover samples of raw ground beef that had been purchased at the butcher shop on five dates during December 21 -29 and served in different homes. These samples were cultured for Salmonella sp.; all grew tartrate-negative Salmonella Typhimurium. On December 30,1994, staff of the Meat Safety and Inspection Bureau (MSIB), Wisconsin Department of Agriculture, Trade, and Consumer Protection (WDATCP), informed the proprietor of the butcher shop of a potential problem with consumption of raw ground beef from the shop and the need to properly label meat products. During the winter holiday season, the butcher shop sold both seasoned and unseasoned raw ground beef that had a warning label regarding safe handling of poultry. On January 2, 1995, inspectors from MSIB examined sanitary conditions in the butcher shop, obtained invoices indicating the origin and the quantity of the meat used to prepare the ground beef, and inspected the raw ground beef production method and selling practice in the butcher shop.

Meat from approximately 35 carcasses obtained from three different suppliers had been ground in the shop from December 21 through January 4. Leftover product was reported to have been discarded each day and not carried over for sale the next day. All parts of the meat grinder except for the auger housing were disassembled and individually cleaned and sanitized at the end of each day. This type of grinder allowed easy disassembly of the auger and other smaller parts; the auger housing was attached to the grinder with nuts and bolts and required a wrench for removal. However, the cleaning staff had not received instructions regarding removal of the auger housing and had cleaned only surfaces of the tunnel-like space for the auger with a brush.

Meat remnants were present in the auger housing when the grinder was disassembled. Twenty environmental swabs of the equipment and the areas related to the production of the ground beef were obtained for bacterial culture; all were negative for Salmonella sp. Stool specimens obtained from all five butchers at the shop were cultured; one was positive for tartrate-negative Salmonella Typhimurium. Although this butcher denied illness, he had eaten raw ground beef at the shop during the outbreak interval.

Reported by: PA Frazak, MPH, Public Health Unit, Dodge County Human Svcs and Health Dept, JJ Kazmierczak, DVM, ME Proctor, PhD, JP Davis, MD, State Epidemiologist for Communicable Diseases, Bur of Public Health, Wisconsin Div of Health; J Larson, R Loerke, Meat Safety and Inspection Bur, Wisconsin Dept of Agriculture, Trade, and Consumer Protection. Div of Bacterial and Mycotic Diseases, National Center for Infectious Diseases; Div of Field Epidemiology, Epidemiology Program Office, CDC.

Editorial Note: The investigation of this outbreak implicated consumption of contaminated raw ground beef as the source of Salmonella infection. Inadequate cleaning and sanitization of the meat grinder probably resulted in ongoing contamination of ground beef over many production days. The outbreak occurred during the winter holiday season, and some patients reported that consumption of raw ground beef during these holidays was a practice brought from Europe by their ancestors. The decline of cases after the holidays may have occurred because ground beef from the implicated butcher shop was no longer consumed raw or because the grinder was cleaned more thoroughly after WDATCP personnel spoke with the proprietor of the butcher shop on December 30. The five persons who became ill butdid not report eating raw ground beef may not have remembered eating the raw ground beef, may have eaten undercooked ground beef or food that was contaminated from the raw ground beef, or may have become ill through person-to-person transmission.

Raw ground beef previously has been implicated as a vehicle for transmission of Salmonella (1,2), and undercooked ground beef is the most frequently recognized vehicle for Escherichia coli O157:H7 infection (3). The prevalence of Salmonella in beef ranges from 1% for raw beef carcasses (4) to 5%-7% for ground beef (U.S. Department of Agriculture, Food Safety and Inspection Service, unpublished data, 1994). Prevention measures include warning consumers of the health risks associated with eating raw ground beef and encouraging them to thoroughly cook ground beef and to adhere to safe foodhandling guidelines. Safe cooking and handling labels on raw or partially cooked meat and poultry are now required by the U.S. Department of Agriculture (USDA). However, the presence of safe foodhandling labels does not ensure adherence to safe practices. For example, an investigation of risk factors for sporadic E. coli O157:H7 infection indicated that of 43 food preparers who reported reading the safe foodhandling label on meat packages, 33 (77%) admitted to practices specifically discouraged on the label (5). The investigation in Dodge County underscores that knowledge of health risks is not consistently associated with desirable changes in behavior. Despite public health warnings and publicity about related outbreaks, some consumers in Dodge County and elsewhere have continued to eat raw or undercooked foods of animal origin. For example, a telephone survey of a national sample of adults conducted by the Center for Food Safety and Applied Nutrition, Food and Drug Administration (FDA), during December 1992-February 1993 indicated that 53% consumed raw eggs; 23%, undercooked hamburgers; 17%, raw clams or oysters; 8%, raw sushi or ceviche; and 5%, steak tartare (raw hamburger meat) (6). Consumer advisories can be more effective if targeted to specific cultural or ethnic groups with such high-risk dietary practices, and WDATCP is planning two press releases this winter holiday period to warn consumers of the risks associated with eating raw ground beef.

In addition to consumer advisories, interventions to reduce the risks associated with the consumption of ground beef include the needs for

1. producers of ground beef to emphasize employee education and training on the recommended methods of cleaning and sanitizing meat-grinding equipment;
2. manufacturers to design meatgrinding equipment that is easily accessible for cleaning and sanitization; and
3. state regulatory and inspection authorities to adopt and enforce FDA's Food Code model requirements, which offer specific recommendations for handling, cooking, and storing raw meat; cleaning and sanitizing equipment and utensils; designing and constructing equipment; and advising consumers about the risks associated with consumption of raw or undercooked food of animal origin (7).

The USDA's Food Safety and Inspection Service also has proposed changes in the meat and poultry inspection system to improve assessment and control of microbial pathogens in raw meat and poultry (8). Consumers can obtain more information on safe meat handling from the USDA's Meat and Poultry Hotline (telephone[800] 535-4555).

References
1. Fontaine RE, Arnon S, Martin WT, et al. Raw hamburger: an interstate common source of human salmonellosis. Am J Epidemiol 1978;107:36-45.
2. CDC. Salmonella Typhimurium Minnesota, Wisconsin, Michigan. MMWR 1972;21:411,416.
3. Griffin PM. Escherichia coli O157:H7 and other enterohemorrhagic Escherichia coli. In: Blaser MJ, Smith PD, Ravdin Jl, Greenberg HG, Guerrant RL, eds. Infections of the gastrointestinal tract. New York: Raven Press, Ltd, 1995.
4. Food Safety and Inspection Service. US Department of Agriculture. Nationwide beef microbiological baseline data collection program: steers and heifers (October 1992-September 1993). Washington, DC: US Department of Agriculture, January 1994.
5. Mead PS, Finelli L, Spitalny K, et al. Risk factors for sporadic infection with Escherichia coli O157:H7 [Abstract. In: 44th Annual Epidemic Intelligence Service Conference, March 27-31, 1995. Atlanta, Georgia: US Department of Health and Human Services, Public Health Service, CDC, 1995.
6. Klontz KC, Timbo B, Fein S, Levy A. Prevalence of selected food consumption and preparation behaviors associated with increased risks of food-borne disease. Journal of Food Protection 1995;58:927-30.
7. Public Health Service. Food code, 1995. Washington, DC: US Department of Health and Human Services, Public Health Service, Food and Drug Administration, 1995.
8. Food Safety and Inspection Service, US Department of Agriculture. Pathogen reduction: hazard analysis and critical control point (HACCP) systems; proposed rule. Federal Register 1995; 60:6774-889.

MMWR 44(42):1995 Oct 27

Outbreak of Salmonellosis Associated With Beef Jerky--New Mexico, 1995

In February 1995, the New Mexico Department of Health (NMDOH) was notified of cases of salmonellosis in two persons who had eaten beef jerky. An investigation by the New Mexico Environment Department determined that these cases were associated with beef jerky processed at a local plant. An investigation by NMDOH identified 91 additional cases. This report summarizes the investigation of this outbreak.

On January 26, 1995, two men presented to the emergency department of a local hospital after onset of diarrhea and abdominal cramps. On January 24, the men had purchased and consumed carne seca, a locally produced beef jerky. Cultures of leftover beef jerky and stool obtained from one patient grew Salmonella. On February 7, NMDOH identified both isolates as Salmonella serotype Montevideo.

NMDOH initiated efforts to determine whether other cases of salmonellosis associated with beef jerky had occurred. On February 8, NMDOH issued a news release advising the public not to eat the implicated brand of beef jerky and to contact the local health department if illness had occurred after eating the product. Cases also were identified through a review of NMDOH records for isolates matching those identified in jerky samples. A confirmed case of beef jerky-related salmonellosis was defined as isolation of Salmonella from a stool sample obtained from a person who had consumed the implicated jerky. A probable case was defined as onset of diarrhea, abdominal cramps, vomiting, and/or nausea in a person who had consumed the implicated jerky.

Illness in 93 persons met the probable or confirmed case definitions. 111 persons reported purchasing the jerky at the local processing plant and eating the jerky during January 21-February 7; onset of symptoms occurred during January 22-February 11 (Figure 1). Incubation periods for most (89%) persons were less than or equal to 3 days. The median age of ill persons was 22 years (range: 2-65 years); 56 (60%) were male. Symptoms of the 93 persons included diarrhea (93%), bloody diarrhea (13%), abdominal cramps (87%), headache (74%), fever (61%), vomiting (43%), and chills (40%). The median duration of illness was 7 days (range: 1-40 days). Five persons (5.4%) were hospitalized.

Of the 93 cases, 40 were culture-confirmed. From the stool samples of these 40 ill persons,three Salmonella serotypes were isolated: Salmonella Typhimurium (31 persons), Salmonella Montevideo

(12), and Salmonella Kentucky (11). Stool samples from 12 persons yielded two serotypes, and the sample from one patient contained all three serotypes. Samples of leftover beef jerky were obtained from five ill persons and from the manufacturer; 11 of the 12 samples tested contained one or more of the three Salmonella serotypes isolated from the patients. Each of the Salmonella Typhimurium isolates obtained from 31 persons with culture-confirmed cases and from the beef jerky were the same uncommon phenotypic variant.

The processing plant that manufactured the contaminated beef jerky was inspected by state authorities on January 31. However, because the plant was not in production, processing-stage temperatures could not be obtained. The owner of the plant described the processing to include placement of slices of partially frozen beef on racks in a drying room at 140 F (60 C) for 3 hours, then holding the meat at 115 F (46 C) for approximately 19 hours; however, temperatures of the meat were never measured. After processing, the jerky was placed in uncovered plastic tubs for sale to the public. The plant owner, who performed all the work in the plant, denied a history of recent gastrointestinal illness but declined to provide a stool specimen. The plant voluntarily closed permanently on February 10. Salmonella was not isolated from environmental swabs taken from 20 surfaces within the plant on February 20.

Reported by: FH Crespin, MD, B Eason, K Gorbitz, T Grass, C Chavala, Public Health Div, PA Gutierrez, MS, J Miller, LJ Nims, MS, Scientific Laboratory Div, M Tanuz, M Eidson, DVM, E Umland, MD, P Ettestad, DVM, Div of Epidemiology, Evaluation and Planning, CM Sewell DrPH, State Epidemiologist, New Mexico Dept of Health; T Madrid, K Smith, C Hennessee, Div of Field Svcs, New Mexico Environment Dept Foodborne and Diarrheal Diseases Br, Div of Bacterial and Mycotic Diseases, National Center for Infectious Diseases; Div of Field Epidemiology, Epidemiology Program Office, CDC.

Editorial Note: Although beef jerky and other processed meat products are considered to be ready-to-eat and, therefore, are expected to be pathogen-free, some recent foodborne disease outbreaks have been associated with ready-to-eat meat products, including salami and sausage (1,2). In the outbreak described in this report, the isolation of the same Salmonella serotype from leftover beef jerky and the stool specimen of an ill person who reported eating the jerky warranted the rapid intervention initiated by NMDOH. Isolation of the combination of uncommon Salmonella serotypes from leftover jerky and the stool specimen of one patient confirmed beef jerky as the source of the outbreak. In addition to this outbreak, NMDOH investigated five outbreaks of salmonellosis associated with locally produced beef jerky during 1966-1988 (3,4) and one outbreak of staphylococcal food poisoning in 1982; none of the beef jerky implicated in these outbreaks had been shipped to other states.

To determine whether consumption of jerky had been associated with foodborne outbreaks in other states during 1976-1995, NMDOH and CDC during May-August 1995 conducted an electronic mail survey with telephone follow-up of all other state health departments. Of the 47 state health departments that responded, 24 (51%) reported that processors of beef jerky were located within their state; however, only four states reported foodborne disease outbreaks associated with locally produced or homemade jerky during 1976-1995, and these outbreaks were caused by Trichinella spiralis and nitrite poisoning. In addition to beef, jerky implicated in these outbreaks had been produced from meat obtained from cougar and bear. Potential explanations for the larger number of jerky-related cases in New Mexico include higher prevalences of consumption of beef jerky, enhanced surveillance for outbreaks, and differences in production methods. This outbreak underscores the risk for foodborne disease associated with consumption of locally produced beef jerky and the need for preventive measures. Conditions recommended for the prevention of bacterial growth during jerky production include rapid drying at high temperatures (i.e., initial drying temperature >155 F [68.3 C] for 4 hours, then >140 F [60 C] for an additional 4 hours) and decreased water activity (i.e., a_w=0.86) (5,6). In 1989, because of several beef jerky-related foodborne outbreaks, the New Mexico Environment Department promulgated regulations regarding the commercial production of jerky made from meat or poultry. The outbreak described in this report is the first jerky-related outbreak to be recognized in New Mexico since the regulations were implemented. As a result of this outbreak, the New Mexico Environment Department plans to evaluate the production processes, including temperatures of meat during drying, of all jerky processors in New Mexico and to assist processors in implementing changes necessary to comply with the regulations.

References
1. CDC. Escherichia coli O157:H7 outbreak linked to commercially distributed dry-cured salami--Washington and California, 1994. MMWR 1995;44:157-60.
2. CDC. Community outbreak of hemolytic uremic syndrome attributable to Escherichia coli O111:NM--South Australia, 1995. MMWR 1995;44:550-1,557-8.
3. CDC. Salmonellosis associated with carne seca--New Mexico. MMWR 1985;34:645-6.
4. CDC. Salmonellosis--New Mexico. MMWR 1967;16:70.
5. Holley RA. Beef jerky: viability of food-poisoning microorganisms on jerky during its manufacture and storage. Journal of Food Protection 1985;48:100-6.
6. Holley RA. Beef jerky: fate of Staphylococcus aureus in marinated and corned beef during jerky manufacture and 2.5 C storage. Journal of Food Protection 1985;48:107-71.

MMWR 43(40):1994 Oct 14

Outbreak of *Salmonella enteritidis* Associated with Nationally Distributed Ice Cream Products--Minnesota, South Dakota, and Wisconsin, 1994

From September 19 through October 10, 1994, a total of 80 confirmed cases of Salmonella enteritidis (SE) infection were reported to the Minnesota Department of Health (MDH); in comparison, 96 cases were reported statewide during all of 1993. Cases were characterized by diarrhea, abdominal cramps, and fever. Recent increases in SE cases also were reported from South Dakota (14 cases during September 6-October 7, compared with 20 cases during all of 1993) and Wisconsin (48 cases during September 6- October 7, compared with 187 during all of 1993). This report summarizes preliminary findings from the outbreak investigation.

On October 5 and 6, to assess potential risk factors for infection, the MDH conducted a case-control study of 15 cases and 15 age- and neighborhood-matched controls. A case was defined as culture-confirmed SE in a person with onset of illness during September. Eleven case-patients (73%) and two controls (13%) reported consumption of Schwan's ice cream within 5 days of illness onset for case-patients and a similar period for controls (odds ratio=10.0; 95% confidence interval=1.4-434.0).

On October 7 and 9, the MDH issued press releases informing the public of this problem and advising persons who had been ill since September 1 and who had consumed Schwan's ice cream to contact the health department. During October 8-11, a total of 2014 persons who had consumed suspected products and had been ill with diarrhea contacted the MDH by telephone. Samples of ice cream from households of ill persons grew SE.

Ill persons reported eating all types and flavors of ice cream products produced at the Schwan's plant in Marshall, Minnesota, including ice cream, sherbet, frozen yogurt, and ice cream sandwiches and cones; these products had production dates in August and September. The implicated products are distributed nationwide, primarily by direct delivery to homes, and are sold only under the Schwan's label. Investigations to examine the extent and causes of the outbreak are under way.

On October 7, the company voluntarily stopped distribution and production at the Marshall plant pending further findings from these investigations.

Reported by: Acute Disease Epidemiology Section, Minnesota Dept of Health. South Dakota Dept of Health. Wisconsin Dept of Health and Social Svcs. Center for Food Safety and Applied Nutrition, Food and Drug Administration. Foodborne and Diarrheal Diseases Br, Div of Bacterial and Mycotic Diseases, National Center for Infectious Diseases, CDC.

Editorial Note: Gastroenteritis caused by Salmonella is characterized by abdominal cramps and diarrhea, vomiting, fever, and headache. Antimicrobial therapy is not indicated in uncomplicated gastroenteritis, which typically resolves within 1 week. Persons at increased risk for infection or more severe disease include infants; the elderly; persons with achlorhydria; those receiving immunosuppressive therapy; persons who may have received antimicrobials for another illness; and those persons with sickle-cell anemia, cancer, or acquired immunodeficiency syndrome (1). Complications include meningitis, septicemia, Reiter syndrome, and death (1).

Salmonella sp. are second only to Campylobacter as a cause of bacterial diarrheal illness in the United States, causing an estimated 2 million illnesses annually (2). Among the more than 2000 Salmonella serotypes, SE has ranked first or second in frequency of isolation from humans since 1988 and accounted for 21% of reported isolates in 1993. Each year, an average of 55 outbreaks of SE infections are reported to CDC; approximately 11% of patients are hospitalized, and 0.3% die (3).

Preliminary findings from this outbreak indicate that the number of persons exposed to contaminated products may be substantial. Approximately 400,000 gallons of the implicated products are produced weekly and are distributed throughout the contiguous United States. Previous investigations have established the potential for large-scale outbreaks of foodborne salmonellosis; for example, in 1985, pasteurized milk produced at one dairy plant caused up to 197,000 Salmonella infections (4).

Consumers should discard or return any Schwan's ice cream products. Persons who have become ill since September 1 with diarrhea and who have consumed Schwan's ice cream products are urged to contact their state health departments.

References

1. Pavia AT, Tauxe RV. Salmonellosis: nontyphoidal. In: Evans AS, Brachman PS, eds. Bacterial infections in humans: epidemiology and control. 2nd ed. New York: Plenum Medical Book Company, 1991:573- 91.
2. Helmick CG, Griffin PM, Addiss DG, Tauxe RV, Juranek DD. Infectious diarrheas. In: Everheart JE, ed. Digestive diseases in the United States: epidemiology and impact. Washington, DC: US Department of Health and Human Services, Public Health Service, National Institutes of Health, National Institute of Diabetes and Digestive and Kidney Diseases, 1994:85-123; DHHS publication no. (NIH)94-1447.
3. CDC. Outbreaks of Salmonella enteritidis gastroenteritis--California, 1993. MMWR 1993; 42:793-7.
4. Ryan CA, Nickels MK, Hargrett-Bean NT, et al. Massive outbreak of antimicrobial-resistant salmonellosis traced to pasteurized milk. JAMA 1987;258:3269-74.

MMWR 43(36):1994 Sep 16

Outbreak of *Salmonella enteritidis* Associated with Homemade Ice Cream--Florida, 1993

On September 7, 1993, the Epidemiology Program of the Duval County (Florida) Public Health Unit was notified about an outbreak of acute febrile gastroenteritis among persons who attended a cook-out at a psychiatric treatment hospital in Jacksonville, Florida. This report summarizes the outbreak investigation.

On September 6, seven children (age range: 7-9 years) and seven adults (age range: 29-51 years) attended the cookout at the hospital. A case of gastroenteritis was defined as onset of diarrhea, nausea or vomiting, abdominal pain, or fever within 72 hours of attending the cookout. Among the 14 attendees, 12 cases (in five of the children and all seven adults) were identified. The median incubation period was 14 hours (range: 7-21 hours); the mean duration of illness was 18 hours (range: 8-40 hours). Predominant symptoms were diarrhea (93%), nausea or vomiting (86%), abdominal pain (86%), and fever (86%). All ill persons were examined by a physician. Salmonella enteritidis (SE) (phage type 13a) was isolated from stool of three of the seven patients from whom specimens were obtained.

Eleven of the 12 ill persons had eaten homemade ice cream served at the cookout. No other food item was associated with illness. Testing of a sample of ice cream revealed contamination with SE (phage type 13a).

The ice cream was prepared at the hospital on September 6 using a recipe that included six grade A raw eggs. An electric ice cream churn was used to make the ice cream approximately 3 hours before the noon meal. The ice cream had been properly cooled, and no food-handling errors were identified. The person who prepared the ice cream was not ill before preparation; however, she became ill

13 hours after eating the ice cream. Her stool specimen was one of the three stools positive for SE (phage type 13a).

The U.S. Department of Agriculture's (USDA) Animal and Plant Health Inspection Service attempted to trace the implicated eggs back to the farm of origin. The hospital purchased eggs from a distributor in Florida. However, the traceback was termin- ated because the implicated eggs from the distributor had been purchased from two suppliers--one of whom bought and mixed eggs from many different sources. Current USDA Salmonella regulations limit testing of flocks to one clearly implicated flock.

Reported by: P Buckner, MPH, D Ferguson, HRS Duval County Public Health Unit, F Anzalone, MD, D Anzalone, DrPH, College of Health, Univ of North Florida, Jacksonville; J Taylor, Office of Lab Svcs, WG Hlady, MD, RS Hopkins, MD, State Epidemiologist, State Health Office, Florida Dept of Health and Rehabilitative Svcs. Foodborne and Diarrheal Diseases Br, Div of Bacterial and Mycotic Diseases, National Center for Infectious Diseases, CDC.

Editorial Note: The outbreak described in this report represents the fourth SE outbreak in Florida since 1985; this outbreak is the first in the state to implicate eggs. In the United States, the number of sporadic and outbreak-associated cases of SE infection has increased substantially since 1985; much of the increase can be attributed to consumption of raw or undercooked eggs (1-3). During 1983-1992, the proportion of reported Salmonella isolates that were SE increased from 8% to 19%. During 1985-1993, a total of 504 SE outbreaks were reported to CDC and resulted in 18,195 cases, 1978 hospitalizations, and 62 deaths (Table 1). Of the 233 outbreaks for which epidemiologic evidence was sufficient to implicate a food vehicle, 193 (83%) were associated with eggs. Of these 193 outbreaks, 14 (7%) were associated with consumption of homemade ice cream. No outbreaks have been associated with pasteurized egg products.

After eggs are identified by public health officials as the cause of an SE outbreak, USDA attempts to trace the implicated eggs back to the farm of origin to conduct serologic and microbiologic assessments of the farm. If SE is detected on the source farm, the eggs are diverted to pasteurization, or the flocks are destroyed. Under current regulations, USDA can pursue the traceback only if one farm is identified as the source. During 1990-1993, the success rate of USDA tracebacks to the source farm declined from 86% (19/22 outbreaks) in 1990 to 17% (3/21 outbreaks) in 1993. The rate declined primarily because eggs increasingly have been marketed in shipments containing eggs from multiple sources.

Although 0.01% of all eggs contain SE and, therefore, pose a risk for infection with SE (4), raw or undercooked eggs are consumed frequently. Based on the Food and Drug Administration (FDA) Food Safety Survey conducted in 1993, 53% of a nationally representativesample of 1620 respondents reported ever eating foods containing raw eggs; of these, 50% had eaten cookie batter, and 36% had eaten ice cream containing raw eggs (S. Fein, FDA, personal communication, September 9, 1994). Many persons may eat raw or undercooked eggsbecause they are unaware that eggs are a potential source of Salmonella (3) and that certain foods (e.g., homemade ice cream, cookie batter, Caesar salad, and hollandaise sauce) contain raw eggs.

Consumers should be informed that eating undercooked eggs may result in Salmonella infection. In addition, eggs should be refrigerated to prevent proliferation of Salmonella if present and should be cooked thoroughly to kill Salmonella. Because most serious illnesses and deaths associated with salmonellosis occur among the elderly and immunocompromised persons, these persons in particular should not eat foods containing raw or undercooked eggs. Hospitals, nursing homes, and commercial kitchens should use pasteurized egg products for all recipes requiring pooled eggs or raw or undercooked eggs and should refrigerate all eggs and egg products.

References
1. St. Louis ME, Morse DL, Potter ME, et al. The emergence of grade A eggs as a major source of Salmonella enteritidis infections: new implications for the control of salmonellosis. JAMA 1988;259:2103-7.
2. Mishu B, Koehler J, Lee LA, et al. Outbreaks of Salmonella enteritidis infections in the United States, 1985-199J Infect Dis 1994;169:547-5
3. Hedberg CW, David MJ, White KE, MacDonald KL, Osterholm MT. Role of egg consumption in sporadic Salmonella enteritidis and Salmonella typhimurium infections in Minnesota. J Infect Dis 1993;167:107-1
4. Mason J, Ebel E. APHIS Salmonella enteritidis Control Program [Abstract]. In: Snoeyenbos GH, eds Proceedings of the Symposium on the Diagnosis and Control of Salmonella. Richmond, Virginia: US Animal Health Association. 1992:78.

MMWR 42(41):1993 Oct 22

Outbreaks of *Salmonella enteritidis* Gastroenteritis--California, 1993

Foodborne infections cause an estimated 6.5 million cases of human illness and 9000 deaths annually in the United States (1). Salmonella is the most commonly reported cause of foodborne outbreaks, accounting for 28% of such outbreaks of known etiology and 45% of outbreak-associated cases during 1973-1987 (2). During 1985-1992, state and territorial health departments reported 437 Salmonella enteritidis (SE) outbreaks (Table 1), which accounted for 15,162 cases of illness, 1734 hospitalizations, and 53 deaths. This report describes three SE outbreaks in California during a 4-month period in 1993.

OUTBREAK 1: LOS ANGELES COUNTY
In January 1993, routine surveillance for salmonellosis identified four unrelated persons with gastroenteritis and stool cultures yielding SE who recently had eaten at a local restaurant; one person had been hospitalized. The mean period from eating at the restaurant to onset of illness was 20 hours (range: 11-24 hours); duration of symptoms ranged from 1 to 14 days. All four isolates were phage type 13a and plasmid profile type 2 (36 and 3.7 megadalton plasmids), an unusual pattern among SE isolates. All four ill persons reported having eaten an egg-based dish (omelette, scrambled eggs, or egg salad) at the restaurant during December 26, 1992-January 6, 1993.

An investigation by the Los Angeles County Department of Health Services involved the four reported cases, five well meal companions, and 100 restaurant patrons identified through credit card receipts; two additional cases were identified. A case was defined as onset of diarrhea (three or more loose stools in a 24-hour period) plus fever, abdominal cramps, nausea, and/or vomiting within 3 days after eating at the restaurant. Five of the six case-patients had eaten an egg-based dish, compared with 16 (16%) of 103 well persons (odds ratio {OR}=27.2; 95% confidence interval {CI}=2.7-1300); no other food was associated with illness.

Inspection of the restaurant revealed that egg salad was stored on a cold table at a holding temperature of 60 F (15.5 C), a temperature that allows growth of Salmonella. For pooled egg dishes, 22-30 dozen extra-large grade AA eggs were pooled several times daily and stored in a walk-in refrigerator. A 2-quart container of pooled eggs was stored in a reach-in refrigerator. The temperature of the pooled eggs in the reach-in refrigerator was 50 F (10 C); California regulations require eggs to be refrigerated at less than or equal to 45 F (less than or equal to 7.2 C).

In February, cultures of swabs of utensils used for pooling and storing the eggs were negative for SE, and rectal swabs obtained from all 43 food handlers at the restaurant also were negative. No eggs from the implicated shipment remained. Eggs from a later shipment from the same distributor, delivered February 9, did not yield SE.

The U.S. Department of Agriculture (USDA) Salmonella enteritidis Control Program and the California Department of Food and Agriculture (CDFA) attempted to trace the implicated eggs back to the farm of origin. However, the traceback was terminated because the eggs were purchased from a distributor who bought and mixed eggs from many different suppliers. Current USDA Salmonella regulations limit the testing of flocks to a single, clearly implicated flock.

OUTBREAK 2: SAN DIEGO COUNTY
In February 1993, 23 persons who had eaten at a local restaurant on February 16 developed abdominal cramps and diarrhea; two were hospitalized. The mean period from eating at the restaurant to onset of illness was 20 hours (range: 3.5-77.0 hours); duration of symptoms ranged from 2 to 14 days. Stool cultures from 11 of 13 ill persons tested yielded SE; all isolates were phage type 13a and plasmid profile type 2, indistinguishable from the SE strains in outbreak 1.

An investigation by the San Diego County Department of Health Services involved the 23 reported cases and 24 well meal companions. A case was defined as onset of diarrhea (three or more loose stools in a 24-hour period) within 5 days after eating at the restaurant. Eighteen (78%) of the 23 case-patients had eaten an entree served with hollandaise or bearnaise sauce, compared with three (13%) of 24 well persons (OR=25.2; 95% CI=4.4-170.7).

The hollandaise sauce, also used as a base for the bearnaise sauce, was prepared with 12 pooled raw egg yolks. A new batch was prepared at the beginning of each meal shift and placed in a clean dispenser. The dispenser was kept under a heat lamp for up to 3-1/2 hours at approximately 100 F-120 F (37.8 C-48.9 C).

Traceback of implicated eggs by USDA and CDFA indicated they had been purchased from the same distributor that had provided eggs to the restaurant involved in outbreak 1. Again, traceback was terminated.

OUTBREAK 3: SANTA CLARA COUNTY
In March 1993, 22 persons who had eaten at a local sandwich shop during February 28-March 4 developed diarrhea, fever, and abdominal cramps; none were hospitalized. Stool cultures from all 22 ill persons yielded SE; all isolates were phage type 13a and plasmid profile type 2, indistinguishable from the SE strains in outbreaks 1 and 2. Preliminary findings of a case-control study conducted by the Santa Clara County Health Department implicated sandwiches as the vehicle of transmission; no other food was associated with illness. Further investigation revealed that mayonnaise was the only food ingredient containing a raw product of animal origin and was common to all sandwiches eaten by ill persons. None of the implicated mayonnaise remained at the time of the investigation, but unrefrigerated eggs from the implicated shipment obtained from the sandwich shop were cultured in five pools of 10 eggs each; one of the pools yielded SE. This isolate was phage type 13a and plasmid profile type 2. Traceback of implicated eggs by USDA and CDFA indicated they had been purchased from the same distributor that had provided eggs to the two restaurants in outbreaks 1 and 2. Again, traceback was terminated.

Reported by: D Ewert, MPH, N Bendana, MS, M Tormey, L Kilman, L Mascola, MD, Los Angeles County Dept of Health Svcs, Los Angeles; LS Gresham, MPH, MM Ginsberg, MD, PA Tanner, ME Bartzen, S Hunt, RS Marks, CR Peter, PhD, San Diego County Dept of Health Svcs, San Diego; J Mohle-Boetani, MD, M Fenstersheib, MD, J Gans, K Coy, MS, S Liska, DrPH, Disease Control and Prevention Unit, Santa Clara County Health Dept, San Jose; S Abbott, R Bryant, L Barrett, DVM, K Reilly, DVM, M Wang, PhD, SB Werner, MD, RJ Jackson, MD, GW Rutherford, III, State Epidemiologist, California Dept of Health Svcs. Animal and Plant Health Inspection Svc, US Dept of Agriculture. Foodborne and Diarrheal Diseases Br, Div of Bacterial and Mycotic Diseases, National Center for Infectious Diseases, CDC.

Editorial Note: Although most reported SE outbreaks have occurred in the New England and Mid-Atlantic states (3), an increasing proportion of outbreaks has been reported from other areas (Table 1). From 1976 through 1991, the proportion of reported Salmonella isolates in the United States that were SE increased from 5% to 20%; SE was second only to S. typhimurium, except in 1989 and 1990, when SE was the most frequently reported serotype. In California, only four SE outbreaks had been reported since 1985, when active surveillance for SE outbreaks began; the proportion of reported Salmonella isolates that were SE increased from 5% to 13% from 1985 through 1992 and to 21% for the first half of 1993.

An estimated 0.01% of all shell eggs contain SE, although this percentage may be higher in the northeastern United States (4). Consequently, foods containing raw or undercooked eggs (e.g., homemade mayonnaise, hollandaise sauce, and runny omelettes) pose a slight risk for infection with SE. In contrast, commercial mayonnaise is made with pasteurized eggs and is safe. Outbreaks of salmonellosis--some of substantial magnitude--may occur when commercial kitchens serve foods made with contaminated shell eggs that have not been sufficiently cooked to kill Salmonella. This is particularly likely when refrigeration is inadequate or holding temperatures are too low and when eggs are pooled, whereby a single contaminated egg can contaminate a large pool. However, egg-handling practices of affected restaurants may be similar to the routine practices of many other restaurants (5). In August 1990, FDA issued recommendations to state agencies that directly regulate commercial establishments concerning the proper handling of shell eggs by restaurants, grocery stores, caterers, institutional feeders, and vending operators. These recommendations include guidelines on refrigeration, cooking, pooling, and substitution with pasteurized eggs.

The temporal clustering of the outbreaks in this report and the same unusual combination of phage type and plasmid profile type common to all three outbreaks suggest that one farm supplied contaminated eggs to all three restaurants. However, because eggs are distributed nationwide and 70% of eggs sold by the distributor in California were obtained or purchased from other states, the source farm may have been outside California. During most egg- associated traceback efforts, the outbreak strain of SE is almost always found on the source farm (6).

Most SE infections occur as sporadic cases or in limited family outbreaks, rather than as part of large common-source outbreaks. Such sporadic cases also are often associated with eating undercooked eggs (7). The risk for infection acquired through consumption of contaminated foods prepared in the kitchens of private homes can be reduced through improved education of consumers regarding the risks of eating raw or undercooked eggs and through increased availability of pasteurized eggs in the retail marketplace. Because most serious illnesses and deaths associated with salmonellosis occur among infants, the elderly, and immunocompromised persons (8,9), persons in these groups should not be served foods containing raw or undercooked eggs. In addition, hospitals, nursing homes, and commercial kitchens should use pasteurized egg products for all recipes requiring pooled eggs or lightly cooked eggs and should refrigerate all eggs and egg products.

On October 27, 1992, the USDA Agricultural Marketing Service published a proposed rule on requirements for storage and transport temperatures of eggs and for carton labeling aimed at increasing the safety of raw shell eggs nationwide. * The comment period for this proposed rule ended March 29, 1993; final regulations are pending and subject to revised legislation. In addition, on August 2, 1993, the USDA Animal and Plant Health Inspection Service (APHIS) published a proposed rule that would revise current USDA regulations concerning chicken infection caused by SE. ** These proposed changes will improve control of the spread of SE in commercial egg-type chicken flocks and include a provision that allows identification of more than one flock as the probable source of eggs causing an SE outbreak. The comment period for this proposed rule has been extended to November 15, 1993. Additional information is available from Dr. John Mason, Director Salmonella enteritidis Control Program, Veterinary Services, APHIS, USDA, Room 205, Presidential Building,6525 Belcrest Road, Hyattsville, MD 20782; telephone (301) 436-4363.

References
1. Bennett JV, Holmberg SD, Rogers MF, Solomon SL. Infectious and parasitic diseases. In: Amler RW, Dull HB, eds. Closing the gap: the burden of unnecessary illness. Am J Prev Med 1987;3(suppl):102-14.
2. Bean NH, Griffin PM. Foodborne disease outbreaks in the United States, 1973-1987: pathogens, vehicles and trends. J Food Protect 1990;53:804-17.
3. St. Louis ME, Morse DL, Potter ME, et al. The emergence of grade A eggs as a major source of Salmonella enteritidis infections: new implications for the control of salmonellosis. JAMA 1988;259:2103-7.
4. Mason J, Ebel E. APHIS Salmonella enteritidis control program {Abstract}. In: Snoeyenbos GH, ed. Proceedings of the Symposium on the Diagnosis and Control of Salmonella. Richmond, Virginia: US Animal Health Association, 1992:78.
5. Vugia DJ, Mishu B, Smith M, Tavris DR, Hickman-Brenner FW, Tauxe RV. Salmonella enteritidis outbreak in a restaurant chain: the continuing challenges of prevention. Epidemiol Infect 1993;119:49-61.
6. Altekruse S, Koehler J, Hickman-Brenner FW, Tauxe RV, Ferris K. A comparison of Salmonella enteritidis phage types from egg-associated outbreaks and implicated laying flocks. Epidemiol Infect 1993;110:17-22.
7. Hedberg CW, David MJ, White KE, MacDonald KL, Osterholm MT. Role of egg consumption in sporadic Salmonella enteritidis and Salmonella typhimurium infections in Minnesota. J Infect Dis 1993;167:107-11.
8. Levine WC, Buehler JW, Bean NH, Tauxe RV. Epidemiology of nontyphoidal Salmonella bacteremia during the human immunodeficiency virus epidemic. J Infect Dis 1991;164:81-7.
9. Levine WC, Smart JF, Archer DL, Bean NH, Tauxe RV. Foodborne disease outbreaks in nursing homes, 1975 through 1987. JAMA 1991;266:2105-9.

* 58 FR 48569-48575. ** 58 FR 41048-41061.

MMWR 42(26):1993 Jul 09

Salmonella Serotype Tennessee in Powdered Milk Products and Infant Formula--Canada and United States, 1993

Since May 1993, three cases of infection with Salmonella serotype Tennessee in infants in Canada and the United States have been linked to consumption of contaminated powdered infant formula. This report summarizes preliminary data on isolation of this organism from powdered milk products and alerts laboratories to the possibility that, because this strain may ferment lactose, it may not be identified as Salmonella.

Following the isolation of Salmonella serotype Tennessee from the stools of two infants in Canada who had consumed Soyalac Powder(R) infant formula in May, the Food and Drug Administration (FDA) isolated Salmonella Tennessee from production equipment at the Minnesota plant where the product had been dried, and from cans of the powdered infant formula. In June 1993, one case of infection with Salmonella Tennessee occurred in Illinois in an infant who consumed Soyalac Powder(R). From November 4, 1992, through June 29, 1993, 48 cases of infection with Salmonella Tennessee have been reported to CDC; when annualized, this number is not substantially different from the mean of 120 cases reported annually from 1981 through 1991.

On June 28, 1993, FDA ordered a recall of all Soyalac Powder(R) infant formula produced on or after November 4, 1992. FDA has identified additional products that are spray-dried at this plant; these products include Sumacal(R) medical food supplement, Propac(R) protein supplement, canned Medibase(R) medical meal replacement, Kresto Denia(R) powdered milk, Enercal(R) diet beverage, Enercal Plus(R), and Promil(R) weaning formula. No cases of illness have been linked to these products. FDA is working with plant officials to determine whether any other products were dried or packaged at this plant during this time. No spray-dried products have been distributed from this plant since June 7, 1993. FDA has requested recall of all products spray-dried at this plant since November 4, 1992. More detailed product information is available from the Division of Emergency and Epidemiological Operations, FDA, telephone (301) 443-1240.

Reported by: KK Louie, REHO, Boundary Health Unit, Surrey, British Columbia; AM Paccagnella, WD Osei, British Columbia Center for Disease Control, Vancouver; H Lior, MSc, Chief, National Laboratory for Enteric Pathogens, Laboratory Center for Disease Control, Ottawa, Ontario, Canada. BJ Francis, MD, State Epidemiologist, Illinois Dept of Public Health. MT Osterholm, PhD, State Epidemiologist, Minnesota Dept of Health. Minneapolis District, Center for Food Safety and Applied Nutrition, Food and Drug Administration. Foodborne and Diarrheal Diseases Br, Div of Bacterial and Mycotic Diseases, National Center for Infectious Diseases, CDC.

Editorial Note: Outbreaks of salmonellosis caused by powdered milk products have been reported in the United States (1) and elsewhere (2,3). The isolates of Salmonella Tennessee that were identified from the three infants described in this report are atypical of salmonellae because most colonies ferment lactose and, therefore, may not be detected by clinical laboratories that use media or methods that identify salmonellae based on absence of lactose fermentation.

To isolate this organism, plating media that include an indicator of hydrogen sulfide (H2S) production, such as bismuth sulfite (BS) agar, Hektoen enteric (HE) agar, or xylose-lysine-deoxycholate (XLD) agar, should be used. BS does not contain lactose, so typical H2S-producing (black) colonies can be selected from this medium. Both HE and XLD contain an indicator of H2S production, as well as lactose; selection of colonies from these media should be based on H2S production rather than absence of lactose fermentation. At CDC, H2S production by this strain was detected more easily on HE than on XLD. Use of either BS or HE is recommended for recovery of this strain. XLD agar should be used only if other media are not available.

To screen colonies selected from isolation plates, lysine-iron agar (LIA) is recommended because the reaction produced by lactose-fermenting salmonellae in this medium is typical and because H2S produced by lactose-fermenting organisms can be detected. Triple sugar iron agar (TSI) or other media that depend on lactose fermentation to identify suspect salmonellae should not be used. H2S production may not be detected on TSI because of acidic conditions caused by fermentation of lactose. Automated test systems should be used with caution, since lactose-fermenting salmonellae tested at CDC in several such systems were sometimes identified incorrectly. This particular strain was correctly identified as Salmonella by the Analytab Products' API 20E(R) * system.

CDC requests that health-care providers and public health departments continue routine reporting to the Salmonella surveillance system; that all Salmonella serogroup C1 (of which Salmonella Tennessee is a member) isolates be serotyped; that persons infected with Salmonella Tennessee be questioned specifically about consumption of powdered milk products or infant formula; and that, until August 15, 1993, new cases of infection with Salmonella Tennessee, whether lactose fermenting or nonlactose fermenting, be reported promptly to the state health department.

References
1. Collins RN, Treger MD, Goldsby JB, Boring JR III, Coohon DB, Barr RN. Interstate outbreak of Salmonella newbrunswick infection traced to powdered milk. JAMA 1968;203:838-44.
2. Weissman JB, Deen RMAD, Williams M, Swanston N, Ali S. An island-wide epidemic of salmonellosis in Trinidad traced to contaminated powdered milk. West Indian Med J 1977; 26:135-43.
3. Rowe B, Begg NT, Hutchinson DN, et al. Salmonella Ealing infections associated with consumption of infant dried milk. Lancet 1987;2:900-3.

* Use of trade names and commercial sources is for identification only and does not imply endorsement by the Public Health Service or the U.S. Department of Health and Human Services.

MMWR 37(32):1988 Aug 19

Update: *Salmonella enteritidis* Infections and *Salmonella enteritidis* Grade A Shell Eggs-- United States

Salmonella enteritidis (SE) continues to be an important cause of outbreaks of gastroenteritis. This report describes recent outbreaks of SE infections that have been associated with Grade A eggs.

FORT MONMOUTH, NEW JERSEY.
From May 3 to May 9, 1988, 88 (47%) of 188 students in a New Jersey college preparatory school developed febrile gastroenteritis. Symptoms included diarrhea, abdominal pain, headache, and fever. Twenty-seven (31%) of the ill students were hospitalized, and all recovered; stool cultures from each ill patient yielded SE. An epidemiologic investigation indicated that homemade ice cream prepared with Grade A raw eggs only 2 hours before consumption was the source of the outbreak. A culture of the implicated ice cream yielded SE. The ice cream had been properly cooled, and no food handling errors were identified.

ASBURY PARK, NEW JERSEY.
An outbreak of SE infections was reported in a group of 100 service organization trainees who had stayed at the same hotel in Asbury Park, New Jersey. Forty-seven (60%) of 78 trainees interviewed reported having had onset of gastrointestinal illness from June 13 to June 16, 1988. Two were hospitalized and recovered; seven stool cultures were taken, and all yielded SE. Epidemiologic data implicated scrambled eggs served on June 11 and 12. In addition, culture of a pooled egg mixture obtained at the hotel yielded SE. Neither the clinical isolates nor the isolate from the eggs were lysine-positive. Since most SE isolates are lysine-positive, a relation-ship between the SE strains found in the patients and in the eggs seems probable. The implicated Grade A eggs were traced to a farm in Pennsylvania.

LIVONIA, NEW YORK.
In late May 1988, an outbreak of gastrointestinal illness occurred among patrons of a restaurant in Livonia, New York. Twelve (38%) of 32 persons who attended a brunch on May 22 reported diarrhea, nausea, vomiting, or abdominal cramps. Stool cultures from four patients yielded SE. Egg omelets made from pooled Grade A eggs were the only food statistically associated with illness.

Investigation did not identify improper food handling practices, such as cross-contamination or inadequate storage, which could have played a role in this outbreak. None of the food handlers were ill, and none had stool cultures that yielded Salmonella. The implicated Grade A eggs were traced to a Maryland farm.

Reported by: GC Taylor, MPH, Fort Monmouth; MA Meadows, LW Jargowsky, MPH, Monmouth County Dept of Health; K Pilot, MJ Teter, DO, J Brook, MD, ME Petrone, MD, KC Spitalny, MD, State Epidemiologist, New Jersey State Dept of Health. SB Spitz, MS, Monroe County Dept of Health; RJ Davin, J Ellison, Livingston County Dept of Health; SF Kondracki, JJ Guzewich, MPH, JK Fudala, MS, JG Debbie, DVM, DL Morse, MD, State Epidemiologist, New York State Dept of Health. Enteric Diseases Br, Div of Bacterial Diseases, Center for Infectious Diseases, CDC.

Editorial Note: A total of 6390 SE isolates were reported for 1987 (16% of total reported Salmonella isolates). SE is the second most common Salmonella serotype reported. National surveillance data for 1987 indicate continued high isolation rates of SE in the northeast, mid-Atlantic, and south Atlantic regions (Figure 1). Recent isolation rates of SE have also increased in the east north central, mountain, and Pacific regions of the country. The outbreaks described in this report confirm the continuing association between eggs and outbreaks of SE infections (1). Of the 19 outbreaks caused by SE with a known vehicle reported to CDC in 1987, 15 (79%) were associated with Grade A shell eggs. No vehicle of transmission was known for 11 other reported outbreaks of SE infections in 1987. An examination of data from 1973 to 1987 reveals that most outbreaks caused by SE occur during the summer months (Figure 2). Warm temperatures may provide opportunities for SE to multiply and survive in the eggs during production, transport, storage, or use.

Although food handling errors can contribute to outbreaks of Salmonella infections, the outbreaks in Fort Monmouth, New Jersey (ice cream), and Livonia, New York (egg omelet), demonstrate that SE infections can occur even when acceptable food preparation techniques have been used.

An SE control program is being developed by state health departments, poultry scientists, the egg industry, the U.S. Department of Agriculture, the Food and Drug Administration, and CDC. Long-term control of SE may depend on the elimination of infected flocks or use of pasteurized egg products. Proper handling and cooking of eggs can minimize the risk of salmonellosis (2); thorough cooking kills Salmonella.*

Clinicians are encouraged to report cases of salmonellosis to local and state health departments. Salmonella isolates can be serotyped by most state public health laboratories to aid in epidemiologic investigations.

References
1. St. Louis ME, Morse DL, Potter ME, et al. The emergence of grade A eggs as a major source of Salmonella enteritidis infections: new implications for the control of salmonellosis. JAMA 1988;259:2103-7.
2. CDC. Update: Salmonella enteritidis infections in the northeasternUnited States. MMWR 1987;36:204-5.
*Further information about proper food preparation with eggs can be obtained through local county extension home economists or by calling the USDA Meat and Poultry Hotline (800) 535- 4555.

MMWR 39(50):1990 Dec 21

Update: *Salmonella enteritidis* Infections and Grade A Shell Eggs--United States, 1990

Salmonella enteritidis (SE) remains an important cause of outbreaks and sporadic cases of gastroenteritis in the United States. This report summarizes three outbreaks in 1989 that were associated with Salmonella-contaminated Grade A eggs.

SUFFOLK COUNTY, NEW YORK.
An outbreak of gastroenteritis occurred among 21 of 24 persons who attended a baby shower on

July 1. Severe diarrhea, vomiting, fever, and cramps occurred a median of 9 hours (range: 5.5-57 hours) after the shower. Twenty ill persons sought medical care, and 18 were hospitalized. One attendee who was 38 weeks pregnant delivered while ill; the infant subsequently developed SE septicemia and required prolonged hospitalization. Additional secondary cases occurred in two household members of primary case-patients. SE was isolated from stool or rectal-swab cultures of all 21 primary and three secondary case-patients.

All 21 ill attendees, but none of the three attendees who remained well, reported eating a homemade baked ziti pasta dish consisting of one raw egg and ricotta cheese combined in a large baking pan with cooked tomato meat sauce and refrigerated overnight. The ziti dish was baked for 30 minutes at 350 F (176.7 C) immediately before serving. Several attendees commented that the center of the ziti was still cold when served. SE was isolated from samples of the leftover baked ziti and from a pool of seven eggs from the original carton. The eggs were supplied by a New Jersey egg producer; SE was isolated from several flocks tested at the farm.

CARBON COUNTY, PENNSYLVANIA.
The Pennsylvania State Department of Health was notified of gastroenteritis in 12 of 32 persons who attended an office party on August 24. Symptoms included diarrhea (100%), headache (58%), abdominal pain (42%), nausea (42%), fever (25%), and vomiting (17%). The median incubation period was 27.5 hours (range: 7-72 hours). Of three persons who were hospitalized, two recovered. The third person, a 40-year-old previously healthy man, experienced severe diarrhea and high fever, was admitted to an emergency room on the fourth day of illness, and died within 2 hours. Clinical course and autopsy findings were compatible with acute salmonellosis; postmortem blood, urine, and stool cultures yielded SE.

The only food and beverages served at the party were six pies (two fruit-based pies and four egg-based custard pies), coffee, and juice. Illness was associated with consumption of the custard pies (relative risk=8.6; 95% confidence interval=1.3-58.6). All the pies had been prepared August 23 by a commercial bakery and held without refrigeration for approximately 21 hours before consumption. Two other cases of Salmonella (identified as group D, which includes SE) infection were reported in persons who did not attend the office party but who ate custard pie prepared by this bakery on the same day. The source of the eggs is unknown.

KNOX COUNTY, TENNESSEE.
An outbreak of SE gastroenteritis occurred among persons who patronized a restaurant on April 8. Twenty-seven cases were reported to the county health department; stool cultures from 23 persons all yielded SE. At least 24 ill persons reported onset of fever, abdominal cramps, and diarrhea within 48 hours after eating at the restaurant; 11 were hospitalized. All had eaten either Hollandaise or Bernaise sauce on April 8. Ten meal companions of ill persons were contacted; none had developed illness or had eaten Hollandaise or Bernaise sauce (p less than 0.01). Both sauces were prepared with Grade A extra-large eggs that were heated but not thoroughly cooked. No other food item was consumed by greater than 20% of those who became ill. The eggs were traced to a farm in Indiana.

Reported by: L Steinert, D Virgil, Brookhaven Memorial Hospital, Patchogue; E Bellemore, Stonybrook Univ Hospital, Stonybrook; B Williamson, E Dinda, D Harris, MD, D Scheider, L Fanella, V Bogacki, F Liska, Suffolk County Dept of Health Svcs; GS Birkhead, MD, JJ Guze wich, MPH, JK Fudala, SF Kondracki, M Shayegani, PhD, DL Morse, MD, State Epidemiologist, New York State Dept of Health. DT Dennis, MD, B Healey, DR Tavris, MD, State Epidemiologist, Pennsylvania State Dept of Health. M Duffy, MD, Knox County Health Dept; K Drinnen, RH Hutcheson, MD, State Epidemiologist, Tennessee Dept of Health and Environment. Div of Field Svcs, Epidemiology Program Office; Enteric Diseases Br, Div of Bacterial Diseases, Center for Infectious Diseases, CDC.

Editorial Note: Since 1979, isolation rates of SE have increased dramatically in New England and, more recently, in the mid-Atlantic states (Figure 1) (1). As of October 31, 1989, 49 SE outbreaks had been reported for 1989; these outbreaks have been associated with 1628 cases and 13 deaths (including 12 deaths in nursing homes). From 1985 through 1988, state health departments reported 140 SE outbreaks associated with 4976 ill persons (of whom 896 were hospitalized) and 30 deaths (Table 1). Contaminated food was implicated in 89 (64%) outbreaks; Grade A shell eggs were implicated in 65 (73%) of these. From 1985 to 1989, the proportion of outbreaks from outside New England and the mid-Atlantic regions increased from 5% to 43%.

Foods containing a single SE-contaminated egg can cause outbreaks of severe illness (2). Salmonellosis can be especially severe in infants less than 3 months of age, in the elderly, and in persons who are immunocompromised. Most SE-associated deaths occur in nursing home residents, but salmonellosis can be fatal in otherwise healthy hosts when ingested in sufficient doses (3).

Thorough cooking kills Salmonella. Contaminated eggs that are liquid or runny after light cooking can contain Salmonella. When eggs are heavily contaminated, standard cooking methods for many egg-containing foods (including Hollandaise and Bernaise sauces, meringue, and scrambled and soft-boiled eggs) may not kill all Salmonella (4,5). If raw or incompletely cooked eggs are held at room temperature for greater than 2-4 hours, the risk of outbreaks of Salmonella infections may increase because Salmonella can grow to high concentrations under such conditions.

In regions where egg-associated salmonellosis has been identified, the public should be advised to not eat raw or undercooked eggs. In addition, consumers should avoid eating foods that contain raw eggs, such as Caesar salad, homemade eggnog, and homemade mayonnaise. Foods made with pasteurized eggs (e.g., commercial eggnog, ice cream, and mayonnaise) are safe to eat. In hospitals and nursing homes, where high-risk patients may be exposed, the risk for outbreaks can be reduced by use of pasteurized egg products in recipes that require pooled eggs and by proper preparation and storage of foods containing eggs. Bulk-quantity pasteurized egg products are available commercially for use in food-service establishments.

Clinicians and microbiologists are encouraged to report cases of salmonellosis to local and state health departments. Because some strains of SE isolated from patients in the northeast are reported to produce minimal H((2))S, suspect isolates that otherwise resemble Salmonella should not be discarded on this basis alone. To help characterize sporadic cases and to assist in epidemiologic investigations, Salmonella isolates can be serotyped by state public health laboratories. When eggs are implicated, investigation of outbreaks and notification of state agriculture departments and the U.S. Department of Agriculture are crucial in efforts to identify sources of contaminated eggs and to develop and implement control measures.

Information on cooking and handling eggs safely is available from the U.S. Department of Agriculture Meat and Poultry Hotline ([800] 535-4555) and from county extension home economists.

References
1. St Louis ME, Morse DL, Potter ME, et al. The emergence of grade A egs as a major source of Salmonella enteritidis infections: new implications for the control of salmonellosis. JAMA 1988;259:2103-7.
2. CDC. Update: Salmonella enteritidis infections and Grade A shell egg. MMWR 1988;37:490,495-6.
3. Taylor DN, Bopp CA, Birkness K, Cohen ML. An outbreak of Salmonella associated with a fatality in a healthy child: a large dose and severe illness. Am J Epidemiol 1984;119:907-12.
4. Baker RC, Hogarty S, Poon W, et al. Survival of Salmonella typhimurium and Staphylococcus aureus in eggs cooked by different methods. Poultry Sci 1983;62:1211-6.
5. Humphrey TJ, Greenwood M, Gilbert RJ, Rowe B, Chapman PA. The survival of Salmonella in shell eggs cooked under simulated domestic conditions. Epidemiol Infect 1989;103:35-45.

MMWR 44(11):1995 Mar 24

Foodborne Botulism--Oklahoma, 1994

On July 2, 1994, the Arkansas Department of Health and the Oklahoma State Department of Health were notified about a possible case of foodborne botulism. This report summarizes the investigation, which implicated consumption of improperly stored beef stew.

On June 30, 1994, a 47-year-old resident of Oklahoma was admitted to an Arkansas hospital with sub-acute onset of progressive dizziness, blurred vision, slurred speech, difficulty swallowing, and nausea. Findings on examination included ptosis, extraocular palsies, facial paralysis, palatal weakness, and impaired gag reflex. The patient also had partially healed superficial knee wounds incurred while laying cement. He developed respiratory compromise and required mechanical ventilation.

Differential diagnoses included wound and foodborne botulism, and botulism antitoxin was administered intravenously. Electromyography demonstrated an incremental response to rapid repetitive stimulation consistent with botulism. Anaerobic culture of the wounds were negative for Clostridium. However, analysis of a stool sample obtained on July 5 detected type A toxin, and culture of stool yielded C. botulinum. The patient was hospitalized for 49 days, including 42 days on mechanical ventilation, before being discharged.

The patient had reported that, during the 24 hours before onset of symptoms, he had eaten home-canned green beans and a stew containing roast beef and potatoes. Although analysis of the leftover green beans was negative for botulism toxin, type A toxin was detected in the stew. The stew had been cooked, covered with a heavy lid, and left on the stove for 3 days before being eaten without reheating. No other persons had eaten the stew.

Reported by: W Knubley, MD, Cooper Clinic, Fort Smith; TC McChesney, DVM, State Epidemiologist, Arkansas Dept of Health. J Mallonee, MPH, Acting State Epidemiologist, Oklahoma State Dept of Health. Foodborne and Diarrheal Diseases Br, Div of Bacterial and Mycotic Diseases, National Center for Infectious Diseases; Div of Field Epidemiology, Epidemiology Program Office, CDC.

Editorial Note: Botulism is a paralytic illness resulting from a potent toxin produced under anaerobic conditions by C. botulinum. Although foodborne botulism is rare in the United States (34 cases reported in 1994 [CDC, unpublished data, 1995]), manifestations can be severe and can progress rapidly. Because of the potential severity of disease and the possibility for exposure of many persons to contaminated products, foodborne botulism is a public health emergency requiring rapid investigation.

When botulism is suggested by clinical manifestations, (e.g., descending neuroparalysis, ptosis, and extraocular palsies), physicians should obtain a thorough food history to assist in the diagnosis and in identifying and obtaining potentially contaminated leftover food. In the case described in this report, heat-resistant C. botulinum spores either survived the initial cooking or were introduced afterwards; the spores subsequently germinated and produced toxin. The lid of the pot or the gravy of the stew most likely provided the anaerobic environment necessary for toxin production. Previous cases with similar features have resulted from consumption of commercial pot pies (1) and onions sauteed in margarine (2), both of which were left at room temperature for hours after cooking.

Most outbreaks of foodborne botulism in the United States result from eating improperly preserved home-canned foods (3); vegetables (especially asparagus, green beans, and peppers) account for most outbreaks caused by home-canning (CDC, unpublished data, 1995). A pressure cooker must be used to home-can vegetables safely because it can reach temperatures necessary to kill botulism spores (substantially more than 212F [more than 100C] for 10 minutes); however, specific times and pressures needed vary for different foods (4). Jams and jellies can be safely home-canned without a pressure cooker because their high sugar content will not support the growth of C. botulinum. Instructions for home-canning are available from county extension offices. Cooked foods should not be held at temperatures 40 F-140 F (4 C 60 C) for hours (5). Boiling food for 10 minutes before eating destroys any toxin present.

CDC provides epidemiologic consultation and laboratory diagnostic services for suspected botulism cases and authorizes release of botulism antitoxin to state health departments and physicians in the United States. These services are available 24 hours a day from CDC through state health departments.

References
1. CDC. Botulism and commercial pot pie--California. MMWR 1983;32:390,45.
2. MacDonald KL, Spengler RF, Hatheway CL, Hargrett NT, Cohen ML. Type botulism from sauteed onions: clinical and epidemiologic observations. JAMA 1985;253:1275-8.
3. St. Louis ME. Botulism. In: Evans AS, Brachman P, eds. Bacterial infections of humans: epi-

demiology and control. 2nd ed. New York: Plenum Publishing, 1991:115-31.
4. Extension Service, US Department of Agriculture. Complete guide to home canning. Washington, DC: US Department of Agriculture, Extension Service, September 1994. (Agriculture information bulletin no. 539).
5. Food and Drug Administration. Food code, 1993. Washington, DC: US Department of Health and Human Services, Public Health Service, Food and Drug Administration, 1993.

MMWR 44(2):1995 Jan 20

Type B Botulism Associated with Roasted Eggplant in Oil--Italy, 1993

In August and October 1993, public health officials in Italy were notified of seven cases of type B botulism from two apparently unrelated outbreaks in different communites. Investigations were initiated by the Regional Health Observatory of Campania and the Italian National Institute of Health. This report summarizes the outbreak investigations, which indicated that illness was associated with eating commercially prepared roasted eggplant in oil.

OUTBREAK 1
On August 14, two waitresses working in a sandwich bar in Santa Maria di Castellabate were admitted to a local hospital with dysphagia, diplopia, and constipation; a clinical diagnosis of botulism was made. On August 12, the waitresses had prepared and eaten ham, cheese, and eggplant sandwiches. A third waitress also ate the sandwiches and developed dyspepsia for which vomiting was induced; she did not have neurologic symptoms. The owner of the bar, who had tasted a small piece of eggplant from the same jar later on August 12, remained asymptomatic. The cook had initially opened the jar of commercially prepared sliced roasted eggplant in oil and had tasted its contents on August 11 and developed diarrhea. Both the cook and the owner reported that the eggplant tasted spoiled.

Botulism was presumptively diagnosed in the two hospitalized patients; both were treated with trivalent botulism antitoxin and gradually improved. No food samples were available for testing. No botulism toxin was detected in the serum of the two hospitalized patients. However, cultures of their stools subsequently yielded type B Clostridium botulinum.

OUTBREAK 2
During October 5 6, four of nine members of an extended family who had dined together on October 2 were hospitalized in Naples with suspected botulism. The meal consisted of green olives, prosciutto, bean salad, green salad, mozzarella cheese, sausages, and commercially prepared roasted eggplant in oil. Based on an investigation and analysis of food histories, the eggplant was implicated as the probable source (relative risk=undefined; p less than 01). All of the patients were treated with trivalent botulism antitoxin and gradually improved. Investigation indicated that on September 27, another family member had opened and dipped a fork into the implicated jar of eggplant; although he did not eat any eggplant, he used the fork for other food items. On September 28, he had developed vomiting, dysphagia, and double vision but was not hospitalized; his symptoms resolved spontaneously. On October 8, he was asymptomatic but was hospitalized and treated with trivalent botulism antitoxin after botulism was diagnosed in other family members.

One of the hospitalized patients developed respiratory muscle weakness and required mechanical ventilation. A serum specimen from one patient was negative for botulism toxin. Cultures of stool specimens from three patients yielded proteolytic type B C. botulinum. No eggplant was available for testing.

FOLLOW-UP
The commercially prepared eggplant suspected of causing both outbreaks was produced by one company and sold only in Italy. The company reported preparing the eggplant in the following manner: eggplant slices were washed and soaked overnight in a solution of water, vinegar, and salt; roasted in an oven; and subsequently placed in glass jars. Garlic, peppers, oregano, and citric acid were added. The mixtures then were covered with sunflower oil and sealed with screw-on lids; after being filled, the jars were boiled in water for 30 minutes. The pH of the product was not consistently

monitored. A total of 119 jars of eggplant from the same lot that caused the outbreaks was tested; neither C. botulinum spores nor botulism toxin were detected. The pH of the product varied from 3.9 to 5.1; the pH was less than 4.6 in 24 (20%) jars tested.

Public health officials issued a national warning and recalled unused jars of eggplant. No additional cases of botulism associated with this product were reported.

Reported by: P D'Argenio, MD, F Palumbo, MD, R Ortolani, MD, R Pizzuti, MD, M Russo, DBiol, Regional Health Observatory of Campania, R Carducci, MD, Cardarelli Hospital, M Soscia, MD, Contugno Hospital, Naples; P Aureli, DBiol, L Fenicia, DBiol, G Franciosa, DBiol, National Institute of Health, Rome; A Parella, MD, Public Health Dept, Eboli; V Scala, MD, Public Health Dept, Vallo della Lucania, Italy. Foodborne and Diarrheal Diseases Br, Div of Bacterial and Mycotic Diseases, National Center for Infectious Diseases; Global EIS Program, International Br, Div of Field Epidemiology, Epidemiology Program Office, CDC.

Editorial Note: Foodborne botulism is a paralytic illness caused by the ingestion of botulism toxin, a neurotoxin produced by the ubiquitous spore-forming bacterium C. botulinum. Although botulism toxin in food can be destroyed by heating to boiling (212 F [100 C]) for 10 minutes, the spores are heat resistant and can survive prolonged boiling. To destroy C. botulinum spores, food must be heated under pressure to temperatures substantially greater than 212 F. Certain environmental conditions, such as absence of oxygen (anaerobic conditions), pH less than 4.6, warm temperatures (generally less than 39 F [less than 4 C]), high moisture content (water activity), and lack of competing bacterial flora promote production of botulism toxin in foods contaminated with C. botulinum spores. The process used to produce the eggplant epidemiologically implicated in the outbreaks in Italy probably failed to remove C. botulinum spores and may have provided such conditions.

Covering foods in oil may provide the anaerobic conditions required for the production of botulism toxin. Outbreaks of botulism in the United States and Canada have been caused by covering vegetables with oil or grease. For example, in 1983, onions covered in grease and left overnight on a grill caused a large outbreak of botulism in Illinois (1), and commercially processed garlic in oil caused outbreaks in1985 and 1989 (2,3). As a result of these outbreaks, the Food and Drug Administration recommended the addition of antimicrobial growth inhibitors or acidifying agents to such products (4) .

In Italy, approximately 50 cases of botulism are reported annually (National Institute of Statistics, unpublished data, 1994), compared with approximately 20 cases annually in the United States (CDC, unpublished data, 1994). However, because the clinical and laboratory diagnoses of botulism can be difficult, these counts of incident cases probably underestimate the actual occurrence. In the United States, surveillance for botulism is linked to the release of botulism antitoxin for treatment of suspected cases. CDC maintains supplies of botulism antitoxin at quarantine stations nationwide for rapid release at the request of state health officials who report suspected cases. This centralized system for controlling antitoxin supplies results in reporting of botulism cases and a well-maintained, reliable source of antitoxin; no such system exists in Europe.

In Italy, as in the United States, outbreaks of botulism associated with commercial products are uncommon; most result from eating improperly preserved home-canned foods (National Institute of Statistics, unpublished data, 1994). However, two previous outbreaks in Italy have been linked to commercial products--mushrooms in oil and pickled olives (5). The outbreaks described in this report probably resulted from a commercial process that was inadequate to prevent contamination of the final product with C. botulinum spores. Another potential explanation is that the jars of eggplant may have been contaminated after they were opened; however, this is less likely because both outbreaks were caused by the same commercial product. Once contaminated with spores, the pH, oil covering, and lack of refrigeration probably provided conditions conducive to the production of botulism toxin. Strict control of the commercial processes used to manufacture such products and the addition of antimicrobial growth inhibitors or acidifying agents could assist in preventing such outbreaks. Persons who prepare roasted vegetables in oil at home should be aware that this practice may be hazardous, especially if such foods are allowed to remain above refrigerator temperature (generally less than 39 F [less than 4 C]) .

References
1. MacDonald KL, Spengler RF, Hatheway CL, Hargrett NT, Cohen ML. Type A botulism from sauteed onions: clinical and epidemiologic observations. JAMA 1985;253:1275-8.

2. St. Louis ME, Shaun HS, Peck MB. Botulism from chopped garlic: delayed recognition of a major outbreak. Ann Intern Med 1988;108:363-8.
3. Morse DL, Pickard LK, Guzewich JJ, Devine BD, Shayegani M. Garlic-in-oil associated botulism: episode leads to product modification. Am J Public Health 1990;80:1372-3.
4. Food and Drug Administration. Press release no. P89-20. Washington, DC: US Department of Health and Human Services, Public Health Service, Food and Drug Administration, April 17,1989.
5. Fenicia L, Ferrini AM, Aureli P, Padovan MT. Epidemic of botulism caused by black olives [Italian]. Industrie Alimentari 1992;31:3078.

MMWR 36(49):1987 Dec 18

International Outbreak of Type E Botulism Associated With Ungutted, Salted Whitefish

On November 2, 1987, a 39-year-old Russian immigrant and his 9-year-old son were admitted to a suburban New York hospital with symptoms indicative of botulism. The father's stool specimen contained type E botulinum toxin. On October 23, the father had purchased a whole, ungutted, salted, air-dried whitefish known as either ribyetz or kapchunka from a delicatessen in Queens, New York City. He and his son had eaten the fish on October 30 and 31. On November 3, 1987, CDC received a report from the Ministry of Health, Jerusalem, Israel, of five additional cases suspected to be botulism; one case was fatal. The patients had eaten ribyetz purchased in a grocery in Brighton Beach, Brooklyn, New York City, on October 17 and taken to Israel. The fish as well as a serum sample from one surviving patient subsequently yielded type E botulinum toxin.

The implicated fish was distributed in the New York City area by Gold Star Smoked Fish Inc., a firm in Brooklyn. On November 3, the New York City Department of Health issued an embargo on the sale and distribution of ribyet zor kapchunka and removed the implicated product from the shelves of stores selling Gold Star products. The public was alerted through news releases, and acute care hospitals in New York City and surrounding areas were notified. No additional cases have been identified in New York. However, one additional laboratory-confirmed case of botulism has been reported in Israel. On November 13, the patient, a 17-year-old female, had eaten whitefish that had been purchased on October 18 at the same delicatessen in Queens associated with the original patients.

Reported by: S Kotev, MD, Hadassah Univ Hospital, Jerusalem; A Leventhal, MD, MPH, A Bashary, RN, H Zahavi, RN, Jerusalem DistHealth Office; A Cohen, National Botulism Reference Lab; P Slater, MD, MPH, Ministry of Health, Israel. A Ruston, MD, E Baron, PhD, B Farber, MD, J Greenspan, MD, M Tenenbaum, MD, R vanAmerongen, MD, North Shores Univ Hospital, Manhasset; V Tulumello, J Lynch, Nassau County Health Dept; S Schultz, MD, C Reisberg, S Shahidi, PhD, S Joseph, MD, New York City Dept of Health; L Crowell, DVM, J Ferrara, New York State Dept of Agriculture and Markets; J Guzewich, M Shayegani, PhD, G Hannett, DL Morse, MD, MS, State Epidemiologist, New York State Dept of Health. US Food and Drug Administration. Div of Field Svcs, Epidemiology Program Office; Enteric Diseases Br, Div of Bacterial Diseases, Center for Infectious Diseases, CDC.

Editorial Note: Ribyetz, or kapchunka, is an ethnic food consumed in this country primarily by Russian immigrants. It has been implicated as a vehicle for botulism twice in recent years. In 1981, a California man became ill (1), and, in 1985, two Russian immigrants died in New York City after eating the fish (2,3). Type E botulism is typically associated with foods of marine origin (4). The mechanism of contamination of the ribyetz has not been established. However, Clostridium botulinum spores can be found in the intestinal contents of fish, and the fact that the fish were uneviscerated may have been important (5).

The whitefish implicated in this outbreak was produced by one firm an distributed only in New York City. In addition to halting the distribution of the fish, officials in New York City and New York State are developing regulations that would in effect prohibit the production and sale of such uneviscerated whitefish. Although refrigeration is recommended, some consumers may be storing the fish unrefrigerated before eating it uncooked. Persons who purchased ribyetz in New York City in October should dispose of any remaining fish in such a way as to make it inaccessible to others.

Public health personnel should be aware of the potential problem, especially for people in ethnic groups known to eat this product. Guidance in treating botulism and testing serum and stool samples for botulinal toxin can be obtained through state or city health departments. Requests for testing specimens of ribyetz can be made through the district offices of the Food and Drug Administration (FDA) or the FDA Division of Emergency and Epidemiological Operations, Rockville, Maryland 20857; telephone number (301) 443-1240.

References
1. California Department of Health Services. Alert: botulism associated with commercially pro-duced, dried, salted whitefish. California Morbidity, November 6, 1981;(suppl 43).
2. Centers for Disease Control. Botulism associated with commercially distributed kapchunka--New York City. MMWR 1985;34:546-7.
3. Badhey H, Cleri DJ, D'Amato RF, et al. Two fatal cases of type E adut food-borne botulism with early symptoms and terminal neurologic signs. J Clin Microbiol 1986;23:616-8.
4. Centers for Disease Control. Botulism in the United States, 1899-197: handbook for epidemiolo-gists, clinicians, and laboratory workers. Atlanta: US Department of Health, Education, and Welfare, Public Health Service, 1979.
5. Bott TL, Deffner JS, McCoy E, Foster EM. Clostridium botulinum type in fish from the Great Lakes. J Bacteriol 1966;91:919-24.

MMWR 40(7):1991 Feb 22

Campylobacter enteritis --New Zealand, 1990

In August-September 1990, an outbreak of Campylobacter enteritis occurred at a camp near Christchurch, New Zealand. This report provides a preliminary summary of the investigation of this outbreak by the New Zealand Communicable Disease Centre and the Canterbury Area Health Board.

The outbreak occurred at a modern camp and convention center (which hosts greater than 15,000 visitors each year) located approximately 19 km (12 miles) from Christchurch. The facility caters to schools and church and youth groups and provides meals, housing, and indoor and outdoor recre-ation for visitors. Water at the camp, obtained from three springs on the premises, was neither chlori-nated nor filtered before use. On September 4, the Canterbury Area Health Board received reports that two persons who lived at the camp had been hospitalized with Campylobacter enteritis and that a number of children who had visited the camp during the week of August 27-31 had become ill with vomiting and headaches.

All persons at the camp during August 27-31 (58 visiting children (age range: 9-12 years), 19 camp leaders, and 39 staff and their family members) were interviewed to identify cases of Campylobacter enteritis and risk factors for infection with Campylobacter. Because of concerns about the accuracy of information provided by children who had attended the camp, analysis of food and water con-sumption was limited to camp leaders and staff.

Based on completed interviews with 99 (85%) of the 116 persons, 44 (44%) had developed a gas-trointestinal illness that met the case definition for Campylobacter enteritis,* with onset from August 9 through September 7 (Figure 1). Predominant manifestations included abdominal pain (80%), diar-rhea (75%), headache (61%), nausea (60%), fever (59%), and vomiting (55%). The 44 case- patients ranged in age from 3 to 51 years (median: 11 years); 30 (68%) were male. Stool specimens from 11 of 14 symptomatic persons yielded C. jejuni. The pattern of clinical illness in persons with culture-confirmed Campylobacter enteritis was similar to that in persons whose illness was not culture-con-firmed.

Investigation determined that case-patients drank more unboiled water than did persons who were not ill (median: 4 cups vs. 2 cups each day; p=0.03, Kruskal-Wallis test) and were more likely to drink water obtained from one particular spring (40/44 (90%) vs. 38/55 (69%); p less than 0.01, Fisher's exact test).** Coliform counts of water specimens from all three springs (collected at taps from staff houses and the camp kitchen) indicated fecal contamination. Water was not examined specifically for Campylobacter.

Private farmland adjacent to the camp is grazed by sheep and cattle. During the investigation, runoff from the surrounding pasture was noted to enter two springs through the basin covers. Torrential rains during the middle of August may have facilitated the seepage of surface contamination into the spring water.

Control efforts were initiated on September 11 and included 1) using rainwater and potable water supplied by tanker and boiling the water used in staff households until a water-treatment system was installed, 2) installing a water-treatment system, 3) conducting a complete water and sanitation survey, and 4) implementing an informal surveillance system to monitor illness among visitors and staff at the camp.

Since implementation of these control measures, no further cases of enteritis have been reported from the camp.

Reported by: J Stehr-Green, MD, New Zealand Communicable Disease Centre, Porirua; P Mitchell, MB BS, C Nicholls, RGON, S McEwan, Dip Home Science, A Payne, BSc (Hons), Canterbury Area Health Board, Christchurch, New Zealand. Enteric Diseases Br, Div of Bacterial Diseases, Center for Infectious Diseases, CDC.

Editorial Note: C. jejuni is the most common bacterial cause of gastroenteritis in the developed world (1). C. jejuni is frequently cultured from stool samples from patients with diarrhea in Africa, Australia, Europe, and North America and has been isolated from patients with diarrhea more frequently than Salmonella and Shigella combined (2). During 1989, in the United States, state health departments reported 7970 isolates of C. jejuni through CDC's Campylobacter Surveillance System; in New Zealand, Campylobacter infections accounted for 67% of reported gastrointestinal illnesses (3).

Most outbreaks of C. jejuni enteritis have been associated with consumption of raw milk or contaminated water (4). In the first known outbreak of waterborne campylobacteriosis, approximately 3000 persons in Bennington, Vermont, developed C. jejuni enteritis after the town's water system became contaminated with water from an unfiltered source (5). As in the New Zealand outbreak, boiling of water and other interim control measures were effective in stemming the outbreak. Waterborne outbreaks of C. jejuni infection reported to CDC from 1978 through 1986 were all associated with consumption of untreated surface water or inadequately chlorinated water. No reported outbreaks of Campylobacter enteritis have been associated with treated water.

Although other outbreaks such as that in Christchurch have been reported, most Campylobacter infections occur as sporadic cases (6). As with Salmonella, foods of animal origin are the most important sources of Campylobacter. In the United States, poultry is the most common source of sporadic infections (7,8). Epidemiologic investigations have also implicated raw milk (9), eggs, beef (6), contaminated water (5), and contact with infected animals, including cats and puppies (7,10).

References
1. Blaser MJ, Hopkins JA, Vasil ML. Campylobacter enteritis. N Engl J Med 1984;305:1444-52.
2. Blaser MJ, Wells JG, Feldman RA, Pollard RA, Allen JR, the Collaborative Diarrhea Disease Study Group. Campylobacter enteritis in the United States: a multicenter study. Ann Intern Med 1983;98:360-5.
3. New Zealand Communicable Disease Centre. Communicable Disease New Zealand: annual supplement. Porirua, New Zealand: New Zealand Communicable Disease Centre, 1989:16.
4. CDC. Campylobacter isolates in the United States, 1982-1986. MMWR 1988;37(no. SS-2):1-13.
5. Vogt RL, Sours HE, Barett T, et al. Campylobacter enteritis associated with contaminated water. Ann Intern Med 1982;96:292-6.
6. Finch MJ, Blake PA. Foodborne outbreaks of campylobacteriosis: the United States experience, 1980-1982. Am J Epidemiol 1985;122:262-8.
7. Deming MS, Tauxe RV, Blake PA, et al. Campylobacter enteritis at a university: transmission from eating chicken and from cats. Am J Epidemiol 1987;126:526-34.
8. Seattle-King County Department of Public Health. Surveillance of the flow of Salmonella and Campylobacter in a community. Seattle: Seattle-King County Department of Public Health, Communicable Disease Control Section, 1984.
9. Schmid GP, Schaefer RE, Pilkaytis BD, et al. A one-year study of endemic campylobacteriosis in a midwestern city: association with consumption of raw milk. J Infect Dis 1987;156: 218-22.
10. Blaser MJ, Cravens J, Powers BW, Wang WL. Campylobacter enteritis associated with canine infection. Lancet 1978;2:979-81.

* The following in a person who had been at the camp: either a stool culture positive for C. jejuni, a history of diarrhea lasting greater than or equal to 2 days, or four of the following signs/symptoms-- diarrhea for 1 day, nausea, vomiting, abdominal pain, fever, headache, myalgia, and malaise.

** Analysis based on total sample of 99 persons because spring source was known for all persons interviewed.

MMWR 39(45):1990 Nov 16

Yersinia enterocolitica Infections during the Holidays in Black Families--Georgia

During the 1988-89 winter holidays (i.e., Thanksgiving through New Year's Day), an outbreak of gastroenteritis caused by raw chitterlings (i.e., pork intestines, a traditional winter holiday food in some black families) contaminated with Yersinia enterocolitica 0:3 occurred among 15 children in metropolitan Atlanta (1). All the children were black, and 11 were enrolled in the Women, Infants, and Children (WIC) Program. Chitterlings had been prepared in 12 of 13 case households and five of 26 control households (p less than 0.001). The infecting organism was primarily transferred from the raw chitterlings to the children through contact with the hands of the foodhandlers. Of child-caretakers enrolled in the Fulton County (the county where most of the cases occurred) WIC Program, nearly half reported household preparation of chitterlings for a Thanksgiving, Christmas, or New Year's Day meal.

To increase community awareness about the potential risk for acquiring yersiniosis from raw chitter-lings, particularly among WIC Program participants, a supplementary lesson plan was developed and incorporated from October 1989 to January 1990 into an existing Fulton County WIC Program group nutrition education program. The lesson included a lecture and discussion that informed moth-ers, grandmothers, and other child-caretakers about 1) the signs and symptoms of yersiniosis in chil-dren; 2) the transmission of Y. enterocolitica infections to children through direct and indirect con-tact with contaminated raw chitterlings; 3) the need for special care when handling raw chitterlings because of potential contamination with bacteria; and 4) the prevention of Y. enterocolitica infec-tions. Means of preventing illness discussed with each group included 1) careful handwashing by persons cleaning chitterlings before touching a child or anything used by a child (e.g., a toy or bot-tle) and 2) not allowing children to touch raw chitterlings. All WIC Program enrollees who attended classes or obtained vouchers during the winter holidays were also given an educational flyer summa-rizing key points of the lesson plan; enrollees were encouraged to share the flyer with other house-hold foodhandlers (Figure 1).

Reported by: MW Monroe, MS, PE McCray, MS, RJ Finton, MSPH, WR Elsea, MD, Fulton County Health Dept, Atlanta; JD Smith, Georgia Dept of Human Resources. Enteric Diseases Br, Div of Bacterial and Mycotic Diseases, Center for Infectious Diseases, CDC.

Editorial Note: Y. enterocolitica causes an enteric infection with fever, diarrhea, and abdominal pain. The recent emergence of Y. enterocolitica 0:3 infections in the United States appears to have been accompanied by the establishment of a widely distributed swine reservoir: chitterlings from many regions of the country harbor Y. enterocolitica 0:3 (1). Because chitterlings are a common traditional food in some black households, particularly during the winter holidays, they probably represent an important vehicle for transmitting infections to children.

Yersiniosis should be suspected in black infants and children with febrile diarrheal illnesses during the winter holidays. During the winter, hospitals with large black pediatric populations should con-sider routinely culturing all stool specimens on cefsulodin-irgasan-novobiocin (CIN) agar, a medium selective for Yersinia (2).

Cleaning raw chitterlings is a labor-intensive and time-consuming process that may expose house-hold members to potentially infectious agents. Because the potential for transmission of the agent is strongest from foodhandlers to children, someone other than the foodhandler should care for the children while chitterlings are being prepared.

The efforts of the Fulton County Health Department indicate that educational messages can be incorporated into existing WIC educational programs; these messages can provide information to child-caretakers about transmission and prevention of Y. enterocolitica infections due to contaminated chitterlings. Information on the lesson plan and a copy of the educational flyer is available from the WIC Program Office, Fulton County Health Department; telephone (404) 730-1441.

References
1. Lee LA, Gerber AR, Lonsway DR, et al. Yersinia enterocolitica 0:3 infections in infants and children, associated with the household preparation of chitterlings. N Engl J Med 1990;322:984-7.
2. Farmer JJ III, Wells JG, Griffin PM, Wachsmuth IK. "Enterobacteriacea infections" in diagnostic procedures for bacterial infections. 7th ed. Washington, DC: American Public Health Association, 1987:285-96.

May 22, 1998 / 47(19);389-391

Cholera Outbreak among Rwandan Refugees -- Democratic Republic of Congo, April 1997

In April 1997, a cholera outbreak occurred among 90,000 Rwandan refugees residing in three temporary camps between Kisangani and Ubundu, Democratic Republic of Congo (formerly Zaire). Medecins Sans Frontieres (MSF) established two referral medical centers and a cholera treatment center in these camps. Personnel from MSF, Zairean nongovernmental organizations (NGOs), and the Office of the United Nations High Commissioner for Refugees (UNHCR) implemented morbidity and mortality surveillance to monitor refugee health status. This report presents the findings of the surveillance system and indicates this outbreak was characterized by a higher death rate than that observed in previous cholera outbreaks in refugee populations.

The daily number of deaths in the camps was obtained from Zairean Red Cross Society volunteers, who were responsible for burying bodies in mass graves. During March 30-April 20, 1997, a total of 1521 deaths were recorded, most of which occurred outside of health-care facilities. The daily crude mortality rate (CMR) ranged from seven to 14 per 10,000 population; the average daily CMR during this period was 9.9 per 10,000 population.

Active identification and referral for treatment of cholera cases was initiated by hiring Rwandan community health workers who were familiar with the refugees in their section of the camps. Cholera was defined as sudden onset of watery diarrhea resulting in dehydration. Clinical characteristics included vomiting (60% of patients), moderate to severe dehydration (50%-70%), and fever greater than 99.5 F (greater than 37.5 C) (less than 20%).

During April 4-19, 1997, a total of 545 persons with cholera were admitted to the cholera treatment center (attack rate: 0.9%); 67 (12.3%) died. Most deaths in the treatment center occurred during the night when MSF health-care workers were absent. According to MSF personnel, most patients with cholera were severely malnourished and suffered from concurrent health problems (e.g., malaria or acute respiratory illnesses). Most (80%) persons with cholera were aged greater than or equal to 5 years. Cholera cases also occurred among health-care workers at the cholera-treatment center. Three of seven stool specimens tested from patients with watery diarrhea were positive for Vibrio cholerae O1, biotype El Tor, serotypes Inaba or Ogawa.

Cholera-control interventions included filtration and chlorination of the camps' water systems, health education, and construction and maintenance of latrines. Treatment of cholera patients by intravenous and oral rehydration therapy was instituted by MSF (1,2). The overall evaluation of cholera control measures was not possible because of the dispersion of the refugees by unidentified armed forces on April 21, 1997.

Reported by: F Matthys, Medecins Sans Frontieres Belgium, Brussels, Belgium. S Male, Z Labdi, Office of the United Nations High Commissioner for Refugees, Geneva, Switzerland. International Emergency and Refugee Health Program, National Center for Environmental Health; and an EIS Officer, CDC.

Editorial Note: The findings in this report indicate that the implementation of a rapid surveillance system facilitated recognition of the need for increased health-care services and appropriate intervention strategies. Timely surveillance using simple case definitions is crucial to targeting interventions during the emergency phase of refugee situations.

During emergency situations, CMR (normally less than 0.5 per 10,000 population per day in developing countries) is the most specific indicator of health status in refugee populations (3). The CMR among refugees in this outbreak was 9.9. This rate was substantially higher than that in Tingi-Tingi (a temporary settlement of Rwandan refugees in the Democratic Republic of Congo) in 1997 (2.5 per 10,000 per day) (4); lower than in Goma in July 1994 (34-54 per 10,000 per day) (5); and similar to those in refugee camps in Thailand in 1979 (10.6 per 10,000 per day) and Somalia in 1980 (10.1 per 10,000 per day) (3).

The situation in the Democratic Republic of Congo demonstrates the importance of immediate and unrestricted access to displaced populations by the international community if local authorities do not have the means or the political will to assist in emergency situations. The case-fatality ratio for cholera in this outbreak was substantially higher than that observed in previous outbreaks of cholera in refugee camps (3,4). Case-fatality ratios of less than or equal to 1% are expected if adequate rehydration services are available (1).

Several factors accounted for the high mortality among the refugees in this outbreak. First, the refugees had been without adequate food, shelter, or access to health care during the preceding 5 months. In addition, the location of the camps assigned by local authorities was far from the nearest villages (4-50 miles {7-82 km} from Kisangani) and the only transport available for relief personnel and supplies was a railway line controlled by the military. As a result, relief workers were required to take a ferry across the Congo River, then travel to the camps by off-road vehicles; these transfers required up to 6 hours in both directions, leaving only 4 hours daily for building treatment facilities and for patient care. Finally, the camps were moved during the outbreak, requiring relocation of ill patients, rebuilding of cholera treatment facilities, and delaying the proper construction of water-treatment and sanitation facilities.

As in the refugee crisis in Goma (5), active identification of cholera cases with the assistance of Rwandan community health-care workers may have prevented the deaths of many refugees outside of treatment centers. Other intervention strategies included health education of refugees, provision of clean water, construction of latrines, and training health workers in aggressive rehydration therapy using a standardized treatment algorithm. Although these measures may have been effective in preventing the further spread of cholera, they abruptly stopped when the 90,000 refugees were dispersed by unidentified armed forces on April 21, 1997; only 37,000 were repatriated to Rwanda by May 1997.

References
1. World Health Organization. The management and prevention of diarrhoea: practical guidelines Geneva, Switzerland: World Health Organization, 1993
2. Medecins sans Frontieres. Clinical guidelines -- diagnostic and treatment manual. 3rd ed. Paris, France: Hatier, 1993.
3. CDC. Famine-affected refugee and displaced populations: recommendations for public health issues. MMWR 1992;41(no. RR-13).
4. Nabeth P, Vasset B, Guerin P, Doppler B, Tectonidis M. Health situation of refugees in eastern Zaire {Letter}. Lancet 1997;349:1031-2.
5. Goma Epidemiology Group. Public health impact of Rwandan refugee crisis: what happened in Goma, Zaire, in July, 1994? Lancet 1995;345:339-44.

May 26, 1995 / 44(20);385-386

Cholera Associated with Food Transported from El Salvador -- Indiana, 1994

Since the onset of the cholera epidemic in Latin America in 1991, most cases of cholera in the

United States have occurred among persons traveling to the United States from cholera-affected areas or who have eaten contaminated food brought or imported from these areas. In December 1994, a cluster of cholera cases occurred among persons in Indiana who had shared a meal of contaminated food brought from El Salvador. This report summarizes the investigation of the cases conducted by the Indiana State Department of Health (ISDH) in collaboration with the local health departments in Jasper and Newton counties (Indiana), the Illinois Department of Public Health, and the DeWitt-Piatt (Illinois) Bi-County Health Department.

On December 30, 1994, a 56-year-old male resident of Illinois who was visiting relatives in Indiana had onset of severe watery diarrhea, nausea, and vomiting. On December 31, he was evaluated at a local hospital and admitted because of dehydration and hypothermia. Culture of a stool sample obtained from the patient on admission yielded toxigenic Vibrio cholerae O1, serotype Ogawa, biotype El Tor. The culture was confirmed by ISDH, the Kentucky Department for Health Services, and CDC. He was treated with intravenous rehydration and antibiotics and was discharged on January 7, 1995. The patient's 51-year-old wife also had onset of watery diarrhea on December 30. She was evaluated at the same hospital on December 31 and again on January 2, 1995. Stool cultures obtained on both occasions were negative for bacterial pathogens but were not cultured specifically for V. cholerae on thiosulfate-citrate-bile salts-sucrose (TCBS) agar.

During the month preceding onset of their illnesses, these persons had neither traveled outside the United States nor eaten raw shellfish. On December 29, while visiting their 26-year-old daughter in Indiana, they shared a meal with her and their 18-year-old son. The meal comprised palm fruit, bread, and white cheese, all of which had been brought from El Salvador to Indiana 2 days earlier by a relative. Neither their daughter nor son reported diarrhea.

To determine the number of persons infected with V. cholerae O1, serum was obtained from the four persons who shared the meal and from the 28-year-old son-in-law who did not eat any of the food items from El Salvador. Vibriocidal antibody titers greater than or equal to 640, indicating recent infection with V. cholerae O1, were detected in the four persons who had shared the meal but not in the son-in-law. Although the relative who brought the food had returned to El Salvador before he could be interviewed, family members reported that he had had no diarrheal illness while in the United States. The methods of preparation of the foods in El Salvador could not be determined; however, the palm fruit was reportedly home-canned in a salt and vinegar solution. No food items were available for testing. Reported by: N Bailey, M Louck, MD, Jasper County Health Dept, Rensselaer; D Hopkins, J Parker, MD, Newton County Health Dept, Morocco; A Oglesby, D Ewert, MPH, B Barrett, K Laurie, E Muniz, MD, State Epidemiologist, Indiana State Dept of Health. N Wade, DeWitt-Piatt Bi-County Health Dept, Clinton; P Piercy, MSPH, BJ Francis, MD, State Epidemiologist, Illinois Dept of Public Health. T Maxson, DrPH, M Russell, R Finger, MD, State Epidemiologist, Dept for Health Svcs, Kentucky Cabinet for Human Resources. Foodborne and Diarrheal Diseases Br, Div of Bacterial and Mycotic Diseases, National Center for Infectious Diseases, CDC.

Editorial Note: Although most recent cases of cholera in the United States have been associated with international travel (1,2), three U.S. outbreaks have been linked to consumption of food transported from other countries: two associated with crab meat transported in suitcases from Ecuador (3,4) and one associated with commercial frozen coconut milk imported from Thailand (5). The investigation of the cases in Indiana did not implicate a specific contaminated food item; however, of the three food items transported from El Salvador, canned palm fruit is more likely to support the growth of V. cholerae than dry foods, such as bread or cheese.

Since the introduction of cholera into Latin America in 1991, approximately 1 million cases and 9000 associated deaths have been reported to the Pan American Health Organization (PAHO) (2). In 1994, El Salvador and 12 other countries in Latin America reported cholera cases to PAHO (2). Travelers to Latin America and cholera-affected areas in Asia and Africa should eat only foods that have been cooked and are still hot and should drink only beverages that are carbonated or made from boiled or chlorinated water. Travelers also should be advised not to transport food from cholera-affected areas.

The health-care providers who evaluated and treated the patients in this report initially did not suspect cholera because the patients had had no history of recent travel. Patients with severe diarrhea or suspected cholera should be asked about histories of recent travel and consumption of foods transported from another country. Stool samples obtained from persons with suspected cholera should be cultured on TCBS agar because other media routinely used for stool cultures may not support the growth of V. cholerae. Isolates of V. cholerae should be sent to a state public health laboratory for serogrouping; isolates that are serogroup O1 or O139 should subsequently be referred to CDC for toxin testing.

References
1. CDC. Cholera associated with international travel, 1992. MMWR 1992;41:664-7.
2. CDC. Update: Vibrio cholerae O1 -- western hemisphere, 1991-1994, and V. cholerae O139 -- Asia, 1994. MMWR 1995;44:215-9.
3. CDC. Cholera -- New York, 1991. MMWR 1991;40:516-8.
4. Finelli L, Swerdlow D, Mertz K, Ragazzoni H, Spitalny K. Outbreak of cholera associated with crab brought from an area with epidemic disease. J Infect Dis 1992;166:1433-5.
5. Taylor JL, Tuttle J, Pramukul T, et al. An outbreak of cholera in Maryland associated with imported commercial frozen fresh coconut milk. J Infect Dis 1993;167:1330-5.

March 24, 1995 / 44(11);215-219

Update: Vibrio cholerae O1 -- Western Hemisphere, 1991-1994, and V. cholerae O139 -- Asia, 1994

The cholera epidemic caused by Vibrio cholerae O1 that began in January 1991 has continued to spread in Central and South America Figure_1. In southern Asia, the epidemic caused by the newly recognized strain V. cholerae O139 that began in late 1992 also has continued to spread Figure_2. This report updates surveillance findings for both epidemics.

From the onset of the V. cholerae O1 epidemic in January 1991 through September 1, 1994, a total of 1,041,422 cases and 9642 deaths (overall case-fatality rate: 0.9%) were reported from countries in the Western Hemisphere to the Pan American Health Organization. In 1993, the numbers of reported cases and deaths were 204,543 and 2362, respectively Table_1. From January 1 through September 1, 1994, a total of 92,845 cases and 882 deaths were reported. In 1993 and 1994, the number of reported cases decreased in some countries but continued to increase in several areas of Central America, Brazil, and Argentina (1-3).

The epidemic of cholera caused by V. cholerae O139 has affected at least 11 countries in southern Asia. V. cholerae O139 produces severe watery diarrhea and dehydration that is indistinguishable from the illness caused by V. cholerae O1 (4) and appears to be closely related to V. cholerae O1 biotype El Tor strains (5). Specific totals for numbers of V. cholerae O139 cases are unknown because affected countries do not report infections caused by O1 and O139 separately; however, greater than 100,000 cases of cholera caused by V. cholerae O139 may have occurred (6).

In the United States during 1993 and 1994, 22 and 47 cholera cases were reported to CDC, respectively. Of these, 65 (94%) were associated with foreign travel. Three of these were culture-confirmed cases of V. cholerae O139 infection in travelers to Asia. Reported by: Cholera Task Force, Diarrheal Disease Control Program, World Health Organization, Geneva. Expanded Program for the Control of Diarrheal Diseases, Special Program on Maternal and Child Health and Population, Pan American Health Organization, Washington, DC. Foodborne and Diarrheal Diseases Br, Div of Bacterial and Mycotic Diseases, National Center for Infectious Diseases, CDC.

Editorial Note: Cholera is transmitted through ingestion of fecally contaminated food and beverages. Because cholera remains epidemic in many parts of Central and South America, Asia, and Africa, health-care providers should be aware of the risk for cholera in persons traveling in cholera-affected countries -- particularly those persons who are visiting relatives or departing from the usual tourist routes because they may be more likely to consume unsafe foods and beverages.

Persons traveling in cholera-affected areas should not eat food that has not been cooked and is not hot (particularly fish and shellfish) and should drink only beverages that are carbonated or made from boiled or chlorinated water. The licensed parenteral cholera vaccine provides only limited and brief protection against V. cholerae O1, may not provide any protection against V. cholerae O139, and has a high cost-benefit ratio (7); therefore, the vaccine is not recommended for travelers (8). New oral cholera vaccines are being developed and provide more reliable protection, although still at a high cost per case averted. None of these vaccines have attained the combination of high efficacy, long duration of protection, simplicity of administration, and low cost necessary to make mass vaccination feasible in cholera-affected countries.

The diagnosis of cholera should be considered in patients with watery diarrhea who have recently (i.e., within 7 days) returned from cholera-affected countries (9). Patients with suspected cholera should be reported immediately to local and state health departments. Treatment of cholera includes rapid fluid and electrolyte replacement with adjunctive antibiotic therapy. Stool specimens should be cultured on thiosulfate-citrate-bile salts-sucrose (TCBS) agar. Clinical isolates of non-O1 V. cholerae should be referred to a state public health laboratory for testing for O139 if the patient traveled in an O139-affected area, has life-threatening dehydration typical of severe cholera, or has been linked to an outbreak of diarrhea.

References
1. CDC. Update: cholera -- Western hemisphere, 1992. MMWR 1993;42:89-
3. Wilson M, Chelala C. Cholera is walking south. JAMA 1994;272:1226-7.
4. Tauxe R, Seminario L, Tapia R, Libel M. The Latin American epidemic. In: Wachsmuth I, Blake P, Olsvik O, eds. Vibrio cholerae and cholera: molecular to global perspectives. Washington, DC: ASM Press, 1994:321-44.
5. CDC. Imported cholera associated with a newly described toxigenic Vibrio cholerae O139 strain -- California, 1993. MMWR 1993;42:501-3.
6. Popovic T, Fields P, Olsvik O, et al. Molecular subtyping of toxigenic Vibrio cholerae O139 causing epidemic cholera in India and Bangladesh, 1992-1993. J Infect Dis 1995;171:122-7.
7. Cholera Working Group, International Center for Diarrheal Diseases Research, Bangladesh. Large epidemic of cholera-like disease in Bangladesh caused by Vibrio cholerae O139 synonym Bengal. Lancet 1993;342:387-90.
8. MacPherson D, Tonkin M. Cholera vaccination: a decision analysis. Can Med Assoc J 1992; 146:1947-52.
9. CDC. Cholera vaccine. MMWR 1988;37:617-8,623-4.
10. Besser RE, Feikin DR, Eberhart-Phillips JE, Mascola L, Griffin PM. Diagnosis and treatment of cholera in the United States: are we prepared? JAMA 1994;272:1203-5.

August 27, 1993 / 42(33);636-639

Surveillance for Cholera -- Cochabamba Department, Bolivia, January-June 1992

Following the epidemic spread of cholera in Peru (1), in April 1991, health officials in neighboring Bolivia established a surveillance system to detect the appearance and monitor the spread of cholera in their country. The first confirmed case in Bolivia was reported on August 26, 1991; by December 31, 1991, a total of 206 cases had been reported, and 21,324 probable and confirmed cases were reported during 1992. This report summarizes cholera surveillance in Cochabamba department (1992 population: 1,070,000) in central Bolivia (Figure 1) for January-June 1992; the assessment was one element of the Data for Decision Making (DDM) Project conducted by the Child and Community Health Project, Bolivia's Ministry of Social Security and Public Health (MSSPH), the U.S. Agency for International Development (USAID), and CDC.

In April 1991, the MSSPH established three categories of case definitions for cholera surveillance: 1) suspected -- acute diarrhea in a person living in an area where Vibrio cholerae O1 had not been reported previously (stool cultures were obtained from patients with suspected cases); 2) probable -- diarrhea with dehydration, vomiting, and leg cramps in a person living in an area with reported cholera cases or related epidemiologically to another person with cholera (stool cultures were not recommended for patients with probable cases); and 3) confirmed -- diarrhea in a person with a stool culture positive for V. cholerae O1. A two-page case-report form was designed for tabulating and investigating each case and was distributed to all health units in the country. In July 1992, the two-page cholera surveillance form was replaced by a quarter-page surveillance form that collected data on fewer variables.

Cases reported during January 1-June 30, 1992, were analyzed. During this period, 4087 cholera cases in residents of Cochabamba department were reported to the MSSPH; surveillance forms were submitted for 2962 (72%) and oral reports for 1125 (28%) cases. Data about the 2962 cases reported on the surveillance form were used to evaluate the form and to characterize the epidemiology of cholera in Cochabamba department. Of the forms received, data on patient's age, sex, address, and outcome were available for 97% of reported cases; however, information on signs and symptoms of illness was reported for approximately 63% of cases.

The 2962 reported cases included 2667 classified as probable and 295 classified as confirmed and represented an incidence of 2.8 per 1000 population in Cochabamba department. Of the 2962 persons, 1527 (52%) were male (Table 1); 2539 (86%) were aged greater than or equal to 15 years, and 157 (5%) were aged less than 5 years. A total of 1621 (55%) cases occurred in residents of urban areas and 1341 (45%) in residents of rural areas. Of 2878 patients for whom hospitalization status was known, 2449 (85%) were hospitalized; hospitalization rates were similar in urban (83%) and rural (87%) areas. Forty-three persons died (overall case-fatality rate {CFR}=1.4%). Thirteen deaths occurred among all urban cases (CFR=0.8% for urban areas), and 30 deaths occurred among 1328 reported rural cases (CFR=2.2% for rural areas).

Reported by: G Pereira, MD, Ministry of Social Security and Public Health, La Paz; Child and Community Health Project, La Paz; J Flores, MD, Chief of Epidemiology, R Agudo, Health Unit, Cochabamba; US Agency for International Development, Bolivia. US Agency for International Development, Washington, DC. Pan American Health Organization, Washington, DC. Data for Decision Making Project, International Br, Div of Field Epidemiology, Epidemiology Program Office; Foodborne and Diarrheal Diseases Br, Div of Bacterial and Mycotic Diseases, National Center for Infectious Diseases; Div of Field Svcs, International Health Program Office, CDC.

Editorial Note: Features of the cholera epidemic in Bolivia have been similar to those in neighboring countries: the disease has predominantly affected adults in both rural and urban areas (1). The overall CFR for cholera in Latin America has been approximately 1% (2) -- lower than that in other epidemics (3). The CFR has been higher in rural areas of Latin America (as demonstrated in Cochabamba department), reflecting factors such as lack of access to health care, inadequate distribution of oral rehydration salts, and delays in providing prevention and treatment education outside urban areas (4).

The challenges associated with cholera surveillance in Bolivia are similar to those in other Latin American countries that initiated cholera prevention and control programs after the epidemic began in Peru. For example, surveillance systems established to detect and investigate the earliest cases initially were effective; however, as the number of cases increased, available resources for reporting were strained because 1) complex case definitions constrained reporting and interpretation of data; 2) lengthy and detailed surveillance forms that were useful in investigating the earliest cases were subsequently unnecessary and cumbersome (in Cochabamba department, reporting using the two-page form was considered incomplete, inefficient, and was often delayed for cases in rural areas; essential data elements could be listed on the quarter-page form, and since its introduction, all cholera cases reported to the MSSPH have been reported with the form); and 3) laboratories in areas of intense cholera activity were inundated by requests for cultures to confirm suspect cases. CDC and the Pan American Health Organization have recommended measures to simplify cholera surveillance and facilitate rapid dissemination of surveillance information for Latin America and the Caribbean (see box) (5).

Analysis of surveillance information at levels below the national level provides health authorities with more immediate information on local disease activity, allowing appropriate decisions to be made regarding the distribution of treatment supplies and/or support personnel. The evaluation of cholera surveillance in Cochabamba department for January-June 1992 is a component of the DDM Project in Bolivia. The USAID-funded DDM Project, in which Bolivia is one of five countries collaborating with CDC, aims to increase data-based decision making in public health for formulating health policies and for program planning, monitoring, and evaluation. In 1992, the MSSPH requested assistance from USAID/Bolivia and CDC to provide training to 41 national, regional, and district program managers, epidemiologists, and other health officials in applied epidemiology, management, biostatistics, and communication skills. The evaluation of cholera surveillance in Cochabamba department was one of the 41 applied epidemiology projects conducted as part of this training program. The results of the evaluation described in this report have been used to strengthen cholera surveillance efforts and prevention activities in Bolivia.

References
1. Pan American Health Organization. Cholera situation in the Americas: an update. Epidemiol Bull 1991;12:11.
2. CDC. Update: cholera -- Western Hemisphere, 1992. MMWR 1993;42:89-3.
4. Glass RI, Black RE. The epidemiology of cholera. In: Barua D, Greenbough WB III, eds. Cholera. New York: Plenum, 1992.
5. Quick RE, Vargas R, Moreno D, et al. Epidemic cholera in the Amazon: the challenge of preventing death. Am J Trop Med Hyg 1993;48:597-602.

July 09, 1993 / 42(26);501-503

Imported Cholera Associated with a Newly Described Toxigenic Vibrio cholerae O139 Strain -- California, 1993

Epidemics of cholera-like illness caused by a previously unrecognized organism occurred recently in southern Asia (1). This report documents the first case of cholera imported into the United States that was caused by this organism, the newly described toxigenic Vibrio cholerae O139 strain.

On February 5, 1993, a 48-year-old female resident of Los Angeles County sought care at a local outpatient health-care facility for acute onset of watery diarrhea and back pain. A few hours before seeking medical care, she had returned to the United States from a 6-week visit with relatives in Hyderabad, India.

Her diarrheal illness began in India on February 4 and increased in severity while she traveled to the United States. She reported a maximum of 10 watery stools per day but no vomiting, visible blood or mucous in her stools, or documented fever. The patient was prescribed trimethoprim-sulfamethoxazole without rehydration treatment and recovered uneventfully. Duration of illness was approximately 4 days. No secondary illness occurred among family members.

When the patient sought medical care, the physician suspected cholera, and a culture of a stool specimen obtained from the patient at that time yielded colonies suspected of being V. cholerae. This was confirmed by the Los Angeles County Public Health Laboratory. The isolate was identified as V. cholerae non-O1. The isolate produced cholera toxin by Y-1 adrenal cell assay and latex agglutination in the California State Public Health Laboratory. Testing at CDC identified the isolate as toxigenic V. cholerae serogroup O139, resistant to trimethoprim-sulfamethoxazole.

Before this illness, the patient had been in good health. In Hyderabad, she stayed with relatives and did not travel outside the city. Although the source of her infection was not confirmed, on January 30, the patient had eaten fried shrimp and prawns purchased from a local market and prepared by relatives. She also recalled drinking a half glass of unbottled water in Hyderabad on February 3.

Reported by: M Tormey, MPH, L Mascola, MD, L Kilman, Los Angeles County Dept of Health Svcs, Los Angeles; P Nagami, MD, Southern California Permanente Medical Group, Los Angeles; E DeBess, DVM, S Abbott, GW Rutherford, III, MD, State Epidemiologist, California Dept of Health Svcs. Foodborne and Diarrheal Diseases Br, Div of Bacterial and Mycotic Diseases, National Center for Infectious Diseases, CDC.

Editorial Note: In October 1992, an epidemic of cholera-like illness began in Madras, India, associated with an atypical strain of V. cholerae (2). In early 1993, similar epidemics began in Calcutta (with more than 13,000 cases) and in Bangladesh (with more than 10,000 cases and 500 deaths) caused by similarly atypical strains of V. cholerae (3,4). These strains could not be identified as any of the 138 known types of V. cholerae and have been designated as a new serogroup, O139 (5). Although the extent of the ongoing epidemic in southern Asia is unclear, this strain is now associated with epidemic cholera-like illness along a 1000-mile coastline of the Bay of Bengal (from Madras, India, to Bangladesh) and appears to have largely replaced V. cholerae O1 strains in affected areas.

The emergence of this new cause of epidemic cholera represents an important shift in the epidemiology of this infectious disease (6). Until 1993, the only recognized causes of epidemic cholera were V. cholerae strains that were part of serogroup O1. V. cholerae isolates from other serogroups (i.e., non-O1) were recognized as causes of sporadic diarrheal and invasive infections but were not considered to have epidemic potential. The relation of the new non-O1 serogroup to typical O1 strains is unclear; except for the presence of O1 antigen, the strains are nearly identical in most characteristics.

Descriptions of the symptoms associated with V. cholerae O139 infection suggest it is indistinguishable from cholera caused by V. cholerae O1 and should be treated with the same rapid fluid replacement (7). Although the illness may be severe, it is treatable with oral and intravenous rehydration therapy. The new organism has been susceptible to tetracycline, which is the recommended antibiotic for treatment of cholera. However, the organism is reportedly resistant to trimethoprim-sulfamethoxazole and furazolidone, other antibiotics used to treat cholera.

Health-care providers should consider the new strain as a possible cause of cholera-like illness in persons returning from the Indian subcontinent. Although previous cases were reported from Madras and Calcutta in India and from Bangladesh, this report suggests that Hyderabad, India -- which is inland -- is also affected. Because of effective sewerage and water treatment, further spread of this strain is unlikely in the United States. However, the potential for epidemic cholera caused by V. cholerae O139 exists for much of the developing world, and further spread to other parts of Asia is probable.

The emergence of this new strain has at least three other major public health implications. First, it expands the definition of cholera beyond the illness caused exclusively by toxigenic V. cholerae of serogroup O1. Because it appears to cause the same illness and to have similar epidemic potential, the World Health Organization has asked all nations to report illnesses caused by this strain as cholera (1). In the United States, clinicians, laboratorians, and public health authorities should report infections with toxigenic V. cholerae O139 as cholera, in addition to cases of toxigenic V. cholerae O1 infection.

Second, the rapid spread of the V. cholerae O139 epidemic in southern Asia, even among adults previously exposed to cholera caused by V. cholerae O1, suggests that preexisting immunity to toxigenic V. cholerae O1, whether the result of natural infection or cholera vaccine, offers little or no protective benefit. Travelers to areas affected by this epidemic should exercise particular care in selecting food and drink and should not assume that cholera vaccination is protective against the V. cholerae O139 strain.

Third, laboratory identification methods for V. cholerae O1 depend on detection of the O1 antigen on the surface of the bacterium, and therefore do not identify this new strain. A specific diagnostic antiserum for V. cholerae O139 is being prepared for use in U.S. public health laboratories and will be distributed soon. Without such antiserum, this strain might be confused with other non-O1 V. cholerae isolates unrelated to the newly described O139 strain that occasionally cause infections in the United States.

In 1989, a pilot surveillance effort in four states determined that the reported infection rate for non-O1 V. cholerae was 1 per 1 million population (8). Although non-O1 strains can cause illness, non-O1 strains other than the newly described O139 have not been implicated as a cause of epidemics and are not considered a major public health problem. Accordingly, CDC recommends that:

1. Sporadic clinical isolates of non-O1 V. cholerae should be referred to a state public health laboratory for further characterization if there is an epidemiologic link to areas of the world known to be affected by O139 (currently India and Bangladesh); if the disease is typical of severe cholera (i.e., watery diarrhea with life-threatening dehydration); or if the isolate has been linked to an outbreak (i.e., more than one linked case) of diarrheal illness.

2. Physicians should ask that specimens from persons with suspected cholera be cultured on thio-sulfate-citrate-bile salts-sucrose (TCBS) medium for isolation of V. cholerae. All cases of suspected cholera should be reported immediately to local and state health departments.

References
1. World Health Organization. Epidemic diarrhea due to Vibrio cholerae non-O1. Wkly Epidemiol Rec 1993;68:141-2.
2. Ramamurthy T, Garg S, Sharma R, et al. Emergence of novel strain of Vibrio cholerae with epidemic potential in southern and eastern India {Letter}. Lancet 1993;341:703-4.
3. Albert MJ, Siddique AK, Islam MS, et al. Large outbreak of clinical cholera due to Vibrio cholerae non-O1 in Bangladesh {Letter}. Lancet 1993;341:704.
4. Bhattacharya MK, Bhattacharya SK, Garg S, et al. Outbreak of Vibrio cholerae non-O1 in India and Bangladesh {Letter}. Lancet 1993;341:1346-7.
5. Shimada T, Balakrish Nair G, Deb BC, et al. Outbreak of Vibrio cholerae non-O1 in India and Bangladesh {Letter}. Lancet 1993;341:1347.
6. CDC. Emerging infectious diseases -- introduction. MMWR 1993;42:257.
7. Swerdlow DL, Ries AA. Cholera in the Americas: guidelines for the clinician. JAMA 1992;267:1495-9.
8. Levine WC, Griffin PM, Gulf Coast Vibrio Working Group. Vibrio infections on the Gulf Coast: results of first year of regional surveillance. J Infect Dis 1993;167:479-83.

August 27, 1993 / 42(33);639

Recommended Measures for Cholera Surveillance and Rapid Surveillance Information Dissemination for Latin America and the Caribbean

Case definitions. Two categories should be used in case definitions in areas with epidemic cholera: clinical and laboratory-confirmed. A clinical case should be defined as acute, watery diarrhea in a person aged greater than or equal to 5 years; a laboratory-confirmed case, as culture-confirmed Vibrio cholerae O1 infection in a person with diarrhea.

Report forms. Lengthy surveillance forms should not be used. Basic data (e.g., age, sex, address, date of onset or treatment, hospitalization, and outcome) can be collected using short forms and kept for analysis at the local level.

Laboratory confirmation of cases. Cultures should be performed for clinical cholera cases in a cholera-threatened area. After cholera has become established in an area, stool cultures should be performed at a reduced frequency (e.g., 10 cultures per month) to confirm the continuing presence of V. cholerae O1 and to monitor antimicrobial resistance.

Surveillance during an evolving epidemic. In areas threatened by cholera, acute dehydrating diarrhea in persons aged greater than or equal to 5 years should be investigated and cultured. When small numbers of cases are being confirmed, only laboratory-confirmed cases should be reported. When the number of laboratory-confirmed cases increases, the clinical case definition should be used for reporting, and culturing should be used only on a limited basis to confirm the continuing presence of cholera. As the number of cholera cases decreases, the definition for clinical cases should be used for at least 1 year to detect seasonal recurrences of the epidemic. To determine routes of cholera transmission and the potential for prevention, case-control investigations should be conducted at outbreak sites.

Analysis and communication of surveillance data. Surveillance data (e.g., numbers of cases, hospitalizations, and deaths) should be transmitted weekly to the central level and analyzed in a timely manner. Summary reports should be disseminated regularly to all components of and levels within the surveillance system and to the Pan American Health Organization.

February 12, 1993 / 42(05);91-93

Isolation of Vibrio cholerae O1 from Oysters -- Mobile Bay, 1991- 1992

On July 2, 1991, during routine monitoring, the Food and Drug Administration (FDA) isolated toxigenic Vibrio cholerae O1, serotype Inaba, biotype El Tor from oysters and intestinal contents of an oyster-eating fish taken from closed oyster beds in Mobile Bay (1). This isolate was indistinguishable from the Latin American epidemic strain and differed from the strain of V. cholerae O1 that is endemic to the Gulf Coast. This report summarizes the public health response to this isolation of V. cholerae O1.

On July 18, Gulf Coast residents were advised by the Mobile County Health Department and the Alabama Department of Public Health (ADPH) to wash their hands after handling raw seafood and to eat seafood well cooked. FDA and ADPH initiated biweekly sampling of oysters from Mobile Bay, and on July 22 and September 16, 1991, the Latin American strain was again isolated from oysters. The Mobile Bay oyster beds, initially closed on May 31, 1991, remained closed to harvesting until November 4, 1991. On June 15, 1992, toxigenic V. cholerae O1 was again isolated from a sample of oysters from a restricted shellfish-growing area, and the adjacent growing areas were closed to harvesting. On August 19, 1992, the oyster beds were reopened after samples were repeatedly negative. No toxigenic vibrios have been isolated since June 1992.

Toxigenic vibrios have not been isolated from Moore swabs that were placed in effluent from sewage treatment plants in the Mobile Bay area after each isolation of V. cholerae O1 from oysters. FDA and ADPH continue monitoring of shellfish obtained or harvested from the Mobile Bay area, and ADPH maintains surveillance for cases of cholera.

Reported by: BH Eichold, II, MD, JR Williamson, MPH, Mobile County Health Dept, Mobile; CH Woernle, MD, State Epidemiologist, Alabama Dept of Public Health. RM McPhearson, ScD, Food and Drug Administration. Enteric Diseases Br, Div of Bacterial and Mycotic Diseases, National Center for Infectious Diseases, CDC.

Editorial Note: No cases of cholera have been identified in Alabama in recent decades. Surveillance for clinical cases has increased since the beginning of the Latin American outbreak in 1991, and many clinical laboratories now routinely culture diarrheal stool specimens on culture media appropriate for isolation of V. cholerae. The strain responsible for the epidemic in Latin America can be distinguished in the laboratory from the endemic V. cholerae O1 strain that is unique to the U.S. Gulf Coast (2).

The isolation of the Latin American strain of V. cholerae O1 from Gulf Coast oysters during two successive summers illustrates the potential for this organism to be repeatedly introduced or to persist in the environment at least transiently after a single introduction. However, there have been no recognized cases of cholera in the United States caused by the Latin American strain as a result of consumption of seafood harvested from the Gulf of Mexico. It is unknown how the Latin American strain was introduced into Mobile Bay. Surveillance using Moore swabs would have detected clinical cases and asymptomatically infected shedders of V. cholerae O1 (3). However, repeatedly negative Moore swabs indicate that municipal sewage was probably not the source of the strain.

Introduction of toxigenic vibrios into Mobile Bay may have resulted from discharge of contaminated ballast water from freighter vessels. To control buoyancy, ships take on large volumes of ballast water in a harbor and discharge it in other locations. This process may have been responsible for the introduction of other harmful species such as the zebra mussel in the Great Lakes (4). In 1991, the FDA isolated toxigenic V. cholerae O1 from the ballast tanks of ships that had originated from Latin American ports and arrived at Mobile Bay (5). To reduce the risk of introducing harmful organisms through contaminated freighter ballast water, the International Maritime Organization has recommended that freighters empty and refill their ballast water tanks twice on each voyage while in international waters (6). The efficacy of ballast water exchanges in reducing the level of contamination of ballast water has not been assessed. Although ballast water exchanges may decrease the risk of introduction of V. cholerae O1 from other ports into U.S. harbors, this approach would not eliminate the strain already endemic in U.S. Gulf Coast waters.

Since 1973, 91 cases of cholera have occurred in the United States that were unrelated to international travel. Most of these followed consumption of raw or undercooked seafood harvested from the U.S. Gulf Coast contaminated with the Gulf Coast strain of V. cholerae O1. The risk for transmission of cholera can be reduced by avoiding consumption of raw or undercooked seafood.

References
1. DePaola A, Capers GM, Motes ML, et al. Isolation of Latin American epidemic strain of Vibrio cholerae O1 from U.S. Gulf Coast {Letter}. Lancet 1992;339:624.
2. Wachsmuth IK, Bopp CA, Fields PA. Difference between toxigenic Vibrio cholerae O1 from South America and U.S. Gulf Coast {Letter}. Lancet 1991;337:1097-8.
3. Barrett TJ, Blake PA, Morris GA, et al. Use of Moore swabs for isolating Vibrio cholerae from sewage. J Clin Microbiol 1980;11:385-8.
4. Roberts L. Zebra mussel invasion threatens U.S. waters. Science 1990;249:1370-2.
5. McCarthy SA, McPhearson RM, Guarino AM, Gaines JL. Toxigenic Vibrio cholerae O1 and cargo ships entering Gulf of Mexico {Letter}. Lancet 1992;339:624-5.
6. Coast Guard, US Department of Transportation. International Maritime Organization ballast water control guidelines. Federal Register 1991;56:6483.

September 11, 1992 / 41(36)

Cholera Associated with International Travel, 1992

Approximately one case of cholera per week is being reported in the United States. Most of these cases have been acquired during international travel and involve persons who return to their homelands to visit family or foreign nationals visiting relatives in the United States. This report summarizes case reports from four states during 1992. Connecticut

On January 8, the Connecticut Department of Health Services was notified about suspected cholera in two persons. The first, a 43-year-old woman born in Ecuador, traveled with her daughters, aged 13 and 16 years, to Guayaquil, Ecuador, to visit relatives during the Christmas holidays. On January 3, the mother ate raw clams, and the 16-year-old ate cooked shrimp. The following evening, the mother ate cooked crab and lobster, and the 16-year-old ate cooked crab. The 13-year-old ate no seafood during the trip. On January 5, approximately 16 hours after the second meal, the mother had onset of vomiting, cramps, and diarrhea. On January 6, about 48 hours after the second meal and during the return flight to Connecticut, the 16-year-old developed similar symptoms.

Both persons were treated as outpatients at an emergency room in Connecticut with intravenous fluids and oral antimicrobials. Toxigenic Vibrio cholerae O1, biotype El Tor, serotype Inaba, was recovered from stool cultures of both persons. In addition, both Shigella and Campylobacter were isolated from the 16-year-old's stool. The 13-year-old daughter remained well. Florida

On June 8, the Florida Department of Health and Rehabilitative Services was notified of suspected cholera in a 48-year-old man born in Ecuador. The man and his brother traveled by air on June 4 from Guayaquil, Ecuador, to the United States to visit relatives in Miami. Before leaving Guayaquil, he ate ceviche at the airport restaurant. His brother had a different meal.

On the morning of June 6, the patient awoke with severe diarrhea and was hospitalized in Miami. He recovered and was discharged on the 5th hospital day. Culture of the patient's stool yielded toxigenic V. cholerae O1, biotype El Tor, serotype Ogawa. The patient's brother remained well. Hawaii

On July 30, the Hawaii Department of Health was notified about suspected cholera in a 58-year-old male traveler from the Philippines. On July 28, the man boarded a flight in Manila for Honolulu and Panama, where he was employed. Approximately 90 minutes into the flight, he developed severe diarrhea that continued for the duration of the 10.5-hour flight to Honolulu. No oral rehydration therapy was available on the airliner. Shortly before arrival in Honolulu, he had onset of nausea, vomiting, and dizziness.

On arrival, the patient was met by a CDC quarantine officer and was taken by ambulance to a hospital, where he was admitted to the intensive care unit in hypovolemic shock. A stool culture yielded toxigenic V. cholerae O1, biotype El Tor, serotype Ogawa. The patient received intravenous antimicrobials and approximately 10-12 liters of intravenous fluids daily for 5 days. He recovered and was discharged on the 7th hospital day. Texas

On April 29, the Texas Department of Health was notified of suspected cholera in a 40-year-old Hispanic male resident of Brownsville. On April 27, the man and his brother from Houston traveled by automobile to Tampico, Mexico, to visit their father. That evening, they ate fried shrimp and boiled crab at a restaurant in Tampico. The two men returned to Brownsville on April 28. Shortly after midnight, the man had onset of severe vomiting, diarrhea, and confusion; he was hospitalized at 6 a.m.

The emergency room physician suspected cholera. Motile vibrios were visible on a wet preparation of stool examined by darkfield microscopy, and toxigenic V. cholerae O1, biotype El Tor, serotype Inaba, was isolated from a stool sample. The isolate was similar to the Latin American strain by multilocus enzyme testing at CDC.

The man received 13 liters of fluid intravenously during the first day of hospitalization; he recovered and was discharged after 2 days. His brother reported mild diarrhea after the trip. His serum, obtained approximately 2 weeks after his illness, had no detectable vibriocidal antibodies, indicating that he had not had cholera.

Reported by: G Cooper, JL Hadler, MD, State Epidemiologist, Connecticut State Dept of Health Svcs. S Barth, PhD, RC Mullen, MPH, WG Hlady, MD, RS Hopkins, MD, State Epidemiologist, Florida Dept of Health and Rehabilitative Svcs. J Kelly, Philippine Airlines (Honolulu station); S Castillo, FD Pien, MD, Straub Clinic and Hospital, Inc; HY Higa, VY Goo, MS, Div of Microbiology, LK Inouye, PhD, M Sugi, MPH, Div of Epidemiology, EW Pon, MD, State Epidemiologist, Hawaii Dept of Health. L Pelly, MD, Brownsville Medical Center, J Trevino, City of Brownsville Health Dept; GR Garza, A Calderin, South Texas Hospital Laboratory, Harlingen; NL Shelton, MPH, Houston Health and Human Svcs Dept; K Williams, Div of Microbiology, B Ray, K Hendricks, MD, DM Simpson, MD, State Epidemiologist, Texas Dept of Health. Div of Quarantine, National Center for Prevention Svcs; Enteric Diseases Br, Div of Bacterial and Mycotic Diseases, National Center for Infectious Diseases, CDC.

Editorial Note: In 1991, 26 cases of cholera were reported in the United States; 18 were associated with travel to Latin America. Of these, 11 were related to crabs brought back in suitcases (1). Although no further domestic cholera cases associated with souvenir crab have occurred, the number of travel-associated cholera cases is increasing.

Since January 1, 1992, 96 cholera cases have been reported in the United States (with one death), more than in any year since CDC began cholera surveillance in 1961. Of these, 95 were travel-associated. In comparison, from 1961 to 1981, only 10 travel-associated cholera cases were reported (2). Of the 96 cases, 75 were associated with an outbreak on board an Aerolineas Argentinas flight between Argentina and Los Angeles (3). Of the remaining 21 cases, 14 have been linked with travel between the United States and Latin America and six with travel between the United States and Asia. The source of one patient's infection remains unknown. None of the 20 travel-associated cases occurred on typical tourist itineraries. Twelve of the 14 cases associated with travel to Latin America occurred in U.S. residents who were visiting relatives in Latin America; two occurred in residents of Latin America who were ill in the United States. Similarly, five of the six cases associated with travel to Asia occurred in persons visiting relatives.

Most persons infected with V. cholerae O1 have no symptoms, and attempts to prevent the introduction of cholera through restriction of travel have not been successful (4). Because immigrants or foreign nationals may not speak English and are unlikely to obtain pretravel medical advice, they may be difficult to reach with cholera-prevention messages. In addition, these persons may be exposed to cholera while staying in the households of relatives in their homelands.

The report of the Filipino traveler illustrates how a cholera strain could be introduced into another part of the world. Infected travelers can easily move from one part of the world affected by cholera to another where sanitary conditions may permit spread of cholera.

Although spread of cholera on an aircraft is unlikely if routine sanitary measures are followed, cabin crew of commercial aircraft traveling to and from areas affected by cholera should be prepared to treat passengers who develop symptoms of cholera. Most persons with cholera can be treated with oral rehydration solution (ORS) which can be kept on board in dehydrated packets. CDC has advised domestic and foreign airlines serving the western hemisphere and the International Air Transport Association to stock ORS and instructions in its use on flights to and from cholera-affected areas. With prompt and appropriate replacement of fluids, dehydration in persons with severe ongoing fluid losses can be prevented. Regardless of treatment en route, any patient suspected of having cholera should seek medical assistance immediately on arrival.

Risk for cholera and traveler's diarrhea can be reduced by following the general rule "boil it, cook it, peel it, or forget it" (5). In particular, travelers should not consume 1) unboiled or untreated water and ice made from such water; 2) food and beverages from street vendors; 3) raw or partially cooked fish and shellfish, including ceviche; and 4) uncooked vegetables. Cold seafood salads may be particularly risky. Travelers should eat only foods that are cooked and hot, or fruits they peel themselves. Carbonated bottled water and carbonated soft drinks are usually safe if no ice is added (6). Persons planning travel to cholera-affected areas may call the pretravel hotline made available through CDC in English ((404) 332-4559) and Spanish ((404) 330-3132).

References
1. CDC. Cholera -- New Jersey and Florida. MMWR 1991;40:287-9.
2. Synder JD, Blake PA. Is cholera a problem for U.S. travelers? JAMA 1982;247:2268-9.
3. CDC. Cholera associated with an international airline flight, 1992. MMWR 1992;41:134-5.
4. World Health Organization. Guidelines for cholera control. Geneva: World Health Organization, Programme for Control of Diarrhoeal Disease, 1992; publication no. WHO/CCD/SER/80.4, rev. 4.
5. Kozicki M, Steffen R, Schar M. "Boil it, cook it, peel it, or forget it": does this rule prevent travellers' diarrhoea? Int J Epidemiol 1985;14:169-72.
6. CDC. Health information for international travel, 1992. Atlanta: US Department of Health and Human Services, Public Health Service, 1992; DHHS publication no. (CDC)92-8280.

December 13, 1991 / 40(49);844-845

Cholera Associated with Imported Frozen Coconut Milk -- Maryland, 1991

During August 1991, three cases of cholera in Maryland were associated with the consumption of frozen coconut milk imported from Asia. Following an investigation, the product was recalled, and no other cases have been reported.

On August 19, a woman residing in Maryland had onset of severe watery diarrhea and vomiting and, on August 22, was hospitalized with dehydration. Vibrio cholerae O1, serotype Ogawa, biotype El Tor, and Plesiomonas shigelloides were isolated from the stool specimen obtained from the patient; the V. cholerae O1 isolate was confirmed at the Maryland State Department of Health and Mental Hygiene (MDHMH) and CDC and was toxigenic.

The patient had neither traveled outside the United States nor eaten raw shellfish during the preceding month. She and five other persons had attended a private party on August 17. Two of the other persons also had onset of an acute diarrheal illness after the party; incubation periods were 6 hours and 14 hours. Vibriocidal antibody titers were elevated, indicating recent infection with V. cholerae O1. One asymptomatic person also had an elevated vibriocidal antibody titer. Thus, four persons attending the party had laboratory evidence of recent infection, and three of the four had symptoms of cholera. None of the four reported recent foreign travel or cholera vaccination.

Food served at the party included steamed crabs and a homemade Thai-style rice pudding served with a topping made from frozen coconut milk. All six persons ate crabs and rice pudding with coconut milk. However, crabs left over from this party were served at a second party held later on August 17 at the same site; the coconut milk topping was not served. One of 20 persons at the second party had onset of mild diarrhea; specimens obtained from this person and 14 others were negative for vibriocidal antibodies when tested 12-26 days after the party.

The Food and Drug Administration's (FDA) Baltimore District Laboratory cultured unopened packages of the same brand of frozen coconut milk (but a different shipment) as that served at the party. Toxigenic V. cholerae O1, serotype Ogawa, biotype El Tor, was isolated from one of six bags tested. In addition, V. cholerae non-O1, V. fluvialis, V. alginolyticus, Aeromonas species, and group B, E1, and E2 Salmonella were isolated from this product, with coliform counts measuring up to 11,000 most probable number per gram.

No secondary cases of cholera were identified among contacts of the affected persons. In addition, surveillance through emergency rooms failed to identify additional cases in the area. The MDHMH placed Moore swabs in four central sewage collection points in the Baltimore metropolitan and Montgomery County areas as a surveillance measure for the presence of V. cholerae O1 infection in the general population; swabs collected from September 11 through October 3 did not yield V. cholerae O1.

The implicated product in this outbreak was Asian Best brand of frozen coconut milk, produced in Thailand and exported by a Bangkok trading company to a Maryland distributor. Nineteen shipments, totaling 36,160 8-ounce bags, had been imported since January 1, 1991. On September 20, the distributor issued a voluntary product recall, and FDA halted all further importation of this product. The Thai Ministry of Public Health reported that the manufacturer of this brand was not licensed by the Thai FDA and shipped the product only to the United States. Reported by: C Lacey, Montgomery County Health Dept; R Talbot, Howard County Health Dept; J Taylor, MPH, D Dwyer, MD, B Jolbitado, C Morrison, E Butler-Senkel, S Strauss, MD, D Murphy-Baxam, MS, J Libonati, PhD, E Israel, MD, State Epidemiologist, Maryland State Dept of Health and Mental Hygiene. N Ridley, MS, Massachusetts Dept of Public Health. M Smith, MD, State Epidemiologist, New Hampshire State Dept of Health and Human Svcs. J Zingeser, DVM, Vermont Dept of Health. G Miller Jr, MD, State Epidemiologist, Virginia Dept of Health. K Ungchusak, MD, Thai Ministry of Public Health. Baltimore District; Div of Emergency and Epidemiological Operations, Food and Drug Administration. Enteric Diseases Br, Div of Bacterial and Mycotic Diseases, National Center for Infectious Diseases, CDC.

Editorial Note: Of the 24 cases of cholera reported in the United States during 1991, 16 were exposed during travel to South America (1-3); all 16 patients were infected with V. cholerae O1, serotype Inaba, the strain epidemic in Latin America. Three were exposed during travel to Asia; two of the three were infected with serotype Ogawa, the serotype identified in the patient from Maryland.

The source of infection of the coconut milk implicated in the Maryland cholera outbreak remains under investigation. This product, marketed primarily for home use (distribution to restaurants was limited), is usually consumed well-cooked in ethnic curries and desserts. In this outbreak, the heating of the coconut milk was apparently insufficient to kill cholera organisms, and prolonged holding time at room temperature was sufficient to allow the organisms to multiply to infectious levels (4). The risk for cholera infection to the general public by this product is minimal given its limited distribution and usual preparation procedure. However, this outbreak illustrates the potential for global dissemination of cholera in a frozen food product. Canned coconut milk is safe because heat treatment during the standard canning process is sufficient to kill vibrios.

References
1. CDC. Importation of cholera from Peru. MMWR 1991;40:258-9.
2. CDC. Cholera--New Jersey and Florida. MMWR 1991;40:287-9.
3. CDC. Cholera--New York, 1991. MMWR 1991;40:516-8.
4. Kolvin JL, Roberts D. Studies on the growth of Vibrio cholerae biotype El Tor and biotype classical in foods. J Hyg Epidemiol Microbiol Immunol 1982;89:243-52.

Epidemiologic Notes and Reports Cholera -- New York, 1991

Through June 26, 1991, cholera has been reported from seven countries in the Western Hemisphere: Brazil, Chile, Colombia, Ecuador, Mexico, Peru, and the United States. In the United States, a total of 14 confirmed cases of epidemic-associated cholera have been reported among persons in Florida (one) (1), Georgia (one) (2), New Jersey (eight) (1), and New York (four). This report summarizes information regarding the four cases reported in New York and describes a new laboratory procedure used to confirm the vehicle of transmission in this outbreak.

On April 26, 1991, a 57-year-old man (patient B) was hospitalized in New York City with a 2-day history of diarrhea; stool culture yielded Vibrio cholerae O1. An investigation by the New York City Department of Health identified additional cases among his family and friends. The first person to become ill was a man (patient A) who had returned from Ecuador on April 21 and had onset of watery diarrhea April 22. Although he sought care from a physician, he was not hospitalized, and a stool culture was not obtained.

On April 24, three other persons (patients B, C, and D) had onset of diarrhea. All patients had laboratory evidence of infection with V. cholerae O1. A stool culture from patient C, a woman, yielded V. cholerae O1. Convalescent phase blood samples from patient D, a woman, and patient A had vibriocidal antibody titers greater than or equal to 1:640, indicating recent V. cholerae O1 infection. The New York City Department of Health Laboratory and CDC identified the isolates as toxigenic V. cholerae O1, biotype El Tor, serotype Inaba--the serotype that is causing epidemic cholera in South America.

Patients B, C, and D had not recently visited South America. However, on the evening of April 22 they had eaten a salad containing crab meat from crabs that had been brought from Ecuador by patient A. The crabs had been purchased by patient A at a pier in Guayaquil, Ecuador, on April 20, then boiled and shelled; meat and claws were then stored in a plastic bag in a freezer. On April 21, when patient A returned to New York, he carried the bag in his suitcase; on arrival, the meat and claws were still frosted and were placed in a freezer overnight. On April 22, the crab meat was thawed in a double-boiler for 15-20 minutes. Two hours later, without further cooking, the crab was served in a crab salad and as cold crab in the shell. The crab was consumed during a 6-hour period by patients B, C, and D and by four persons who remained well. Patient A had onset of diarrhea before eating the crab meat but ate after patients B, C, and D had eaten; he did not assist in preparing the food.

Four samples of crab were obtained for culture, including a claw, two pieces of meat that had remained in the plastic bag, and juice saved when the crab meat was thawed for the crab salad. Standard culture procedures were negative for V. cholerae O1 at the New York City Department of Health and CDC. However, use of the polymerase chain reaction (PCR) technique with primers recently constructed at CDC enabled dection of the V. cholerae O1 toxin gene in one of the pieces of crab meat from the plastic bag. Reported by: R Roman, MPH, M Middleton, S Cato, E Bell, KR Ong, MD, Commission on Disease Intervention; R Gruenewald, PhD, A Ferguson, MS, A Ramon, MD, Bur of Laboratories, New York City Dept of Health. Enteric Diseases Br, Div of Bacterial and Mycotic Diseases, National Center for Infectious Diseases, CDC.

Editorial Note: Epidemic cholera had not been reported in South America during the 1900s until January 1991, when cholera was reported from several locations in Peru (3). As of July 24, 1991, 257,399 probable cholera cases and 2697 cholera-associated deaths have been reported to the Pan American Health Organization from seven countries (Table 1). The cases in New York bring to 14 the total number of confirmed cholera cases in the United States associated with the epidemic in South America; in addition, these cases are the second episode of transmission of V. cholerae O1 associated with crabs brought back by a traveler from South America (1). The Food and Drug Administration monitors seafood imported into the United States; no cases of cholera in the continental United States have been linked to commercially imported food products.

Patient A probably became infected with V. cholerae O1 while in Ecuador because he had onset of illness within 24 hours of returning to New York. Patients B, C, and D were probably infected by eating the crab. Patient A was unlikely to have contaminated the crab because his illness began after the crab had been cooked, frozen, and packaged, and he touched the crab meat again only after the others had eaten. Secondary spread from patient A is unlikely because person-to-person spread of cholera is infrequent--especially in settings where adequate access to water for washing and sanitation facilities exist. Since 1961, more than 100 domestically acquired and imported cases of cholera have been reported to CDC; none of these cases has been associated with person-to-person spread (CDC, unpublished data).

Crabs are a likely vehicle for transmission of cholera and may be contaminated with V. cholerae O1 before or after harvest. Vibrios can survive in crabs boiled for up to 8 minutes (4), and undercooked crabs have caused several previous outbreaks (2,4). V. cholerae O1 biotype El Tor strains multiply rapidly at room temperature in cooked shellfish (5). In this report, vibrios that survived boiling in Ecuador or that contaminated the meat during shelling may have multiplied during transport or while the crab salad was held at ambient temperature.

Standard culture procedures can detect only viable organisms; in contrast, PCR can detect DNA from nonviable organisms. Because of the freezing and thawing, V. cholerae O1 organisms in the crab may not have been viable. However, PCR analysis indicated that the crabs from the outbreak had been contaminated with toxigenic V. cholerae O1. The PCR procedure and other new laboratory tests are potentially important tools for investigating outbreaks of cholera.

The cholera outbreaks in New Jersey (1) and New York prompted an ongoing educational campaign to discourage travelers from returning from infected areas (including Peru, Ecuador, and Colombia) with perishable seafood and other high-risk food items. This campaign includes publication by CDC of a travel advisory in English and Spanish and the distribution of letters to airline passengers traveling to and returning from these countries. Newspapers and radio and television stations in the New Jersey/New York area have also helped publicize this message. No additional cases of cholera associated with food brought back from South America have been reported.

A CDC ``travelers' hot line'' is available in English and Spanish for persons planning travel to Central and South America: the telephone numbers are (404) 332-4559 (English) and (404) 330-3132 (Spanish).

References
1. CDC. Cholera--New Jersey and Florida. MMWR 1991;40:287-9.
2. CDC. Importation of cholera from Peru. MMWR 1991;40:258-9.
3. CDC. Cholera--Peru, 1991. MMWR 1991;40:108-10.
4. Blake PA, Allegra DT, Snyder JD, et al. Cholera--a possible endemic focus in the United States. N Engl J Med 1980;302:305-9.
5. Kolvin JL, Roberts D. Studies on the growth of Vibrio cholerae biotype El Tor and biotype classical in foods. J Hygiene 1982;89:243-52.

May 03, 1991 / 40(17);287-289

Epidemiologic Notes and Reports Cholera -- New Jersey and Florida

Through April 30, 1991, epidemic cholera has been reported from five countries in South America: Brazil, Chile, Colombia, Ecuador, and Peru. In addition, in the United States a total of 10 confirmed cases of epidemic-associated cholera have been reported in Georgia (1), New Jersey, and Florida. This report summarizes information regarding the cases reported in New Jersey and Florida.

NEW JERSEY
From March 31 through April 3, eight residents of Hudson and Union counties developed profuse watery diarrhea after eating crab meat transported from South America. Five of the patients also reported vomiting, and at least three had severe leg cramps; five were hospitalized. Ingestion of the crab meat was statistically associated with illness; of the 11 persons who attended the two meals where the crab was served, all eight who ate the crab meat became ill; the three who did not remained well (p less than 0.01). Each of the patients had onset of symptoms within 3 days of

ingesting the crab meat. Stool samples from four of the eight patients yielded toxigenic Vibrio cholerae O1, serotype Inaba, biotype El Tor, the same serotype responsible for the epidemic in South America. In convalescent serum specimens obtained from the four patients who were culture negative, vibriocidal antibody titers were greater than or equal to 1:1280, indicating recent V. cholerae infection.

The crab was purchased in a fish market in Ecuador, then boiled, shelled, and wrapped in foil. On March 30, it was transported into the United States, unrefrigerated, in a plastic bag on an airplane. It was delivered to a private residence, refrigerated overnight, then served in a salad on March 31 and April 1. No crab meat was available for culture.

All eight patients have fully recovered. No cases of secondary transmission have been reported.

FLORIDA
On April 6, a woman with severe watery diarrhea was admitted to a Dade County hospital on her return from Ecuador. Although stool cultures were negative for V. cholerae O1, testing of acute and convalescent blood samples detected a 32-fold rise in vibriocidal antibody titers, indicating recent infection with V. cholerae O1.

The patient had traveled in Ecuador from March 27 through April 6. She reported eating raw oysters in Salinas Beach, Ecuador, on March 29 and ceviche on March 30; she also consumed ice during her stay. On April 2, she developed watery diarrhea with 30-40 stools per day. On return to the United States, she was admitted to the hospital. The patient recovered, and no cases of secondary transmission have been identified. Reported by: H Ragazzoni, DVM, K Mertz, MD, L Finelli, DrPH, C Genese, MBA, Div of Epidemiology, FJ Dunston, MD, State Commissioner of Health, New Jersey State Dept of Health. B Russell, MPH, W Riley, PhD, E Feller, MD, Baptist Hospital, Miami; MB Ares, MD, M Fernandez, MD, E Sfakianaki, MD, Dade County Public Health Unit; JA Simmons, MD, RS Hopkins, MD, State Epidemiologist, Florida Dept of Health and Rehabilitative Svcs. Enteric Diseases Br, Div of Bacterial and Mycotic Diseases, Center for Infectious Diseases, CDC.

Editorial Note: Epidemic cholera had not been reported in South America this century (2) until January 1991, when cholera appeared simultaneously in several coastal cities of Peru. As of April 29, 169,255 probable cholera cases and 1244 deaths in Peru had been reported to the Pan American Health Organization; cholera had also been reported in Ecuador (3898 cases and 140 deaths), Chile (26 cases), Colombia (176 cases), and Brazil (four cases). The cases reported in Florida and New Jersey bring to 10 the total number of confirmed cases in the United States associated with the epidemic in South America.

No reported cases of cholera have been linked to commercially imported food products. In New Jersey, the confirmed V. cholerae O1 infections resulted from consumption of noncommercial crab meat that had been grossly mishandled and illegally transported into the United States. Although it is unclear how the crab meat became contaminated, contamination may have occurred at harvest, at purchase, or after cooking. V. cholerae O1 can survive in contaminated crabs that are boiled for less than 10 minutes (3). Because V. cholerae biotype El Tor strains multiply rapidly at room temperature in cooked shellfish (4), the lack of refrigeration during transport may have permitted growth of vibrios.

Previous cases acquired in the United States have been associated with undercooked crabs or raw oysters harvested domestically in the Gulf of Mexico (3,5). In the United States, secondary transmission from imported or domestic cases is unlikely because of the availability of safe drinking water and proper treatment of sewage.

The risk for cholera to tourists in affected areas is considered extremely low (6). Although it cannot be determined whether the source of infection in the traveler to Ecuador was consumption of raw oysters, ceviche, or contaminated ice or some other vehicle of infection, this case illustrates the need for travelers to areas with epidemic cholera to follow scrupulously the precautions described for prevention of travelers' diarrhea (7). The general rule "boil it, cook it, peel it, or forget it" has been proposed for preventing travelers' diarrhea (8). In particular, travelers to Colombia, Ecuador, and Peru should not consume 1) unboiled or untreated water and ice made from such water; 2) food and beverages from street vendors; 3) raw or partially cooked fish and shellfish, including ceviche; and 4) uncooked vegetables. Travelers should eat only foods that are cooked and hot, or fruits they peel themselves. Carbonated bottled water and carbonated soft drinks are usually safe if no ice is added. Cholera vaccination, which protects approximately 50% of vaccinated persons for 3-6 months, is not recommended for travelers and is not a substitute for scrupulously choosing food and drink.

V. cholerae may not be isolated from stool samples of cholera patients if the samples are collected late in illness or after antimicrobial therapy is begun. Vibriocidal antibody titers peak 10-21 days after infection and can be used to confirm V. cholerae infection (9).

Travelers who develop severe watery diarrhea, or diarrhea and vomiting, during or within 1 week after travel to an area with known cholera should seek medical attention immediately. Physicians should request that specimens from suspected cases be cultured on media designed for isolation of V. cholerae and should report suspected cases of cholera to their local and state health departments.

References

1. CDC. Importation of cholera from Peru. MMWR 1991;40:258-9.
2. CDC. Cholera--Peru 1991. MMWR 1991;40:108-10.
3. Blake PA, Allegra DT, Snyder JD, et al. Cholera--a possible endemic focus in the United States. N Engl J Med 1980;302:305-9.
4. Kolvin JL, Roberts D. Studies on the growth of Vibrio cholerae biotype El Tor and biotype classical in foods. J Hygiene 1982;89:243-52.
5. Pavia AT, Campbell JF, Blake PA, Smith JD, McKinley TW, Martin DL. Cholera from raw oysters shipped interstate. JAMA 1987;285:2374.
6. Snyder JD, Blake PA. Is cholera a problem for US travelers? JAMA 1982;247:2268-9.
7. CDC. Health information for international travel, 1990. Atlanta: US Department of Health and Human Services, Public Health Service, 1990; DHHS publication no. (CDC)90-8280.
8. Kozicki M, Steffen R, Schar M. Boil it, cook it, peel it or forget it: does this rule prevent travellers' diarrhoea? Int J Epidemiol 1985;14:169-72.
9. Feeley JC, DeWitt WE. Immune response to Vibrio cholerae. In: Rose NR, Friedman H, eds. Manual of clinical immunology. Washington, DC: American Society for Microbiology, 1976:289-95.

April 19, 1991 / 40(15);258-259

Epidemiologic Notes and Reports Importation of Cholera from Peru

On April 9, 1991, a U.S. physician attending a conference in Lima, Peru, had onset of diarrhea. He reported a maximum of eight watery stools in 24 hours and experienced no other symptoms except moderate weakness. The diarrhea lasted 5 days. After arriving in Peru on April 5, he had eaten all his meals, including a cold crab meat appetizer 2 days before onset of illness, in his hotel or at events catered solely for the conference participants. He also consumed ice and municipal water that the hotel reported had been purified. Culture of a stool sample obtained on April 11, after his return to the United States, yielded toxin-producing Vibrio cholerae O1, serotype Inaba, biotype El Tor. His family did not accompany him to Peru and has remained well. Reported by: JA Wilber, MD, State Epidemiologist, Georgia Dept of Human Resources. Enteric Diseases Br, Div of Bacterial and Mycotic Diseases, Center for Infectious Diseases, CDC.

Editorial Note: An epidemic of cholera is occurring in Peru, Ecuador, and Colombia, and there is potential for spread to other countries. Although the risk for cholera is small for U.S. residents traveling in cholera-infected areas (1), some U.S. travelers nonetheless may become infected (2). The best protection is provided by scrupulous adherence to recommendations to prevent traveler's diarrhea (3,4); particularly, raw seafood and potentially contaminated water should be avoided. Optimally, travelers should drink only water that they have treated (e.g., by adding iodine or boiling) themselves. In addition, ice, which may be made from contaminated water, should be avoided. Commercially bottled water has transmitted cholera (5), but carbonated bottled water has a low pH and permits only brief survival of V. cholerae O1.

Most V. cholerae O1 infections cause no symptoms or only mild to moderate diarrhea, but in a small proportion of cases the illness can be life-threatening. Travelers who develop severe watery diarrhea or diarrhea and vomiting during or following travel to an area with known cholera should seek medical attention immediately. Treatment of cholera with proper oral and, if indicated, intravenous rehy-dration is simple and highly effective.

The risk for secondary transmission of cholera in the United States is extremely small (2).

References
1. Snyder JD, Blake PA. Is cholera a problem for US travelers? JAMA 1982;247:2268-9.
2. CDC. Update: cholera outbreak--Peru, Ecuador, and Colombia. MMWR 1991;40:225-7.
3. CDC. Cholera--Peru, 1991. MMWR 1991;40:108-10.
4. CDC. Health information for international travel, 1990. Atlanta: US Department of Health and Human Services, Public Health Service, 1990; DHHS publication no. (CDC)90-8280.
5. Blake PA, Rosenberg ML, Florencia J, Bandeira Costa J, Quintino LDP, Gangarosa EJ. Cholera in Portugal, 1974. II. Transmission by bottled mineral water. Am J Epidemiol 1977;105:344-8.

February 15, 1991 / 40(6);108-110

International Notes Cholera -- Peru, 1991

On January 29, 1991, the General Office of Epidemiology, Ministry of Health (MOH) in Lima, Peru, received reports of an increase in gastroenteritis in Chancay, a coastal district approximately 1 hours by road north of Lima (Figure 1). On January 30, teams from the Field Epidemiology Training Program (FETP), Division of Epidemiology, MOH, traveled to Chancay to investigate this problem.

Investigation identified an outbreak of diarrheal illness that had begun on January 23. Illness in initial cases was characterized by voluminous watery diarrhea, vomiting, and to a lesser extent, severe muscle cramping. Vibrio cholerae O1, Inaba, biotype El Tor, was isolated from patients' stools from Chancay and Chimbote by the National Institute of Health, MOH; Cayetano Heredia University; and the Navy Army Medical Research Institute Detachment and was confirmed by CDC. Additional cases of gas-troenteritis have been reported from the cities of Chimbote, Piura, Trujillo, and Chiclayo along the northern coast of Peru (Figure 1).

Active surveillance and a national laboratory network have been implemented throughout the country. From January 24 through February 9, 1859 persons with gastroenteritis who required hospitalization and 66 deaths were reported to the MOH. Epidemiologic investigations are being carried out by FETP resi-dents to further define the extent of the epidemic and the mode of transmission. As a result of the epi-demic, a national permanent Committee of Epidemiologic Surveillance has been established. The gener-al population has been alerted to ongoing activity, and information on preventive measures has been widely disseminated through the media. The MOH has recommended 1) the exclusive use of boiled water for drinking, 2) careful cleaning of fruit and vegetables, and 3) avoidance of raw or inadequately cooked fish or seafood. Reported by: C Vidal Layseca, MD, Minister of Health and Social Services, Lima; C Carrillo Parodi, MD, Director, National Institutes of Health, Lima; L Seminario Carrasco, MD, Director, General Office of Epidemiology, Lima; Field Epidemiology Training Program, Lima; Laboratory of Cayetano Heredia Univ, Lima, Peru. Navy Army Medical Research Institute Detachment, Lima, Peru. Enteric Diseases Br, Div of Bacterial Diseases, Center for Infectious Diseases; Global EIS Program, International Br, Div of Field Epidemiology, Epidemiology Program Office, CDC.

Editorial Note: The appearance of cholera in several towns along the Peruvian seacoast represents the first time this century that epidemic cholera has been identified in South America. During the 19th cen-tury, epidemic cholera affected the Americas in several pandemic waves. The pandemic of cholera that began in Southeast Asia in 1961 affected many areas of Asia, the Middle East, Europe, Oceania, and Africa but apparently did not reach the American continents. An endemic focus of a unique Western Hemisphere strain exists along the coast of Louisiana and Texas, and possibly northern Mexico (1). Isolates from Peru are being examined to determine their relation to the pandemic or Western Hemisphere strains.

Following its introduction in sub-Saharan Africa in 1970, cholera was initially confined to coastal regions but spread following rivers and the routes of traders and travelers (2). The El Tor pandemic strain grows in many foods and can persist in aquatic environments. After initial outbreaks, cholera can disappear or become endemic and remain a public health threat. High attack rates are more common in areas with poor sanitation and inadequate water supplies. In previous epidemics, documented vehicles of transmission have included contaminated water, raw or undercooked shellfish and other seafood, moist-grain gruels, and leftover rice.

When the profuse watery diarrhea and vomiting associated with severe cholera are not treated, patients may die from dehydration in hours. Treatment with oral and, if necessary, intravenous rehydration can decrease death rates of severe cholera from 50% to 1%-2%. Therapeutic antibiotics can decrease the volume of stool produced. Mass chemoprophylaxis, vaccination, and quarantine have proven ineffective and can divert valuable resources from efforts to ensure adequate treatment of cases and control of transmission (3).

The impact of epidemic cholera can be diminished by organized control efforts. Public health officials should establish surveillance networks in areas with cholera, or at risk for cholera, and establish oral rehydration facilities throughout the country. Epidemiologic investigations, such as that being conducted by the Peruvian FETP (4,5) of the MOH, can help control efforts by determining the extent and source of outbreaks.

The risk to U.S. travelers of acquiring cholera in endemic areas is low. During the first 20 years of the current pandemic, only 10 cases of cholera in U.S. travelers were reported to CDC--representing a risk of acquiring a reported case of cholera of less than one per 500,000 returning travelers (6). Cholera vaccination confers only brief and incomplete protection and is not recommended. The usual precautions to prevent traveler's diarrhea should be observed carefully (7); particularly, raw seafood and potentially contaminated water should be avoided. A traveler who develops severe watery diarrhea, or diarrhea and vomiting, during or following travel to an area with known cholera should seek medical attention immediately.

References
1. Blake PA, Wachsmuth K, Davis BR, Bopp CA, Chaiken BP, Lee JV. Toxigenic Vibrio cholerae O1 strain from Mexico identical to United States isolates. Lancet 1983;2:912.
2. Goodgame RW, Greenough WB. Cholera in Africa: a message for the West. Ann Intern Med 1974;82:101-6.
3. World Health Organization. Guidelines for cholera control. Geneva: World Health Organization, 1986; publication no. WHO/CDD/SER/80.4 Rev 1.
4. Music SI, Schultz MG. Field Epidemiology Training Programs: new international health resources. JAMA 1990;263:3309-10.
5. Malison MD, Dayrit MM, Limpakarnjanarat K. The Field Epidemiology Training Programmes. Int J Epidemiol 1989;18:995-6.
6. Snyder JD, Blake PA. Is cholera a problem for US travelers? JAMA 1982;247:2268-9.
7. CDC. Health information for international travel, 1990. Atlanta: US Department of Health and Human Services, Public Health Service, 1990; DHHS publication no. (CDC)90-8280.

April 05, 1991 / 40(13);225-227

International Notes Update: Cholera Outbreak -- Peru, Ecuador, and Colombia

The following report is reprinted from the Weekly Epidemiological Record of the World Health Organization (WHO). The editorial note was prepared by CDC.

SMALL RISK OF CHOLERA TRANSMISSION BY FOOD IMPORTS WHO has no documented evidence of a cholera outbreak occurring as a result of the importation of food across international borders.

* Dried, acidic and pickled foods, fruit juices: cholera organisms are sensitive to drying and to acidity (pH less than 4.5); therefore, these foods and juices are unlikely to cause infection.
* Coffee, cereals: same as for dried foods above.

* Frozen foods: freezing below -20 centigrade will reduce, but may not completely eliminate, cholera organisms from food.
* Canned foods: canned foods produced according to the relevant Codex standard* are free of cholera organisms even if the raw product was contaminated.
* Irradiated foods: irradiated foods produced according to the relevant Codex standard** and which have received a dose of at least 1kGy are free of cholera organisms even if the raw product was contaminated.
* Fresh sea food: sea food from shallow coastal waters (such as prawns and shellfish) may be contaminated. It should be properly cooked as shown below. Deep sea fish are unlikely to have been infected in their habitat, but could become contaminated during subsequent handling.
* Fresh vegetables and fruit: these may be surface contaminated and may remain so up to a maximum of 10 days.
* Animal feeds: since there is no known reservoir of cholera in poultry or livestock, animal feeds, and in particular dried fish meal, do not pose a risk of transmission.

Cholera transmission through food can be eliminated by thorough cooking (core temperature 70 centigrade), and by prevention of contamination of cooked foods by contact with raw foods or infected food handlers. Refrigeration prevents multiplication of the cholera organism but may prolong its survival. Fruit from which the peel can be removed should also be safe.

If national authorities are concerned about the importation of any product, they are urged to consult with the World Health Organization, Food Safety unit, 1211 Geneva 27, Switzerland, or with the Pan American Health Organization, Program Coordinator, HST, 525 Twenty-Third Street, NW, Washington, DC 20037-2897, United States of America (Fax (202) 223-5971).

Countries are reminded that cholera vaccine is not recommended as a measure for prevention or control and they should not require it from persons entering or leaving infected countries. On no account should the travel of people across frontiers be restricted because of cholera. Reprinted from: World Health Organization. Weekly Epidemiological Record 1991;66:55-6. Reported by: Enteric Diseases Br, Div of Bacterial and Mycotic Diseases, Center for Infectious Diseases, CDC.

Editorial Note: In late January 1991, cholera appeared in South America for the first time this century (1). It was first identified in Peru and has now spread to Ecuador and Colombia (Figure 1). Because all three countries export food, there has been concern that food from these countries might infect consumers. WHO published the above report to respond to this concern; CDC concurs with the statements in the WHO report.

The duration of survival of Vibrio cholerae O1 in food is affected by several factors, including pH, humidity, temperature, and inoculum size. Many vibrios are needed to infect a person who has normal gastric acidity. Most imported foods, even if contaminated with V. cholerae O1 at the point of origin, pose minimal risk to the consumer. Bivalve molluscan shellfish eaten raw are the food most likely to carry V. cholerae O1 and infect consumers. In 1988, raw oysters shipped from the U.S. coast on the Gulf of Mexico caused single cases of cholera in six states (2). The U.S. Food and Drug Administration's (FDA) National Shellfish Sanitation Program does not sanction any imports of bivalve molluscan shellfish from Peru, Ecuador, and Colombia.

Cholera may spread to additional countries in South America, and a small number of U.S. residents may acquire the disease during travel or by eating imported food. Treatment of cholera is simple and highly effective, with case-fatality rates of less than 1% when proper oral and/or intravenous rehydration therapy is given. Sanitation in the United States is adequate to make the risk of continued transmission extremely small; none of the cholera cases imported into the United States since 1961 have resulted in secondary transmission (3). Sporadic cases of cholera that have occurred in the United States since 1973 have been associated with consumption of seafood from the Gulf coast; only one outbreak of cholera (on a floating oil rig) has been traced to fecal contamination (4).

In response to the situation in Peru, in mid-February, FDA substantially increased its surveillance of food imports to safeguard consumers in the United States by increasing sampling and testing of crustaceans, finfish, and produce from Peru for V. cholerae O1.

The risk of cholera to tourists is extremely low (3), and cholera vaccine is not recommended for persons traveling to affected countries. Careful selection of safe foods and beverages is paramount (5).

References
1. CDC. Cholera--Peru, 1991. MMWR 1991;40:108-10.
2. CDC. Toxigenic Vibrio cholerae O1 infection acquired in Colorado. MMWR 1989;38:19-20.
3. Snyder JD, Blake PA. Is cholera a problem for US travelers? JAMA 1982;247:2268-9.
4. Johnston JM, Martin DL, Perdue J, et al. Cholera on a Gulf Coast oil rig. N Engl J Med 1983;309:523-6.
5. CDC. Health information for international travel, 1990. Atlanta: US Department of Health and Human Services, Public Health Service, 1990:141-3; DHHS publication no. (CDC)90-8280.

* Recommended International Code of Practice for Low-Acid and Acidified Low-Acid Canned Food, Codex Alimentarius Vol. G, FAO/WHO 1983.
** Codex General Standard for Irradiated Foods, and Recommended International Code of Practice for the Operation of Radiation Facilities used for the Treatment of Foods, Codex Alimentarius Vol. XV, FAO/WHO 1984.

January 20, 1989 / 38(2);19-20

Epidemiologic Notes and Reports Toxigenic Vibrio cholerae O1 Infection Acquired in Colorado

On August 17, 1988, a 42-year-old man was treated for profuse watery diarrhea, vomiting, and dehydration at an emergency room in Rifle, Colorado. On August 15, he had eaten approximately 12 raw oysters from a new oyster-processing plant in Rifle. Approximately 36 hours after eating the oysters, he had sudden onset of symptoms and passed 20 stools during the day before seeking medical attention. Stool culture subsequently yielded toxigenic Vibrio cholerae O1, biotype El Tor, serotype Inaba. The patient had no underlying illness, was not taking medications, and had not traveled outside the region during the month before onset.

The oysters had been harvested on August 8, 1988, in a bay off the coast of Louisiana. Approximately 1000 bushels (200,000 oysters) arrived by refrigerator truck at the plant in Rifle on August 11. The patient purchased three dozen of these oysters on August 15.

During a 6-day period, eight other persons shared the oysters purchased by the patient. None became ill. Although one of seven tested had a vibriocidal antibody titer of 1:640, none had elevated antitoxic antibody titers, and none had V. cholerae O1 isolated from stool. Physicians and local health departments were asked to notify the Colorado Department of Health about similar cases, but none were reported.

The oyster-processing plant in Rifle began operation in May 1988 and functioned as a wet-storage unit. The Gulf oysters were reportedly harvested from approved waters, trucked to Colorado, and placed in recirculating disinfected artificial seawater baths for a variable number of days before packaging for market. These oysters were probably the vehicle of infection for the case of cholera. Reported by: M Doran, P Shillam, RE Hoffman, MD, State Epidemiologist, Colorado Dept of Health. LM McFarland, DrPH, Louisiana Dept of Health and Hospitals. Div of Field Svcs, Epidemiology Program Office; Enteric Diseases Br, Div of Bacterial Diseases, Center for Infectious Diseases, CDC.

Editorial Note: VcA-3 phage typing showed that the organism is identical to all others associated with an endemic focus known to have been present in the Gulf of Mexico since 1973 (1-3). This is the third reported case of toxigenic V. cholerae O1 apparently acquired from oysters shipped interstate in the United States (4) and is the first case known to have been acquired in Colorado during this century.

This report suggests that V. cholerae O1 may persist in oysters for many days after harvest. Several different Vibrio species previously have been associated with infections related to consumption of raw oysters (5). Since this case occurred, five additional oyster-related cases of cholera have been reported by five other states from August to October 1988. Thorough cooking remains he best method to prevent acquisition of infectious diseases from raw shellfish.

References
1. Blake PA, Allegra DT, Snyder JD, et al. Cholera--a possible endemic focus in the United States. N Engl J Med 1980;302:305-9. 2.CDC. Toxigenic Vibrio cholerae O1 infections--Louisiana and Florida. MMWR 1986;35:606-7. 3.CDC. Cholera in Louisiana--update. MMWR 1986;35:687-8. 4.Pavia AT, Campbell JF, Blake PA, Smith JDL, McKinley TW, Martin DL. Cholera from raw oysters shipped interstate (Letter). JAMA 1987;258:2374. 5.Blake PA. Vibrios on the half shell: what the walrus and the carpenter didn't know. Ann Intern Med 1983;99:558-9.

September 26, 1986 / 35(38);606-7

Epidemiologic Notes and Reports Toxigenic Vibrio cholerae 01 Infections -- Louisiana and Florida

Four cases of cholera acquired in Louisiana and one case acquired in Florida have been detected since mid-August 1986. All five patients were hospitalized with severe diarrhea and had stool cultures yielding toxigenic Vibrio cholerae 01, serotype Inaba.

The four Louisiana patients had onset between August 8 and September 9; they lived in New Orleans and three towns south and west of New Orleans. The single confirmed case in Florida occurred in a woman from California who arrived in Miami on August 18 and became ill with diarrhea on August 24. The patients had no known common source exposures, and the vehicles of transmission are still under investigation, but all had eaten seafood within 5 days before onset of symptoms. The Louisiana patients had eaten crabs and shrimp from multiple sites along the Louisiana coast of the Gulf of Mexico. The Florida patient had eaten raw oysters; their source is being traced. Reported by L McFarland, DrPH, Chief, Epidemiology Section, J Mathison, MD, State Epidemiologist, HB Bradford, PhD, Director of Laboratory Sciences, Louisiana Dept of Health and Human Resources; MH Wilder, MD, Acting State Epidemiologist, W Riley, PhD, Miami Regional Laboratory, C Shank, Jacksonville Central Laboratory, Florida Dept of Health and Rehabilitative Svcs; Enteric Diseases Br, Div of Bacterial Diseases, Center for Infectious Diseases, Div of Field Svcs, Epidemiology Program Office, CDC.

Editorial Note: Although the vehicles of transmission for these cases are as yet unknown, the patients' exposures to seafood have been of particular interest because seafood has been an important vehicle for cholera in several countries, including the United States. Toxigenic V. cholerae 01, serotype Inaba, biotype El Tor, appears to have an environmental reservoir on the U.S. Gulf Coast, and domestically acquired cases of cholera with identical organisms were detected in 1973, 1978, 1981, and 1984 (1-4). Almost all of the cases occurred during the summer and fall. The vehicles of transmission implicated in those instances were boiled or steamed crabs and rice that was contaminated after being boiled. V. cholerae 01 has been shown to survive in crabs boiled for 8 minutes, but not in crabs boiled for 10 minutes (1).

Toxigenic V. cholerae 01 multiplies readily in a variety of foods. Foodborne cholera is prevented by the same measures that are routinely stressed in prevention of other bacterial foodborne diseases: thorough cooking of any possibly contaminated foods, preventing contamination of foods after cooking, and storing foods that are not eaten soon after cooking at temperatures too low (below 4 C) or too high (above 60 C) to permit multiplication of the organism. Waterborne cholera is prevented by chlorination of water supplies. The disease is rarely, if ever, spread by person-to-person contact.

Cholera can be confirmed by stool culture, preferably on thiosulfate-citrate-bile salts-sucrose (TCBS) agar. If V. cholerae is isolated from stool, the isolate should be serogrouped and assayed for cholera toxin production; arrangements for this testing can be made through state public health laboratories. Cases of cholera should be reported immediately to state epidemiologists.

References
1. Blake PA, Allegra DT, Snyder JD, et al. Cholera--a possible endemic focus in the United States. New Engl J Med 1980;302:305-9.
2. Shandera WX, Hafkin B, Martin DL, et al. Persistence of cholera in the United States. Am J Trop Med Hyg 1983;32:812-7.
3. Johnston JM, Martin DL, Perdue J, et al. Cholera on a Gulf Coast oil rig. New Engl J Med 1983;309:523-6.
4. Lin FYC, Morris JG, Kaper JB, et al. Persistence of cholera in the United States: Isolation of Vibrio cholerae 01 from a patient with diarrhea in Maryland. J Clin Microbiol 1986;23:624-6.

November 07, 1986 / 35(44);687-8

Cholera in Louisiana -- Update

Since mid-August 1986, 12 cases of cholera have been identified among residents of Louisiana. The cases occurred in nine families living in New Orleans and in other towns in six parishes (Jefferson, LaFourche, Assumption, St. Mary, Iberia, and Jefferson Davis) within a 200-mile radius to the south and west of New Orleans. None of the patients had traveled abroad within the past year.

Onset of symptoms occurred between August 8 and October 1. Ten of the patients had severe diarrhea, seven required hospitalization, and four required treatment in an intensive care unit for hypotension. All patients recovered following intravenous fluid therapy. Seven patients had stool cultures yielding toxigenic Vibrio cholerae 01, biotype El Tor, serotype Inaba. The remaining five patients did not have stool cultures performed but had vibriocidal antibody titers greater than or equal to 1280, suggesting recent infection with V. cholerae 01.

Sewer system surveillance using Moore swabs has detected toxigenic V. cholerae 01 in sewage in eight separate sites in southern Louisiana (three in Jefferson Parish, one in Orleans Parish, one in St. Tammany Parish, one in Iberia Parish, and two in Jefferson Davis Parish). Five of these sites are in towns without a clinically identified case of cholera.

Although no common source has been identified, eleven of the patients reported eating crabs or shrimp within 5 days before the onset of symptoms. The seafoods were harvested from multiple sites in a wide area along the Louisiana coast of the Gulf of Mexico. Surveillance is continuing, and further epidemiologic studies are underway. Reported by L McFarland, DrPH, HB Bradford, PhD, J Mathison, MD, State Epidemiologist, Louisiana Dept of Health and Human Resources; Enteric Diseases Br, Div of Bacterial Diseases, Center for Infectious Diseases, Div of Field Svcs, Epidemiology Program Office, CDC.

Editorial Note: Thirteen cases of domestically acquired cholera (one involving a Florida patient (1)) have been detected near the U.S. Gulf coast so far during 1986. Past studies of El Tor V. cholerae infections in both endemic and non-endemic countries indicate that many mild or clinically inapparent infections occur for every hospitalized patient (2). The detection of toxigenic V. cholerae 01 in the sewer systems of several towns with no identified cases suggests that undetected cases have occurred in Louisiana.

The source of infection, as in 1978 in Louisiana (3), appears to be crustacea. Because seafood from the Gulf Coast is shipped to many states, even physicians located far from the Gulf should consider the possibility of cholera when a patient has severe, watery diarrhea. Diagnosis is confirmed by the isolation of V. cholerae 01 from stool culture, preferably on thiosulfate-citrate-bile salts-sucrose (TCBS) agar. Isolates of V. cholerae should be serotyped and tested for toxin production through state public health laboratories, and all cases should be reported immediately to the state epidemiologist.

In this outbreak, inadequate cooking or improper handling of crustacea appeared to play a significant role in the development of V. cholerae 01 infection. Thoroughly cooking potentially contaminated food and then carefully handling and storing cooked food will prevent foodborne cholera. (V. cholerae 01 has been shown to survive in crabs boiled for 8 minutes, but not in crabs boiled for 10 minutes (3)).

Vigorous rehydration (preferably with Ringer's lactate) and careful correction of electrolyte and acid-base disturbances are the mainstays of therapy and result in very low mortality rates among hospitalized patients. Tetracycline shortens the duration of symptoms and the period of fecal shedding of the organism (4).

References
1. CDC. Toxigenic Vibrio cholerae O1 infections--Louisiana and Florida. MMWR 1986;35:606-7.
2. Harris JR, Holmberg SD, Parker RDR, et al. Impact of epidemic cholera in a previously uninfected island population: evaluation of a new seroepidemiologic method. Am J Epidemiol 1986;123:424-30.
3. Blake PA, Allegra DT, Snyder JD, et al. Cholera--a possible endemic focus in the United States. N Engl J Med 1980;302:305-9.
4. Greenbough WB. In: Mandell GL, Douglas RG Jr, Bennett JE, eds. Principles and practice of infectious diseases. 2nd ed. New York: John Wiley and Sons, 1985:1208-18.

March 24, 1995 / 44(11);215-219

Update: Vibrio cholerae O1 -- Western Hemisphere, 1991-1994, and V. cholerae O139 -- Asia, 1994

The cholera epidemic caused by Vibrio cholerae O1 that began in January 1991 has continued to spread in Central and South America Figure_1. In southern Asia, the epidemic caused by the newly recognized strain V. cholerae O139 that began in late 1992 also has continued to spread Figure_2. This report updates surveillance findings for both epidemics.

From the onset of the V. cholerae O1 epidemic in January 1991 through September 1, 1994, a total of 1,041,422 cases and 9642 deaths (overall case-fatality rate: 0.9%) were reported from countries in the Western Hemisphere to the Pan American Health Organization. In 1993, the numbers of reported cases and deaths were 204,543 and 2362, respectively Table_1. From January 1 through September 1, 1994, a total of 92,845 cases and 882 deaths were reported. In 1993 and 1994, the number of reported cases decreased in some countries but continued to increase in several areas of Central America, Brazil, and Argentina (1-3).

The epidemic of cholera caused by V. cholerae O139 has affected at least 11 countries in southern Asia. V. cholerae O139 produces severe watery diarrhea and dehydration that is indistinguishable from the illness caused by V. cholerae O1 (4) and appears to be closely related to V. cholerae O1 biotype El Tor strains (5). Specific totals for numbers of V. cholerae O139 cases are unknown because affected countries do not report infections caused by O1 and O139 separately; however, greater than 100,000 cases of cholera caused by V. cholerae O139 may have occurred (6).

In the United States during 1993 and 1994, 22 and 47 cholera cases were reported to CDC, respectively. Of these, 65 (94%) were associated with foreign travel. Three of these were culture-confirmed cases of V. cholerae O139 infection in travelers to Asia. Reported by: Cholera Task Force, Diarrheal Disease Control Program, World Health Organization, Geneva. Expanded Program for the Control of Diarrheal Diseases, Special Program on Maternal and Child Health and Population, Pan American Health Organization, Washington, DC. Foodborne and Diarrheal Diseases Br, Div of Bacterial and Mycotic Diseases, National Center for Infectious Diseases, CDC.

Editorial Note: Cholera is transmitted through ingestion of fecally contaminated food and beverages. Because cholera remains epidemic in many parts of Central and South America, Asia, and Africa, health-care providers should be aware of the risk for cholera in persons traveling in cholera-affected countries -- particularly those persons who are visiting relatives or departing from the usual tourist routes because they may be more likely to consume unsafe foods and beverages.

Persons traveling in cholera-affected areas should not eat food that has not been cooked and is not hot (particularly fish and shellfish) and should drink only beverages that are carbonated or made from boiled or chlorinated water. The licensed parenteral cholera vaccine provides only limited and brief protection against V. cholerae O1, may not provide any protection against V. cholerae O139, and has a high cost-benefit ratio (7); therefore, the vaccine is not recommended for travelers (8). New oral cholera vaccines are being developed and provide more reliable protection, although still at a high cost per case averted. None of these vaccines have attained the combination of high efficacy, long duration of protection, simplicity of administration, and low cost necessary to make mass vaccination feasible in cholera-affected countries.

The diagnosis of cholera should be considered in patients with watery diarrhea who have recently (i.e., within 7 days) returned from cholera-affected countries (9). Patients with suspected cholera should be reported immediately to local and state health departments. Treatment of cholera includes rapid fluid and electrolyte replacement with adjunctive antibiotic therapy. Stool specimens should be cultured on thiosulfate-citrate-bile salts-sucrose (TCBS) agar. Clinical isolates of non-O1 V. cholerae should be referred to a state public health laboratory for testing for O139 if the patient traveled in an O139-affected area, has life-threatening dehydration typical of severe cholera, or has been linked to an outbreak of diarrhea.

References
1. CDC. Update: cholera -- Western hemisphere, 1992. MMWR 1993;42:89-
2.
3. Wilson M, Chelala C. Cholera is walking south. JAMA 1994;272:1226-7.
4. Tauxe R, Seminario L, Tapia R, Libel M. The Latin American epidemic. In: Wachsmuth I, Blake P, Olsvik O, eds. Vibrio cholerae and cholera: molecular to global perspectives. Washington, DC: ASM Press, 1994:321-44.
5. CDC. Imported cholera associated with a newly described toxigenic Vibrio cholerae O139 strain -- California, 1993. MMWR 1993;42:501-3.
6. Popovic T, Fields P, Olsvik O, et al. Molecular subtyping of toxigenic Vibrio cholerae O139 causing epidemic cholera in India and Bangladesh, 1992-1993. J Infect Dis 1995;171:122-7.
7. Cholera Working Group, International Center for Diarrheal Diseases Research, Bangladesh. Large epidemic of cholera-like disease in Bangladesh caused by Vibrio cholerae O139 synonym Bengal. Lancet 1993;342:387-90.
8. MacPherson D, Tonkin M. Cholera vaccination: a decision analysis. Can Med Assoc J 1992; 146:1947-52.
9. CDC. Cholera vaccine. MMWR 1988;37:617-8,623-4.
10. Besser RE, Feikin DR, Eberhart-Phillips JE, Mascola L, Griffin PM. Diagnosis and treatment of cholera in the United States: are we prepared? JAMA 1994;272:1203-5.

October 08, 1982 / 31(39);538-9

Epidemiologic Notes and Reports Non-O1 Vibrio cholerae Gastroenteritis--New Hampshire

In September 1981, an isolated case of non-O1 Vibrio cholerae gastroenteritis occurred in a Laconia, New Hampshire, resident following consumption of raw clams harvested from New England coastal waters. The patient was a previously healthy 40-year-old woman; her recent travel and personal-contact histories were unremarkable. Within 26 hours after eating the clams, she developed acute abdominal cramps, followed by fever and bloody diarrhea. She was treated symptomatically with rest and oral hydration and recovered without sequelae. Her stool culture grew V. cholerae (Smith serotype 361) and no other enteric pathogens. Studies for production of heat-labile and heat-stable toxins were negative. The asymptomatic family members had also eaten the clams; their stool cultures grew only normal flora. Subsequent cultures of shellfish harvested from the same coastal area were negative for vibrio organisms.

The market where the clams were purchased provided names of eight restaurants it routinely supplies, none of which reported any recent gastrointestinal illness among customers or employees. A retrospective review of hospital emergency-room records identified 36 other patients who had presented with gastrointestinal symptoms during the week the index case occurred. Only one had had a stool culture, which grew Campylobacter jejuni. All patients were sent food-history questionnaires; none of the 14 respondents reported eating raw shellfish before onset of symptoms or eating at any of the restaurants supplied by the market. After the index case was reported, prospective surveillance

189

was initiated for patients presenting with diarrheal disease at the local emergency room and at a regional medical clinic. Stool cultures were obtained from these patients and screened on thiosulfate-citrate bile salts sucrose (TCBS) agar, a selective medium for vibrio species. No further cases were identified.

This represents the first reported case of non-O1 V. cholerae gastroenteritis apparently caused by shellfish from New England waters. Reported by S MacRae, Dept of Microbiology, T Clements, Dept of Infection Control, Lakes Region General Hospital, J Cournoyer, New Hampshire State Dept of Health and Welfare; Field Svcs Div, Epidemiology Program Office, CDC.

Editorial Note: Isolated cases of non-O1 V. cholerae gastroenteritis have been reported previously in the United States (1,2), and outbreaks of intestinal illness caused by this organism have occurred elsewhere (3-5). Investigations of recent isolated cases in the United States have demonstrated a statistically significant association between eating raw shellfish and development of disease (1,2). Most of these cases have been associated with Gulf Coast oysters.

Environmental studies have demonstrated that the organisms can be found in brackish surface waters and are more numerous during warmer summer months (1). A Food and Drug Administration study of 790 samples of randomly selected oysters collected between June 1979 and May 1980 revealed non-O1 V. cholerae in 111 samples (14%) (6). Some investigators have demonstrated an association between fecal contamination of water and presence of the organism (1), but non-O1 V. cholerae has been found in waters free of fecal contamination and thus may be a constituent of normal marine flora (7).

Although no outbreaks of illness due to this organism have been reported in the United States, it is possible that common-source exposures have occurred in which milder cases have gone undetected and unreported. Non-O1 V. cholerae should be included in the differential diagnosis of acute gastroenteritis following ingestion of raw seafood. Diagnosis can be facilitated by culture of stool specimens on TCBS medium.

References
1. Wilson R, Lieb S, Roberts A, et al. Non-O group 1 Vibrio cholerae gastroenteritis associated with eating raw oysters. Am J Epidemiol 1981;114:293-8.
2. Morris JG Jr, Wilson R, Davis BR, et al. Non-O group 1 Vibrio cholerae gastroenteritis in the United States: clinical, epidemiologic, and laboratory characteristics of sporadic cases. Ann Intern Med 1981;94:656-8.
3. Aldova E, Laznickova K, Stepankova E, Leitava J. Isolation of nonagglutinable vibrios from an enteritis outbreak in Czechoslovakia. J Infect Dis 1968;118:25-31.
4. Dakin WP, Howell DJ, Sutton RG, O'Keefe MF, Thomas P. Gastroenteritis due to non-agglutinable (non-cholera) vibrios. Med J Aust 1974;2:487-90.
5. CDC. Outbreak of Vibrio cholerae non O-1 gastroenteritis--Italy. MMWR 1981;30:374-5.
6. Twedt RM, Madden JM, Hunt JM, et al. Characterization of Vibrio cholerae isolated from oysters. Appl Environ Microbiol 1981;41:1475-8.
7. Colwell RR, Kaper J, Joseph SW. Vibrio cholerae, Vibrio parahaemolyticus, and other vibrios: occurrence and distribution in Chesapeake Bay. Science 1977;198:394-6.

January 29, 1999 / 48(03);48-51

Outbreak of Vibrio parahaemolyticus Infection Associated with Eating Raw Oysters and Clams Harvested from Long Island Sound -- Connecticut, New Jersey, and New York, 1998

During July-September 1998, an outbreak of Vibrio parahaemolyticus infections associated with consumption of oysters and clams harvested from Long Island Sound occurred among residents of Connecticut, New Jersey, and New York. This is the first reported outbreak of V. parahaemolyticus linked to consumption of shellfish harvested from New York waters. This report summarizes the investigation of this outbreak.

On August 10, 1998, a New York City resident with toxigenic V. cholerae O1 infection who had not traveled recently was reported to the New York City Department of Health (NYCDOH). NYCDOH initiated an investigation to determine the most likely source of the infection. Using a broadcast facsimile, NYCDOH contacted all Queens County laboratories on August 12 and, on August 26, asked selected infectious diseases physicians and all New York City hospitals and laboratories to consider V. cholerae as a potential cause of diarrhea and to report any confirmed or suspected Vibrio infections to the NYCDOH. Although no additional V. cholerae infections were reported, 23 culture-confirmed cases of V. parahaemolyticus were reported among residents of Connecticut, New Jersey, and New York. Dates of illness onset ranged from July 21 through September 17 (Figure_1).

An investigation coordinated by the New York State Department of Health determined that 22 of 23 ill persons had eaten or handled oysters, clams, or crustaceans: 16 ate raw oysters or clams, two ate steamed crabs, one ate crab cakes, one ate boiled crabs and lobsters, one ate lobster roll, and one handled live crabs. The median onset of illness following consumption of shellfish was 19 hours (range: 12-52 hours). Clinical histories were available for 19 of the 23 ill persons; 17 (89%) had gastroenteritis and two (11%) had bloodstream infections with lower extremity edema and bullae. Among patients with gastroenteritis, reported clinical symptoms included diarrhea (100%), abdominal cramps (94%), nausea (94%), vomiting (82%), fever (47%), bloody stools (29%), headache (24%), and myalgia (24%). Median duration of gastrointestinal illness was 5 days.

Traceback investigations by local and state health departments identified the site of harvest for oysters or clams eaten by 11 of the 16 patients. Oysters or clams eaten by eight patients were harvested from Oyster Bay, off New York's Long Island Sound, during August 4-27. Shellfish tags from oysters and clams eaten by the other three persons indicated harvest areas elsewhere off Long Island or, in one case, Washington state (1) *.

During the outbreak period, mean surface water temperature measurements from 15 Oyster Bay stations was 77.2 F (25.1 C), compared with cooler 1997 and 1996 measurements (74.1 F {23.4 C} and 69.4 F {20.7 C}, respectively). On September 10, the New York State Department of Environmental Conservation (NYSDEC) closed Oyster Bay to harvesting of shellfish and recalled shellfish harvested from that area after August 10.

Laboratory testing of 12 V. parahaemolyticus clinical isolates, including the eight traced to Oyster Bay, identified O3:K6 serotype. Pulsed-field gel electrophoresis (PFGE) performed on four clinical isolates at the New York City Bureau of Labs indicated that three isolates epidemiologically linked to Oyster Bay had indistinguishable PFGE patterns, and the other isolate not linked to Oyster Bay had a distinctly different pattern. Oysters harvested on five occasions from Oyster Bay during September 11-October 14 contained V. parahaemolyticus at less than or equal to 120 colony forming units {cfu} per gram of oyster meat. None of these environmental isolates matched the outbreak strain or other clinical isolates by PFGE. On the basis of these results and a decline in water temperature to 63.5 F (17.5 C), NYSDEC reopened Oyster Bay to commercial shellfish harvesting on October 22. No additional culture-confirmed cases of V. parahaemolyticus infection have been reported.

Reported by: E Wechsler, C D'Aleo, VA Hill, J Hopper, D Myers-Wiley, E O'Keeffe, J Jacobs, F Guido, A Huang, MD, Westchester County Health Dept, New Rochelle; SN Dodt, B Rowan, M Sherman, A Greenberg, MD, Div of Disease Control, Nassau County Dept of Health, Mineola; D Schneider, B Noone, L Fanella, BR Williamson, E Dinda, M Mayer, MD, Suffolk County Dept of Health Svcs, Hauppauge; M Backer, A Agasan, MD, Enteric Pathogens Laboratory, L Kornstein, PhD, Environmental Microbiology Laboratory, New York City Bur of Laboratories; F Stavinsky, Bur of Environmental Investigations; B Neal, D Edwards, M Haroon, D Hurley, L Colbert, J Miller, MD, B Mojica, MD, New York City Dept of Health; E Carloni, B Devine, M Cambridge, Bur of Community Sanitation and Food Protection; T Root, D Schoonmaker, M Shayegani, Wadsworth Laboratories, Albany; W Hastback, New York State Dept of Environmental Conservation; B Wallace, MD, S Kondracki, Bur of Communicable Disease Control; P Smith, MD, State Epidemiologist, New York State Dept of Health. S Matiuck, K Pilot, M Acharya, Bur of Labs; G Wolf, W Manley, C Genese, J Brooks, MD, Acting State Epidemiologist, New Jersey Dept of Health. Z Dembek, PhD, J Hadler, MD, State Epidemiologist, Connecticut Dept of Public Health. Center for Food Safety and Applied Nutrition, Food and Drug Administration. Fish and Wildlife Svc, US Dept of Agriculture. Foodborne and Diarrheal Diseases Br, Div of Bacterial and Mycotic Diseases, National Center for Infectious Diseases; State Br, Div of Applied Public Health Training, Epidemiology Program Office; and EIS officers, CDC.

Editorial Note: This is the fourth multistate outbreak of V. parahaemolyticus infections in the United States since 1997, and the first associated with shellfish harvested from the northeast Atlantic Ocean. Before 1997, foodborne outbreaks caused by V. parahaemolyticus had been infrequently reported in the United States (1). During 1997-1998, multistate outbreaks of V. parahaemolyticus were associated with consumption of raw or undercooked oysters harvested from the Pacific Northwest and Texas (2; CDC, unpublished data, 1998).

V. parahaemolyticus is a halophilic, gram-negative bacterium that naturally inhabits marine and estuarine waters. V. parahaemolyticus infections are usually acquired by persons who eat raw or undercooked shellfish, particularly oysters, or whose skin wounds are exposed to warm seawater. The most common clinical manifestation of infection is self-limited gastroenteritis, but infections may result in septicemia that can be life threatening (3,4). The concentration of V. parahaemolyticus in seawater increases with increasing water temperature and corresponds with a seasonal increase in sporadically occurring cases in warmer months (4). This outbreak and the recent outbreaks of V. parahaemolyticus infections in the Pacific Northwest and Texas occurred during summer months.

To reduce the risk for V. parahaemolyticus and other shellfish-associated infections, persons should avoid eating raw or undercooked shellfish, particularly during warmer months. Monitoring of environmental conditions, such as water temperature and salinity, may help determine when shellfish harvesting areas should be closed and re-opened to harvesting.

Guidelines regulating the harvesting of oysters and clams rely on quantitative measurement of V. parahaemolyticus levels in oyster or clam meat. However, data from recent outbreaks may require revision of these guidelines. The recommended action level of V. parahaemolyticus per gram of oyster meat that must be detected in the absence of human illness before closing oyster beds is greater than 10,000 cfu/g. Oyster samples that were harvested from implicated beds in the Pacific Northwest in 1997 and Oyster Bay in 1998 yielded less than 200 V. parahaemolyticus cfu/g of oyster meat, indicating that human illness can occur at levels much lower than the current action level.

Infection with V. parahaemolyticus is not a notifiable condition in most states, including New York. This outbreak was detected only coincidentally because of enhanced surveillance during an investigation of a case of V. cholerae O1. Health-care providers treating patients with gastroenteritis who have a history of recent ingestion of raw or undercooked shellfish should consider Vibrio infection and request a stool culture specifically for Vibrio. Clinical laboratories should use thiosulfate-citrate-bile salts-sucrose agar (TCBS), a selective medium for culturing for Vibrio spp., when culturing stool specimens for Vibrio and should consider using TCBS for routine screening of all stools specimens, at least during summer months.

CDC coordinates a passive Gulf Coast Vibrio surveillance system and the Foodborne Diseases Active Surveillance Network (FoodNet) to monitor the incidence of Vibrio infections. Because of these multistate outbreaks, all states should consider making infections with V. parahaemolyticus and other vibrioses reportable, with referral of clinical isolates to public health laboratories for confirmation and strain subtyping.

References
1. Bean NH, Goulding JS, Lao C, Angulo FJ. Surveillance for foodborne-disease outbreaks -- United States, 1988-1992. MMWR 1996;45(no. SS-5).
2. CDC. Outbreak of Vibrio parahaemolyticus infections associated with eating raw oysters -- Pacific Northwest, 1997. MMWR 1998;47;457-62.
3. Levine WC, Griffin PM, and the Gulf Coast Vibrio Working Group. Vibrio infections on the Gulf Coast: results of first year of regional surveillance. J Infect Dis 1993;167:479-83.
4. Hlady WG, Klontz KC. The epidemiology of Vibrio infections in Florida, 1981-1993. J Infect Dis 1996;173:1176-83.

The shipper that provided the oysters harvested elsewhere in Long Island also had received oysters from Oyster Bay at approximately the same time. Although comingling of shellfish is against state regulations, it is known to occur.

June 12, 1998 / 47(22);457-462

Outbreak of Vibrio parahaemolyticus Infections Associated with Eating Raw Oysters -- Pacific Northwest, 1997

During July-August 1997, the largest reported outbreak in North America of culture-confirmed Vibrio parahaemolyticus infections occurred. Illness in 209 persons was associated with eating raw oysters harvested from California, Oregon, and Washington in the United States and from British Columbia (BC) in Canada; one person died. This report summarizes the investigations of the outbreak, which suggest that elevated water temperatures may have contributed to increased cases of illness and highlights the need for enhanced surveillance for human infections.

BRITISH COLUMBIA

During July 1-19, the BC Provincial Laboratory received isolates of V. parahaemolyticus from nine patients, more than twice the expected number for July. Because of the high number of isolates identified, the BC Center for Disease Control (BCCDC) conducted interviews with the eight patients who could be contacted; seven had eaten raw oysters during the 24 hours before illness onset, and one had eaten crabs. On July 30, the BC Ministry of Health (BCMOH) issued a public health alert advising that molluscan shellfish (e.g., oysters, clams, mussels, and scallops) should not be eaten raw or undercooked. On July 31, the Vancouver/Richmond Health Board banned the sale of raw molluscan shellfish in restaurants in the cities of Vancouver and Richmond, BC. These actions were followed by a rapid decline in the number of new cases. On August 19, the Federal Department of Fisheries and Oceans (DFO) closed all BC coastal waters to the harvesting of oysters.

The BCMOH continued to interview BC residents with culture-confirmed V. parahaemolyticus infections; information was obtained from 42 of the 51 persons with illness reported during July 1-September 26. Of the 42, a total of 39 (93%) had eaten molluscan shellfish and 35 (83%) had eaten raw or undercooked oysters during the 4 days before onset of illness; 28 had eaten oysters purchased at restaurants or other food establishments in BC; and seven had eaten oysters they had harvested. Oysters eaten by ill persons were traced by BCCDC, the Canadian Food Inspection Agency (CFIA), and BCMOH to harvesting areas along the BC coast. Samples of oysters harvested from these areas contained multiple V. parahaemolyticus serotypes at less than 200 colony-forming units (CFU) per gram of oyster tissue. No additional outbreak-related illnesses were reported in BC residents after DFO closed the coastal waters to the harvesting of oysters. The closure remained in effect until September 12, after which no additional cases were reported.

WASHINGTON

On July 18, on the basis of reports of illness received from local health departments and from ill persons, the Washington Department of Health (WDOH) issued an advisory that persons eat only thoroughly cooked oysters. On August 14, after additional cases had been reported, the WDOH advised commercial harvesters to refrigerate oysters within 4 hours after harvesting, and on August 20, advised the public to thoroughly cook molluscan shellfish from both commercial and noncommercial sources. On August 23, the U.S. Food and Drug Administration (FDA) also issued a statement regarding proper procedures for cooking oysters (1).

WDOH interviewed 54 of the 56 persons who had culture-confirmed V. parahaemolyticus during May 26-September 9. Of the 54, a total of 48 (89%) had eaten molluscan shellfish before becoming ill; 42 (88%) reported eating oysters. Product traceback by the WDOH's Shellfish Program determined that 35 case-patients had eaten molluscan shellfish harvested in Washington. On August 20, members of the Pacific Coast Oyster Growers Association voluntarily halted shipments of shell oysters from Washington, and on August 28, WDOH closed oyster beds in major shellfish harvesting areas. The oyster beds were reopened on September 15, and no additional illnesses were reported.

OREGON

On August 21, the Oregon Health Division (OHD) requested that local county health departments and microbiology laboratories provide immediate notification of illnesses associated with or isolations of V. parahaemolyticus. The request was prompted by an increased number of V. parahaemolyticus cases detected by the Foodborne Disease Active Surveillance Network (FoodNet) (a collaboration between CDC, the U.S. Department of Agriculture, FDA, and seven states for surveillance of foodborne diseases and related epidemiologic studies) and simultaneous reports from BC and Washington of a V. parahaemolyticus outbreak associated with eating raw or undercooked shellfish.

OHD interviewed the 13 persons reported with culture-confirmed V. parahaemolyticus infections with onsets during July 19-September 27. Twelve had eaten molluscan shellfish; 10 (77%) had eaten raw oysters. Traceback of the oysters that had been eaten indicated they had been harvested in waters near BC (four cases), Washington (four), Oregon (one), and California (one). On August 26, the implicated oyster harvest bed in Oregon was closed by the Oregon Department of Agriculture; only oysters to be cooked could be harvested. On August 28, OHD, in conjunction with the Food Safety Division of the Oregon Department of Agriculture, issued a press release warning persons not to eat raw molluscan shellfish harvested along the Pacific Northwest coast.

After closure of the implicated oyster harvest bed in Oregon, no additional cases associated with eating raw oysters harvested from Oregon waters were reported. The sale of oysters to be eaten raw was reestablished on September 30.

CALIFORNIA
During May-July, the City and County of San Francisco Department of Public Health reported 11 culture-confirmed V. parahaemolyticus infections to the California Department of Health Services (CDHS). On the basis of these cases, on August 18, San Francisco health officials issued a health advisory recommending that persons not eat raw shellfish and advising restaurants not to serve raw oysters, clams, or mussels. On August 19, CDHS issued a warning about eating raw oysters, clams, and mussels harvested off the coasts of BC and Washington. CDHS interviewed each of the 83 persons reported with culture-confirmed V. parahaemolyticus infections with onset during June 9-December 9. Of the 83, a total of 68 (82%) reported eating oysters during the week before onset of illness. Although 59 persons ate oysters identified through traceback as having been harvested off the coast of Washington and BC, nine persons with culture-confirmed illness ate oysters harvested from Tomales Bay, California (40 miles north of San Francisco).

SUMMARY FINDINGS
During July 20-August 24, culture-confirmed cases of V. parahaemolyticus infections associated with eating shellfish harvested from Washington or BC also were reported to the state health departments of Utah (three), Alaska (one), Maryland (one), and Hawaii (one). A total of 209 culture-confirmed V. parahaemolyticus infections were reported throughout North America during this outbreak. Dates of illness onset ranged from May 26 through December 9 (median: August 8) (Figure_1). V. parahaemolyticus isolates from ill persons included many different serotypes, some of which matched serotypes found in oysters. The median age of patients was 39 years (range: 12-85 years); 141 (67%) were male. Clinical histories were available for 196 persons with culture-confirmed infection: 194 (99%) reported diarrhea; 172 (88%), abdominal cramps; 101 (52%), nausea; 77 (39%), vomiting; 64 (33%), fever; and 24 (12%), bloody diarrhea. Of 137 persons providing information on underlying illnesses, 17 (12%) reported an underlying illness. Two patients were hospitalized; one with V. parahaemolyticus isolated from her bloodstream died.

Mean Pacific coastal sea surface temperatures recorded by the U.S. Navy ranged from 54 F-66 F (12 C-19 C) during May 13-September 9, 1997 (B. McKenzie, U.S. Navy, personal communication, 1998). These temperatures were 2 F-9 F (1 C-5 C) above temperatures from the same period in 1996.

Oysters from implicated harvest sites contained V. parahaemolyticus, but the number of organisms per gram was often less than 200 CFU. The highest levels were greater than 11,000 CFU in samples tested by CFIA.

Reported by: M Fyfe, MD, Communicable Disease Epidemiology; MT Kelly, MD, Provincial Laboratory, British Columbia Center for Disease Control; ST Yeung, MBBS, Field Epidemiology Training Program, Health Canada; P Daly, MD, Vancouver/Richmond Health Board; K Schallie, Canadian Food Inspection Agency; S Buchanan, Food Protection Programs, British Columbia Ministry of Health. P Waller, MS; J Kobayashi, MD, Communicable Disease Epidemiologist; N Therien, MPH, M Guichard, MS, S Lankford, Public Health Laboratories; P Stehr-Green, DrPH, State Epidemiologist, Washington Dept of Health. R Harsch, MD, Oregon Health Sciences Univ, Portland; E DeBess, DVM, M Cassidy, T McGivern, S Mauvais, D Fleming, MD, State

Epidemiologist, State Health Div, Oregon Dept of Human Resources. M Lippmann, Communicable Disease Control Unit; L Pong, Environmental Health Management Section, City and County of San Francisco Dept of Public Health. RW McKay, Food Safety Div, Dept of Agriculture; DE Cannon, Environmental Health, Shellfish Program; SB Werner, MD; S Abbott, Div of Communicable Disease Control; M Hernandez, C Wojee, J Waddell, Div of Food, Drug, and Radiation Safety, S Waterman, MD, State Epidemiologist, California Dept of Health Svcs. J Middaugh, MD, State Epidemiologist, State of Alaska Dept of Health and Social Svcs. D Sasaki, DVM, Epidemiology Br, P Effler, MD, State Epidemiologist, Hawaii Dept of Health. C Groves, MS, N Curtis, Maryland State Epidemiology and Disease Control, D Dwyer, MD, State Epidemiologist, Maryland State Dept of Health and Mental Hygiene. G Dowdle, MSPH, Communicable Disease Control, C Nichols, MPA, State Epidemiologist, Utah Dept of Health. Center for Food Safety and Applied Nutrition, US Food and Drug Administration. Foodborne and Diarrheal Diseases Br, Div of Bacterial and Mycotic Diseases, National Center for Infectious Diseases, CDC.

Editorial Note: The last large outbreak of V. parahaemolyticus infections reported in North America occurred in 1982 and resulted in 10 culture-confirmed cases. Although V. parahaemolyticus outbreaks are rare, sporadic cases are not infrequent. Most infections are associated with ingestion of raw or undercooked shellfish harvested from both the Gulf of Mexico and the Pacific Ocean.

V. parahaemolyticus is a gram-negative bacterium that naturally inhabits U.S. and Canadian coastal waters and is found in higher concentrations during the summer (2,3). The outbreak described in this report may have been associated with elevated water temperatures. Because V. parahaemolyticus concentrations in oysters and shellfish increase with warmer temperatures, enhanced surveillance at the beginning of summer may lead to earlier recognition and appropriate public health action. Water temperature monitoring may help determine when oyster beds should be closed to harvesting to prevent further outbreaks (4).

Epidemiologic and microbiologic studies conducted during this outbreak primarily implicated eating raw oysters. On the basis of studies suggesting that the infectious dose of V. parahaemolyticus might be greater than or equal to 100,000 CFU (5), the United States and Canada allow the sale of oysters if there are less than 10,000 CFU of V. parahaemolyticus per gram of oyster. However, adherence to these guidelines did not prevent this outbreak. Closure of implicated shellfish beds by health officials was useful; in Canada, additional human illness rapidly declined following a federally mandated suspension of harvesting of shellfish from BC waters in September. In the United States, shellfish-associated infections continued to occur into December.

The mean incubation period for V. parahaemolyticus is 15 hours (range: 4-96 hours). In immunocompetent persons, V. parahaemolyticus causes a mild to moderate gastroenteritis with a mean duration of illness of 3 days. Infection can cause serious illness in persons with underlying disease (e.g., persons who use alcohol excessively or have diabetes, pre-existing liver disease, iron overload states, compromised immune systems, or gastrointestinal problems) (2,6). During this outbreak, most ill persons had no underlying illness. To reduce the risk for V. parahaemolyticus and other shellfish-associated infections, persons should avoid eating raw or undercooked shellfish. If persons who eat raw or undercooked shellfish develop gastroenteritis within 4 days of ingestion, they should consult a health-care provider and request a stool culture. Only three states (California, Florida, and Louisiana) require visible posting of alerts regarding the risks associated with eating raw oysters at point of retail sale (2,7,8). Although assessment of these regulatory educational strategies have indicated compliance is variable (7), other states might consider posting such alerts.

V. parahaemolyticus is not a reportable disease in all states. During this outbreak, public health officials in Washington and California and in BC promptly became aware of the outbreak through routine reporting; in Oregon, although V. parahaemolyticus is not reportable, the outbreak was detected through an active surveillance program. All states should consider making V. parahaemolyticus and other vibrioses reportable; standard forms are available from CDC's Foodborne and Diarrheal Diseases Branch, Division of Bacterial and Mycotic Diseases, National Center for Infectious Diseases, telephone (404) 639-2206; fax (404) 639-2205.

References
1. US Department of Health and Human Services. HHS news: statement advising consumers about oysters from the Pacific. World-Wide Web site http://vm.cfsan.fda.gov/~lrd/hhsoyst.html. Accessed

June 8, 1998.

2. Hlady WG, Klontz KC. The epidemiology of Vibrio infections in Florida, 1981-1993. J Infect Dis 1996;173:1176-83.

3. Morris JG Jr, Black RE. Cholera and other vibrioses in the United States. N Engl J Med 1985;312:343-9.

4. Shapiro R, Altekruse S, Hutwagner L, et al. The role of Gulf Coast oysters harvested in warmer months in Vibrio vulnificus infections in the United States, 1988-1996. J Infect Dis 1998 (in press).

5. Sanyal SC, Sen PC. Human volunteer study on the pathogenicity of Vibrio parahaemolyticus. In: Fujino T, Sakaguchi G, Sakazaki R, Takeda Y, eds. International Symposium on Vibrio para-haemolyticus. Tokyo, Japan: Saikon Publishing Co., Ltd., 1974:227-30.

6. Blake PA, Merson MH, Weaver RE, Hollis DG, Heublein PC. Disease caused by a marine vibrio: clinical characteristics and epidemiology. N Engl J Med 1979;300:1-5.

7. Mouzin E, Mascola L, Tormey MP, Dassey DE. Prevention of Vibrio vulnificus infections: assessment of regulatory educational strategies. JAMA 1997;278:576-8.

8. CDC. Vibrio vulnificus infections associated with raw oyster consumption -- Florida, 1981-1992. MMWR 1993;42:405-7.

July 26, 1996 / 45(29);621-624

Vibrio vulnificus Infections Associated with Eating Raw Oysters -- Los Angeles, 1996

Of all foodborne infectious diseases, infection with Vibrio vulnificus is one of the most severe; the case-fatality rate for V. vulnificus septicemia exceeds 50% (1,2). In immunocompromised hosts, V. vulnificus infection can cause fever, nausea, myalgia, and abdominal cramps 24-48 hours after eating contaminated food; because the organism can cross the intestinal mucosa rapidly, sepsis and cutaneous bullae can occur within 36 hours of the initial onset of symptoms. Cases are most commonly reported during warm-weather months (April-November), and often are associated with eating raw oysters. During April 1993-May 1996, a total of 16 cases of V. vulnificus infection were reported in Los Angeles County. Fifteen (94%) of these patients were primarily Spanish-speaking, 12 (75%) had preexisting liver disease (associated with alcohol use or viral hepatitis), all were septicemic, and all had eaten raw oysters 1-2 days before onset of symptoms. In May 1996, three deaths related to V. vulnificus infection among primarily Spanish-speaking persons were reported to the Los Angeles County Department of Health Services (LACDHS). This report summarizes the findings of the investigations of these fatal cases and illustrates the importance of prevention strategies for persons with preexisting liver disease. Case Investigations

Case 1. On May 1, 1996, a 38-year-old man had onset of fever, chills, nausea, and myalgia. On April 29, he had eaten at home raw oysters purchased from a retail store. On May 2, he was admitted to a hospital because of a fever of 102 F (39 C) and two circular necrotic lesions on the left leg. He reported a history of regular beer consumption (36-72 oz per day) and insulin-dependent diabetes. Sepsis and possible deep-vein thrombosis were diagnosed, and the patient was transferred to the intensive-care unit (ICU). In the ICU, therapy was initiated with ticarcillin/clavulanic acid, gentamicin, vancomycin, and ceftazidime. On May 3, V. vulnificus was isolated from the blood sample obtained from the patient on admission, and ciprofloxacin was added to his therapy. On May 4, he died. Traceback of the oysters by environmental health inspectors indicated they originated from a lot harvested in Galveston Bay, Texas, on April 27.

Case 2. On May 10, a 46-year-old man had onset of fever, sweats, and nausea. On May 9, he had eaten at home raw oysters purchased from a retail store. On May 11, he was admitted to a hospital because of a fever of 101.5 F (38.5 C), jaundice, and ascites. He reported a history of heavy alcohol use (72 oz of beer per day) and alcoholic liver disease; in 1995, he had had jaundice for 1 month and had cirrhosis diagnosed. In the hospital, sepsis of unknown etiology was diagnosed, and he was transferred to the ICU; therapy was initiated with piperacillin and gentamicin. On May 12, he died. V. vulnificus was isolated from samples of blood and peritoneal fluid obtained on admission. Traceback of the oysters by environmental health inspectors indicated they originated from a lot harvested in Galveston Bay on May 4; however, harvesters associated with case 1 were different from those for case 2.

Case 3. On May 20, a 51-year-old woman had onset of fever, nausea, and muscle aches. On May 19, she had eaten raw oysters served at a party. On May 21, she was admitted to a hospital because of a fever of 105 F (40.5 C) and bilateral leg cellulitis. In 1982, she had had breast cancer diagnosed and in 1986, chronic hepatitis C. Following the cellulitis, hemorrhagic bullous lesions developed, then septic shock, and the patient was transferred to the ICU. Therapy was initiated with ticarcillin/clavulanic acid and one dose each of ciprofloxacin and doxycycline. On May 22, she died. V. vulnificus was isolated from blood and wound cultures obtained on admission. Traceback of the oysters by environmental health inspectors indicated they originated from a lot harvested in Eloi Bay, Louisiana, on May 14. Follow-Up Investigation

During the investigation of cases 1-3, no implicated oysters were available for analysis. Because V. vulnificus is present in up to 50% of oyster beds with the water conditions that prevail in the Gulf of Mexico during warm months (i.e., temperature greater than 68 F {greater than 20 C} and salinity of less than 16 parts per thousand) (3), no oysters from these waters were obtained for analysis following the tracebacks. Other than ingestion of oysters, no other known source of exposure to V. vulnificus (e.g., ingestion of other raw shellfish or skin exposure to seawater or shellfish) was identified for the three case-patients, and no cases of V. vulnificus-associated illness were identified among the persons who ate raw oysters with the case-patients.

As a result of these three deaths, LACDHS initiated an educational campaign to inform health-care providers and public health professionals about prevention of V. vulnificus infection. Brochures published in English and Spanish also were distributed to immunocompromised persons, including persons with liver disease, to warn them about the hazards of eating raw shellfish.

Reported by: L Mascola, MD, M Tormey, MPH, D Dassey, MD, Acute Communicable Disease Control, L Kilman, S Harvey, PhD, Public Health Laboratory, A Medina, A Tilzer, Consumer Product Div, Food and Milk Inspection Program, Los Angeles County Dept of Health Svcs; S Waterman, MD, State Epidemiologist, California State Dept of Health Svcs. Foodborne and Diarrheal Diseases Br, Div of Bacterial and Mycotic Diseases, National Center for Infectious Diseases; State Br, Div of Applied Public Health Training (proposed), Epidemiology Program Office, CDC.

Editorial Note: V. vulnificus is a gram-negative bacterium that causes septicemia, wound infections, and gastroenteritis. Transmission occurs through ingestion of contaminated raw or undercooked seafood, especially raw oysters, or through contamination of a wound by seawater or seafood drippings. Persons with liver disease are at particularly high risk for fatal septicemia following ingestion of contaminated seafood; immunocompromised persons also are at increased risk (1,4,5).

The findings in this report suggest that these three fatal cases of V. vulnificus infection were associated with eating contaminated raw oysters. Three factors support this conclusion: 1) V. vulnificus infection previously has been associated only with sea-water, brackish water, or shellfish; 2) ingestion of raw oysters was the only known source of exposure for these three cases; and 3) the implicated oysters were harvested in waters in which V. vulnificus is commonly present during warm months.

Although there is no national surveillance system for V. vulnificus infections, the Gulf Coast states, in collaboration with CDC, conduct regional Vibrio surveillance; Alabama, Florida, Louisiana, and Texas have participated since 1988 and Mississippi, since 1989. From 1988 through 1995, CDC received reports of 302 V. vulnificus infections from the Gulf Coast states; of these, 141 (47%) were associated with eating contaminated seafood, 128 (42%) with wound infections, and 33 (11%) with unknown sources. Of the 141 persons with V. vulnificus infections associated with ingestion, 136 (96%) had eaten raw oysters. Among the 242 persons for whom outcome was known, 86 (36%) died (CDC, unpublished data, 1996).

V. vulnificus thrives in warm sea water (3). The organism is frequently isolated from shellfish from the Gulf of Mexico (3) and from shellfish harvested from U.S. Pacific (6) and Atlantic (7) coastal waters. Although oysters can be harvested legally only from waters devoid of fecal contamination, even legally harvested oysters can be contaminated with V. vulnificus because the bacterium is naturally present in marine environments. V. vulnificus contamination does not alter the appearance, taste, or odor of oysters. Regulations in California and other states requiring oyster lot tagging, labeling, and record retention have facilitated traceback investigations. From 1990 through 1995, the Food and Drug Administration (FDA) and state officials traced oysters eaten by 26 patients who acquired V. vulnificus infections in states outside the Gulf Coast region; among oysters that could be traced to the harvest site (19 cases), all had been harvested in the Gulf of Mexico (FDA, unpublished data, 1996). Timely, voluntary reporting of V. vulnificus infections to CDC and regional FDA shellfish specialists enhances ongoing collaborative efforts to improve investigation and control of these infections. Regional FDA specialists with expert knowledge about shellfish assist state officials with tracebacks of shellfish and, when notified rapidly about cases, are often able to identify and sample harvest waters.

In California, Florida, and Louisiana, warning notices are required to be posted at sites of raw oyster sales. However, these states do not require notices in languages other than English; this policy may decrease the effectiveness of warning notices in areas such as Los Angeles where use of languages other than English is common. For example, the three persons described in this report were fluent in Spanish and spoke English as a second language. Information about consumption of raw oysters is available 24 hours a day in English and Spanish from FDA's Seafood Hotline, telephone (800) 332-4010 or (202) 205-4314.

Because of the high case-fatality rate of V. vulnificus infections in persons with preexisting liver disease or immunocompromising conditions, these persons especially should be informed about the health hazards associated with consumption of raw or undercooked seafood, particularly oysters (2,8,9); the need to avoid contact with sea water during the warm months; and the importance of using protective clothing (e.g., gloves) when handling shellfish (8). Health-care providers should consider V. vulnificus infection in the differential diagnosis of fever of unknown etiology. In addition, providers should ask about a history of raw oyster ingestion or sea water contact when persons with preexisting liver disease or immunocompromising conditions present with fever (especially when bullae, cellulitis, or wound infection is also present) and should promptly administer appropriate antibiotic therapy (tetracycline or a third-generation cephalosporin {e.g., ceftazidime or cefotaxime}) when indicated.

References
1. Tacket CO, Brenner F, Blake PA. Clinical features and an epidemiological study of Vibrio vulnificus infections. J Infect Dis 1984;149:558-61.
2. CDC. Vibrio vulnificus infections associated with raw oyster consumption -- Florida, 1981-1992. MMWR 1993;42:405-7.
3. Kelly MT. Effect of temperature and salinity on Vibrio (Beneckea) vulnificus occurrence in a Gulf Coast environment. Appl Environ Microbiol 1982;44:820-4.
4. Johnston JM, Becker SF, McFarland LM. Vibrio vulnificus: man and the sea. JAMA 1985;253: 2850-3.
5. Blake PA, Merson MH, Weaver RE, Hollis DG, Heublein PC. Disease caused by a marine Vibrio: clinical characteristics and epidemiology. N Engl J Med 1979;300:1-5.
6. Kelly MT, Stroh EM. Occurrence of Vibrionaceae in natural and cultivated oyster populations in the Pacific Northwest. Diagn Microbiol Infect Dis 1988;9:1-5.
7. Tilton RC, Ryan RW. Clinical and ecological characteristics of Vibrio vulnificus in the northeastern United States. Diagn Microbiol Infect Dis 1987;6:109-17.
8. Whitman CM, Griffin PM. Preventing Vibrio vulnificus infection in the high-risk patient. Infectious Diseases Clinical Practice 1993;2:275-6.
9. Food and Drug Administration. If you eat raw oysters, you need to know. Washington, DC: US Department of Health and Human Services, Public Health Service, 1995; DHHS publication no. (FDA)95-2293.

June 04, 1993 / 42(21);405-407

Vibrio vulnificus Infections Associated with Raw Oyster Consumption -- Florida, 1981-1992

Vibrio vulnificus is a gram-negative bacterium that can cause serious illness and death in persons with preexisting liver disease or compromised immune systems. From 1981 through 1992, 125 persons with V. vulnificus infections, of whom 44 (35%) died, were reported to the Florida Department of Health and Rehabilitative Services (HRS). This report summarizes data on these cases and presents estimates of the at-risk population in Florida.

The infections generally occurred each year from March through December and peaked from May through October. Seventy-two persons (58%) had primary septicemia, 35 (28%) had wound infections, and 18 (14%) had gastroenteritis. In patients with primary septicemia, 58 infections (81%) occurred among persons with a history of raw oyster consumption during the week before onset of illness. The mean age of these persons was 60 years (range: 33-90 years; standard deviation: 12.9 years); 51 (88%) were male. Fourteen (78%) of the patients with gastroenteritis also had raw oyster-associated illness. Their mean age was 49 years (range: 19-89 years; standard deviation: 25.7 years); seven (50%) were male.

Of the 40 deaths caused by septicemia, 35 (88%) were associated with raw oyster consumption. Nine of these deaths occurred in 1992. The case-fatality rate from raw oyster-associated V. vulnificus septicemia among patients with pre-existing liver disease was 67% (30 of 45) compared with 38% (5 of 13) among those who were not known to have liver disease.

Results of the 1988 Florida Behavioral Risk Factor Survey (BRFS) were used to estimate the proportions of the Florida population who ate raw oysters, and the proportion of the population who ate raw oysters and who believed they had liver disease (e.g., cirrhosis). These estimates were used in conjunction with case reports and population data from the Florida Office of Vital Statistics to estimate the risk for illness and death associated with V. vulnificus (1).

BRFS and state population data indicate that approximately 3 million persons in Florida eat raw oysters; of these, 71,000 persons believe they have liver disease. Based on the number of cases reported to the Florida HRS during 1981-1992, the annual rate of illness from V. vulnificus infection for adults with liver disease who ate raw oysters was 72 per 1 million adults -- 80 times the rate for adults without known liver disease who ate raw oysters (0.9 per 1 million). The annual rate of death from V. vulnificus for adults with liver disease who ate raw oysters was 45 per 1 million -- more than 200 times greater than the rate for persons without known liver disease who ate raw oysters (0.2 per 1 million).

Reported by: WG Hlady, MD, RC Mullen, MPH, RS Hopkins, MD, State Epidemiologist, Florida Dept of Health and Rehabilitative Svcs. Foodborne and Diarrheal Diseases Br, Div of Bacterial and Mycotic Diseases, National Center for Infectious Diseases, CDC.

Editorial Note: V. vulnificus was first described as a cause of human illness in 1979 (2). Although there is no national surveillance for infections caused by this pathogen, regional surveillance in four states along the Gulf Coast indicates an annual incidence for V. vulnificus infections of at least 0.6 per 1 million persons and a case-fatality rate of 22% (3).

V. vulnificus, a free-living bacterium, occurs naturally in the marine environment, rather than as a result of pollution by human or animal fecal waste. This organism is commonly found in estuarine waters of the Gulf of Mexico, where it may contaminate oysters and other shellfish. Legal harvesting of oysters is limited to areas free of fecal contamination; however, V. vulnificus is ubiquitous in warm ocean waters, and oysters harvested from approved sites may be contaminated. Therefore, regardless of the source of the oysters, the potential for infection exists whenever raw oysters are consumed.

Ingestion of raw or undercooked shellfish contaminated with V. vulnificus can lead to primary septicemia or gastroenteritis. In addition, V. vulnificus can cause infection by directly contaminating open wounds during swimming, shellfish cleaning, and other marine activities.

The findings in this report are consistent with other studies indicating that persons with liver disease are at increased risk for infection with V. vulnificus and death (2,4). Persons with compromised immune systems (e.g., chronic renal insufficiency, cancer, diabetes, steroid-dependent asthma, and chronic intestinal disease) or iron overload states (e.g., thalassemia and hemochromatosis) may also be at increased risk for infection with V. vulnificus and death (2,5).* Whether persons with acquired immunodeficiency syndrome are at increased risk for V. vulnificus infections is unknown.

A previous study in north Florida indicated that less than 15% of high-risk patients were aware of the risks associated with raw oyster consumption (6). To increase awareness of risks for infection with this pathogen, the Florida HRS has issued press releases to inform the general public and has provided gastroenterologists in the state with clinical references and information for their patients with liver disease. California and Louisiana both require written consumer alerts regarding the risk of raw oyster consumption be visible where raw oysters are sold at retail food establishments. The Florida HRS also is working with other agencies in the state to establish labeling requirements for raw oysters that would inform consumers at all points of sale of the risk for serious illness for persons with liver disease or compromised immune systems who consume raw oysters. The wording of such labeling will be similar to the label already required by the Florida Department of Natural Resources for all wholesale shellstock and shucked products: "Consumer Information -- There is a risk associated with consuming raw oysters or any raw animal protein. If you have chronic illness of the liver, stomach, or blood or have immune disorders, you are at a greater risk of serious illness from raw oysters and should eat oysters fully cooked. If unsure of your risk, consult a physician."

References
1. Desenclos JA, Klontz KC, Wolfe LE, Hoercherl S. The risk of Vibrio illness in the Florida raw oyster eating population, 1981-1988. Am J Epidemiol 1991;134:290-7.
2. Blake PA, Merson MH, Weaver RE, Hollis DG, Heublein PC. Disease caused by a marine vibrio: clinical characteristics and epidemiology. N Engl J Med 1979;300:1-4.
3. Levine WC, Griffin PM, the Gulf Coast Vibrio Working Group. Vibrio infections on the Gulf Coast: the results of a first year of regional surveillance. J Infect Dis 1993;167:479-83.
4. Tacket CO, Brenner F, Blake PA. Clinical features and an epidemiological study of Vibrio vulnificus infections. J Infect Dis 1984;149:558-61.
5. Johnston JM, Becker SF, McFarland LM. Vibrio vulnificus: man and the sea. JAMA 1985;253:2850-3.
6. Johnson AR, Anderson CR, Rodrick GE. A survey to determine the awareness of hazards related to raw seafood ingestion in at risk patient groups. In: Proceedings of the 13th annual conference of the Tropical and Subtropical Fisheries Technology Society of the Americas. Gulf Shores, Alabama: Tropical and Subtropical Fisheries Technology Society of the Americas, October 1988.

*   The Food and Drug Administration (FDA) publishes brochures on seafood safety, including ones with special information for patients with liver diseases, immune disorders, gastrointestinal disorders, or diabetes mellitus. Free brochures are available to patients and their physicians from the FDA's 24-hour Seafood Safety Hotline, (800) 332-4010 ({800} FDA-4010); in the Washington, D.C., area the number is (202) 205-4314.

March 04, 1994 / 43(08);137-138,143-144

Clostridium perfringens Gastroenteritis Associated with Corned Beef Served at St. Patrick's Day Meals -- Ohio and Virginia, 1993

Clostridium perfringens is a common infectious cause of outbreaks of foodborne illness in the United States, especially outbreaks in which cooked beef is the implicated source (1,2). This report describes two outbreaks of C. perfringens gastroenteritis following St. Patrick's Day meals in Ohio and Virginia during 1993. Ohio

On March 18, 1993, the Cleveland City Health Department (CCHD) received telephone calls from 15 persons who became ill after eating corned beef purchased from one delicatessen. After a local newspaper article publicized this problem, 156 persons contacted CCHD to report onset of diarrheal illness within 48 hours of eating food from the delicatessen on March 16 or March 17. Symptoms included abdominal cramps (88%) and vomiting (13%); no persons were hospitalized. The median incubation period was 12 hours (range: 2- 48 hours). Of the 156 persons reporting illness, 144 (92%) reported having eaten corned beef; 20 (13%), pickles; 12 (8%), potato salad; and 11 (7%), roast beef.

In anticipation of a large demand for corned beef on St. Patrick's Day (March 17), the delicatessen had purchased 1400 pounds of raw, salt-cured product. Beginning March 12, portions of the corned beef were boiled for 3 hours at the delicatessen, allowed to cool at room temperature, and refrigerated. On March 16 and 17, the portions were removed from the refrigerator, held in a warmer at 120 F (48.8 C), and sliced and served. Corned beef sandwiches also were made for catering to several groups on March 17; these sandwiches were held at room temperature from 11 a.m. until they were eaten throughout the afternoon.

Cultures of two of three samples of leftover corned beef obtained from the delicatessen yielded greater than or equal to 105 colonies of C. perfringens per gram.

Following the outbreak, CCHD recommended to the delicatessen that meat not served immediately after cooking be divided into small pieces, placed in shallow pans, and chilled rapidly on ice before refrigerating and that cooked meat be reheated immediately before serving to an internal temperature of greater than or equal to 165 F (greater than or equal to 74 C). Virginia

On March 28, 1993, 115 persons attended a traditional St. Patrick's Day dinner of corned beef and cabbage, potatoes, vegetables, and ice cream. Following the dinner, 86 (76%) of 113 persons interviewed reported onset of illness characterized by diarrhea (98%), abdominal cramps (71%), and vomiting (5%). The median incubation period was 9.5 hours (range: 2-18.5 hours). Duration of illness ranged from 1 hour to 4.5 days; one person was hospitalized.

Corned beef was the only food item associated with illness; cases occurred in 85 (78%) of 109 persons who ate corned beef compared with one of four who did not (relative risk=3.1; 95% confidence interval=0.6-17.1). Cultures of stool specimens from eight symptomatic persons all yielded greater than or equal to 106 colonies of C. perfringens per gram. A refrigerated sample of leftover corned beef yielded greater than or equal to 105 colonies of C. perfringens per gram.

The corned beef was a frozen, commercially prepared, brined product. Thirteen pieces, weighing approximately 10 pounds each, had been cooked in an oven in four batches during March 27-28. Cooked meat from the first three batches was stored in a home refrigerator; the last batch was taken directly to the event. Approximately 90 minutes before serving began, the meat was sliced and placed under heat lamps.

Following the outbreak, Virginia health officials issued a general recommendation that meat not served immediately after cooking be divided into small quantities and rapidly chilled to less than or equal to 40 F (less than or equal to 4.4 C), and that precooked foods be reheated immediately before serving to an internal temperature of greater than or equal to 165 F (greater than or equal to 74 C). Follow-Up Investigation

The results of the epidemiologic and laboratory investigations suggest that the two outbreaks in this report were not related. Traceback of the corned beef in both of these outbreaks indicated that the meat had been produced by different companies and sold through different distributors. Serotyping was performed on C. perfringens isolates recovered from the stool samples in Virginia and on an isolate from a food sample obtained in Ohio. Six of the seven Virginia stool isolates were serotype PS86; however, the food isolate from Ohio could not be serotyped using available antisera.

Reported by: J Zimomra, MPA, T Wenderoth, A Snyder, R Russ, Div of Environmental Health, Cleveland City Health Dept; ED Peterson, R French, MPA, TJ Halpin, MD, State Epidemiologist, Div of Preventive Medicine, Ohio Dept of Health. JE Florance, MD, A Adkins, J Andrew, M Burkgren, K Crisler, T Fagen, L Fass, JM Galloway, S Haines, RH Hinton, C Jackson, NS Rivera, EL Testor, C Williams, Prince William Health District; AA DiAllo, PhD, DR Patel, Virginia Div of Consolidated Laboratory Svcs, Dept of General Svcs; CW Armstrong, MD, D Woolard, MPH, GB Miller, MD, State Epidemiologist, Virginia Dept of Health. Div of Field Epidemiology, Epidemiology Program Office; Foodborne and Diarrheal Diseases Br, Div of Bacterial and Mycotic Diseases, National Center for Infectious Diseases, CDC.

Editorial Note: C. perfringens is a ubiquitous, anaerobic, gram-positive, spore-forming bacillus and a frequent contaminant of meat and poultry (3). C. perfringens food poisoning is characterized by onset of abdominal cramps and diarrhea 8-16 hours after eating contaminated meat or poultry (4). By sporulating, this organism can survive high temperatures during initial cooking; the spores germinate during cooling of the food, and vegetative forms of the organism multiply if the food is subsequently held at temperatures of 60 F-125 F (16 C-52 C) (3). If served without adequate reheating, live vegetative forms of C. perfringens may be ingested. The bacteria then elaborate the enterotoxin that causes the characteristic symptoms of diarrhea and abdominal cramping (4).

Laboratory confirmation of C. perfringens foodborne outbreaks requires quantitative cultures of implicated food or stool from ill persons. Both outbreaks described in this report were confirmed by the recovery of greater than or equal to 105 organisms per gram of epidemiologically implicated food (5). Cultures of stool samples from persons affected in Virginia also met the alternate criterion of a median of greater than or equal to 106 colonies per gram (6). Serotyping is not useful for confirming C. perfringens outbreaks and, in general, is not available (7).

Corned beef is a popular ethnic dish that is commonly served to celebrate St. Patrick's Day. The errors in preparation of the corned beef in these outbreaks were typical of those associated with previously reported foodborne outbreaks of C. perfringens (8). Improper holding temperatures were a contributing factor in most (97%) C. perfringens outbreaks reported to CDC from 1973 through 1987 (2). To avoid illness caused by this organism, food should be eaten while still hot or reheated to an internal temperature of greater than or equal to 165 F (greater than or equal to 74 C) before serving (9).

References
1. Shandera WX, Tacket CO, Blake PA. Food poisoning due to Clostridium perfringens in the United States. J Infect Dis 1983;147:167-70.
2. Bean NH, Griffin PM. Foodborne disease outbreaks in the United States, 1973-1987: pathogens, vehicles, and trends. Journal of Food Protection 1990;53:804-17.
3. Hall HE, Angelotti R. Clostridium perfringens in meat and meat products. Appl Microbiol 1965;13:352-7.
4. Hughes JM, Tauxe RV. Food-borne disease. In: Mandell GL, Douglas RG Jr, Bennett JE, eds. Principles and practice of infectious diseases. 3rd ed. New York: Churchill Livingstone Inc, 1990;893-
5.
6. Hauschild WAH. Criteria and procedures for implicating Clostridium perfringens in food-borne outbreaks. Can J Public Health 1975;66:388-92.
7. Hauschild WAH, Desmarchelier P, Gilbert RJ, Harmon SM, Vahlefeld R. ICMSF methods studies: XII. Comparative study for the enumeration of Clostridium perfringens in feces. Can J Microbiol 1979;25:953-63.
8. Hatheway CL, Whaley DN, Dowell VR Jr. Epidemiological aspects of Clostridium perfringens foodborne illness. Food Technology 1980;34:77-9.
9. Loewenstein MS. Epidemiology of Clostridium perfringens food poisoning. N Engl J Med 1972;286:1026-8.
10. Bryan FL. What the sanitarian should know about Clostridium perfringens foodborne illness. Journal of Milk and Food Technology 1969;32:381-9.

March 18, 1994 / 43(10);177-8

Epidemiologic Notes and Reports Bacillus cereus Food Poisoning Associated with Fried Rice at Two Child Day Care Centers -- Virginia, 1993

Bacillus cereus, an infectious cause of foodborne illness, accounted for 2% of outbreaks with confirmed etiology that were reported to CDC during 1973-1987 (1). On July 21, 1993, the Lord Fairfax (Virginia) Health District received reports of acute gastrointestinal illness that occurred among children and staff at two jointly owned child day care centers following a catered lunch. This report summarizes the investigation of this outbreak.

The catered lunch was served on July 21 to 82 children aged less than or equal to 6 years and to nine staff; dietary histories were obtained for 80 persons. Staff and all children aged greater than or equal to 4 years were interviewed directly; staff and parents were questioned for children aged less than 4 years.

Of the 80 persons, 67 ate the catered lunch. A case was defined as vomiting by a person who was present at either day care center on July 21. Fourteen (21%) persons who ate the lunch became ill, compared with none of 13 who did not. Symptoms included nausea (71%), abdominal cramps or pain (36%), and diarrhea (14%). Twelve of the 14 cases occurred among children aged 2.5-5 years, and two occurred among staff. The median incubation period was 2 hours (range: 1.5-3.5 hours). Symptoms resolved a median of 4 hours after onset (range: 1.5-22 hours).

Chicken fried rice prepared at a local restaurant was the only food significantly associated with illness; illness occurred in 14 (29%) of 48 persons who ate chicken fried rice, compared with none of 16 who did not (relative risk=undefined; lower confidence limit=1.7); three persons who were not ill were uncertain if they had eaten the rice. B. cereus was isolated from leftover chicken fried rice (greater than 10 superscript 6 organisms per gram) and from vomitus from one ill child (greater than 10 superscript 5 organisms per gram) but not from samples of leftover milk. Other food items (peas and apple rings) were not available for analysis.

The rice had been cooked the night of July 20 and cooled at room temperature before refrigeration. On the morning of the lunch, the rice was pan-fried in oil with pieces of cooked chicken, delivered to the day care centers at approximately 10:30 a.m., held without refrigeration, and served at noon without reheating.

Following the outbreak, health officials from the Lord Fairfax Health District recommended to day care staff and restaurant food handlers that the practice of cooling rice or any food at room temperature be discontinued, food be maintained at proper temperatures (i.e., below 41 F {5 C} or above 140 F {60 C}), and a thermometer be used to verify food temperatures.

Reported by: M Khodr, MD, S Hill, L Perkins, S Stiefel, C Comer-Morrison, S Lee, Lord Fairfax Health District, Winchester; DR Patel, D Peery, Virginia Div of Consolidated Laboratory Svcs, Dept of General Svcs; CW Armstrong, MD, GB Miller, Jr, MD, State Epidemiologist, Virginia Dept of Health. Div of Field Epidemiology, Epidemiology Program Office; Foodborne and Diarrheal Diseases Br, Div of Bacterial and Mycotic Diseases, National Center for Infectious Diseases, CDC.

Editorial Note: B. cereus, a ubiquitous, spore-forming bacteria, causes two recognized forms of foodborne gastroenteritis: an emetic syndrome resembling that caused by Staphylococcus aureus and characterized by an incubation period of 1-6 hours and a diarrheal illness characterized by an incubation period of 6-24 hours (2). Fever is uncommon with either syndrome. The emetic syndrome -- which occurred in the outbreak described in this report -- is mediated by a highly stable toxin that survives high temperatures and exposure to trypsin, pepsin, and pH extremes; the diarrheal syndrome is mediated by a heat- and acid-labile enterotoxin that is sensitive to proteolytic enzymes (3).

The diagnosis of B. cereus food poisoning can be confirmed by the isolation of greater than or equal to 105 B. cereus organisms per gram from epidemiologically implicated food. Underreporting of such outbreaks is likely because illness associated with B. cereus is usually self-limiting and not severe. In addition, findings of a recent survey about culture practices for outbreaks of apparent foodborne illness indicate that 20% of state public health laboratories do not make B. cereus testing routinely available (South Carolina Department of Health and Environmental Control and CDC, unpublished data, 1991).

Fried rice is a leading cause of B. cereus emetic-type food poisoning in the United States (1,4). B. cereus is frequently present in uncooked rice, and heat-resistant spores may survive cooking. If cooked rice is subsequently held at room temperature, vegetative forms multiply, and heat-stable toxin is produced that can survive brief heating, such as stir frying (4). In the outbreak described in this report, vegetative forms of the organism probably multiplied at the restaurant and the day care centers while the rice was held at room temperature.

The day care staff and restaurant food handlers in this report were unaware that cooked rice was a potentially hazardous food. This report underscores the ongoing need to educate food handlers about basic practices for safe food handling.

References
1. Bean NH, Griffin PM. Foodborne disease outbreaks in the United States, 1973-1987: pathogens, vehicles, and trends. Journal of Food Protection 1990;53:804-17.

2. Benenson AS, ed. Control of communicable diseases in man. 15th ed. Washington, DC: American Public Health Association, 1990:177-8.

3. Kramer JM, Gilbert RJ. Bacillus cereus and other Bacillus species. In: Doyles MP, ed. Foodborne bacterial pathogens. New York: Marcel Dekker, Inc, 1989:21-70.

4. Terranova W, Blake PA. Bacillus cereus food poisoning. N Engl J Med 1978;298:143-4.

June 27, 1986 / 35(25);408-10

Bacillus cereus -- Maine

On September 22, 1985, the Maine Bureau of Health was notified of a gastrointestinal illness among patrons of a Japanese restaurant. Because the customers were exhibiting symptoms of illness while still on the restaurant premises, and because uncertainty existed as to the etiology of the problem, the local health department, in concurrence with the restaurant owner, closed the restaurant at 7:30 p.m. that same day.

Eleven (31%) of the approximately 36 patrons reportedly served on the evening of September 22 were contacted in an effort to determine the etiology of the outbreak. Those 11 comprised the last three dining parties served on September 22. Despite extensive publicity, no additional cases were reported.

A case was defined as anyone who had vomiting or diarrhea within 6 hours of dining at the restaurant. All 11 individuals were interviewed for symptoms, time of onset of illness, illness duration, and foods ingested. All 11 reported nausea and vomiting; nine reported diarrhea; one reported headache; and one reported abdominal cramps. Onset of illness ranged from 30 minutes to 5 hours (mean 1 hour, 23 minutes) after eating at the restaurant. Duration of illness ranged from 5 hours to several days, except for two individuals still symptomatic with diarrhea 2 weeks after dining at the restaurant. Ten persons sought medical treatment at local emergency rooms on September 22; two ultimately required hospitalization for rehydration.

Analysis of the association of food consumption with illness was not instructive, since all persons consumed the same food items: chicken soup; fried shrimp; stir-fried rice; fried zucchini, onions, and bean sprouts; cucumber, cabbage, and lettuce salad; ginger salad dressing; hibachi chicken and steak; and tea. Five persons ordered hibachi scallops, and one person ordered hibachi swordfish. However, most individuals sampled each other's entrees.

One vomitus specimen and two stool specimens from three separate individuals yielded an overgrowth of Bacillus cereus organisms. The hibachi steak was also culture-positive for B. cereus, although an accurate bacterial count could not be made because an inadequate amount of the steak remained for laboratory analysis. No growth of B. cereus was reported from the fried rice, mixed fried vegetables, or hibachi chicken.

According to the owner, all meat was delivered 2-3 times a week from a local meat supplier and refrigerated until ordered by restaurant patrons. Appropriate-sized portions for a dining group were taken from the kitchen to the dining area and diced or sliced, then sauteed at the table directly in front of restaurant patrons. The meat was seasoned with soy sauce, salt, and white pepper, open containers of which had been used for at least 2 months by the restaurant. The hibachi steak was served immediately after cooking.

The fried rice served with the meal was reportedly customarily made from leftover boiled rice. It could not be established whether the boiled rice had been stored refrigerated or at room temperature. Reported by J Vandeloski, Portland City Health Dept, KF Gensheimer, MD, State Epidemiologist, Maine Dept of Human Svcs; Enteric Diseases Br, Div of Bacterial Diseases, Center for Infectious Diseases, CDC.

Editorial Note: B. cereus is an aerobic, spore-forming, gram-positive rod with a ubiquitous distribution in the environment. Spores of B. cereus have been found in a wide variety of cereals, pulses, vegetables, spices, and pasteurized fresh and powdered milk. Food-poisoning can result from toxins elaborated by germinating organisms, which most commonly follows from inadequate refrigeration and subsequent reheating of foods that have already been cooked.

Two different clinical syndromes appear to be associated with B. cereus food poisoning, which correspond to two different toxins elaborated by the bacteria. A diarrheal syndrome similar to Clostridium perfringens food poisoning with an average incubation period of 10-12 hours has been associated with a heat-labile toxin elaborated by B. cereus. An emetic syndrome similar to staphylococcal food poisoning, with an average incubation period of 1-6 hours, has been associated with a heat-stable toxin from B. cereus (1).

The emetic syndrome has almost always been associated with fried rice served in Oriental restaurants. The common practice of storing boiled rice at room temperature for subsequent preparation of fried rice has generally been implicated in such outbreaks. However, a recent, well-documented outbreak of the emetic syndrome of B. cereus in a British prison implicated beef stew (2). This was thought to be caused by adding to the stew vegetables that were cooked a day earlier.

Fresh meat cooked rapidly, then eaten immediately, seems an unlikely vehicle for B. cereus food poisoning. The laboratory finding of B. cereus in a foodstuff without quantitative cultures and without accompanying epidemiologic data is insufficient to establish its role in the outbreak. A negative culture of fried rice eaten with the meal does not exclude the obvious vehicle; reheating during preparation may eliminate the bacteria in the food without decreasing the activity of the heat-stable toxin. While the question of the specific vehicle remains incompletely resolved, the clinical and laboratory findings substantially support B. cereus as the cause of the outbreak.

Most episodes of food poisoning undoubtedly go unreported, and in most of those reported, the specific pathogens are never identified. Alert recognition of the clinical syndrome and appropriate laboratory work permitted identification of the role of B. cereus in this outbreak.

References
1. Terranova W, Blake PA. Bacillus cereus food poisoning. N Engl J Med 1978;298:143-4.
2. CDC. Communicable disease report, no. 21, May 25, 1984:3.

May 25, 1990 / 39(20);334-335,341

Aeromonas Wound Infections Associated with Outdoor Activities -- California

Aeromonas species are associated with gastroenteritis and with wound infections, particularly wounds incurred in outdoor settings. On May 1, 1988, isolates of Aeromonas became reportable in California, the first state to mandate reporting of isolates of and infections with these organisms. Surveillance data for 1988 and 1989 represent the first population-based estimates of both the occurrence and public health impact of Aeromonas infections in the United States and provide a basis for assessing the need for further surveillance of these organisms.

From May 1, 1988, through April 30, 1989, clinicians and clinical laboratories in California reported 225 Aeromonas isolates from 219 patients. Cases were reported on Confidential Morbidity Report cards to local health departments, which then conducted case investigations and forwarded their reports to the California Department of Health Services. Of the 225 isolates, 178 (79.1%) were recovered from stool, 19 (8.4%) from wounds, 11 (4.9%) from blood, and 17 (7.6%) from other sites. A. caviae was recovered from seven stool cultures; A. sobria was recovered from two stool cultures and one vaginal culture. All other cultures were reported as A. hydrophila or Aeromonas unspecified.

Based on reported cases, the incidence of Aeromonas wound infections in California was 0.7 per million population. Of the 19 patients with wound infections, 13 were injured outdoors (Table 1). Six of these patients required hospitalization for their injuries and/or infections. One patient had a mixed infection including Aeromonas, Proteus, and Pseudomonas species. The number of infections peaked in the summer months with three cases each in July and August. The cases reported among persons aged 30-39 years represented the highest incidence rate for all age groups (1.4 per million). Reported by: SB Werner, MD, Infectious Disease Br, GW Rutherford, III, MD, State Epidemiologist, California Dept of Health Svcs. Div of Field Svcs, Epidemiology Program Office;

Enteric Diseases Br, Div of Bacterial Diseases, and Epidemiology Br, Hospital Infections Program, Center for Infectious Diseases, CDC.

Editorial Note: Aeromonas species are gram-negative, facultatively anaerobic bacteria found in soil and fresh and brackish water worldwide (1). Although Aeromonas species were recognized in 1891 as colonizers and pathogens of cold-blooded animals, especially fish (2), they were not identified as human pathogens until 1968 (3). Since then, they have been associated with a wide spectrum of human diseases (especially in immunocompromised patients), most commonly gastroenteritis (4) and soft tissue infections (5).

The taxonomy of Aeromonas species requires further clarification. Three species, A. hydrophila, A. sobria, and A. caviae, have been associated with human disease (4), but DNA hybridization analyses support seven or more distinct genotypes (6). Because many clinical laboratories are unable to perform precise identification, many aeromonad isolates are reported as

A. hydrophila or A. hydrophila complex. Although the California surveillance data provide limited information about the morbidity of the wound infections reported, they suggest that the public health impact of these soft tissue infections is low and may be determined more by the nature of the underlying injury than by the presence of Aeromonas organisms.

The California data do not provide information on case management. However, one reported case series (7) suggests that surgical debridement is an important component of treatment and has enabled resolution of the infection when either no antibiotics or ineffective antibiotics (i.e., antibiotics to which the organisms were resistant) were used. These findings, as well as the occurrence of Aeromonas organisms in mixed infections, suggest that in some cases Aeromonas species may be colonizers in wounds rather than pathogens.

References
1. Von Graevenitz A. Aeromonas and Plesiomonas. In: Lennette EH, Ballows A, Hausler WJ, Shadomy HJ, ed. Manual of clinical microbiology. Washington, DC: American Society for Microbiology, 1985:278-81.
2. Ewing WH, Hugh R, Johnson JG. Studies on the Aeromonas group. Atlanta, Georgia: US Department of Health, Education, and Welfare, Public Health Service, Communicable Disease Center, 1961.
3. Von Graevenitz A, Mensch AH. The genus Aeromonas in human bacteriology: report of 30 cases and review of the literature. N Engl J Med 1968;278:245-9.
4. Holmberg SD, Schell WL, Fanning GR, et al. Aeromonas intestinal infections in the United States. Ann Intern Med 1986;105:683-9.
5. Janda JM, Duffey PS. Mesophilic Aeromonads in human disease: current taxonomy, labo ratory identification and infectious disease spectrum. Rev Infect Dis 1988;10:980-97.
6. Popoff MY, Coynault C, Kiredjian M, Lemelin M. Polynucleotide sequence relatedness among motile Aeromonas species. Curr Microbiol 1981;5:109-14.
7. Isaacs RD, Paviour SD, Bunker DE, Land SDR. Wound infection with aerogenic Aeromonas strains: a review of twenty-seven cases. Eur J Clin Microbiol Infect Dis 1988;7:355-60.

May 22, 1998 / 47(19);394-396

Plesiomonas shigelloides and Salmonella serotype Hartford Infections Associated with a Contaminated Water Supply -- Livingston County, New York, 1996

On June 24, 1996, the Livingston County (New York) Department of Health (LCDOH) was notified of a cluster of diarrheal illness following a party on June 22, at which approximately 30 persons had become ill. This report summarizes the findings of the investigation, which implicated water contaminated with Plesiomonas shigelloides and Salmonella serotype Hartford as the cause of the outbreak.

The party was held at a private residence on June 22 and was attended by 189 persons. Food was provided by a local convenience store that sells gasoline, packaged goods, sandwiches, and pizza and prepares food for catered events. The convenience store had not catered any parties during the preceding 5 days but catered two parties on June 23. LCDOH contacted the organizers of these events and found no other reports of illness.

To determine the source and extent of the outbreak and mechanism of contamination, LCDOH conducted a cohort study, an environmental investigation, and micro-biologic examinations of stool specimens, leftover food items, and water samples. A menu and guest list were obtained and guests were interviewed by telephone. A probable case was defined as diarrhea (greater than 3 loose stools during a 24-hour period) in a person who attended the party and became ill within 72 hours. Persons with a confirmed case had either Plesiomonas shigelloides or Salmonella serotype Hartford or both isolated from stool. The caterer and facility employees were interviewed to obtain information on food preparation, and the water source was inspected.

Of the 189 attendees, 98 (52%) were interviewed. Sixty persons reported illness; 56 (57%) of 98 respondents had illnesses meeting the case definition. The mean age for case-patients was 41 years (range: 2-85 years), and 32 (57%) were male. Stool specimens were obtained from 14 ill attendees: nine yielded only P. shigelloides, three only Salmonella serotype Hartford, and two had both organisms. One person with culture-confirmed Salmonella serotype Hartford was hospitalized. The clinical profiles of the culture-confirmed (n=14) and probable (n=42) cases were similar.

Twenty food and beverage items were served at the party. Three food items were associated with illness: macaroni salad, potato salad, and baked ziti. Of 56 attendees who ate macaroni salad, 43 (77%) became ill, compared with 17 (40%) of 42 who did not eat macaroni salad (relative risk {RR}=2.6; 95% confidence interval {CI}=1.5-4.4). Of 49 guests who ate potato salad, 36 (73%) became ill, compared with 20 (44%) of 45 who did not eat potato salad (RR=2.1; 95% CI=1.2-3.6). Of 46 attendees who ate baked ziti, 36 (78%) became ill, compared with 20 (42%) of 48 that did not eat baked ziti (RR=2.7; 95% CI=1.5-4.9).

Leftover food samples of these three items were collected on June 25 and sent for microbiologic examination. Salmonella serotype Hartford was isolated from the macaroni salad and baked ziti. Both Salmonella serotype Hartford and P. shigelloides were isolated from the potato salad. Escherichia coli was isolated from a water sample collected on June 27 from the tap in the store. Water samples collected on July 8 from the well that supplied water to the store contained both Salmonella serotype Hartford and P. shigelloides.

Preparation of the salads and the baked ziti began on June 21, and prepared food items were stored in a walk-in cooler overnight. On June 22, the ziti was prepared by heating the tomato sauce, pouring it over the meat and pasta, and heating in an oven for 50 minutes at an unknown temperature. The ziti remained in the oven with the heat off until it and the salads were transported to the party.

All foodhandlers denied gastrointestinal illness with onset before June 22. However, three foodhandlers reported illness beginning after June 22; all three reported having eaten foods prepared for the party. P. shigelloides was recovered from stool specimens from these three workers only.

The New York State Department of Agriculture and Markets found nine sanitary violations at the caterer's facilities. The water source, an unprotected dug well approximately 10 feet deep, served only the store. The well was fed by shallow ground water and may have received surface runoff from surrounding tilled and manured farm land and water from adjacent streams. A small poultry farm was located approximately 1600 feet upstream of the well. Farm field drainage systems discharged into the source water stream just above the well. A water sample collected at the store on June 27 showed no chlorine residual, indicating that the pellet chlorinator was off-line at the time of the event. The pellet chamber was empty and the system did not contain any filtration mechanism. Well water used for food preparation (i.e., rinsing pasta used in salads, mixing ingredients, cooking food items, and cleaning equipment) was probably contaminated as a result of rainfall on June 19 and June 20 that transported pathogens from the surrounding farmland. The improperly maintained chlorinator allowed these pathogens to reach the food preparation area. After the outbreak, the store was prohibited from preparing food until an adequate water-treatment system that met drinking water standards could be provided. Store employees and the public were instructed not to drink the water.

Reported by: R Van Houten, D Farberman, J Norton, J Ellison, Livingston County Dept of Health, Mt. Morris; J Kiehlbauch, PhD, T Morris, MD, P Smith, MD, State Epidemiologist, New York State

Dept of Health. Foodborne and Diarrheal Diseases Br, Div of Bacterial and Mycotic Diseases, National Center for Infectious Diseases, CDC.

Editorial Note: The findings in this report implicated a deficient water supply system as the cause of an outbreak of diarrheal illness caused by Salmonella serotype Hartford and P. shigelloides. Unfiltered, untreated surface water led to contamination of food during its preparation.

Most infections with P. shigelloides have been associated with drinking untreated water, eating uncooked shellfish, or with travel to developing countries (1-3). P. shigelloides (previously Aeromonas shigelloides) are ubiquitous, facultatively anaerobic, flagellated, gram-negative rods (3). Although few are widespread in the environment, few waterborne or foodborne outbreaks have been reported (4). P. shigelloides have been isolated from a variety of sources, including wild and domestic animals (2). Infection is characterized by self-limited diarrhea with blood or mucus, abdominal cramps, and vomiting or fever (5). Symptoms usually occur within 48 hours of exposure. Fecal leukocytes and erythrocytes have been found on stool smears (1); however, the exact mechanism of the diarrhea (secretory versus inflammatory) is unknown.

Salmonella serotype Hartford is a rare serotype that has been isolated from porcine and bovine sources. In May 1995, freshly squeezed, unpasteurized commercial orange juice was implicated as the cause of an outbreak (6). Contamination was thought to have originated from inadequate sanitization of the exterior surfaces of oranges.

In this outbreak, the well water most likely became contaminated with both P. shigelloides and Salmonella serotype Hartford through runoff from nearby farms. The outbreak could have been prevented if effective public health measures had been in place. Routine testing of well water for total fecal coliform bacteria, turbidity, and chlorine residual may enable early detection of fecal contamination and rapid decontamination. Filtration and chlorination of potable water systems have substantially reduced waterborne outbreaks and subsequent morbidity and mortality. Where possible, water sources subject to contamination from agricultural runoff should not be used for drinking or food preparation. Disinfection and filtration of water from any source can further reduce the risk for waterborne illness.

References
1. Soweid AM, Clarkston WK. Plesiomonas shigelloides: An unusual cause of diarrhea. Am J Gastroenterol 1995;90:2235-6.
2. Jeppesen C. Media for Aeromonas spp., Plesiomonas shigelloides and Pseudomonas spp. food and environment. Int J Food Microbiol 1995;26:25-41.
3. San Joaquin VH. Aeromonas, Yersinia, and miscellaneous bacterial enteropathogens. Pediatr Ann 1994;23:544-8.
4. Schofield GM. Emerging foodborne pathogens and their significance in chilled foods. J Appl Bacteriol 1992;72:267-73.
5. Holmberg SD, Wachsmuth IK, Hickman-Brenner FW, Blake PA, Farmer JJ. Pleisiomonas enteric infections in the United States. Ann Intern Med 1986;105:690-4.
6. Cook KA, Swerdlow D, Dobbs T, et al. Fresh-squeezed Salmonella: an outbreak of Salmonella Hartford associated with unpasteurized orange juice -- Florida {Abstract}. EIS Conference Abstract 1996;38-9.

September 15, 1989 / 38(36);617-619

Epidemiologic Notes and Reports Aquarium-Associated Plesiomonas shigelloides Infection -- Missouri

In July 1988, a community hospital in southeastern Missouri reported isolating Plesiomonas shigelloides from the stool of a 14-month-old girl with watery diarrhea (no blood or mucus) and fever. Her highest recorded rectal temperature was 102 F (38.9 C). Her stool was negative for Campylobacter, Salmonella, Shigella, Yersinia, Aeromonas, and rotavirus. The child was treated with trimethoprim/sulfamethoxazole, and her illness resolved after 5 days.
The child had consumed no shellfish and had never traveled more than 80 miles from her home. She

had consumed water only from the municipal system and recently had waded in two area lakes. She attended a day-care center, but no other children in her age group were reported ill. The child did not have an aquarium or other close association with animals. However, 1 evening each week, the child stayed in the home of a babysitter who kept piranhas in an aquarium. When the aquarium was cleaned, the water was poured into the bathtub. The child routinely was bathed in the bathtub before going home. The babysitter reported that the child could have been bathed immediately after the aquarium water had been poured into the bathtub.

P. shigelloides was isolated from samples of aquarium water submitted to the State Public Health Laboratory. However, plasmid studies were not performed, and it was not determined whether the bacterial strain isolated from the child's stool was identical to that isolated from the babysitter's aquarium.

To estimate the prevalence of P. shigelloides in tropical fish tanks, investigators from the Missouri Department of Health (MDH) surveyed aquarium water samples from several sites in Missouri (Table 1). Samples were taken from 18 aquariums, including at least two tanks from each of Missouri's six regional health districts. P. shigelloides was isolated from four (22%) of the 18 tanks. The four tanks were located in three different pet shops: two in central Missouri and one in eastern Missouri. Employees of the three pet shops reported no health problems in the fish in the culture-positive tanks.

MDH advised managers of all surveyed pet shops to have employees wash hands after contact with aquarium water or fish. No special precautions were recommended to managers of shops from which P. shigelloides was isolated. In addition, the baby sitter was advised to clean the tub thoroughly using chlorine bleach after discarding the aquarium water and before using the tub for bathing. Reported by: PS Tippen, A Meyer, EC Blank, DrPH, State Public Health Laboratory, HD Donnell, Jr, MD, State Epidemiologist, Missouri Dept of Health. Div of Field Svcs, Epidemiology Program Office, CDC. Editorial Note: P. shigelloides, a gram-negative bacterial rod, is an opportunistic pathogen in the immunocompromised host and has been suspected to cause diarrheal illness in normal hosts (1,2). However, the organism failed to produce illness in volunteer feeding studies, and its role as an enteric pathogen remains unproven (1). Persons with P. shigelloides infection typically describe a self-limited diarrhea, sometimes with blood and mucus in the stool; appropriate antibiotic therapy appears to shorten the duration of illness (3,4). P. shigelloides can also cause cellulitis and septicemia.

This organism has been isolated from surface water, the gut of freshwater fish, and many animals (including dogs and cats) and is particularly common in tropical and subtropical habitats (5). In humans, most isolates have been from stools of patients with diarrhea who live in tropical and subtropical regions of Asia, Africa, and Australia; isolations from Europe and the United States have been rare and usually associated with foreign travel or consumption of raw oysters (3,6).

Although no other P. shigelloides gastrointestinal infections associated with aquarium water have been reported, the frequency of P. shigelloides in pet shop aquariums reported here suggests this could be a source of this rarely recognized infection. Basic precautions, such as handwashing after contact with aquarium water and preventing the contamination of potable or bathing water by aquarium water, should decrease transmission of potentially pathogenic microorganisms from aquarium water.

References
1. Herrington DA, Tzipori S, Robins-Browne RM, Tall BD, Levine MM. In vitro and in vivo pathogenicity of Plesiomonas shigelloides. Infect Immun 1987;55:979-85.
2. Nolte FS, Poole RM, Murphy GW, Clark C, Panner BJ. Proctitis and fatal septicemia caused by Plesiomonas shigelloides in a bisexual man. J Clin Microbiol 1988;26:388-91.
3. Holmberg SD, Wachsmuth IK, Hickman-Brenner FW, Blake PA, Farmer JJ III. Plesiomonas enteric infections in the United States. Ann Intern Med 1986;105:690-4.
4. Kain KC, Kelly MT. Clinical features, epidemiology, and treatment of Plesiomonas shigel loides diarrhea. J Clin Microbiol 1989;27:998-1001.
5. von Graevenitz A. Aeromonas and Plesiomonas. In: Lennette EH, Balows A, Hausler WJ Jr, Shadomy HJ, eds. Manual of clinical microbiology. 4th ed. Washington, DC: American Society for Microbiology, 1985:278-81.
6. Reinhardt JF, George WL. Plesiomonas shigelloides-associated diarrhea. JAMA 1985;253: 3294-5.

April 16, 1999 / 48(14);285-9

Outbreaks of Shigella sonnei Infection Associated with Eating Fresh Parsley -- United States and Canada, July-August 1998

In August 1998, the Minnesota Department of Health reported to CDC two restaurant-associated outbreaks of Shigella sonnei infections. Isolates from both outbreaks had two closely related pulsed-field gel electrophoresis (PFGE) patterns that differed only by a single band. Epidemiologic investigations implicated chopped, uncooked, curly parsley as the common vehicle for these outbreaks. Through inquiries to health departments and public health laboratories, six similar outbreaks were identified during July-August (in California {two}, Massachusetts, and Florida in the United States and in Ontario and Alberta in Canada). Isolates from five of these outbreaks had the same PFGE pattern identified in the two outbreaks in Minnesota. This report describes the epidemiologic, traceback, environmental, and laboratory investigations, which implicated parsley imported from a farm in Mexico as the source of these outbreaks.

UNITED STATES

Minnesota. On August 17, the Minnesota Department of Health received reports of shigellosis in two persons who ate at the same restaurant during July 24-August 17 (Figure_1). S. sonnei subsequently was isolated from stool samples of 43 ill restaurant patrons; an additional 167 persons had probable shigellosis (diarrhea {three or more loose stools during a 24-hour period} lasting greater than or equal to 3 days or accompanied by fever). Eight (18%) of 44 restaurant employees had a similar illness; five had laboratory-confirmed S. sonnei infection. In a case-control study of 172 ill and 95 well restaurant patrons, five items were associated with illness: water (odds ratio {OR}=1.9; 95% confidence interval {CI}=1.0-3.8), ice (OR=3.7; 95% CI=1.6-8.6), potatoes (OR=2.6; 95% CI=1.5-4.6), uncooked parsley (OR=4.3; 95% CI=2.4-8.0), and raw tomato (OR=1.9; 95% CI=1.0-3.9). In a multivariate analysis, only uncooked parsley (OR=4.3; p less than 0.01) and ice (OR=6.9; p less than 0.01) remained significantly associated with illness.

California. On August 5, the Los Angeles County Department of Health Services was notified of two persons with shigellosis who ate at the same restaurant on July 31. Stool samples from six ill restaurant patrons yielded S. sonnei; an additional three had probable shigellosis (diarrhea {three or more loose stools during a 24-hour period}, or any loose stools accompanied by fever). All 27 foodhandlers denied illness and had stool samples that were negative for S. sonnei. In an unmatched comparison with 10 well dining companions, ill patrons were significantly more likely to have eaten foods sprinkled with chopped, uncooked parsley (OR=32.0; 95% CI=1.8-1381.4).

Massachusetts. On August 11, the Massachusetts Department of Health was notified of six persons who reported illness after eating at a restaurant lunch party on July 30. Stool samples from three persons yielded S. sonnei; an additional three had probable shigellosis (diarrhea within 4 days of the July 30 meal). Chopped, uncooked parsley was served on chicken sandwiches and in cole slaw served at the lunch. In a cohort study of 23 lunch attendees, illness was significantly associated with eating chicken sandwiches (relative risk {RR}=10.0; 95% CI=2.7-37.2) or eating uncooked parsley with any item (RR=10.0; 95% CI=1.4-70.2). All restaurant employees except one submitted a stool sample for culture; all were negative for S. sonnei.

CANADA

On August 10, the Ontario Ministry of Health was notified of a family of three persons with S. sonnei infection who attended a food fair during July 31-August 3. Laboratory-based surveillance identified 32 additional persons with S. sonnei infection who had eaten at a specific kiosk at the fair or at the restaurant that had supplied the kiosk. Of the 35 persons, 20 were questioned about food history; all reported eating a smoked salmon and pasta dish made with fresh chopped parsley. Stool samples from six (38%) of 16 foodhandlers, including the four who handled the parsley, were negative for S. sonnei. One child who had eaten at the kiosk was the index patient at a day care center, from which five secondary cases of shigellosis were reported.

OTHER INVESTIGATIONS

In addition to these four outbreaks, four additional restaurant-associated outbreaks of S. sonnei were identified, involving an additional 218 persons with culture-confirmed or probable shigellosis. Of the 111 persons interviewed, 106 (96%) reported eating chopped, uncooked, curly parsley. Isolates from three of these outbreaks (in Minnesota and California in the United States and in Alberta in Canada) matched the outbreak PFGE pattern. In the fourth outbreak (in Florida), one culture-confirmed case was identified; the isolate was not available for PFGE testing.

TRACEBACK AND ENVIRONMENTAL INVESTIGATIONS

To determine the source(s) of parsley for the seven outbreaks linked by PFGE, state and provincial health departments, CDC, the Food and Drug Administration (FDA), and the Canadian Food Inspection Agency conducted traceback investigations. Farm A in Baja California, Mexico, was a possible source of parsley served in six of the seven outbreaks; four farms in California were possible sources of parsley in two to four of the seven outbreaks.

Field investigations of farm A by FDA and CDC found that the municipal water that supplied the packing shed was unchlorinated and vulnerable to contamination. This water was used for chilling the parsley in a hydrocooler immediately after harvest and for making ice with which the parsley was packaged for transport. Because the water in the hydrocooler was recirculated, bacterial contaminants in the water supply or on the parsley could have survived in the absence of chlorine and contaminated many boxes of parsley. Farm workers and village residents served by this water system reported drinking bottled water or water from other sources. Workers had limited hygiene education and limited sanitary facilities available on the farm at the time of the outbreak.

Foodhandlers at six (75%) of the eight implicated restaurants reported washing parsley before chopping it. Usually parsley was chopped in the morning and left at room temperature, sometimes until the end of the day, before it was served to customers.

LABORATORY INVESTIGATIONS

The Minnesota Department of Health laboratory, which has tested isolates of S. sonnei by PFGE routinely since 1995, identified a previously unrecognized PFGE pattern of S. sonnei and a closely related pattern that differed by a single band associated with the two outbreaks in Minnesota. The pattern was distributed to other laboratories through PulseNet, the national molecular subtyping network for foodborne disease. In Minnesota and at CDC, strains from all seven outbreaks for which isolates were available for PFGE testing had the outbreak PFGE pattern. Isolates from the seven outbreaks were resistant to ampicillin, trimethoprim-sulfamethoxazole, tetracycline, sulfisoxazole, and streptomycin.

Investigators at the University of Georgia Center for Food Safety and Quality Enhancement conducted studies to determine the effects of temperature and handling on the growth and survival of S. sonnei on parsley. Colony-forming units of S. sonnei per gram (cfu/g) decreased by approximately 1 log per week on parsley, whether chopped or whole, under refrigeration (39 F {4 C}). In contrast, S. sonnei counts increased on parsley kept at room temperature (70 F {21 C}). On whole parsley, the increase was limited to 1 log cfu/g during the first 1-2 days, but on chopped parsley a 3 log cfu/g increase was observed within 24 hours.

Reported by: L Crowe, W Lau, L McLeod, Calgary Regional Health Authority; CM Anand, Provincial Laboratory of Southern Alberta; B Ciebin, C LeBer, Ontario Ministry of Health; S McCartney, Ottawa-Carleton Health Unit; R Easy, C Clark, F Rodgers, National Enterics Laboratory, Health Canada; A Ellis, Health Canada; A Thomas, L Shields, B Tate, A Klappholz, I LaBerge, Canadian Food Inspection Agency. R Reporter, H Sato, E Lehnkering, L Mascola, Los Angeles County Dept of Health Svcs, Los Angeles; J Waddell, S Waterman, State Epidemiologist, California Dept of Health Svcs. J Suarez, Miami-Dade County Health Dept, Miami; R Hammond, R Hopkins, State Epidemiologist, Florida Dept of Health. P Neves, Massachusetts Div of Food and Drugs; MS Horine, P Kludt, A DeMaria, Jr, State Epidemiologist, Massachusetts Dept of Public Health. C Hedberg, J Wicklund, J Besser, D Boxrud, B Hubner, M Osterholm, State Epidemiologist, Minnesota Dept of Health. FM Wu, L Beuchat, Center for Food Safety and Quality Enhancement, Univ of Georgia, Athens, Georgia. Food and Drug Administration. Epidemiology Br, Div of Parasitic Diseases; Hospital Environment Laboratory Br, Hospital Infections Program; Foodborne and Diarrheal Diseases Br, Div of Bacterial and Mycotic Diseases, National Center for Infectious Diseases; and EIS officers, CDC.

Editorial Note: S. sonnei is a common cause of gastroenteritis, accounting for 10,262 (73%) of the 14,071 laboratory-confirmed Shigella infections reported to CDC in 1996 (1). Humans and other primates are the only reservoirs for S. sonnei, and transmission occurs through the fecal-oral route. As few as 10-100 organisms can cause infection, enabling person-to-person transmission where hygienic conditions are compromised. In the United States, S. sonnei primarily infects young children and is a common cause of diarrheal outbreaks in child care centers (2). Although reported infrequently, food-borne outbreaks of shigellosis have been associated with raw produce, including green onions (3), iceberg lettuce (4-7), and uncooked baby maize (8).

Before the outbreak described in this report, PFGE was not used routinely by most state public health laboratories to subtype isolates of S. sonnei, making it difficult to detect clusters or outbreaks. This investigation demonstrated how the routine use of PFGE and PulseNet can link clusters of S. sonnei infections in widely dispersed geographic areas. This same technology is now used widely for comparing isolates of Escherichia coli O157:H7. CDC, in consultation with the Minnesota Department of Health, is developing a standard protocol for PFGE subtyping of S. sonnei isolates by PulseNet laboratories.

In the outbreak described in this report, isolates were resistant to many antimicrobial agents, including ampicillin and trimethoprim-sulfamethoxazole, which are commonly used to treat shigellosis. This highly resistant pattern is seen more frequently in countries other than the United States. During 1985-1995, antimicrobial resistance among Shigella increased substantially in the United States (9): resistance to ampicillin increased from 32% to 67%, resistance to trimethoprim-sulfamethoxazole increased from 7% to 35%, and resistance to both agents increased from 6% to 19%. A history of international travel was the strongest risk factor for Shigella infection resistant to trimethoprim-sulfamethoxazole (9).

The findings in this report indicate that several changes in food storage and food preparation procedures are needed. In restaurants, foodhandling practices such as pooling large batches of parsley for chopping and holding chopped parsley at room temperature increase the risk that sporadic low-level bacterial contamination will lead to outbreaks of gastrointestinal illness. When fresh produce is chopped, the release of nutrients may provide a favorable medium for bacterial growth. The risk for outbreaks can be reduced by storing chopped parsley for shorter times, keeping it refrigerated, and chopping smaller batches (10). Changes in parsley production on the farm (e.g., the use of adequately chlorinated water for chilling and icing parsley, education of farm workers on proper hygiene, and possibly the use of post-harvest control measures such as irradiation) may be necessary to ensure that produce is not contaminated with pathogens.

References
1. Foodborne and Diarrheal Diseases Branch. Shigella surveillance: annual tabulation summary, 1996. Atlanta, Georgia: US Department of Health and Human Services, Public Health Service, CDC, National Center for Infectious Diseases, Division of Bacterial and Mycotic Diseases, Foodborne and Diarrheal Diseases Branch, 1997.
2. Mohle-Boetani JC, Stapleton M, Finger R, et al. Communitywide shigellosis: control of an outbreak and risk factors in child day-care centers. Am J Public Health 1995;85:812-6.
3. Cook K, Boyce T, Langkop C, et al. A multistate outbreak of Shigella flexneri 6 traced to imported green onions. Presented at the 35th Interscience Conference on Antimicrobial Agents and Chemotherapy. San Francisco, California, September 1995.
4. Martin DL, Gustafson TL, Pelosi JW, Suarez L, Pierce GV. Contaminated produce -- a common source for two outbreaks of Shigella gastroenteritis. Am J Epidemiol 1986;124:299-305.
5. Davis H, Taylor JP, Perdue JN, et al. A shigellosis outbreak traced to commercially distributed shredded lettuce. Am J Epidemiol 1988;128: 1312-21.
6. Kapperud G, Rorvik LM, Hasseltvedt V, et al. Outbreak of Shigella sonnei infection traced to imported iceberg lettuce. J Clin Microbiol 1995;33:609-14.
7. Frost JA, McEvoy MB, Bentley CA, Andersson Y. An outbreak of Shigella sonnei infection associated with consumption of iceberg lettuce. Emerging Infectious Diseases 1995;1:26-9.
8. Molbak K, Neimann J. Outbreak in Denmark of Shigella sonnei infections related to uncooked "baby maize" imported from Thailand. Eurosurveillance Weekly 1998;2:980813. Available at http://www.outbreak.org.uk/1998/980813.html. Accessed April 2, 1999.
9. Cook K, Boyce T, Puhr N, Tauxe R, Mintz E. Increasing antimicrobial-resistant Shigella infections in the United States. Presented at the 36th Interscience Conference on Antimicrobial Agents and Chemotherapy. New Orleans, Louisiana, September 1996.
10. Wu FM, Doyle MP, Beuchat LR, Mintz E, Swaninathan B. Factors influencing survival and

growth of Shigella sonnei on parsley. Presented at the sixth annual meeting of the Center for Food Safety and Quality Enhancement. Atlanta, Georgia, March 1999.

March 22, 1996 / 45(11);229-231

Shigella sonnei Outbreak Associated with Contaminated Drinking Water -- Island Park, Idaho, August 1995

On August 20, 1995, the District 7 Health Department requested the Idaho Department of Health to assist in investigating reports of diarrheal illness among visitors to a resort in Island Park in eastern Idaho; Shigella sonnei had been isolated from stool cultures of some cases. This report summarizes the findings of the investigation, which implicated contaminated drinking water as the cause of the outbreak.

The resort is located in an area frequented by tourists because of its recreational waters and proximity to a large national park. Facilities include a 36-room motel, conference room, two hot tubs, and 10 hook-ups for recreational vehicles. The resort does not have a restaurant but offers catered meals to groups. To determine the source and extent of the outbreak, persons who had either stayed overnight or eaten at the resort during August 1-21 were telephoned and interviewed; resort staff also were interviewed. Names of visitors were obtained from the resort's records and from interviews with other guests. A probable case was defined as onset of diarrhea (two or more loose stools during a 24-hour period) with either fever or bloody stools while at the resort or within 11 days of leaving the resort. A confirmed case additionally required Shigella sonnei isolated from stool.

Approximately 810 persons stayed or ate at the resort during August 1-21; of these, 222 were contacted, and 221 (99%) agreed to be interviewed. A total of 82 cases (attack rate: 35%) were identified, including 67 probable and 15 confirmed. The median age of case-patients was 31 years (range: 3 months-81 years), and 42 (51%) were male. Onset of illness occurred during August 6-24 (Figure_1). The average duration from time of arrival until onset of diarrhea was 4 days (range: 1-11 days). Fifteen patients (18%) had bloody diarrhea, eight sought treatment in local emergency departments, and five were admitted to local hospitals.

Risk for illness was higher among persons who had drunk tap water or had used ice from the ice machines at the resort than among those who did not (80 {46%} of 175 versus one {3%} of 39; relative risk=17.6; 95% confidence interval=2.5-123.0). Increased risk for illness was not associated with eating or drinking any resort food or beverages (other than water), swimming or fishing in the area recreational waters, using a hot tub, or dining in any local restaurants in Island Park. At least 14 of the case-patients stayed only one night at the resort and had drunk tap water obtained in their rooms but had not eaten food prepared at the resort.

After receiving reports of diarrheal illness among guests at the resort, the District 7 Health Department recommended several prevention measures before initiating the investigation. On August 17, the resort posted warning signs at water taps cautioning against drinking water; on August 19, food service was terminated; and on August 21, bottled water was placed in every room. Resort water is supplied by one well, which was dug in 1993. Samples of water obtained from the well on August 23 were positive for fecal coliform bacteria; however, cultures were negative for Shigella.

During the outbreak investigation, residents in some houses in a new subdivision adjacent to the resort reported acute diarrheal illness. Each house either had a private well or shared a well with a neighbor. S. sonnei was isolated from stool samples from six persons who resided in three of these homes. All six persons denied direct contact with other neighbors or visiting the resort. Fecal coliform bacteria were identified in samples obtained from six of 10 neighborhood wells during August 21-23. However, cultures of water samples from two of these wells were negative for S. sonnei.

The water table in the area was substantially higher than normal because of high rainfall levels during the spring. Initial inspection of a sewer line that had been placed from the subdivision and the resort by a private developer indicated that sewage was draining improperly, although no breaks were identified in selected sections that were excavated for inspection.

Plasmid profiles were performed on Shigella isolates from 15 ill resort visitors, two ill staff members, and five of six ill residents of the neighboring houses; all 22 isolates shared seven identical plasmids. S. sonnei isolates obtained from patients elsewhere in Idaho did not match this pattern.

The District 7 Health Department required that the resort provide bottled or boiled water to visitors and recommended that persons residing in the area have their well water tested and boil all drinking water. Since the investigation, the resort has drilled a new and deeper well.

Reported by: B Arnell, District 7; J Bennett, Southeast District; R Chehey, State Bur of Laboratories; J Greenblatt, MD, State Epidemiologist; Idaho State Dept of Health. Foodborne and Diarrheal Diseases Br, Div of Bacterial and Mycotic Diseases, National Center for Infectious Diseases; Div of Field Epidemiology, Epidemiology Program Office, CDC.

Editorial Note: S. sonnei is a well-recognized cause of gastrointestinal illness and the most common cause of bacillary dysentery in the United States. In addition to diarrhea, common manifestations of shigellosis include fever, abdominal pain, and blood or mucus in the stool. Although most outbreaks of shigellosis have been attributed to person-to-person transmission (1), foodborne (2-4), waterborne (5), and swimming-related (6,7) outbreaks have been reported. Waterborne outbreaks commonly are associated with wells that have been fecally contaminated. However, because Shigella organisms rarely are isolated from water sources, the identification of a waterborne source usually is based on epidemiologic evidence.

The findings of this investigation indicate possible transmission from multiple wells in the same area, suggesting possible contamination and spread of viable Shigella organisms through the groundwater. Plasmid profile analysis confirmed that the outbreak isolates were the same strain that caused illness among persons in the neighboring community. Although investigation of the sewer line continues, the source of the contamination of the well water has not yet been determined.

Routine water-quality testing, including testing for fecal coliform (thermotolerant) bacteria, is the most practical indicator of possible bacterial contamination of drinking water from both community and private water supplies. However, many privately owned wells never are tested for fecal coliform bacteria. In addition, timely testing, reporting, and follow-up in cases of contaminated public water systems often are constrained because of limited resources available to local health departments.

References
1. Mandell GM, Bennett JE, Dolin R. Principles and practice of infectious diseases. 4th ed. New York: Churchill Livingstone Inc, 1995:2033-5.
2. Jewell Ja, Warren RE, Buttery RB. Foodborne shigellosis. Commun Dis Rep CDR Rev 1993;3:R42-R44.
3. Kapperud G, Rorvik LM, Hasseltvedt V, et al. Outbreak of Shigella sonnei infection traced to imported iceberg lettuce. J Clin Microbiol 1995;33:609-14.
4. Hedburg CW, Levine WC, White KE. An international foodborne outbreak of shigellosis associated with a commercial airline. JAMA 1992;268;3208-12.
5. Samonis G, Elting L, Skoulika E, et al. An outbreak of diarrhoeal disease attributed to Shigella sonnei. Epidemiol Infect 1994;112:235-45.
6. Keene WE, McAnulty JM, Hoesly FC, et al. A swimming-associated outbreak of hemorrhagic colitis caused by Escherichia coli 0157:H7 and Shigella sonnei. N Engl J Med 1994;331:579-84.
7. Sorvillo FJ, Waterman SH, Vogt JK, England B. Shigellosis associated with recreational water contact in Los Angeles County. Am J Trop Med Hyg 1988;38:6613-7.

September 09, 1994 / 43(35);657

Outbreak of Shigella flexneri 2a Infections on a Cruise Ship

During August 29-September 1, 1994, an outbreak of gastrointestinal illness occurred on the cruise ship Viking Serenade (Royal Caribbean Cruises, Ltd.) during its roundtrip voyage from San Pedro,

California, to Ensenada, Mexico. A total of 586 (37%) of 1589 passengers and 24 (4%) of 594 crew who completed a survey questionnaire reported having diarrhea or vomiting during the cruise. One death occurred in a 78-year-old man who was hospitalized in Mexico with diarrhea. Shigella flexneri 2a has been isolated from fecal specimens from at least 12 ill passengers. Antimicrobial susceptibility testing of representative isolates indicated resistance to tetracycline and susceptibility to ampicillin and trimethoprim-sulfamethoxazole. The subsequent two cruises of the ship were canceled. Investigation of the mode of transmission is under way.

Additional information is available from the Vessel Sanitation Program, Special Programs Group, National Center for Environmental Health, telephone (305) 539-6730.

Reported by: Communicable Disease Control, Los Angeles County Dept of Health Svcs; Div of Communicable Disease Control, California Dept of Health Svcs. Foodborne and Diarrheal Diseases Br, Div of Bacterial and Mycotic Diseases, National Center for Infectious Diseases; Vessel Sanitation Program, Special Programs Group, National Center for Environmental Health; Div of Field Epidemiology, Epidemiology Program Office, CDC.

*This recommendation allows slightly higher doses of prilocaine when body weight is measured in pounds rather than kilograms (8 mg/kg=3.6 mg/lb).

June 26, 1992 / 41(25);440-442

Shigellosis in Child Day Care Centers -- Lexington-Fayette County, K

In January 1991, the Lexington-Fayette County (Kentucky) Health Department (LFCHD) received three reports of Shigella sonnei infections from the University of Kentucky microbiology laboratory. The infections occurred in children aged 2-3 years, each of whom attended a different child day care center in Lexington-Fayette County (population: 200,000). This report summarizes the findings of an investigation by the LFCHD and the Kentucky Department for Health Services to assess the impact of day care center attendance on communitywide shigellosis.

Public health field nurses obtained stool cultures from family members and day care center contacts of the three children; five contacts tested positive for S. sonnei infection. Despite health education efforts and follow-up by LFCHD, cases continued to occur throughout the community. From January 1 through July 15, 1991, 186 culture-confirmed S. sonnei infections were reported in Lexington-Fayette County.

Investigators attempted to interview an adult member of each family with at least one case of culture-confirmed infection. Questions were asked about the occurrence of diarrhea and child day care center attendance for all household members during January 1 through July 15, 1991. A case of shigellosis was defined as diarrhea (i.e., two or more loose stools per day for 2 or more days) in a person who resided in a household with a person who had culture-confirmed shigellosis. An initial case of shigellosis was defined as the first incidence of diarrhea in a household member.

Of the 186 persons with culture-confirmed infection, 165 (89%) were contacted; these 165 persons represented 109 households, within which 111 initial cases of shigellosis were identified. Of the 64 children aged less than 6 years with initial cases, 57 (89%) attended licensed day care centers, compared with 44 (67%) of the 66 children who were not initial case-patients (odds ratio=4.1; 95% confidence interval=1.5-11.6).

In 1990, approximately 20,000 children aged less than 6 years lived in Lexington-Fayette County; the total capacity of licensed day care centers in the county was 7754 children (Urban Research Institute, University of Louisville, Kentucky, unpublished data, 1992). Among children aged less than 6 years, the rates of initial cases were 7.4 per 1000 children who attended licensed child day care centers and 0.6 per 1000 children of the same age group who did not attend day care centers. The rate of initial cases of shigellosis attributable to child day care center attendance was 6.8 per 1000 children aged less than 6 years, and the attributable risk percentage* was 91%. Thus, 52 (91%) of the 57 initial cases among children aged less than 6 years in licensed child day care and 47% of

215

the 111 initial cases of all ages were attributed to child day care center attendance.

To control shigellosis, in June 1991, LFCHD created a Shigella task force that instituted a diarrhea clinic to facilitate proper diagnosis and treatment, intensified infection-control training and surveillance for shigellosis, and encouraged community-based participation in prevention efforts. Children were monitored in handwashing at day care centers, elementary schools, summer camps, and free-lunch sites. Three weeks after intensive interventions were initiated, the incidence of culture-confirmed cases declined substantially.

Reported by: M Kolanz, J Sandifer, J Poundstone, MD, Shigella Task Force, Lexington-Fayette County; M Stapleton, MSPH, R Finger, MD, State Epidemiologist, Dept for Health Svcs, Kentucky Cabinet for Human Resources. Meningitis and Special Pathogens Br, and Enteric Diseases Br, Div of Bacterial and Mycotic Diseases, National Center for Infectious Diseases, CDC.

Editorial Note: Shigellosis is transmitted by the fecal-oral route; transmission is efficient because the infective dose is low. Minor hygienic indiscretions allow fecal-oral spread from person to person, and many persons with mild illness are in contact with others. As a result, community outbreaks are difficult to control (1).

During 1970-1988, the proportion of young children cared for in licensed centers in the United States increased from 3.5% to 22.0% (2,3). Child day care center attendance increases the risk for diarrheal disease (4). The risk for shigellosis is greatest for children aged less than 6 years (5,6) who are most likely to spread disease to their household members (6). Behavior typical in toddlers, including oral exploration of the environment and suboptimal toileting hygiene, may be associated with this risk (7).

From 1974 through 1990, 26 cases of Shigella infection in Lexington-Fayette County had been the maximum reported in any year. However, a large outbreak with 112 culture-confirmed cases of shigellosis affected the same community in 1972-73 (5). In both outbreaks, child day care center attendance was associated with an increased risk for initial cases in households. Secondary attack rates by age group within households were similar in the two outbreaks: for children aged 1-5 years, rates were 47% in 1972-73 and 53% in 1991. However, in 1991, 51% of the initial cases occurred among children aged less than 6 years who attended a licensed child day care center, compared with 23% in 1972-73. The attributable risk of 91% for day care center attendance among initial cases in young children in 1991 suggests a need for improved infection-control practices in child day care centers.

One of the national health objectives for the year 2000 is to reduce by 25% the number of cases of infectious diarrhea among children who attend licensed day care centers (objective 20.8) (8). To decrease the likelihood of transmission of diarrheal illness in day care centers, facility operators should ensure the following:

* Staff and children should be instructed in rigorous and consistent handwashing practices, including the use of soap and running water.
* Staff and children should wash their hands after using the toilet and changing diapers, and before handling, preparing, serving, and eating food. During an outbreak of diarrheal illness, staff and children should also wash their hands on entry to the day care center.
* If possible, staff who prepare food (including bottles) should not change diapers or assist children in using the toilet. If they perform both functions, they should practice rigorous handwashing before handling food and after using the toilet, changing diapers, and assisting children with toilet use.
* Surfaces, hard-surface toys, and other fomites should be decontaminated regularly; in the setting of a diarrheal outbreak, this should be done at least once per day.
* Children with diarrhea should be excluded from child day care until they are well.
* In the outbreak setting, where feasible, convalescing children should be placed in a separate room with separate staff and a separate bathroom until they have two stool cultures that are negative for Shigella 48 hours or more after completion of a 5-day course of antibiotics (9). If cohorting is not feasible, temporary closure of day care centers may be considered to interrupt disease transmission; however, this policy could increase the likelihood of transmission if children are transferred to other centers (10).

References
1. CDC. Community outbreaks of shigellosis -- United States. MMWR 1990;39:509-13,519.
2. Keyserling MD. Windows on day care. New York: National Council of Jewish Women, 1972:1-3.
3. Dawson DA. Child care arrangements: health of our nation's children--United States, 1988. Hyattsville, Maryland: US Department of Health and Human Services, Public Health Service, CDC, 1990. (Advance data no. 187).
4. Bartlett AV, Moore MD, Gary GW, Starko KM, Erben JJ, Meredith BA. Diarrheal illness among infants and toddlers in child day care centers. J Pediatr 1985;107:495-502.
5. Weissman JB, Schmerler A, Weiler P, Filice G, Godby N, Hansen I. The role of preschool children and day-care centers in the spread of shigellosis in urban communities. J Pediatr 1974;84:797-802.
6. Wilson R, Feldman RA, Davis J, Laventure M. Family illness associated with Shigella infection: the interrelationship of age of the index patient and the age of household members in acquisition of illness. J Infect Dis 1981;143:130-2.
7. Pickering LK. The day care center diarrhea dilemma (Editorial). Am J Public Health 1986;76:623-4.
8. Public Health Service. Healthy people 2000: national health promotion and disease prevention objectives -- full report, with commentary. Washington, DC: US Department of Health and Human Services, Public Health Service, 1991; DHHS publication no. (PHS)91-50212.
9. Tauxe RV, Johnson K, Boase J, Helgerson SD, Blake PA. Control of day care shigellosis: a trial of convalescent day care in isolation. Am J Public Health 1986;76:627-30.
10. Tackett CO, Cohen ML. Shigellosis in day care centers: use of plasmid analysis to assess control measures. Pediatr Infect Dis 1983;2:127-9.

* Incidence among children exposed to day care minus incidence among children not exposed to day care, divided by incidence among children exposed to day care.

June 28, 1991 / 40(25);421,427-428

International Notes Shigella dysenteriae Type 1 -- Guatemala, 1991

On March 14, 1991, physicians at a hospital in Guatemala City reported to the Institute of Nutrition of Central America and Panama (INCAP) that a 2-year-old boy living in an orphanage in Guatemala City had been hospitalized with dysentery; stool cultures yielded Shigella dysenteriae type 1. Another child from the orphanage had recently died from dysentery. During March 18-21, two other young children from the orphanage were diagnosed with S. dysenteriae type 1. On March 21, health officials in Rabinal, in the department of Baja Verapaz, reported more than 100 cases of dysentery to the Division of Epidemiology and Disease Control of the Ministry of Health (MOH). This report summarizes the investigation of these outbreaks. Guatemala City

The orphanage houses approximately 150 children. No new children had been admitted to the orphanage in 1991, and no illness had been reported among staff members. The index patient was treated with trimethoprim-sulfamethoxazole; however, a stool culture yielded S. dysenteriae type 1 that was resistant to trimethoprim-sulfamethoxazole as well as to ampicillin, chloramphenicol, and tetracycline. Stool cultures from the two children who became ill after the index patient also yielded S. dysenteriae type 1 with the same resistance pattern as the initial isolate. Stool cultures from 39 children most likely to have had contact with the index patient were negative, except for one isolate of S. flexneri type 4. No additional cases of dysentery have been reported from the orphanage. Rabinal, Baja Verapaz

On March 21, the MOH received a request from health officials in the department of Baja Verapaz (116 miles (186 km) north of Guatemala City) for drugs to treat suspected amebiasis; the health officials reported that more than 100 cases of dysentery had occurred in residents of Rabinal, a community of approximately 10,000 persons. To determine the cause of the outbreak, INCAP investigators traveled to Rabinal and collected stool specimens in Cary-Blair transport medium from 16 per-sons with dysentery. Eleven samples yielded S. dysenteriae type 1, resistant to chloramphenicol and tetracycline. Based on these results, ill persons were treated with trimethoprim-sulfamethoxazole.

On April 2 and 10, investigators from INCAP and the MOH again visited Rabinal. Surveys done by personnel of the local health post showed that at least 540 persons had developed dysentery since early March; two infants had died. Stool samples were obtained from 46 patients with dysentery; 12 grew S. dysenteriae type 1. For 10 patients, strains were indistinguishable from those obtained in March. Strains from two patients were resistant to ampicillin, chloramphenicol, tetracycline, and trimethoprim-sulfamethoxazole. One of these resistant strains was from a boy who had taken trimethoprim-sulfamethoxazole prophylaxis for respiratory illness in mid-March. By the end of April, local personnel reported that the number of new cases of dysentery was declining. Reported by: JR Cruz, F Cano, L Rodriguez, Program on Infection, Nutrition and Immunology, Div of Nutrition and Health, Institute of Nutrition of Central America and Panama; CA Rios, Hospital for Infectious Diseases, Guatemala City; P Guerra, Z Leonardo, Baja Verapaz Health Area, Ministry of Health. Enteric Diseases Br, Div of Bacterial and Mycotic Diseases, National Center for Infectious Diseases, CDC.

Editorial Note: Pandemic S. dysenteriae type 1 (the Shiga bacillus) affected Central America from 1969 through 1972. In Guatemala, there were more than 112,000 cases and at least 10,000 deaths (1,2). The outbreak spread quickly, with high attack rates in all age groups and the highest incidence and mortality rates in young children (2,3). The case-fatality rate estimated from village surveys was 7.4% (2). Many cases were misdiagnosed as amebiasis, and treatment with antiamebic drugs contributed to the high mortality (2,3). Treatment was further complicated by resistance of the epidemic strain of S. dysenteriae type 1 to sulfathiazole, chloramphenicol, and tetracycline, drugs commonly used at that time to treat dysentery (4).

Since 1972, no major outbreaks of dysentery caused by the Shiga bacillus have occurred in Central America. However, in 1988, the number of these infections reported in the United States increased fivefold over the annual mean from the preceding decade, and most ill persons had recently visited the Yucatan peninsula in Mexico (5). The antimicrobial resistance pattern and plasmid profile were similar to those of the 1969-1972 pandemic strain (4,5). In 1989, the number of imported cases decreased in the United States, and outbreaks of documented Shiga infection have not been reported from Mexico.

Appropriate antimicrobial therapy decreases the severity and duration of dysentery caused by Shigella (6). Nalidixic acid is effective therapy for strains resistant to other antimicrobials; the newer quinolones are also effective, but are costly and have not been approved for use in children (6). Moreover, Shigella can rapidly acquire resistance, and are likely to do so in settings in which antimicrobials are commonly used and shigellosis is endemic (7). The recent cases in Guatemala underscore the need for continued surveillance for enteric pathogens, especially those associated with dysentery. Once Shigella are identified, determination of the antimicrobial resistance pattern and the modes of transmission are important in designing control measures. As during the 1969-1972 pandemic, the recent cases in Rabinal were initially misdiagnosed as amebiasis, a misdiagnosis that may be common in some locations (8). Prompt culturing facilitated the correct diagnosis and appropriate therapy.

The appearance of the Shiga bacillus in two locations separated by more than 100 km suggests this pathogen may be present in other areas of Guatemala. The detection of trimethoprim-sulfamethoxazole-resistant strains early in the outbreak highlights the need for continued monitoring of resistance. The MOH and INCAP have requested that any clusters of bloody diarrhea among persons in Guatemala be reported. Training in techniques to identify S. dysenteriae type 1 has been incorporated into the courses for workers from regional laboratories; these courses were initiated in response to the current cholera epidemic.

References
1. Mata LJ, Gangarosa EJ, Caceres A, Perera DR, Mejicanos ML. Epidemic Shiga bacillus dysentery in Central America. I. Etiologic investigations in Guatemala, 1969. J Infect Dis 1970;122:170-80.
2. Mendizabal-Morris CA, Mata LJ, Gangarosa EJ, Guzman G. Epidemic Shiga-bacillus dysentery in Central America: derivation of the epidemic and its progression in Guatemala, 1968-69. Am J Trop Med Hyg 1971;20:927-33.
3. Gangarosa EJ, Perera DR, Mata LJ, Mendizabal-Morris C, Guzman G, Reller LB. Epidemic Shiga bacillus dysentery in Central America. II. Epidemiologic studies in 1969. J Infect Dis 1970;122:181-90.
4. Reller LB, Rivas EN, Masferrer R, Bloch M, Gangarosa EJ. Epidemic Shiga-bacillus dysentery

in Central America: evolution of the outbreak in El Salvador, 1969-70. Am J Trop Med Hyg 1971;20:934-40.

5. Parsonnet J, Greene KD, Gerber AR, Tauxe RV, Aguilar OCJ, Blake PA. Shigella dysenteriae type 1 infections in US travellers to Mexico, 1988. Lancet 1989;2:543-6.

6. Salam MA, Bennish ML. Antimicrobial therapy for shigellosis. Rev Infect Dis 1991;13 (suppl 4):332-41.

7. Griffin PM, Tauxe RV, Redd SC, Puhr ND, Hargrett-Bean N, Blake PA. Emergence of highly trimethoprim-sulfamethoxazole-resistant Shigella in a Native American population: an epidemiologic study. Am J Epidemiol 1989;129:1042-51.

8. Pelto GH. The role of behavioral research in the prevention and management of invasive diarrheas. Rev Infect Dis 1991;13(suppl 4):255-8.

August 03, 1990 / 39(30);509-513,519

Current Trends Community Outbreaks of Shigellosis -- United States

From 1986 to 1988*, the reported isolation rate of Shigella in the United States increased from 5.4 to 10.1 isolates per 100,000 persons (Figure 1). In 1988, state health departments reported 22,796 isolates of Shigella to CDC, the highest number since national surveillance began in 1965. In addition to the recent increase in Shigella isolation rates, many communitywide shigellosis outbreaks that have been difficult to control have been reported. This report describes four community outbreaks of shigellosis during 1986-1989 in which innovative public health control measures were used.

Kankakee County, Illinois. From October 1986 through February 1987, an outbreak of shigellosis caused by S. sonnei occurred in Kankakee County, Illinois (population: 97,800). Of 191 persons with culture-confirmed shigellosis, 70% were black and 61% were aged 1-10 years. Thirty-one percent of patients were hospitalized. Cases were clustered in low-income areas. An epidemiologic investigation did not identify common sources of exposure in the community; many patients reported having had contact with persons with culture-confirmed shigellosis or symptoms compatible with shigellosis.

To control this outbreak, from December 12 to January 10 the following measures were implemented: 1) information about shigellosis and its prevention was provided to parents of all children in the school district where most of the cases occurred, to child-care centers and preschools, and through schools, churches, and the news media; 2) teachers monitored handwashing by students before lunch; 3) parents assisted in monitoring handwashing in schools in the most severely affected areas; and 4) home-prepared foods were not permitted at any school or child-care events. Although the number of reported cases subsequently decreased, the outbreak did not end until March.

Peoria County, Illinois. From February through September 1987, a shigellosis outbreak caused by S. sonnei occurred in Peoria County, Illinois (Figure 2) (population: 181,500). Of the 513 culture-confirmed cases, 75% were in blacks and 69% were in children aged 1-10 years. Most patients resided in low-income areas. Seven percent of patients were hospitalized. Investigation did not identify a common source of exposure; most patients had a history of contact with a person who had culture-confirmed shigellosis or symptoms compatible with shigellosis.

During April, the following interventions were implemented: 1) child-care center and nursery school employees were informed about shigellosis prevention; 2) school officials in the affected area ensured that warm water, soap, and disposable towels for handwashing were always available for students; 3) in schools, parents and teachers instructed students on proper handwashing and monitored children for symptoms of shigellosis; 4) printed educational material about shigellosis was provided to all persons attending Women, Infants, and Children (WIC) clinics, immunization clinics, community clinics, and hospital emergency rooms; 5) volunteers from the local Urban League and housing authority made door-to-door visits in affected neighborhoods to identify cases and provide printed educational material; 6) religious leaders discussed the Shigella outbreak with their congregations, and church publications included information on shigellosis prevention; and 7) parents taught neighborhood children how to wash their hands and monitored them for symptoms of shigellosis. Although the number of reported cases decreased concurrently with the intervention, the outbreak continued at a lower level until September.

Orange County, New York. From November 29, 1986, to February 28, 1987, 110 culture-confirmed cases of S. sonnei gastroenteritis were reported in residents of a religious community (population: 5200) in Orange County, New York (Figure 3). Cases occurred primarily among school children 2-1/2 - 9 years of age; cases were evenly distributed by sex. An epidemiologic investigation did not identify a point source of exposure; spread of disease was consistent with person-to-person transmission.

Control measures were focused in schools and implemented from January 12 through February 28. The measures included 1) widespread dissemination of information about shigellosis and its prevention (e.g., proper handwashing and diaper changing) in schools and the community child-care center, 2) a program in which older children monitored handwashing by young children in the schools, and 3) periodic health department sanitation inspections of the schools. The number of reported cases of shigellosis declined concurrently with the intervention efforts.

Caddo County, Oklahoma. From August through October 1989, 34 persons with gastroenteritis caused by S. sonnei were identified in Caddo County, Oklahoma (Figure 4) (population: approximately 32,100, including 18% Native Americans). Ninety-one percent of cases were in Native Americans. Seventy-one percent were in children and teenagers. An epidemiologic investigation did not identify a common source of infection but did suggest person-to-person transmission: 37 persons with symptoms compatible with shigellosis became ill after being exposed to a person (usually in their household) with a culture-confirmed Shigella infection. Clusters of cases occurred in persons residing in two Native American housing developments where children regularly played and ate snacks together.

Initial interventions implemented from August 29 to September 13 included 1) efforts to contact families of patients to identify potential exposures and secondary cases and to provide information on hygiene and handwashing, 2) education at child-care centers and other institutions on the importance of hygiene and sanitation in preventing transmission, and 3) encouragement of physicians, hospitals, and clinical laboratories in the area to assist in identifying and reporting new cases. The number of new cases reported initially declined; however, when new cases began to increase again, additional measures were implemented from September 26 to October 4, including dissemination of information on shigellosis and its prevention through 1) assistance of tribal leaders in providing information in tribal newsletters and at informal gatherings, 2) presentations at tribal senior citizen lunches, 3) house-to-house visits by public health officials and other persons in areas where clusters of cases were identified, 4) distribution of take-home handouts to students in child-care centers and schools, 5) press releases to local newspapers and radio stations, 6) puppet shows on handwashing performed at all child-care centers, where informational posters were distributed to attendees, and 7) notification to restaurants and churches of the importance of excluding symptomatic persons from food handling duties. The last confirmed case occurred on October 21. Reported by: C Pate, MS, D Safiran, F Sutton, N Scanlon, E Blanchette, Kankakee County Health Dept; A Kennell, MS, C Marvin, MS, L Esch, P Roberts, Peoria City/County Health Dept; K Kelly, C Langkop, MS, BJ Francis, MD, State Epidemiologist, Illinois Dept of Public Health. A Werzberger, MD, Monroe; S Kondracki, R Gallo, DL Morse, MD, State Epidemiologist, New York State Dept of Health. P Callahan, P Boden, MS, GR Istre, MD, State Epidemiologist, Oklahoma State Dept of Health. R Myers, Indian Health Service. Div of Field Svcs, Epidemiology Program Office; Enteric Diseases Br, Div of Bacterial Diseases, Center for Infectious Diseases, CDC.

Editorial Note: Since 1986, the incidence of shigellosis in the United States has increased in all regions of the country. The highest isolation rates were reported among residents of counties with large proportions of low-income minority residents, among young children, and among women of childbearing age.

Communitywide outbreaks of shigellosis can be difficult to control because of the ease of person-to-person transmission among young children, high secondary attack rates, the frequently extended duration of these outbreaks, and multiple points of exposure. The impact of community interventions can be difficult to measure; however, the outbreaks described in this report suggest that effective control efforts should include the following: 1) communitywide recognition of the problem and participation in the intervention, 2) diversified and culture-specific educational efforts to promote handwashing and hygiene, and 3) supervised handwashing for children. Because community leaders can play a key role in developing interventions and ensuring that these interventions are accepted in the community, they should be actively involved in all control efforts.

Handwashing with soap and running water may be the single most important preventive measure to interrupt transmission of shigellosis (1). Soap and running water should be readily accessible to all persons during community outbreaks of shigellosis. Because young children are most likely to be infected with Shigella and are also most likely to infect others (2), a strict policy of supervised hand-washing for young children after they have defecated and before they eat is crucial. Institutions where hygiene may be suboptimal (e.g., schools, child-care centers, and homeless shelters) can amplify transmission of shigellosis into the community and should be targeted for intensive control efforts. Excluding persons with diarrhea from handling food and limiting use of home-prepared foods at large gatherings will reduce the risk of large outbreaks caused by foodborne transmission.

Antimicrobials have a limited role in the control of epidemic shigellosis and are not a substitute for hygienic measures in reducing the secondary spread of shigellosis. Antimicrobials should be reserved for treatment of patients only when clinically indicated, and the decision to use antimicro-bials to treat patients with mild, self-limiting illness should be weighed against the risk of producing resistant strains of Shigella (3). Prophylactic use of antimicrobials cannot be recommended to pre-vent illness in persons who are exposed but not ill. In addition, using antimicrobials to treat patients with mild shigellosis to reduce the spread of secondary infections is not known to be any more effective in preventing Shigella infections than handwashing with soap and water; moreover, this practice can lead to the development of resistant strains that complicate therapy (4,5). Because resistance patterns may change, antimicrobial selection should be based on ongoing monitoring of local antimicrobial resistance of Shigella strains.

Shigellosis outbreaks can occur at any time of the year but are most common in the summertime (6). Shigella infections should be suspected in communitywide (Continued on page 519)epidemics of diarrheal illness that disproportionately affect young children. Stool specimens should be obtained and state and local health departments informed promptly of culture-confirmed cases so that out-breaks of shigellosis can be recognized and appropriate control measures instituted.

References
1. Kahn MU. Interruption of shigellosis by handwashing. Trans R Soc Trop Med Hyg 1982;76:164-8.
2. Wilson R, Feldman RA, Davis J, LaVenture M. Family illness associated with Shigella infection: the interrelationship of age of the index patient and the age of household members in acquisition of illness. J Infect Dis 1981;143:130-2.
3. Weissman JB, Gangarosa EJ, Dupont HL. Shigellosis: to treat or not to treat? JAMA 1974;229:1215-6.
4. CDC. Multistate outbreak of Shigella sonnei gastroenteritis--United States. MMWR 1987;36:440-2,448-9.
5. Griffin PA, Tauxe RT, Redd SC, Puhr ND, Hargrett-Bean N, Blake P. Emergence of highly trimethoprim-sulfamethoxazole resistant Shigella in a Native American population: an epidemiologic study. Am J Epidemiol 1989;129:1042-51.
6. Black RE, Craun GF, Blake PA. Epidemiology of common-source outbreaks of shigellosis in the United States, 1961-1975. Am J Epidemiol 1978;108:47-52. *The most recent year for which national surveillance data are available.

August 12, 1988 / 37(31);465

Epidemiologic Notes and Reports Shigella dysenteriae Type 1 in Tourists to Cancun, Mexico

From January 1 to August 1, 1988, 17 cases of diarrheal disease caused by Shigella dysenteriae type 1 (Shiga bacillus) were reported to CDC. Three cases were reported to CDC during the same period in 1987. Fifteen of the patients with shigellosis had visited Cancun, Mexico, andd two had visited other areas in Mexico in the weeks before or during onset of their illness. The patients had no com-mon exposures in hotels or restaurants. Thirteen (76%) of the patients required hospitalization ; two patients developed hemolytic-uremic syndrome. Six isolates tested thus far at CDC were resistant to chloramphenicol and tetracycline; two isolates were also resistant to ampicillin and trimethoprim-sulfamethoxazole. An epidemiologic and laboratory investigation is under way in Mexico. Reported by: J Sepulveda Amor, Director General de Epidemiologia, Secretaria de Salud, Mexico. Enteric Diseases Br, Div of Bacterial Diseases, Center for Infectious Diseases, CDC.

Editorial Note: The antimicrobial agents often taken prophylactically and therapeutically by travelers--trimethoprim-sulfamethoxazole and tetracycline--may be ineffective against the S. dysenteriae type 1 strains for which sensitivity data are available. Physicians should consider this diagnosis in persons with severe or bloody diarrheal illness who have recently returned from Mexico, obtain appropriate cultures, and report suspected cases of S. dysenteriae to local and state public health authorities. Laboratories are requested to send isolates of S. dysenteriae to appropriate public health laboratories for serotyping. Travelers to Cancun and other regions with recognized risk for travelers' diarrhea should follow CDC's recommendations for international travel (1).

Reference
1. CDC. Health information for international travel, 1988. Atlanta: US Department of Health and Human Services, Public Health Service, 1988; HHS publication no. (CDC)88-8280.

October 02, 1987 / 36(38);633-4

Nationwide Dissemination of Multiply Resistant Shigella sonnei Following a Common-Source Outbreak

In early July 1987, an outbreak of multiply resistant Shigella sonnei gastroenteritis occurred among persons who attended the annual Rainbow Family gathering in North Carolina (1). Since that time, four clusters of gastroenteritis due to multiply resistant S. sonnei have been reported among persons who had no apparent contact with gathering attendees.

Preliminary results from a survey of gathering attendees showed that 157 (58%) of the 270 respondents experienced acute diarrheal illness. This finding is consistent with previous estimates of a 50% or greater attack rate of acute gastroenteritis among the 12,000 attendees (1). Seventy-five attendees from 26 states* and 14 contacts of these persons who had not attended the gathering have had culture-confirmed infection. The S. sonnei isolates from these patients are resistant to ampicillin, tetracycline, and trimethoprim-sulfamethoxazole--the antibiotics usually used to treat shigellosis.

In July, August, and September, clusters of multiply resistant S. sonnei infection occurred in Missouri and Pennsylvania. Isolates from these cases showed an antimicrobial resistance pattern similar to that of the strain involved in the North Carolina outbreak. Two small clusters were reported from Missouri. A third cluster occurred among patrons and employees of a Pennsylvania restaurant. In a fourth cluster, which has been epidemiologically linked to the third, residents and staff of a nursing home in the same Pennsylvania town became ill. Reported by: JN MacCormack, MD, MPH, State Epidemiologist, North Carolina Dept of Human Resources. RH Hutcheson, MD, State Epidemiologist, Tennessee Dept of Health and Environment. HD Donnell Jr, MD, MPH, State Epidemiologist, Missouri Dept of Health. C Diehl, M Hardin, R David, MD, Acting State Epidemiologist, Pennsylvania Dept of Health. Enteric Diseases Br, Div of Bacterial Diseases, Center for Infectious Diseases; Div of Field Svcs, Epidemiology Program Office, CDC.

Editorial Note: In a national survey of Shigella isolates conducted in 1985 and 1986, approximately 4% of isolates from S. sonnei infections acquired in the United States were resistant to trimethoprim-sulfamethoxazole. None had the same antimicrobial resistance pattern as the North Carolina outbreak strain. The occurrence of these four clusters of infection with multiply resistant S. sonnei underscores the need for sensitivity testing to guide in selecting appropriate antimicrobial therapy. Such testing also permits early identification and prompt reporting of multiply resistant strains to public health authorities so further transmission can be prevented.

Further spread of this resistant strain will likely limit the effectiveness of the usual antimicrobial agents for treating shigellosis. Infections that are caused by this multiply resistant Shigella and that require antimicrobial therapy can be treated with nalidixic acid or norfloxacin. Although studies in other countries suggest that both nalidixic acid and norfloxacin are effective for the treatment of shigellosis (2,3), it is important to note that neither nalidixic acid nor norfloxacin has been approved by the Food and Drug Administration (FDA) for the treatment of bacterial gastroenteritis. Both nalidixic acid and norfloxacin are quinolones, and care should be exercised in prescribing either one for children because of experimental evidence that quinolones can cause arthropathy in young animals (4,5). No such lesions have been reported to the FDA in association with nalidixic acid therapy in humans. Life-threatening infections are rare with S. sonnei but could be treated with gentamicin or chloramphenicol, to which the outbreak strain is sensitive.

Basic hygiene and sanitary precautions remain the cornerstones of control measures for shigellosis outbreaks, including those due to multiply resistant strains (6). Vigorous emphasis on handwashing with soap after defecation and before eating has been shown to reduce secondary transmission of shigellosis (7).

References

1. CDC. Shigellosis--North Carolina. MMWR 1987;36:449-50.
2. Rogerie F, Ott D, Vandepitte J, Verbist L, Lemmens P, Habiyaremye
    I. Comparison of norfloxacin and nalidixic acid for treatment of dysentery caused by Shigella dysenteriae type 1 in adults. Antimicrob Agents and Chemother 1986;29:883-6.
2. DuPont HL, Corrado ML, Sabbaj J. Use of norfloxacin in the treatment of acute diarrheal disease. Am J Med 1987;82(suppl 6B):79-83.
3. Schlduter G. Ciprofloxacin: review of potential toxicologic effects. Am J Med 1987;82 (suppl 4A):91-3.
4. Corrado ML, Struble WE, Chennekatu P, Hoagland V, Sabbaj J. Norfloxacin: review of safety studies. Am J Med 1987;82(suppl 6B):22-6.
5. CDC. Multiply resistant shigellosis in a day care center--Texas. MMWR 1986;35:753-5.
6. Khan MU. Interruption of shigellosis by hand washing. Trans R Soc Trop Med Hyg 1982;76:164-8. *California, Colorado, Connecticut, Florida, Georgia, Illinois, Iowa, Maryland, Massachusetts, Michigan, Missouri, New Hampshire, New Jersey, New Mexico, New York, North Carolina, Ohio, Pennsylvania, Rhode Island, Tennessee, Texas, Utah, Vermont, Virginia, West Virginia, and Wisconsin.

July 17, 1987 / 36(27);440-2,448-9

Multistate Outbreak of Shigella sonnei Gastroenteritis -- United States

CDC has received reports that shigellosis outbreaks have occurred in several states, affecting related religious communities. Dates of onset range from November 1986 through June 1987. The largest outbreak was in New York City, and outbreaks in other states began soon after the Passover holiday in April, when many persons visited relatives in New York. Epidemiologic data are incomplete, but in some of these outbreaks new cases continue to occur. A summary of the outbreaks follows. NEW YORK STATE

New York City. Between December 27, 1986, and May 16, 1987, 1,328 cases of culture-confirmed Shigella sonnei gastroenteritis were reported in Brooklyn, New York (Figure 1). On the basis of a sentinel-physician surveillance system, the actual number of cases is likely to have exceeded 13,000. The vast majority of infected persons were tradition-observant Jews belonging to several religious sects. Of the persons with culture-confirmed cases, 55% were less than5 years and 85% were less than17 years of age; 55% were female.

Since more than 25% of the initial isolates were resistant to ampicillian, trimethoprim-sulfamethoxazole (TMP-SXT) was initially recommended for treatment. One isolate of TMP-SXT-resistant Shigella was identified in early January, and TMP-SXT resistance among tested isolates increased from 2% in January to 12% in March. In mid-March, a recommendation was made that patients with mild symptoms should not be treated with antimicrobials.

Person-to-person transmission was thought to be likely, since investigations did not implicate a common source of food or water. With the cooperation of community and religious leaders and physicians, control efforts were directed toward improved sanitation and personal hygiene in schools and homes. Special efforts were made to encourage handwashing with soap and water. The measures were instituted in late March, in anticipation of a large influx of people into these communities to celebrate the Passover holiday during third week in April.

Culture-confirmed cases decreased after the first week in April, but cases continue to be reported above the expected background rates among these religious groups in Brooklyn. The decline in reported Shigella isolates reflects the implementation of hygienic control measures starting the third week in March, the closing of the religious schools during the Passover holiday, and a reduced number of stool specimens obtained for culture as a result of preparations for and observance of the holidays.

Upstate New York. Outbreaks of S. sonnei infections were also recognized in two other tradition-observant Jewish communities in New York State. Approximately 110 culture-confirmed cases were reported in an Orange County community between November 29, 1986, and February 20, 1987. Two-thirds of the patients were less than 5 years and 95% were less than 17 years of age. Cases decreased sharply after a shigellosis advisory bulletin written in Hebrew was distributed throughout the community and a handwashing campaign was directed at school-aged children. A majority of the 110 isolates tested were resistant to ampicillin, but none were resistant to TMP-SXT. Another out-break with greater than260 culture-confirmed cases began November 22, 1986, in a Rockland County community. Although control efforts similar to those used in Orange County were attempted, cases continue to be reported in Rockland County; one outbreak in early June affected 100 (77%) of 130 persons at a private party.

The New York State Department of Health has notified camp directors and nurses in children's sum-mer camps serving the affected communities of the potential for further Shigella transmission. Recommendations include obtaining cultures from children with diarrhea, isolating or excluding cul-ture-positive children from camp activities, and emphasizing personal hygiene. NEW JERSEY

Between May 2 and June 3, 1987, 45 cases of febrile gastroenteritis occurred at a private Hebrew day school in northeast New Jersey, affecting 30% of the children enrolled; half of the affected chil-dren had bloody diarrhea, and one child was hospitalized because of convulsions. S. sonnei was iso-lated from 33 stool specimens. Interviews suggested that the source of the outbreak was a tradition-observant Jewish community in Brooklyn to which many schoolchildren return on weekends. The first cases occurred at the nursery and kindergarten levels, and person-to-person spread appeared to cause cases among other schoolchildren as well as among family members. It was recommended that any child with two loose bowel movements per day or a positive stool culture should remain at home until he or she was asymptomatic for at least 2 days. Teachers and parents were instructed to teach the children to wash their hands thoroughly after defecation and before handling food and playing with other children. OHIO

Five cases of confirmed S. sonnei infections have occurred at a Jewish Orthodox school in Ohio. The student with the earliest confirmed case had illness onset on May 14, but school attendance records indicate that absenteeism for diarrheal illness began a week after the Passover holiday. Preschool and kindergarten children are now being supervised in handwashing after the use of toilet facilities and before meals. Parents have been advised that children with diarrheal illness will be excluded from school. The outbreak is being investigated to establish a possible relationship to the York City outbreak. MARYLAND

On May 26, 1987, the Baltimore County Health Department was notified that shigellosis cases were occurring among students and families associated with four private Jewish schools in Baltimore County, Maryland. In the period 7-June 14, 42 culture-confirmed and 54 probable cases of S. sonnei gastroenteritis occurred in 33 families residing in northwest Baltimore City and adjacent Baltimore County. Of the 87 persons affected whose age was known, 43% were less than6 years old. Symptoms included diarrhea (98.9%), fever (73.6%), abdominal cramps (62.6%), vomiting (21.8%), and bloody diarrhea (10.3%). Index-case children had attended one day-care center, three day-care homes, four private Jewish schools, and one public school. All S. sonnei isolates tested have been resistant to ampicillin but sensitive to TMP-SXT and tetracycline.

No common source for the Baltimore outbreak has been identified, and person- to-person transmis-sion appears likely. Although visiting with friends and family from New York was commonly report-ed, no cases have been linked directly to confirmed cases in New York. Inspections have been per-formed, and hygiene has been emphasized in schools, day-care facilities, summer camps, pools, restaurants, and food stores in an effort to prevent transmission. Reported by: SK Schulman, MD, Brooklyn, New York; A Werzberger, MD, Monroe, New York; S Schultz, MD, City Epidemiologist, New York City Dept of Health; LD Budnick, MD, DL Morse, MD, MS, State Epidemiologist, New York State Dept of Health. H Ragazzoni, DVM, MPH, M Teter, DO, WE Parkin, DVM, DrPH, State Epidemiologist, New Jersey State Dept of Health. LK Giljahn, MPH, T Kramer, TJ Halpin, MD, MPH, State Epidemiologist, Ohio Dept of Health. D Dwyer, MD, J Baumgardner, D Glasser, MD, MPH, Baltimore City Health Dept; E Hopf, MD, Baltimore County Health Dept; E Israel, MD, MPH, State Epidemiologist, Maryland State Dept of Health and Mental Hygiene. Div of Field Svcs, Epidemiology Program Office; Enteric Diseases Br, Div of Bacterial Diseases, Center for Infectious Diseases, CDC.

Editorial Note: Shigella sonnei has become the predominant serotype that causes community out-breaks of shigellosis (1). Children 1-5 years of age are at highest risk of infection; their lack of hygienic practices, combined with the low infectious dose, the frequency of mild illness, and the acquisit of antimicrobial resistance, all predispose to transmission in day-care and preschool settings and to spread within the community (2,3). The community outbreaks reported here appear to be linked, beginning in New York, extending over an 8-month period, and expanding into several states following the Passover holidays; continued transmission is likely. The long duration of the outbreak and the large proportion of cases involving person-to-person transmission are consistent with person-to-person transmis-sion, although the limited epidemiologic data obtainable do not clearly define the routes of transmis-sion. In a recent investigation of an outbreak of hepatitis A in one of the communities affected by this shigellosis outbreak, many opportunities for person-to-person transmission were identified (4).

Devising successful control measures for shigellosis remains a challenge. Handwashing with soap after defecation and before eating has been shown to reduce secondary transmission of shigellosis (5). Although control strategies that emphasize effective handwashing are often difficult to imple-ment among children and families at highest risk, they may interrupt chains of transmission. Creative interventions should be encouraged, including handwashing protocols, posters, and counsel-ing sensitive to local language and custom. Antimicrobial treatment of persons infected with Shigella has also been used to decrease morbidity and the secondary spread of infection (6). However, the appearance of antimicrobial-resistant strains of Shigella, as observed in New York City, has repeated-ly complicated the use of antimicrobials in controlling shigellosis (7). The decision to use antimicro-bials in treating patients with mild, self-limited illness should be weighed against the risk of produc-ing resistant strains of Shigella.

December 05, 1986 / 35(48);753-5

Epidemiologic Notes and Reports Multiply Resistant Shigellosis in a Day-Care Center -- Texas

Between October 10 and November 6, 1985, 15 children at a day-care center in Diboll, Texas, devel-oped a diarrheal illness. Shigella sonnei was isolated from 10 ill children and from two of 19 asymp-tomatic children who were cultured on November 7. All isolates were colicin type 9, resistant to ampicillin, carbenicillin, streptomycin, cephalothin, and trimethoprim/sulfamethoxazole (TMP/SMX), and sensitive to tetracycline, nalidixic acid, chloramphenicol, and gentamicin. The attack rate was highest among the 12- to 22-month-old group. Family members of this group had the highest secondary attack rate (Table 1). No cases occurred among the 22 staff members. None of the children were hospitalized, but four of the five ill family members were.

The 89 children attending the center were cared for, by age group, in separate rooms. All groups except infants and toddlers had separate toilet and playground facilities. Infants and toddlers shared these facilities.

Symptomatic children were excluded from the center until their diarrhea had resolved. Then they were permitted to return, without treatment or cultures, to their classrooms. Handwashing and hygiene were emphasized; contact between age groups was limited; and the routine policy excluding food preparers from child care, particularly diaper-changing, was reinforced. No further cases were reported at the center after November 7, when this strategy was implemented.

During the following month, statewide surveillance for TMP/SMX-resistant S. sonnei infections detected an outbreak among kindergarteners in a town 100 miles away. Although this outbreak strain had the same colicin type and antimicrobial resistance profile as the Diboll strain, its plasmid content differed, and no direct connection between the two outbreaks was discovered. Reported by M Crowder, MD, W Joyce, J Connors, Public Health Region 7, A Quillian, M Czpiel, Angelina County and Cities Health Dist, J Taylor, MPH, DL Martin, MN, CE Alexander, MD, Bureau of Epidemiology, Texas Dept of Health; Enteric Diseases Br, Div of Bacterial Diseases, Center for Infectious Diseases, CDC.

*Editorial Note:*Shigellosis in day-care centers can be difficult to control. Basic hygiene, exclusion of symptomatic persons, and routine antimicrobial therapy for all infected persons have been advocated as control measures (1). In the Texas outbreak reported here, antimicrobial therapy was not part of the control strategy because the strain was resistant to all drugs commonly used to treat shigellosis in children. Nonetheless, the straightforward control strategy in this well-designed day-care center was associated with the end of the outbreak, even though untreated convalescent children returned to the center and untreated asymptomatic carrier children remained there.

The elements contributing to this apparent success included vigorous emphasis on handwashing among staff and children; routine exclusion of ill children; separate areas and staff for diapering and food-preparation; and separate rooms, toilets, and play-facilities for different age groups. There is some evidence that each element is important. Handwashing has been shown to reduce the incidence of diarrheal illness in day-care centers (2). In day-care centers in Houston, Texas, the incidence of diarrheal illness was significantly associated with the proportion of staff who changed diapers and also served or prepared food (3). The usefulness of separating children by age was suggested by uniform shigellosis attack rates observed across ages 0 to 5 years at a day-care center where the children were grouped together (4). Additional study of the efficacy and utility of these specific control measures is needed (5,6).

Providing day-care in isolation for convalescent children may limit the spread of shigellosis in the community. In one outbreak, in which children with shigellosis were rigidly excluded from a day-care center until negative cultures were obtained, the outbreak strain spread to a day-care center in an adjacent county (7). In another outbreak, at a center where isolation of convalescent children was possible, treated, convalescent children without negative cultures were allowed to return to the day-care center, and there was no further spread of illness in either the center or the community (8). Further evaluation of convalescent day-care, with and without isolation, is needed before specific recommendations can be made.

To help day-care center directors, employees, and parents work with health departments to control disease in day-care centers, CDC has produced a training kit: "What To Do To Stop Disease in Child Day-Care Centers". This kit has been distributed to state health departments and licensing boards for distribution to licensed day-care centers. It also can be purchased for $4.00 from the Government Printing Office, Superintendent of Documents, Washington, D.C., 20402. The GPO Stock Number is 017-023-00172-8.

References
1. Child Day Care Infectious Disease Study Group, CDC. Public health considerations of infectious diseases in child day care centers. J Peds 1984;105:683-701.
2. Black RE, Dykes AC, Anderson KE, et al. Handwashing to prevent diarrhea in day-care centers. Am J Epidemiol 1981;113:445-51.
3. Lemp GF, Woodward WE, Pickering LK, Sullivan PS, DuPont HL. The relationship of staff to the incidence of diarrhea in day-care centers. Am J Epidemiol 1984;120:750-8.
4. Pickering LK, Evans DG, DuPont HL, Vollet JJ 3rd, Evans DJ Jr. Diarrhea caused by Shigella, rotavirus and Giardia in day-care centers: prospective study. J Pediatr 1981;99:51-6.
5. Pickering LK, Bartlett AV, Woodward WE. Acute infectious diarrhea among children in day care: epidemiology and control. Rev Inf Dis 1986;8:539-47.
6. Petersen NJ, Bressler GK. Design and modification of the day care environment. Rev Inf Dis 1986;8:618-21.

7. Tacket CO, Cohen ML. Shigellosis in day care centers: use of plasmid analysis to assess control measures. Pediatr Infect Dis 1983;2:127-30.

8. Tauxe RV, Johnson KE, Boase JC, Helgerson SD, Blake PA. Control of day care shigellosis: a trial of convalescent day care in isolation. Am J Publ Health 1986;76:627-30.

October 04, 1985 / 34(39);600-2

Current Trends Shigellosis -- United States, 1984

In 1984, 12,790 Shigella isolates from humans were reported to CDC. This is a 14.4% decrease from the 14,946 isolates reported in 1983. The number of isolates continues to be less than the 15,334 reported during the peak year, 1978 (Figure 2).

Shigella serotypes were reported for 12,179 of the 12,790 isolates. The most frequently isolated serotype, S. sonnei, comprised 64.4% of all isolates serotyped (Table 3). S. flexneri 1a accounted for 14.1% of all S. flexneri subtyped; 1b, 2.6%; 2a, 28.1%; 3a, 24.3%; and 6, 13.3%.

The number of reported isolates in every serotype decreased, compared with the numbers reported in 1983 (Table 3). S. sonnei decreased 15.3%; S. flexneri, 10.8%; S. boydii, 6.5%; and S. dysenteriae, 3.2%. The decreases were not confined to one state or region.

The age-specific rate of reported isolates per 100,000 population was highest for 2-year- old children, lower for older children, and lowest for adults. The age-specific rate for 20- to 29-year-olds was slightly higher than the rates for the older children and the remaining age groups (Figure 3). In addition, in the 20- to 29-year-age group, a slightly higher rate was reported for females than for males. Rates of reported isolates by patient sex were similar for the remaining age groups.

Since some populations have higher rates than others, data were tabulated separately for patients residing in certain institutions (e.g. nursing homes, facilities for the mentally ill, and other resident-care centers) and on American Indian reservations. Only 2,416 (18.9%) of the reports included data on residence at the time of onset of illness. Of those specified, 22 (0.9%) lived in institutions, and 67 (2.8%), on Indian reservations. Fifteen (68.2%) of the reported isolates from residents of institutions were S. sonnei, and five (22.7%) were S. flexneri. Twenty-four (36.4%) of the reported isolates from Indian reservation residents were S. sonnei, and 42 (63.6%) were S. flexneri. For other known residences, S. sonnei accounted for 1,634 (71.7%); S. flexneri, for 587 (25.8%); S. boydii, for 34 (1.5%); and S. dysenteriae, for 24 (1.1%). Reported by Statistical Svcs Activity, Enteric Diseases Br, Div of Bacterial Diseases, Center for Infectious Diseases, CDC.

Editorial Note: This report is based on CDC's Shigella Surveillance Activity, a passive laboratory-based system that receives reports from the 50 states and the District of Columbia. These reports do not distinguish between clinical or subclinical infections or between chronic or convalescent carriers.

November 02, 1984 / 33(43);616-8

Shigellosis -- United States, 1983

In 1983, 14,946 Shigella isolates from humans were reported to CDC. This is a 10.5% increase from the 13,523 isolates reported in 1982. The number of isolates is still less than the 15,334 reported during the peak year, 1978 (Figure 1).

Shigella serotypes were reported for 14,089 of the 14,946 isolates. The most frequently isolated serotype, S. sonnei, comprised 65.8% of all isolates serotyped (Table 1). When compared with 1982, the number of reported isolates increased notably in all serotypes except for S. flexneri, which remained relatively constant (Table 4). S. flexneri 1a accounted for 13.5% of all S. flexneri subtyped; 1b, 13.2%; 2a, 25.4%; 3a, 18.5%; and 6, 11.3%.

The reported increases in the number of isolates from specific serotypes were not confined to one state or region. However, from 1982 to 1983, reported S. sonnei isolates increased notably in Indiana

(35 to 193), Maryland (85 to 199), Missouri (35 to 217), and New York (134 to 899). The increase in New York was associated with an outbreak in New York City.

The age-specific attack rate for persons from whom isolates were reported was highest for 2-year-old children, lower for older children, and lowest for adults. The age-specific attack rate for 20- to 29-year-olds was slightly higher than the attack rates for the older children and the remaining age groups (Figure 2). In addition, in the 20- to 29-year age group, a slightly higher isolation rate was reported for females than for males. The isolation rates by sex were similar for the remaining age groups.

Since some populations have a higher attack rate than others, data were tabulated separately for patients residing in certain institutions (e.g., nursing homes, facilities for the mentally ill, and other resident-care centers) and on American Indian reservations. Only 4,124 (27.6%) of the reports included data on residence at the time of onset of illness. Of those specified, 49 (1.2%) lived in institutions and 54 (1.3%) on Indian reservations. Forty-eight (97.8%) of the reported isolates from residents of institutions were S. sonnei, and one (2.0%) was S. flexneri. Thirty-four (63.0%) of the reported isolates from Indian reservation residents were S. sonnei and 20 (37%) were S. flexneri. For other known residences, S. sonnei accounted for 2,918 (74.4%); S. flexneri 936 (23.9%); S. boydii for 39 (1.0%); and S. dysenteriae for 27 (0.7%). Reported by Statistical Svcs Activity, Enteric Diseases Br, Div of Bacterial Diseases, Center for Infectious Diseases, CDC. Editoral Note: This report is based on CDC's Shigella Surveillance Activity, a passive laboratory-based system that receives reports from the 50 states and the District of Columbia. These reports do not distinguish between clinical or subclinical infections or between chronic or convalescent carriers.

September 02, 1983 / 32(34);444,449-50

Shigellosis -- United States, 1982

In 1982, 13,523 Shigella isolations from humans were reported to CDC. This represents a 9.9% decrease from the 15,006 isolations reported in 1981. The number of isolations has continued to decline from the 15,334 reported during the peak year, 1978 (Figure 1).

Shigella serotypes were reported for 12,818 of the 13,523 isolates and were distributed by serotype as follows: S. sonnei--8,228 (64.2%), S. flexneri--4,165 (32.5%), S. boydii--294 (2.3%), and S. dysenteriae--131 (1.0%). S. flexneri 1a, 1b, 2a, 3a, and 6 comprised 27.3% of all S. flexneri sub-typed. When compared with 1981, reported S. boydii isolations decreased by 42.4%; S. dysenteriae, by 35.8%; and S. sonnei, by 12.7%. The number of S. isolates and were distributed by serotype as follows: S. sonnei--8,228 (64.2%), S. flexneri--4,165 (32.5%), S. boydii--294 (2.3%), and S. dysen-teriae--131 (1.0%). S. flexneri 1a, 1b, 2a, 3a, and 6 comprised 27.3% of all S. flexneri subtyped. When compared with 1981, reported S. boydii isolations decreased by 42.4%; S. dysenteriae, by 35.8%; and S. sonnei, by 12.7%. The number of S. flexneri isolations remained relatively constant.

The decreases were not confined to one state or region. From 1981 to 1982, S. sonnei decreased notably in Florida (166 to 92), Georgia (349 to 135), Hawaii (128 to 46), Indiana (102 to 35), Louisiana (357 to 147), Missouri (128 to 35), and Virginia (889 to 81); S. boydii, in Arizona (38 to 16) and Texas (82 to 45); and S. dysenteriae, in California (129 to 71).

In the reported age distribution of persons from whom isolates were obtained, the age-specific attack rate was highest for 2-year-old children, markedly lower for older children, and slightly lower for adults, except for a slight increase for 20- to 29-year-olds (Figure 2). In the 20- to 29-year age groups, a slightly higher isolation rate was reported for females than for males. The isolation rates by sex were similar for the remaining age groups. The median ages of persons from whom isolates were reported were S. boydii--13 years, S. dysenteriae--24, S. flexneri--11, and S. sonnei--7.

Since shigellosis is a more important problem for some population groups than for others, data were tabulated separately for patients residing in certain institutions (e.g., nursing homes, facilities for the mentally ill, and other resident-care centers) and on American Indian reservations. Thirty-one per-cent of the reports included data on patient residence at the time of onset of illness. Of those speci-fied, 1.2% lived in institutions and 2.7% on Indian reservations. Sixty-nine percent of the reported isolates from residents of institutions were S. sonnei, and 30.8% were S. flexneri. Fifty-nine percent
228

of the isolates from residents of Indian reservations were S. flexneri, and 40.9% were S. sonnei. S. sonnei accounted for 74.4% of the isolates with known residence; S. flexneri, for 22.9%; S. boydii, for 1.2%; and S. dysenteriae, for 0.5%. Reported by Statistical Svcs Activity and Enteric Diseases Br, Div of Bacterial Diseases, Center for Infectious Diseases, CDC.

Editorial Note: This report is based on CDC's Shigella Surveillance Activity, a passive, laboratory-based system that receives reports from the 50 states and the District of Columbia. These reports do not distinguish between clincial or sub-clinical infections or between chronic or convalescent carriers.

May 20, 1983 / 32(19);250-2

Hospital-Associated Outbreak of Shigella dysenteriae Type 2 -- Maryland

An outbreak of severe dysentery caused by Shigella dysenteriae type 2 recently occurred at the U.S. Naval Hospital, Bethesda, Maryland. Epidemiologic investigation implicated the salad bar in the active-duty staff cafeteria as the source of infection.

In March 1983, 95 (6%) of 1,490 active-duty hospital staff members and 12 other individuals (three hospital inpatients, four visitors, and five food-service workers) became ill with acute dysentery. Onset of illness occurred over an 11-day period, and the epidemic curve was consistent with a com-mon-source outbreak (Figure 1). Patients presented with chills, fever, abdominal cramps, and the abrupt onset of profuse watery or bloody diarrhea. Nausea, vomiting, myalgias, and dehydration were frequently noted. A case was defined as a patient with two or more of the following: 1) fever or chills, 2) nausea or vomiting, 3) watery or bloody diarrhea (more than two non-formed bowel move-ments per day), or 4) abdominal cramps. Twenty-four individuals required hospitalization for intra-venous hydration. The duration of illness for most persons ranged from 3 to 8 days.

S. dysenteriae type 2 was cultured from stool specimens of 36 of the 80 affected individuals who were cultured. All symptomatic individuals were treated with a 5-day course of either ampicillin or trimethoprim sulfamethoxazole, both of which were effective against the organism in vitro. One hun-dred three of 107 known symptomatic individuals and 102 controls matched by job category were interviewed, and food-specific histories were obtained. Eating food prepared in the staff cafeteria, where 900-1,300 persons eat one or more meals daily, was significantly associated with illness (p 0.0001). Analysis of food histories from patients and controls who ate at the cafeteria at any time between February 28 and March 3 showed that patients were significantly more likely than controls to have eaten raw vegetables from the salad bar (p = 0.004). A single batch of salad vegetables pre-pared on February 28 was served at the salad bar through March 3. No single salad item or dressing was specifically implicated. No samples of salad from the days in question were available for cul-ture, and no cultures taken from available food items were positive. No other outbreaks of food-related shigellosis were reported to state health authorities during the outbreak period.

Interviews and stool cultures were obtained from 63 food handlers. Five had illnesses meeting the case definition with onset concurrent with the other cases. Although no index case was identified, gastrointestinal illnesses appeared to be common causes of absenteeism among food handlers during the 3 weeks preceding the outbreak. No food handler had a positive stool culture; none reported recent emigration or foreign travel.

The scope of the outbreak was undoubtedly limited by the exclusion of civilians from eating in the cafeteria and because food for inpatients is prepared in separate areas by different personnel. Although a few staff members continued to work while ill, most did not work after onset of symp-toms. Only ambulatory inpatients or visitors who ate in the staff cafeteria became ill. Two persons with presumed secondary cases who denied eating in the cafeteria were identified, including a civil-ian nurse who reported taking a rectal culture 12 hours before onset of her own symptoms from a hospitalized patient with a case, and a student laboratory technician who had contact with other ill technicians.

Preliminary food histories led to closing the salad bar, the cold-sandwich line, and a self-serve ice cream machine on March 7; these were reopened on March 14. Symptomatic health care workers rapidly improved with antibiotic therapy and were allowed to return to work 48 hours after symptoms had subsided. Symptomatic food handlers were required to have two negative rectal cultures, 24 hours apart, after completion of antibiotic therapy. The impact of this outbreak on the hospital operation was considerable. Well staff members volunteered for additional shifts to care for ill coworkers who were hospitalized on a separate ward. The Naval Hospital was not forced to restrict elective admissions. Reported by R Longfield, Commander MC USN, E Strohmer, R Newquist Commander MC USN, Naval Hospital, J Longfield, MD, J Coberly, Uniformed Svcs University of the Health Sciences, Bethesda, Maryland; G Howell, Captain MC USN, R Thomas, Lt MC USNR, Navy Environmental Health Center, Norfolk, Virginia; Enteric Diseases Br, Div of Bacterial Diseases, Hospital Infections Program, Center for Infectious Diseases, CDC.

Editorial Note: S. dysenteriae type 2 is an uncommon cause of disease in the United States; in 1982, only 0.2% (20/8,939) of the Shigella isolates of known serotype reported to CDC were type 2.

Because the infectious dose of Shigella is relatively small, it is more likely to spread person-to-person than are other enteric pathogens such as Salmonella and Vibrio cholerae 01. Although 6% of the active duty hospital staff members became ill, there was no evidence of secondary spread from staff to patients. The opportunity for spread from staff to patients was undoubtedly decreased.

December 24, 1982 / 31(50);681-682

Current Trends Shigellosis -- United States, 1981

In 1981, 15,006 Shigella isolations from humans were reported to CDC. While this represented a 6% increase over the 14,168 isolates reported in 1980, it remained 2% below the 15,334 reported during the peak year, 1978 (Figure 4).

Shigella serotypes were reported for 14,278 of the 15,006 isolates and were distributed by serotype as follows: S. sonnei --9,423 (66%), S. flexneri --4,141 (29%), S. boydii --510 (3.6%), and S. dysenteriae --204 (1.4%). When compared with 1980, this represented increases of 46% for S. boydii, 39% for S. dysenteriae, and 9% for S. sonnei, and a decrease of 5% for S. flexneri. The increases were not confined to one state or region. From 1980 to 1981 S. sonnei increased notably in Connecticut (67 to 337), Missouri (50 to 128), Virginia (83 to 889), and Washington (161 to 307); S. boydii increased in Texas (43 to 82).

The reported age distribution* of persons from whom isolates were obtained is shown in Figure 5. The rate, highest for 2-year-old children, decreased abruptly for older children and decreased more gradually for adults, except for a slight increase for 20-29 year-olds. Although in the 20-29-year age group a slightly higher isolation rate was reported for women, the isolation rates by sex were similar. The median ages in years of persons from whom isolates were reported were S. boydii--9.0, S. dysenteriae--16.5, S. flexneri--10.0, and S. sonnei--6.0. *Age, sex, and type of residence were unavailable for California.

Since shigellosis is a more significant problem for some population groups than for others, data were tabulated separately for patients residing in certain institutions (e.g., nursing homes, facilities for the mentally ill, and other resident-care centers), and on American Indian reservations. Twenty-nine percent of the reports included data on patient residence at the time of illness onset: 0.9% lived in institutions and 1.5% on Indian reservations. Seventy-four percent of the reported isolates from residents of institutions were S. flexneri, and 26% were S. sonnei. Similarly, 69% of the isolates from residents of Indian reservations were S. flexneri, and 31% were S. sonnei. This contrasts with the remainder of Shigella cases with known residence in which S. sonnei represented 75% of isolates and S. flexneri represented 22.5%. Reported by Statistical Svcs Activity, Enteric Diseases Br, Div of Bacterial Diseases, Center for Infectious Diseases, CDC.

Editorial Note: This report is based on CDC's Shigella Surveillance Activity, a passive, laboratory-based system that receives reports from the 50 states and the District of Columbia. These reports do not distinguish between clinical or sub-clinical infections or between chronic or convalescent carriers.

October 21, 1983 / 32(41);533

Epidemiologic Notes and Reports Gastrointestinal Illness Associated with Imported Brie Cheese -- District of Columbia

Between September 19 and September 27, 1983, three outbreaks of a gastrointestinal illness affecting 45 people were reported to the District of Columbia Department of Human Services. All three outbreaks followed office parties that occurred on September 13 (11 of 16 persons ill), September 16 (27/71), and September 22 (7/7). Four people sought medical attention, but none required hospitalization. The mean incubation period was 44 hours, and the mean duration of symptoms was 4.4 days. Illness was characterized by watery diarrhea (91%), abdominal cramps (80%), nausea (38%), and fever (20%). Vomiting and blood in the stool were each reported by one patient.

Analysis of data from a questionnaire administered to 70 of the 71 people who attended the largest party showed a strong association between eating imported French brie cheese and becoming ill (p 0.001). The same brand of cheese was also served at the other two parties.

The cheese was purchased at two stores in the Washington, D.C., area, which were supplied by the same distributor. The cheese bore the brand name Marcillat and the lot number 20208 and was produced in France on July 21, 1983. No evidence of mishandling or improper refrigeration could be determined. No information is yet available about the manufacturing plant in France.

Washington, D.C., area retailers of the cheese were notified of possible contamination and advised by health officials not to sell cheese of the implicated lot. Retailers in Washington, D.C., voluntarily relinquished the remaining quantities of lot 20208 to health officials. Since the shelf-life of brie cheese is short, none of that lot should be available now. However, health departments in other states in which the cheese was distributed have been contacted--Colorado, Connecticut, Florida, Georgia, Illinois, Louisiana, Massachusetts, Minnesota, New Jersey, New York, North Carolina, Pennsylvania, and Texas.

Cultures performed in the laboratory of the District of Columbia Department of Human Services on one cheese sample and four stool samples did not detect any routine enteric pathogens but did detect Citrobacter freundii in the cheese and in three of the stools. C. freundii is not generally recognized as an enteric pathogen, and its role in these outbreaks is uncertain. Specimens of stool and blood from ill and well persons who attended the parties have been obtained for further evaluation at CDC. The U.S. Food and Drug Administration is testing cheese from the implicated lot and from other lots produced by the same manufacturer. Reported by ME Levy, MD, District Epidemiologist, District of Columbia Dept of Human Svcs; Emergency and Epidemiology Operations Br, US Food and Drug Administration; Enteric Diseases Br, Div of Bacterial Diseases, Center for Infectious Diseases, CDC.

January 20, 1984 / 33(2);16,22

Update: Gastrointestinal Illness Associated with Imported Semi-Soft Cheese

In September 1983, three clusters of gastrointestinal illness associated with eating imported French Brie cheese occurred in the District of Columbia (1). All three outbreaks involved one lot of cheese and one distributor in the Washington, D.C., area. Cases of similar clinical illness have subsequently been identified in four states (Colorado, Georgia, Illinois, and Wisconsin) associated with eating the same brand of semi-soft cheese (either Brie or Camembert). The lots implicated in these states included at least one lot produced approximately 40 days after the cheese that caused the District of Columbia cases. Stool specimens were collected in Illinois and Wisconsin from ill individuals and from well family members who did not eat the cheese. Escherichia coli serotype O27:H20 producing a heat-stable toxin was identified in stool specimens from seven of 15 recently ill persons and from none of eight controls. E. coli O27:H20 organisms were also isolated from single cases in Washington, D.C., and Atlanta, Georgia. Plasmid analysis of the organisms from patients who lived in different locations and who ate different lots of cheese revealed an identical plasmid profile. Attempts to isolate the organism from the cheese are in progress. Control measures included recalling the cheese nationwide and instituting a program of regulatory sampling in cooperation with the French government for the importation of semi-soft cheeses. Reported by BJ Francis, MD, State Epidemiologist, Illinois Dept of Public Health; JP Davis, MD, State Epidemiologist, Wisconsin Dept of Health and Social Svcs; Emergency and Epidemiology Operations Br, US Food and Drug Administration; Div of Field Svcs, Epidemiology Program Office, Enteric Diseases Br, Div of Bacterial Diseases, Center for Infectious Diseases, CDC.

Editorial Note: Enterotoxigenic E. coli organisms commonly cause diarrhea in developing countries (2), and they have also been implicated as a common cause of travelers' diarrhea (3). They are rarely associated with illness acquired in the United States, Canada, or Europe (4-6). This represents the third foodborne outbreak caused by enterotoxigenic E. coli and the first common-source outbreak due to a strain producing heat-stable enterotoxin reported in the United States (7,8).

The association of a single pathogen with illness caused by eating semi-soft cheeses from at least two lots manufactured 1 month apart suggests a continuing common source of contamination; however, no information is yet available from the manufacturing plant in France. In a 1971 diarrheal-disease outbreak in the United States caused by enteroinvasive E. coli contaminating Brie, Camembert, or Coulomiers cheese produced by another French manufacturer, a contaminated water supply was implicated as the source of pathogenic organisms (9).

References
1. CDC. Gastrointestinal illness associated with imported Brie cheese--District of Columbia. MMWR 1983;32:533.
2. Ryder RW, Sack DA, Kapikian AZ, et al. Enterotoxigenic Escherichia coli and reovirus-like agent in rural Bangladesh. Lancet 1976;I;659-62.
3. Merson MH, Morris GK, Sack DA, et al. Travelers' diarrhea in Mexico. A prospective study of physicians and family members attending a congress. N Engl J Med 1976;294:1299-305.
4. Brunton J, Hinde D, Langston C, Gross R, Rowe B, Gurwith M. Enterotoxigenic Escherichia coli in central Canada. J Clin Microbiol 1980;11:343-8.
5. Back E, Blomberg S, Wadstrom T. Enterotoxigenic Escherichia coli in Sweden. Infection 1977;5:2-5.
6. Gangarosa EJ. Epidemiology of Escherichia coli in the United States. J Infect Dis 1978;137:634-8.
7. Taylor WR, Schell WL, Wells JG, et al. A foodborne outbreak of enterotoxigenic Escherichia coli diarrhea. N Engl J Med 1982;306:1093-5.
8. Wood LV, Wolfe WH, Ruiz-Palacios G, et al. An outbreak of gastroenteritis due to a heat-labile enterotoxin-producing strain of Esherichia coli. Infection and Immunity 1983;41:931-4.
9. Marier R, Wells JG, Swanson RC, Callahan W, Mehlman IJ. An outbreak of enteropathogenic Escherichia coli foodborne disease traced to imported French cheese. Lancet 1973;II:1376-8.

August 02, 1996 / 45(30);650-653

Invasive Infection with Streptococcus iniae -- Ontario, 1995-1996

During December 1995-February 1996, four cases of a bacteremic illness (three accompanied by cellulitis and the fourth with infective endocarditis, meningitis, and probable septic arthritis) were identified among patients at a hospital in Ontario. Streptococcus iniae, a fish pathogen not previously reported as a cause of illness in humans (1-3), was isolated from all four patients. All four patients were of Chinese descent and had had a history of preparing fresh, whole fish; three patients for whom information was available had had an injury associated with preparation of fresh, whole fish purchased locally. This report summarizes information about these cases and presents preliminary findings of an ongoing investigation by health officials in Canada (4), which suggests that S. iniae may be an emerging pathogen associated with injury while preparing fresh aquacultured fish. Case Reports

The first three cases occurred during December 15-20, 1995, among previously healthy women who ranged in age from 40-74 years. Each had a history of injury to the hand while preparing fresh, whole, aquacultured fish. The first case-patient reported a puncture wound to her hand with a fish bone while preparing a newly purchased tilapia (Oreochromis species) *, a freshwater fish marketed primarily as whole fish; the second lacerated the skin over her finger with a knife that had just been used to cut and clean a freshwater fish of unknown type; and the third punctured her finger with the dorsal fin while scaling a fresh tilapia.

The period from injury to onset of symptoms for the three cases ranged from 16 hours to 2 days. At the time of hospitalization, physical examination findings included fever (range: 100.4 F {38.0 C} to 101.3 F {38.5 C}) and cellulitis with lymphangitic spread proximate to the site of injury. Leukocyte counts ranged from 12,900/mm3 to 16,900/mm3 with an increased proportion of neutrophils. Blood cultures from all three patients were positive for S. iniae, and treatment with beta-lactam antibiotics or clindamycin resulted in complete resolution of all manifestations of illness.

The fourth patient, a 77-year-old man, was admitted to the hospital on February 1, 1996, because of a 1-week history of increasing knee pain, intermittent sweats, fever, dyspnea, and confusion. Past medical history included diabetes mellitus, hypertension, rheumatic heart disease, chronic renal failure, Paget's disease, and osteoarthritis. Approximately 10 days before admission, he had prepared a fresh tilapia, although it was unknown whether he incurred an injury while preparing the fish. Findings on examination included temperature of 96.1 F (35.6 C) and a large effusion and warmth of the right knee without overlying cellulitis. New murmurs of aortic insufficiency and mitral regurgitation were noted. While in the emergency department, he had a respiratory arrest and was intubated; treatment included administration of a beta-lactam agent and erythromycin. The leukocyte count on admission was 25,200/mm3 with 95% neutrophils. Ten hours following admission, his knee was aspirated, and a lumbar puncture was performed. Analysis of the joint fluid included a leukocyte count of 72,000/mm3 but no evidence of crystals. Analysis of the cerebrospinal fluid (CSF) included a leukocyte count of 87/mm3 (54% neutrophils), a glucose of 14 mg/dL, and a protein of 320 mg/dL. Cultures of samples of synovial fluid and CSF were negative, but blood cultures yielded S. iniae. Based on the clinical and laboratory findings, and a transesophageal echocardiogram that documented a mitral-valve vegetation, S. iniae endocarditis and meningitis were diagnosed. Treatment with beta-lactam antibiotics was continued, and he recovered. Microbiology

Isolates from all patients grew on sheep-blood agar incubated in room air at 95.0 F (35 C), appeared as gram-positive cocci in short chains or pairs, and were catalase-negative. During the first 18 hours of incubation, colonies were alpha-hemolytic and initially were identified as viridans streptococci. Further testing conducted by reference laboratories identified them as S. iniae. Three strains were resistant to bacitracin, and the fourth was susceptible. Pulsed-field gel electrophoresis patterns of chromosomal Sma1 digests of all four isolates were identical. Microbroth-dilution testing for susceptibility indicated that all isolates were susceptible to beta-lactams, macrolides, trimethoprim-sulfamethoxazole, and tetracycline. Follow-Up Investigation

All four patients had prepared fresh, whole fish, three of which were known to be tilapia, that had been purchased from different stores. In two cases, the fish were taken live from holding tanks in different fish markets. Surface cultures were obtained from four fresh tilapia purchased at selected fish markets in the community during March 1996. Cultures from three of the four fish yielded S. iniae; however, pulsed-field gel electrophoresis patterns were different for each, and none matched the outbreak strain. None of the vendors at the markets where the fish were purchased reported that the fish appeared to be sick. Fresh, whole tilapia sold in Ontario were imported from U.S. fish farms.

The ongoing epidemiologic and microbiologic investigation includes the establishment of surveillance for cases of upper-extremity cellulitis in patients visiting the emergency departments of 10 Toronto-area hospitals and use of a standardized questionnaire for interviewing patients. In addition, to better characterize the prevalence of S. iniae in fish, samples from live, aquacultured fish imported into Canada are being collected and tested by Canadian health officials for S. iniae.

Reported by: M Weinstein, MD, DE Low, MD, A McGeer, MD, B Willey, Mount Sinai Hospital and Princess Margaret Hospital, Univ of Toronto, and Canadian Bacterial Diseases Network, Toronto; D Rose, MD, M Coulter, P Wyper, Scarborough Grace Hospital, Scarborough; A Borczyk, MSc, Public Health Laboratory of Ontario, Toronto; M Lovgren, National Reference Center for Streptococcus, Laboratory Center for Disease Control, Edmonton, Alberta, Canada. Childhood and Respiratory Diseases Br, Div of Bacterial and Mycotic Diseases, National Center for Infectious Diseases, CDC.

Editorial Note: Because of recent increases in aquaculture, the occurrence of infections caused by a variety of streptococcal species is increasing among some salt-water and freshwater fish. S. iniae was first recognized in 1972 as a cause of disease in an Amazon freshwater dolphin, Inia geoffrensis. In 1986, S. iniae (reported as S. shiloi) was identified as a cause of meningoencephalitis among tilapia and trout in Israel; the organism was identified subsequently among tilapia in the United States and Taiwan. Infections with S. iniae may be asymptomatic or may cause disease associated with death rates of 30% to 50% in affected fishponds (2).

The first recognized case of S. iniae infection in humans occurred in Texas in 1991, and a second case occurred in Ottawa, Canada, in 1994; however, potential sources for both cases were not determined. The pulsed-field gel electrophoresis digest from the isolates causing both of these infections was identical to the isolates of the cases described in this report, except for a one-band shift.

Whether the recent cases of S. iniae infection represent the emergence of a new human pathogen or previously unrecognized disease is unclear. S. iniae infection may not be recognized because cultures rarely are obtained from patients with wound infections or cellulitis and, if cultured, viridans streptococcus isolates may be considered contaminants and not be further characterized. In addition, it is unclear whether human infections may be caused by any S. iniae strain or whether the strain implicated in all six of the cases is more virulent than other strains. Finally, because all four persons described in this report were of Chinese descent, potential racial/ethnic associations with risk for this infection should be further considered. Additional culture surveys and laboratory studies of tilapia should assist in characterizing the diversity and virulence among S. iniae.

To more clearly define the role of S. iniae as a human pathogen, physicians are encouraged to obtain blood and wound cultures from persons with upper-extremity cellulitis and to seek a history of recently having prepared a fresh, whole fish. Microbiology laboratories should be able to make a preliminary identification of S. iniae based on several distinguishing phenotypic characteristics. ** Possible S. iniae isolates can be confirmed at the CDC Streptococcal Reference Laboratory and tested to determine whether they are the same strain as identified from the six cases of human disease.

References
1. Eldar A, Frelier P, Assenta L, et al. Streptococcus shiloi, the name for an agent causing septicemic infection in fish is a junior synonym of Streptococcus iniae. Int J Syst Bacteriol 1995;45:840-2.
2. Eldar A, Bejerano Y, Bercovier H. Streptococcus shiloi, and Streptococcus difficile: two new streptococcal species causing a meningoencephalitis in fish. Curr Microbiol 1994;28:139-43.
3. Perera R, Johnson S, Collins M, et al. Streptococcus iniae associated with mortality of Tilapia nilotica and T. aurea hybrids. Journal of Aquatic Animal Health 1994;6:335-40.
4. Weinstein M, Low D, McGeer A, et al. Invasive infection due to Streptococcus iniae: a new or previously unrecognized disease -- Ontario, 1995-1996. Canada Communicable Disease Report 1996;22:129- 32.
5. Pier GB, Madin SH. Streptococcus iniae sp. nov., a beta-hemolytic streptococcus isolated from an Amazon freshwater dolphin, Inia geoffrensis. Int J Syst Bacteriol 1976;26:545-53.
6. Pier GB, Madin SH, Al-Nakeeb S. Isolation and characterization of a second isolate of Streptococcus iniae. Int J Syst Bacteriol 1978;28:311-4.

* Tilapia is one of the fastest growing aquaculture industries in the United States and the world. ** S. iniae is beta-hemolytic; however, some strains may appear to be alpha-hemolytic because a narrow zone of beta-hemolysis is surrounded by a larger zone of alpha-hemolysis (5,6). Beta-hemolysis always is observed under anaerobic incubation and in the area of stabs in the agar. S. iniae is nongroupable with Lancefield group A through U antisera. In addition, the pyrrolidonylarylaminase and leucine aminopeptidase tests are positive, the Voges-Proskauer test is negative, and the organism may have variable susceptibility to bacitracin.

October 10, 1986 / 35(40);629-30

Epidemiologic Notes and Reports Group-A, -B Hemolytic Streptococcus Skin Infections in a Meat-Packing Plant -- Oregon

In the period October 17, 1985-January 9, 1986, 44 episodes of pyoderma occurred among 32 workers in an Oregon meat-packing plant. Most of the 44 reports involved impetigo-like lesions on the hand, wrist, and forearm, but six episodes of cellulitis and two of lymphangitis were also reported. The same epidemic strain of Group-A, -B hemolytic Streptococcus (GAS) isolated from skin lesions was also isolated from meat in the plant.

In November 1985, emergency-room personnel in Pendleton, Oregon, reported to the Umatilla County Health Department a cluster of skin infections affecting three employees in a meat-packing plant, all from the same small, family-owned facility. After the Oregon State Health Division was asked to investigate, all 69 persons employed in the plant were interviewed for a history of and examined for the presence of pustular, draining, or inflamed skin lesions.

Seventy lesions were cultured, representing the initial 44 episodes of infection and 14 later sporadic cases. GAS, only, was isolated from 26%; both GAS and Staphylococcus aureus from 54%; and Staph. aureus, only, from 17%. Whereas multiple phage types of Staph. aureus were isolated from patients and meat, a single strain of GAS, MNT T14 SOR+, was identified in 24 group A streptococcal isolates serotyped.

Between October 17, 1985, and January 9, 1986, all but four of the 32 ill meat packers worked at least part-time on the kill floor or on the boning line or both. The attack rate for boners/killers was 74%, compared with 13% for workers who were never involved in killing or boning (relative risk (RR)=5.7, 95% confidence limits (CL)=2.9-11.3).

The epidemic investigation suggested that meat was a vehicle of transmission of GAS between workers. Cultures of two pork loins revealed the same epidemic strain (MNT T14 SOR+) as did isolates from patients. An increased risk for acquiring infection could not be shown for other exposures. Workers who became infected did not share knives or gloves more often than did uninfected workers. Meat packers usually own and maintain their own knives.

Recommendations to the meat-packing plant included an increased emphasis on worker safety; an increased emphasis on worker hygiene, e.g., covering skin lacerations; removal of workers with untreated skin infections from the meat-processing line; and improved surveillance of skin injuries and infections, including modifying sick-leave benefits to encourage reporting. Reported by K Flanagan, St. Anthony's Hospital, Pendleton, S Kline, Umatilla County Health Dept, K Quackenbush, Public Health Laboratory, L Foster, State Epidemiologist, Oregon State Health Div; Div of Field Svcs, Epidemiology Program Office, Respiratory Diseases Br, Div of Bacterial Diseases, Center for Infectious Diseases, CDC.

Editorial Note: This is the second reported outbreak of GAS skin infections among U.S. meat packers. During a similar outbreak in a Vermont meat-packing plant involving 18 of 59 employees, a worker with a chronic impetiginous lesion may have introduced GAS into the plant, and meat was postulated as one mode of transmission (1). Epidemic and sporadic cases of GAS skin infections among meat workers have been recognized in Great Britain since the mid-1970s (2-4). In the Oregon outbreak, it is also likely that meat was the vehicle of transmission after initial contamination by an infected human. Knife use is probably the significant risk shared by killers and boners vs. other meat workers. Bone has also been recognized as a source of skin injury among meat workers (5). GAS might spread from a meat-packing plant outside to non-plant workers, although there is no evidence of such transmission in the Oregon outbreak. In Great Britain, retail butchers and restaurant workers have been infected with epidemic GAS strains during outbreaks in meat-packing plants, presumably by handling contaminated meat (6,7). Improved surveillance of skin infections in the meat-packing industry may document more accurately the occurrence of such outbreaks in the United States.

References
1. CDC. Unpublished data.
2. Fraser CA, Ball LC, Morris CA, Noah ND. Serological characterization of group-A streptococci associated with skin sepsis in meat handlers. J Hyg Camb 1977;78:283-96.
3. Barnham M, Kerby J. A profile of skin sepsis in meat handlers. J Infect 1984;9:43-50.
4. Public Health Laboratory Service Working Group on Streptococcal Infection in Meat Handlers. The epidemiology and control of Streptococcus sepsis in meat handlers. Environ Health (Great

Britain) 1982;10:256-8.
5. Prevention of streptococcal sepsis in meat handlers. Communicable Disease Reports (London) 1983;34:1-4.
6. Streptococcal infection in meat workers. Communicable Disease Reports (London) 1981;16:3-4.
7. Fraser CA, Ball LC, Maxted WR, Parker MT. Streptococcal skin sepsis among meat handlers. Pathogenic streptococci. Proceedings of the VII International Symposium on Streptococci and Streptococcal Disease, September 1978, 1979. Parker MT, ed. 1979:115-6.

November 30, 1984 / 33(47);669-72

**Epidemiologic Notes and Reports Streptococcal Foodborne Outbreaks --
Puerto Rico, Missouri**

Two large outbreaks of foodborne group A streptococcal pharyngitis have been reported to CDC during 1984 in Puerto Rico and Missouri.

Puerto Rico: On August 3, 1984, an outbreak occurred among guests attending a party in a private home in San Juan, Puerto Rico. During that weekend, numerous party attendees became ill with sore throat, myalgia, cervical adenopathy, and fever. Many were seen by physicians and had exudative pharyngitis. One was hospitalized.

The Puerto Rico Department of Health was notified of the outbreak on August 8. Because of the high attack rate and the clustering of cases, the outbreak was presumed to be foodborne. Self-administrated questionnaires were received from 45 (96%) of the 47 party attendees, and 25 questionnaires were received from their household contacts. Throat cultures were obtained from 44 (94%) of party attendees.

Four persons were excluded from the questionnaire analysis--three because of onset of pharyngitis before the party, and one, because of an incomplete questionnaire. Of the 41 remaining persons, 23 (56%) had illness meeting the case definition. The attack rate for persons who ate carrucho, a conch salad, was 70%, compared with 29% for persons who did not eat carrucho (p = 0.013). No other food showed significantly different attack rates. No dose-response effect for persons eating carrucho was demonstrated, nor was a difference in attack rates observed between persons who ate early in the evening and those who ate later in the evening. That carrucho was the vehicle for transmission was further supported by the fact that two of four persons who did not attend the party but who ate carrucho that had been brought home to them became ill with pharyngitis. The secondary attack rate for household contacts who did not eat carrucho was 4%. The incubation period was 12-60 hours (median 24 hours).

Throat cultures from 11 party attendees grew group A streptococci, as did a small sample of carrucho remaining from the party. All cultures were of the same serotype (M nontypable, T12, SOR+).

The carrucho was prepared in a small beachside restaurant outside San Juan. The conch used to make the carrucho came in a torn, unlabeled plastic bag and was allegedly imported from Santo Domingo. None of the uncooked conch remained for testing, but the method of salad preparation, which reportably included boiling the conch for 2 1/2 hours, should have been adequate to kill any streptococci. Seventy pounds of carrucho was made the afternoon of the party. The 25 pounds purchased by the party's host was left in an automobile at ambient temperature for 3 hours before delivery to the party.

Approximately 2,000 persons who ate in the restaurant that weekend were potentially exposed to the 45 pounds of remaining carrucho. Because there was no way to identify individuals who might have eaten there that weekend, four clinical microbiology laboratories serving the San Juan area were surveyed in an attempt to determine if the number of positive throat cultures in August was higher than the number during the same time the previous year; no increase was observed.

All foodhandlers at the restaurant were interviewed and examined for skin lesions, and cultures (pharyngeal, nasal, and hand) were obtained. No cultures were positive, and no histories were obtained of recent pharyngitis or skin lesions. Food prepared at the restaurant, including carrucho, during the week after the party was cultured; all was negative for group A streptococci.

Because party attendees were potentially exposed to streptococci, the Puerto Rico Department of Health recommended that all attendees who developed symptoms of pharyngitis, regardless of culture results, receive antibiotic therapy effective against group A streptococci.

Missouri: Another outbreak occurred among participants from seven states at a meeting held at a Kansas City, Missouri, hotel from May 31, to June 1, 1984. On June 6, the Kansas City Health Department was notified of three cases of group A beta-hemolytic streptococcal pharyngitis occurring in three technicians from one blood bank who had attended the meeting. Other cases were subsequently reported among persons who attended the meeting. Clustering of cases and a high attack rate suggested a foodborne source.

A questionnaire was administered by telephone or mail to 136 (98%) of the 139 persons identified as having attended the conference. Cases were defined as persons with acute onset of sore throat between May 31 and June 5, who had had no antecedent contact to household members with pharyngitis. Severity of illness ranged from minor discomfort to symptoms resulting in several days' absence from work. Positive cultures for group A streptococci were reported for 13 (93%) of 14 individuals from whom throat cultures were obtained. However, none of the cultures were still available for typing or confirmation by the time of investigation. The survey implicated a luncheon held May 31. Sixty (57%) cases among the 106 persons who attended it were identified, compared with no cases among 30 conference attendees who did not attend the luncheon (p 0.0001). Food-specific attack rates suggested macaroni salad or mousse as possible vehicles of transmission. The attack rate for persons who ate macaroni salad was 88%, compared with 47% for those who did not (p 0.0001), but only one-third of persons who were ill gave histories of having eaten macaroni salad. The attack rate for persons who ate mousse was 63%, compared with 39% for persons who did not (p = 0.053), and, since 82% of ill persons reported having eaten the mousse, it was considered more likely if only one vehicle were involved. The incubation period of the illness was 24-36 hours (median 36 hours).

All the food for the luncheon was prepared by five hotel employees. The foodhandlers were interviewed and examined, and cultures were obtained. All were negative for group A streptococci, and no visible skin lesions were found on any worker. One worker claimed to have had a sore throat the day of the luncheon but did not seek medical attention.

The pastry chef had prepared two types of mousse the morning of the luncheon. Although it was refrigerated for 30 minutes during one phase of preparation, the final product was kept at room temperature for 1-2 hours before the luncheon. Reported by JG Rigau, MD, Commonwealth Epidemiologist, Puerto Rico Dept of Health; T Martin, V Gibson, D Giedinghagen, GL Hoff, PhD, Div of Communicable Disease Control, Div of Environmental Health, Kansas City Health Dept, HD Donnell, Jr, MD, State Epidemiologist, Missouri Dept of Social Svcs; Respiratory and Special Pathogens Epidemiology Br, Div of Bacterial Diseases, Center for Infectious Diseases, CDC.

Editorial Note: Before the advent of pasteurization of milk and availability of adequate refrigeration, foodborne streptococcal outbreaks were very common. Outbreaks resulting in epidemics of scarlet fever, rheumatic fever, and suppurative complications were reported. Improvements in sanitation have resulted in foodborne streptococcal outbreaks becoming relatively uncommon (1-3).

These outbreaks show the difficulties involved in recognizing foodborne illness. Foodborne transmission of streptococci, rather than person-to-person transmission, is suggested by a large clustering of cases, a shorter incubation period, and a higher attack rate. Unless disease occurs in a setting where people who are ill are likely to notice the epidemic themselves, it is difficult for public health officials to detect the increased incidence of streptococcal pharyngitis in the community, especially since only a small percentage of persons with sore throats seek medical attention and ultimately receive treatment for the illness. The Puerto Rico outbreak was recognized only because a number of ill people worked in the same office. Initially, the party attendees felt the illness resulted from close person-to-person contact; only when persons who were not at the party ate party food and became ill did the office manager notify the health department. The second outbreak almost escaped detection, since the illness peaked after the conference had ended, and the participants had returned to their homes in seven states.

It is unknown how many cases of endemic streptococcal pharyngitis are caused by foodborne transmission. It is important to recognize that rheumatic fever and glomerulonephritis may result from outbreaks of these infections.

References
1. Hill HR, Zimmerman RA, Reid GV, et al. Food-borne epidemic of streptococcal pharyngitis at the United States Air Force Academy. N Engl J Med 1969;280:917-21.
2. McCormick JB, Kay D, Hayes P, Feldman R. Epidemic streptococcal sore throat following a community picnic. JAMA 1976;236:1039-41.
3. Ryder RW, Lawrence DN, Nitzkin JL, et al. An evaluation of penicillin prophylaxis during an outbreak of foodborne streptococcal pharyngitis. Am J Epidemiol 1977;106:139-44.

October 07, 1983 / 32(39);510,515-6

Group C Streptococcal Infections Associated with Eating Homemade Cheese -- New Mexico

Between July 25 and September 9, 1983, 16 cases of invasive group C streptococcal infection were identified in northern New Mexico. The group C streptococcus was isolated from the blood of 15 patients and the pericardial fluid of one patient. The organism isolated from 14 of the patients has been identified as a group C B-hemolytic streptococcus--species Streptococcus zooepidemicus; the species of the remaining two isolates have not yet been determined. Ages of the 16 patients ranged from 19 to 89 years (median 74); 10 were male. All patients were Hispanic.

In general, the clinical syndrome was characterized by fever, of the patients has been identified as a group C B-hemolytic streptococcus--species Streptococcus zooepidemicus; the species of the remaining two isolates have not yet been determined. Ages of the 16 patients ranged from 19 to 89 years (median 74); 10 were male. All patients were Hispanic.

In general, the clinical syndrome was characterized by fever, chills, and vague constitutional symptoms. However, five patients had localized signs of infection, including pneumonia, endocarditis and meningitis, pericarditis, and abdominal pain that led to a cholecystectomy for one patient and an appendectomy for another patient. Two patients with multiple underlying medical problems died.

A case-control study was undertaken to identify possible risk factors for contracting group C streptococcal infections. Patients and controls were matched for age, sex, ethnicity, and neighborhood of residence. Some of the possible risk factors investigated included underlying illnesses, immunosuppressive medications, animal exposure, group activities, restaurants visited, and food items consumed. Initial questionnaires identified eating "queso blanco," a homemade white cheese, as the only risk factor associated with illness (10 of 15 case patients versus 7 of 45 controls, p 0.001). During subsequent discussions, four of the five patients who did not report in the case-control study that they had eaten the homemade cheese later remembered that they had consumed the cheese before the onset of illness. Thus, only one patient did not recall having eaten the cheese before becoming ill.

The sole source of the homemade cheese consumed by the patients was an ungraded, small (seven cows), family dairy farm in northern New Mexico. At the farm, the cheese was made from raw cows' milk and was not subsequently aged. It was distributed to several stores in northern New Mexico within 24-48 hours after preparation in the family kitchen. Milk samples from the cows and cheese samples from the stores were obtained for microbiologic analysis. Group C B-hemolytic streptococci, species S. zooepidemicus, have been isolated from multiple samples of each.

Public health control measures included closing the dairy operation, removing the cheese from the stores, and advising the public to dispose of any "queso blanco" purchased from the stores that sold the implicated product. No new cases of group C streptococcal infections have been identified since these interventions were implemented. Reported by FH Espinosa, MD, WM Ryan, MD, PL Vigil, MD, Espanola Hospital, Espanola, DF Gregory, MD, RB Hilley, MD, DA Romig, MD, RB Stamm, MD, ED Suhre, MD, PS Taulbee, MD, LH Zucal, MD, St. Vincent Hospital, Santa Fe, RW Honsinger Jr, MD, PJ Lindberg, MD, Los Alamos Medical Center, Los Alamos, M Barcheck, JA Miller, R Mitzelfelt, JM Montes, LJ Nims, OJ Rollag, DVM, State Public Health Veterinarian, N

Weber, JM Mann, MD, State Epidemiologist, New Mexico Health & Environment Dept; Div of Bacterial Diseases, Center for Infectious Diseases, CDC.

Editorial Note: Group C streptococci are a common cause of infection in several animal species but are generally considered to be a rare cause of infection in humans (1). Of the four species of group C streptococci, S. equisimilis has been reported to cause most human illnesses, including bacteremia, endocarditis, meningitis, pneumonia, epiglottitis, puerperal sepsis, and wound infections. However, S. zooepidemicus has been associated with two outbreaks of pharyngitis and nephritis in Europe (2,3). While pharyngitis was not a part of the clinical syndrome in the outbreak reported here, it is too early to tell if poststreptococcal glomerulonephritis will develop.

In both of the European outbreaks, unpasteurized milk was suspected as the source of infection. The outbreak reported here is the first epidemic of group C streptococcal infections in the United States and is the first such reported outbreak in which the vehicle--cheese made from unpasteurized cows' milk--has been epidemiologically implicated. Although S. zooepidemicus and S. equisimilis are rarely reported causes of mastitis in cows, the cause of this outbreak was contaminated milk from cows with mammary infections due to S. zooepidemicus.

Because few laboratories routinely determine the species of group C streptococci, the number of human infections due to S. zooepidemicus is not known. Furthermore, group C streptococci may be mistakenly identified as group A strains if only bacitracin susceptibility testing is done to differentiate group A streptococci from other B-hemolytic streptococci. It is impractical for clinical laboratories to routinely determine the serogroup and species of all B-hemolytic streptococci isolated from all sites. However, serogroups of B-hemolytic streptococci and species of group C strains isolated from blood and other normally sterile sites should be identified if further information is to be gained about the epidemiology of such infections.

References
1. Ghoneim AT, Cooke EM. Serious infection caused by group C streptococci. J Clin Pathol 1980;33:188-90.
2. Duca E, Teodorovici G, Radu C, et al. A new nephritogenic streptococcus. J Hyg, Camb. 1969;67:691-8.
3. Barnham M, Thornton TJ, Lange K. Nephritis caused by Streptococcus zooepidemicus (Lancefield group C). Lancet 1983; 8331:945-8.

ENTEROVIRULENT ESCHERICHIA COLI GROUP (EEC Group)

February 11, 1994 / 43(05);81,87-88

Foodborne Outbreaks of Enterotoxigenic Escherichia coli -- Rhode Island and New Hampshire, 1993

Infections with enterotoxigenic Escherichia coli (ETEC) are a frequent cause of diarrhea in developing countries but not in the United States and other industrialized countries. This report describes two foodborne ETEC outbreaks that occurred in the United States in 1993. Rhode Island

On March 25, the Rhode Island Department of Health was notified of gastrointestinal illness among passengers on an airline flight from Charlotte, North Carolina, to Providence, Rhode Island, on March 21. The flight carried 98 passengers; 47 (64%) of 74 passengers who were interviewed met the case definition of three or more loose stools in 24 hours beginning within 4 days after the flight. Additional symptoms included abdominal cramps (94%), nausea (70%), headache (57%), fever (13%), and vomiting (13%). The only common meal for all ill passengers was dinner served on board the flight. The median incubation period was 41 hours (range: 12-77 hours); two (5%) of 44 persons recovered within 48 hours of onset of illness.

Illness was most strongly associated with eating garden salad made from shredded carrots and iceberg, romaine, and endive lettuce (46 {98%} of 47 ill passengers compared with six {22%} of 27 well passengers; relative risk {RR}=4.4; 95% confidence interval {CI}=2.2-8.9). Investigators from the Food and Drug Administration (FDA) contacted 18 passengers who had traveled on March 21 on a different flight operated by the airline and who had been served the same meal; nine passengers reported gastrointestinal illness. On March 21, approximately 4000 portions of salad had been prepared by one catering service for 40 flights operated by the same airline that day. The FDA traceback determined that all of the salad ingredients were of U.S. origin.

Stool specimens obtained from 20 passengers from the index flight were negative on culture for Salmonella, Shigella, Campylobacter, Yersinia, and Vibrio, and viral particles were not observed in 12 stool specimens examined by electron microscopy at CDC. E. coli isolates from 10 ill passengers were tested for ETEC at CDC. ETEC strains (serotype O6:non-motile {NM}) that produced heat stable (ST) and heat labile (LT) toxins were identified in isolates from three passengers.

FDA inspection of the caterer's facilities did not identify deficiencies in sanitary conditions. In addition, all food handlers denied gastrointestinal illness or recent travel outside the United States. Samples of food collected for culture on March 27 did not yield ETEC. New Hampshire

On April 5, the New Hampshire Division of Public Health Services was notified of gastrointestinal illness in eight persons who ate a buffet dinner served at a mountain lodge on March 31. A total of 202 persons ate the dinner, including 132 guests and 70 lodge employees. A case was defined as diarrhea (three or more loose or watery stools in a 24-hour period) and one other symptom (cramps, fever, headache, nausea, or vomiting) with onset from April 1 through April 7 in a guest or employee who had eaten the dinner. Of the 123 guests and 56 employees who were interviewed, 96 (78%) and 25 (45%), respectively, had illness that met the case definition. Additional symptoms included cramps (92%), nausea (59%), myalgias (50%), headache (49%), fever (22%), and vomiting (11%). Illness began a median of 38 hours after foods from the buffet were eaten (range: 3-159 hours); 60 (65%) of 93 persons for whom information was available reported continuing illness 4-6 days after symptom onset.

Illness among guests was most strongly associated with consumption of tabouleh salad (cases occurred in 78 {94%} of 83 guests who ate the tabouleh and 18 {53%} of 34 guests who did not {RR=1.8; 95% CI=1.3-2.5}). Tabouleh was the only food associated with illness among lodge employees (RR=6.4; 95% CI=2.2-18.8). The tabouleh was prepared from onions, carrots, zucchini, peppers, broccoli, mushrooms, green onions, tomatoes, parsley, bulgur wheat, olive oil, lemon juice, and bottled garlic. All of the produce was of U.S. origin. The salad was prepared the evening before the banquet. All food preparers denied gastrointestinal illness or travel outside the United States the week before the banquet.

Cultures of stool specimens obtained from 14 persons were negative for Salmonella, Shigella, Campylobacter, and Yersinia; neither ova nor parasites were detected in stool specimens from seven ill persons. However, ETEC (serotype O6:NM) that produced LT and ST was isolated from stool specimens from seven of nine ill guests and from one of five well employees. Additional ETEC serotypes also were isolated from six specimens. Follow-up Investigation

Plasmid profiles of the O6:NM strains from the outbreaks in New Hampshire and Rhode Island were identical but differed from those of 10 other serotype O6:NM ETEC strains from other sources. Carrots were the only item common to the tabouleh salad implicated in New Hampshire and the garden salad implicated in Rhode Island. Carrots used in both salads were grown in the same state; however, a traceback conducted by the New Hampshire Division of Public Health Services in collaboration with FDA and CDC did not identify a single source. FDA is investigating the implicated carrot sales agency in the state where the carrots were grown. Reported by: V Benoit, P Raiche, MG Smith, MD, State Epidemiologist, New Hampshire Div of Public Health Svcs. J Guthrie, MD, Univ of Rhode Island Infirmary; EF Donnelly, MPH, EM Julian, PhD, R Lee, MS, S DiMaio, M Rittmann, BT Matyas, MD, State Epidemiologist, Rhode Island Dept of Health. Atlanta District Office and Div of Emergency and Epidemiology Operations, Food and Drug Administration. Div of Field Epidemiology, Epidemiology Program Office; Respiratory and Enterovirus Br, Div of Viral and Rickettsial Diseases; Foodborne and Diarrheal Diseases Br, Div of Bacterial and Mycotic Diseases, National Center for Infectious Diseases, CDC.

Editorial Note: Since 1975, 13 outbreaks of ETEC gastroenteritis in the United States have been reported to CDC; four (31%) of these outbreaks, including the two described in this report, occurred in 1993. Although each of the four outbreaks in 1993 and five outbreaks reported previously were foodborne, ETEC outbreaks associated with waterborne and person-to-person transmission have been described (1,2). At least one foodborne ETEC outbreak in the United States was attributed to spread from an infected food handler (3) and another to imported contaminated food (4). However, none of the recent foodborne outbreaks were associated with these sources. Salads containing raw vegetables have been associated with ETEC infection (5).

Because ETEC is not detected by standard stool culture methods for Salmonella, Shigella, Vibrio, or other enteric bacterial pathogens and because symptoms of ETEC infection are relatively nonspecific, outbreaks caused by ETEC may be incorrectly attributed to a viral etiology. Watery diarrhea is the predominant symptom of ETEC infection, usually reported by more than 90% of patients (3- 5). The diarrhea is often accompanied by abdominal cramps and is generally mild, although severe dehydrating diarrhea has been reported (6). Two percent to 13% of patients report vomiting (3-5).

In contrast to illness caused by ETEC, gastroenteritis from infection with Norwalk virus is usually characterized by vomiting but not by diarrhea (7). Because nausea, headache, and myalgias occur with varying frequency in association with ETEC and Norwalk virus infections, these symptoms are less useful for differentiating the two illnesses (3-5,7). The incubation periods are similar for ETEC and Norwalk gastro- enteritis (range: 24-48 hours) (2-4,7). However, duration of illness is shorter for Norwalk gastroenteritis (usually less than or equal to 3 days) and longer for illness caused by ETEC infection (often greater than 4 days) (1-5,7).

Laboratory identification of ETEC depends on testing E. coli isolates by methods that are not widely available. For well characterized outbreaks of watery diarrheal illness for which no pathogen has been identified during routine bacteriologic examinations, arrangements can be made through local and state health departments to send E. coli isolates to CDC for testing. ETEC previously has been recognized primarily as a cause of traveler's diarrhea. However, the findings in this report indicate that clinicians and microbiologists may need to consider ETEC in patients with diarrheal illness who did not travel (8).

References
1. Rosenberg ML, Koplan JP, Wachsmuth IK, et al. Epidemic diarrhea at Crater Lake from enterotoxigenic Escherichia coli: a large waterborne outbreak. Ann Intern Med 1977;86:714-8.
2. Ryder RW, Wachsmuth IK, Buxton AE. Infantile diarrhea produced by heat-stable enterotoxigenic Escherichia coli. N Engl J Med 1976;295:849-53.
3. Taylor WR, Schell WL, Wells JG, et al. A foodborne outbreak of enterotoxigenic Escherichia coli diarrhea. N Engl J Med 1982;306:1093-5.

4. MacDonald KL, Eidson M, Strohmeyer C, et al. A multistate outbreak of gastrointestinal illness caused by enterotoxigenic Escherichia coli in imported semisoft cheese. J Infect Dis 1985; 151:716-20.
5. Merson MH, Morris GK, Sack DA, et al. Traveler's diarrhea in Mexico: a prospective study of physicians and family members attending a conference. N Engl J Med 1976;294:1299-305.
6. Sack RB, Gorbach SL, Banwell JG, Jacobs B, Chatterjee BD, Mitra RC. Enterotoxigenic Escherichia coli isolated from patients with severe cholera-like disease. J Infect Dis 1971; 123:378-85.
7. Kaplan JE, Gary GW, Baron RC, et al. Epidemiology of Norwalk gastroenteritis and the role of Norwalk virus in outbreaks of acute nonbacterial gastroenteritis. Ann Intern Med 1982; 96:756-61.
8. Osterholm MT, Hedberg CW, MacDonald KL. Prevention and treatment of traveler's diarrhea {Letter}. N Engl J Med 1993;329:1584-5.

October 13, 2000 / 49(40);911-3

Outbreak of Escherichia coli O157:H7 Infection Associated With Eating Fresh Cheese Curds --- Wisconsin, June 1998

On June 15, 1998, the Division of Public Health, Wisconsin Department of Health and Family Services, was notified of eight laboratory-confirmed and four suspected Escherichia coli O157:H7 infections among west-central Wisconsin residents who became ill during June 8--12. This report summarizes the outbreak investigation, which implicated fresh (held <60 days) cheese curds from a dairy plant as the source of infection.

A primary case was defined as the first laboratory-confirmed case in a household; a secondary case was one that occurred 3--8 days after a primary case in the same household. A matched case-control study was conducted to assess potential sources of infection. For the purposes of the case-control study, a case was defined as culture-confirmed illness among residents of Chippewa and Eau Claire counties with illness onset during June 7--18. For each case-patient, two community controls matched by sex and age group (range: from <10 years within 2 years to >10 years within 5 years) were interviewed by telephone. Case-patients and controls were interviewed about food exposures and potential risk factors for E. coli O157:H7 infection within 7 days before onset of illness.

In response to the case-control study, the Wisconsin Department of Agriculture, Trade, and Consumer Protection visited dairy plant A to collect cheese samples, raw ingredients, and packaging materials; to review employee food handling and hygienic practices; and to assess potential sources of contamination from raw milk. Product and environmental samples (e.g., vat surfaces and floor drains) from the dairy plant were screened for phosphatase activity to identify evidence of raw milk.

Fifty-five laboratory-confirmed case-patients were identified, including two from secondary households. Case-patients were from seven Wisconsin counties (27 from Chippewa and 16 from Eau Claire counties); two case-patients were visiting from out of state. Median age was 27 years (range: 15 months--90 years) and 37 (67%) were female. The most frequently reported symptoms included bloody diarrhea (55 [100%]), cramps (50 [91%]), fatigue (39 [71%]), and nausea (38 [69%]). Mean duration of diarrhea was 5.1 and 4.5 days for 25 hospitalized and 30 nonhospitalized case-patients, respectively.

Eating fresh cheese curds during June 1--17 was reported by all 24 case-patients in Chippewa and Eau Claire counties and eight (18%) of 45 controls (matched odds ratio=undefined; 95% confidence interval=20.6--infinity). Illness was not linked to eating other cheese products (e.g., shredded, sliced, block, or string cheese). Of the 43 laboratory-confirmed case-patients whose cheese curd source could be identified, all had eaten fresh cheese curds produced at dairy plant A; 19 had purchased the curds from an unrefrigerated display at plant A, and 24 had purchased them refrigerated from retail stores that received shipments from plant A. Fifteen (50%) of 30 case-patients who recalled the purchase date had bought the curds on June 5 or 6. The median number of curds eaten was eight (range: one--28), the equivalent of approximately 1.6 oz of cheese.

Thirty-five specimens from plant A that were produced during the outbreak were tested: nine environmental samples, 18 unopened cheese samples, six opened retail packages of curds, and two unopened retail packages of curds. Five of the six opened retail packages of curds and four of the 18 unopened cheese samples were positive for nonbacterial phosphatase (Scharer method). E. coli O157:H7 was isolated from an opened package of curds that had been served at a party attended by nine persons with culture-confirmed illness. The contents of this package tested positive for nonbacterial phosphatase. Among 44 E. coli O157:H7 case-patient isolates available for pulsed-field gel electrophoresis, 42 were indistinguishable from each other and from the curd isolate.

Dairy plant A had produced four or five vats of pasteurized cheddar and Colby cheese products 5 days a week since 1977. Each vat yielded approximately 1500 pounds of cheese that was pressed into 40-lb blocks, daisies (rounds of cheese), or was packaged as fresh cheese curds. Dairy plant A also produced unpasteurized (raw milk) cheddar cheese daisies every June as part of Dairy Month. Certain raw milk cheese products can be produced and sold legally as long as the cheese is held at >35 F (>1.7 C) for at least 60 days before it is sold*. Curds are sold fresh (held <60 days); therefore, curds must be made with pasteurized milk. At least one 1500-lbs vat of raw milk cheddar cheese was made on May 27 and June 2--5. These vats were used inadvertently to make fresh curds, which were incorrectly labeled "pasteurized" cheddar cheese curds, and distributed and sold in six Wisconsin counties.

Reported by: J Durch, MPH, T Ringhand, MPH, Chippewa County Dept of Public Health, Chippewa Falls; K Manner, M Barnett, Wisconsin Dept of Agriculture, Trade, and Consumer Protection; M Proctor, PhD, S Ahrabi-Fard, MS, Communicable Disease Epidemiology Section; J Davis, MD, State Epidemiologist for Communicable Disease, Wisconsin Div of Public Health. D Boxrud, Minnesota Health Dept. Foodborne and Diarrheal Diseases Br, Div of Bacterial and Mycotic Diseases, National Center for Infectious Diseases; and an EIS Officer, CDC.

Editorial Note: Cheese is made in vats by coagulating milk with enzymes and/or acids. After whey is drained, the large cheese clumps are removed and milled into curds, salted, and packaged in small plastic bags for sale. Raw milk consumption has been associated with campylobacteriosis, salmonellosis, E. coli O157:H7, yersiniosis, listeriosis, tuberculosis, brucellosis, cryptosporidiosis, and staphylococcal enterotoxin poisoning (1). In 1950, the U.S. Food and Drug Administration (FDA) required manufacturers of soft and fresh cheeses to use pasteurized milk and allowed raw milk to be used only for certain aged cheeses (2). In 1986, E. coli O157:H7 illness was associated with consuming raw milk (3). In 1987, FDA banned the interstate sale of raw milk in retail packages. During 1973--1992, 40 (87%) of 46 raw milk-associated outbreaks occurred in the 28 states that permitted the intrastate sale of raw milk (4). During the same period, 11 of 32 cheese-associated outbreaks were attributed to contamination before distribution (5).

This outbreak investigation illustrates the hazards of using raw milk to produce commercial products that may lead to mislabeling or contaminating pasteurized product by equipment or ingredients. This practice can result in pasteurized products contaminated by equipment or ingredients and in product mislabeling. States that allow the sale of unpasteurized milk or dairy products made from unpasteurized milk should take appropriate steps to reduce the risk for contamination and mislabeling to prevent similar outbreaks.

References
1. Potter ME, Kaufmann AF, Blake PA, Feldman RA. Unpasteurized milk: the hazards of a health fetish. JAMA 1984;252:2048--52.
2. US Food and Drug Administration. Cheeses; processed cheeses; cheese food; cheese spreads, and related foods: definitions and standards of identity; final rule. Federal Register 1950;19:5656--90.
3. Martin ML, Shipman LD, Wells JG, et al. Isolation of Escherichia coli O157:H7 from dairy cattle associated with two cases of haemolytic uraemic syndrome [Letter]. Lancet 1986;8514:1043.
4. Headrick ML, Korangy S, Bean NH, et al. The epidemiology of raw milk=associated foodborne disease outbreaks reported in the United States, 1973 through 1992. Am J Public Health 1998;88:1219--21.
5. Altekruse SF, Timbo BB, Mowbray JC, Bean NH, Potter ME. Cheese-associated outbreaks of human illness in the United States, 1973 to 1992: sanitary manufacturing practices protect consumers. Journal of Food Protection 1998;61:1405--7.

* Code of Federal Regulations Title 21, Part 133.

April 21, 2000 / 49(15);321-4

Escherichia coli O111:H8 Outbreak Among Teenage Campers --- Texas, 1999

In June 1999, the Tarrant County Health Department reported to the Texas Department of Health (TDH) that a group of teenagers attending a cheerleading camp during June 9--11 became ill with nausea, vomiting, severe abdominal cramps, and diarrhea, some of which was bloody. Two teenagers were hospitalized with hemolytic uremic syndrome (HUS), and two others underwent appendectomies. Routine stool cultures from eight ill persons failed to yield a pathogen. Stools subsequently were sent to laboratories at the Texas Department of Health and CDC, where Escherichia coli O111:H8 was isolated from two specimens. This report summarizes the investigation of this outbreak.

To identify additional cases, surveillance for non-O157 Shiga toxin-producing E. coli (STEC) illnesses in Texas was enhanced by alerting all local health departments, hospitals, clinical laboratories, and physicians about the outbreak. A cohort study of all campers attending the 3-day camp was conducted to identify the source of the outbreak and to collect data describing the clinical illness. Illness was defined as either diarrhea (three or more loose stools during any 24-hour period) accompanied by abdominal cramps or bloody diarrhea alone, occurring within 14 days after the start of the camp. Campers were interviewed for demographic information, medical histories, and symptoms and about their food and beverage consumption during the camp. Sanitarians inspected the cafeteria where meals were prepared and served to campers and the plumbing system in the dormitory where campers resided. Foodhandlers and other kitchen staff were interviewed about food preparation practices, menus, and the delivery schedules and suppliers for food items served to campers. Foodhandlers submitted stool specimens and rectal swabs for testing. Several food items from the cafeteria were cultured.

Of the 650 campers composing the cohort, 521 (80%) were interviewed. Of these, 58 (11%) had illnesses that met the case definition. The median age of the 58 ill persons was 16 years (range: 12--53 years), and 95% were female. The median length of illness was 5 days; four (7%) persons were hospitalized. Two persons developed HUS. In addition to diarrhea, reported symptoms included abdominal cramping (100%), nausea (62%), headache (56%), vomiting (38%), bloody diarrhea (37%), and fever with a median temperature of 100 F (38 C) (29%).

Illnesses peaked on the third and final day of camp (Figure 1). Illnesses with bloody diarrhea peaked on the day after the camp ended. No campers reported having a diarrheal illness or contact with a person with diarrhea during the 2 weeks before the start of camp.

One meal (supper on the first day of camp) and 21 other exposures were significantly associated with risk for developing illness. Of these 21 exposures, 19 were specific food items from among 202 foods and beverages served in the cafeteria during the camp and two were more general exposures. Only the two general exposures were significantly and independently associated with illness: consuming any ice from large trash can-style lined barrels that the camp provided in the dormitory lobby for filling water bottles (73% of ill persons versus 43% of nonill persons) (adjusted odds ratio [AOR]=3.4; 95% confidence interval [CI]=1.8--6.3; p=0.0001) and eating any salad from the cafeteria salad bar on at least one occasion (93% of ill persons versus 79% of nonill persons; AOR=3.5; 95% CI=1.4--11.8; p=0.02).

Inspection of the camp's water systems showed no evidence of plumbing cross-connections or failures that might have led to exposures to contaminated water or waste. Coliform testing of ice from the ice machines used to fill the barrels was negative. Campers reported dipping their drink containers and arms, hands, and heads into the ice. They also reported observing floating debris in the ice barrels. Inspection of the cafeteria and kitchen indicated that kitchen staff may have improperly followed cooking times and temperatures recommendations when preparing meals.

The laboratory investigation of stools specimens submitted by 11 ill persons yielded E. coli O111:H8 from two specimens. Three enrichment broths prepared from these 11 specimens had detectable Shiga toxin when screened with a commercial enzyme immunoassay (EIA). Two of these three EIA-positive stool specimens yielded colonies of Shiga toxin-producing E. coli, which were serotyped as E. coli O111:H8. Both isolates contained gene sequences for Shiga toxins 1 and 2 by polymerase chain reaction. E. coli O157:H7 was not isolated from any camper, foodhandler, or food or water sample. Samples of the implicated ice and salad items served during the camp were not available for testing.

Reported by: D Bergmire-Sweat, MPH, L Marengo, MS, P Pendergrass, MD, K Hendricks, MD, M Garcia, R Drumgoole, T Baldwin, K Kingsley, B Walsh, MPH, S Lang, L Prine, T Busby, L Trujillo, D Perrotta, PhD, Texas Dept of Health. A Hathaway, MD, B Jones, DVM, A Jaiyeola, MBBS, Tarrant County Health Dept, Fort Worth, Texas. S Bengtson, DVM, Food Safety Inspection Svc, US Dept of Agriculture. Foodborne and Diarrheal Diseases Br, Div of Bacterial and Mycotic Diseases, National Center for Infectious Diseases; and an EIS Officer, CDC.

Editorial Note: This was the first community outbreak of infections attributable to Shiga toxin-producing E. coli O111 reported in the United States. The findings of the investigation suggest a point-source outbreak. Although primary infection from eating a contaminated salad item and then secondary spread through the barrel ice is a plausible hypothesis, the original source of contamination and its means of spread are unknown.

Identification of non-O157 STEC requires techniques not used routinely by clinical laboratories. In this outbreak, a commercially available EIA kit was used to detect and isolate STEC in stool specimens; isolates were then serotyped at CDC.

STEC cause illness in otherwise healthy persons, including severe abdominal cramping (sometimes confused for appendicitis), bloody diarrhea, and HUS. E. coli O111 was the second most common non-O157 STEC (after E. coli O26) isolated from specimens submitted to CDC for serotyping during 1983--1998 and among isolates from persons with diarrhea collected for an ongoing survey in Minnesota initiated in 1995 (Minnesota Department of Public Health, unpublished data, 2000). STEC cause an estimated 110,000 illnesses each year in the United States, of which >30% may be attributable to non-O157 serotypes such as O111 (1); the burden of disease attributable to non-O157 STEC is unknown.

Most STEC outbreaks in North America have resulted from infection with E. coli O157. A household cluster of E. coli O111 infection was reported in 1990 from Ohio (2), and outbreaks have occurred in Australia, Europe, and Japan (3--7). Despite investigations involving large numbers of persons in well-defined settings, the vehicle of transmission has been epidemiologically implicated and microbiologically confirmed in only one 1995 outbreak in South Australia, which was attributable to mettwurst, a dried fermented sausage (3).

As demonstrated by this outbreak, a commercially available kit could be used to screen stool specimens for Shiga toxin and potential STEC. However, culturing and serotyping the causative organism is critical to identify and better understand these emerging pathogens. To facilitate diagnosis of STEC infections, clinicians should inform health departments about clusters of suspected illnesses that could be attributable to STEC (e.g., bloody diarrhea and HUS). Clinical laboratories should screen stool specimens from persons with either bloody diarrhea or HUS for STEC, routinely or when E. coli O157 is not isolated, and attempt to isolate STEC from stools that are positive by the screening test and refer isolates to public health laboratories for serotyping. States should consider adding STEC infections to their notifiable disease lists.

References
1. Mead PS, Slutsker L, Dietz V, et al. Food-related illness and death in the United States. Emerg Infect Dis 1999;5:60.
2. Banatvala N, Debeaukelaer MM, Griffin PM, et al. Shiga-like toxin-producing Escherichia coli O111 and associated hemolytic-uremic syndrome: a family outbreak. Pediatr Infect Dis J 1996;15:1008--11.
3. CDC. Community outbreak of hemolytic uremic syndrome attributable to Escherichia coli O111:NM---South Australia, 1995. MMWR 1995;44:550--1,557--8.
4. Tanaka H, Ohseto M, Yamashita Y, et al. Bacteriological investigation on an outbreak of acute enteritis associated with verotoxin-producing Escherichia coli O111:H [Japanese]. Kansenshogaku Zasshi Journal of the Japanese Association for Infectious Diseases 1989;63:1187--94.
5. Viljanen MK, Peltola T, Junnila SYT, et al. Outbreak of diarrhea due to Escherchia coli O111:B4 in schoolchildren and adults: association of Vi antigen-like reactivity. Lancet 1990;336:831--4.
6. Caprioli A, Luzzi I, Rosmini F, et al. Community-wide outbreak of hemolytic-uremic syndrome

associated with non-O157 verocytotoxin-producing Escherichia coli. J Infect Dis 1994;169:208--11.
7. Wright JP, Rhodes P, Chapman PA, et al. Outbreaks of food poisoning in adults due to
Escherichia coli O111 and campylobacter associated with coach trips to northern France.
Epidemiology & Infection 1997;119:9--25.

September 17, 1999 / 48(36);803

Public Health Dispatch: Outbreak of Escherichia coli O157:H7 and Campylobacter Among Attendees of the Washington County Fair -- New York, 1999

On September 3, 1999, the New York State Department of Health (NYSDOH) received reports of at
least 10 children hospitalized with bloody diarrhea or Escherichia coli O157:H7 infection in counties
near Albany, New York. All of the children had attended the Washington County Fair, which was
held August 23-29, 1999; approximately 108,000 persons attended the fair during that week.
Subsequently, fair attendees infected with Campylobacter jejuni also were identified. An ongoing
investigation includes heightened case-finding efforts, epidemiologic and laboratory studies, and an
environmental investigation of the Washington County fairgrounds. This report presents the preliminary findings implicating contaminated well water.

To identify additional fair attendees with diarrhea, the NYSDOH issued press releases, conducted
daily press briefings, and contacted emergency departments, laboratories, and infection-control practitioners by fax and telephone. Laboratories were asked to culture all diarrheal stool specimens for E.
coli O157:H7 and subsequently for Campylobacter spp.

As of September 15, 921 persons reported diarrhea after attending the Washington County Fair.
Stool cultures yielded E. coli O157:H7 from 116 persons; 13 of these persons were co-infected with
C. jejuni. Stool cultures from 32 additional persons yielded only Campylobacter. Sixty-five persons
have been hospitalized; 11 children have developed hemolytic uremic syndrome (HUS); and two
persons died: a 3-year-old girl from HUS and a 79-year-old man from HUS/thrombotic thrombocytopenic purpura. Cases of diarrheal illness among fair attendees have been reported from 14 New
York counties and four states.

An environmental investigation of the fairgrounds on September 3 determined that much of the fair
was supplied with chlorinated water. However, in at least one area of the fair, a shallow well supplied unchlorinated water to several food vendors who used the water to make beverages and ice.
Initial cultures of water from this well yielded high levels of coliforms and E. coli.

A case-control study was conducted to determine risk factors for infection. Case-patients were residents of Washington County who developed diarrhea after attending the fair and in whom stool cultures yielded E. coli O157:H7 or Campylobacter. Controls were residents of Washington County randomly selected from the telephone directory who had attended the fair and were frequency-matched
by age group. Thirty-two case-patients and 84 controls were enrolled. Analysis was limited to those
attending the fair at least once during the final 4 days of the fair because all ill persons, including
those attending only once, attended during that period. Drinking water or beverages made with water
from the suspect well was associated with illness. Twenty-six (81%) of 32 case-patients and nine
(16%) of 57 controls had consumed water from this well during the final 4 days of the fair (matched
odds ratio=23.3; 95% confidence interval=6.3-86.9). When controlled for water consumption, other
exposures, such as eating food at the fair and contact with manure, were not significantly associated
with illness.

On September 9, the New York State Public Health Laboratory, the Wadsworth Center, used five different polymerase chain reaction assays to demonstrate the presence of E. coli O157:H7 DNA in
water from the implicated well and subsequently isolated the organism from water samples from the
well and the water distribution system. Pulsed-field gel electrophoresis testing by the Wadsworth
Center showed that the DNA "fingerprints" of E. coli O157:H7 isolates from the well, the water distribution system, and most patients were similar. Water sampling for Campylobacter spp. is ongoing.

To prevent secondary transmission of enteric infection, letters were sent to schools and day care centers emphasizing the need to exclude symptomatic children and practice careful handwashing. Letters also were sent to nursing homes and hospitals with recommendations regarding employees and residents with diarrhea. Information to the public about the outbreak also focused on how to prevent secondary infections. On September 13, the state health commissioner issued an order requiring county fairgrounds to use disinfected water when hosting public events; the commissioner also is reviewing laws and regulations applicable to fairs.

Reported by: County health depts in the Capital District; New York state outbreak investigation team; A Novello, MD, Commissioner, New York State Dept of Health. Foodborne and Diarrheal Diseases Br, Div of Bacterial and Mycotic Diseases, National Center for Infectious Diseases; and EIS officers, CDC.

PARASITIC PROTOZOA and WORMS

June 16, 1989 / 38(23);405-407

Epidemiologic Notes and Reports Common-Source Outbreak of Giardiasis -- New Mexico

In April 1988, the Albuquerque Environmental Health Department and the New Mexico Health and Environment Department investigated reports of giardiasis among members of a church youth group in Albuquerque. The first two members to be affected had onset of diarrhea on March 3 and 4, respectively; stool specimens from both were positive for Giardia lamblia cysts. These two persons had only church youth group activities in common. Routine surveillance identified no other cases associated with the church youth group.

The youth group had dinner once a week at the church; food was prepared by parents of group members. The number of attendees at each meal varied, and no record of who attended was kept. A survey of all families attending the church sought to identify any family members who had eaten at any youth group dinners in March and any who had had diarrhea since February 1, 1988. One hundred forty-eight persons who attended at least one youth group dinner in March were interviewed about food they had eaten at the meal(s); the 42 persons reporting diarrheal illness were interviewed about details of their illness.

A case was defined as diarrhea and/or abdominal cramping with onset after February 1, 1988, lasting greater than 7 days and/or a stool specimen positive for Giardia cysts. Twenty-two (15%) persons met the case definition. Onset of illness occurred from March 3 to March 30 (Figure 1), and illness lasted 1-32 days (median: 20 days). Twenty-one (19%) of 108 persons who ate the youth group dinner on March 2 developed an illness meeting the case definition, compared with one (3%) of 40 who did not eat that meal (relative risk (RR)=7.8, 95% confidence interval (CI)=1.1-55.9, p=0.02).

For the 21 ill persons who had eaten the March 2 dinner, the most frequent symptoms reported were fatigue (95%), diarrhea (91%), abdominal cramps (57%), bloating (57%), and weight loss (67%). Patients ranged in age from 11 to 58 years (median: 39 years); 14 (67%) were female; 15 (71%) sought care from a physician. Fourteen (67%) patients submitted stool specimens for ova and parasite examination; 10 (71%) specimens were positive for Giardia cysts. Seven of the stool specimens were also tested for Shigella, Salmonella, Campylobacter, and Yersinia, and all were negative. One ill person attended a day-care center, one had household contact with a day-care center attendee, and none had consumed surface water. The foods served at the dinner on March 2 included tacos (with meat, onions, tomatoes, lettuce, cheese, salsa, sour cream, and tortillas), corn, peaches, cupcakes, soft drinks, coffee, and tea. No food samples were available for microbiologic testing. Persons who became ill were more likely to have reported eating lettuce (RR=8.1, CI=1.1-57.3), salsa (p less than 0.01), onions (RR=4.2, CI=1.9-9.1), or tomatoes (RR=3.5, CI=1.4-8.8) or drinking tea/coffee (RR=5.5, CI=2.3-13.4). Water consumption was not associated with illness. Lettuce, onions, and tea/coffee were most strongly associated with illness by logistic regression analysis.

Except for the commercially prepared salsa, the implicated foods were prepared in the church kitchen. The lettuce and tomatoes were rinsed at the kitchen's main sink; the outer leaves of the lettuce were removed; and the lettuce, tomatoes, and onions were chopped on the same cutting board, which was not washed between items. The dinner was prepared by eight women whose children were in the youth group; all ate the meal. Although the woman who prepared the lettuce and tomatoes taught preschool and had a child in preschool, neither she nor her child was ill when the meal was prepared. None of the eight food preparers reported symptoms at the time of meal preparation; however, five became ill with diarrhea after March 8. Three had stool specimens positive for Giardia cysts.

The church is on the municipal water system. A survey of possible connections between the church's potable water system and the sanitary sewer system identified five potential cross-connections. However, water samples taken at the time of the cross-connection survey had adequate chlorine levels and were negative for coliform bacteria. On April 4, after the investigation began, the church stopped using municipal water for consumption and began catering meals. After elimination of all cross-connections, every outlet was flushed simultaneously for 3 hours. No new cases occurred after the remediation measures were completed. Reported by: DJ Grabowski, MS, KJ Tiggs, JD Hall, DrPH, HW Senke, AJ Salas, Albuquerque Environmental Health Department; CM Powers, JA Knott,

Bernalillo County District Health Office; LJ Nims, Scientific Laboratory Div; CM Sewell, DrPH, Acting State Epidemiologist, New Mexico Health and Environment Dept. Div of Field Svcs, Epidemiology Program Office, CDC.

Editorial Note: In this apparent point-source outbreak of giardiasis, the most likely vehicle of transmission was taco ingredients. Although all the ill persons ate the commercially prepared salsa, salsa was unlikely to have transmitted Giardia cysts because the cysts would not remain viable after the pasteurization and canning processes. Two explanations for the contamination are possible. First, if the potable water was contaminated, the lettuce and tomatoes could have been contaminated when washed. Because the lettuce, tomatoes, and onions were all cut on the same board, cross-contamination could have occurred. However, because plumbing changes were made before completion of the epidemiologic investigation, this hypothesis could not be tested. Second, if the woman who prepared the lettuce and tomatoes was infected and excreting Giardia cysts, she could have contaminated the vegetables during preparation. However, this mode is less likely because this woman had acute onset of diarrhea 10 days after the meal, suggesting a new infection at that time.

Only two reported outbreaks of giardiasis have been associated with food: canned salmon (1) and noodle salad (2). In both outbreaks, contamination occurred when food was mixed with bare hands. Waterborne outbreaks of Giardia are well documented, and persons consuming untreated surface water are at increased risk for developing giardiasis (3). Person-to-person transmission is also well known in day-care and institutional settings (4). Public health officials should consider foodborne transmission when investigating outbreaks of giardiasis.

References
1. Osterholm MT, Forfang JC, Ristenen TL, et al. An outbreak of foodborne giardiasis. N Engl J Med 1981;304:24-8.
2. Petersen LR, Cartter ML, Hadler JL. A food-borne outbreak of Giardia lamblia. J Infect Dis 1988;157:846-8.
3. Craun GF. Waterborne giardiasis in the United States: a review. Am J Pub Health 1979;69:817-9.
4. Pickering LK, Woodward WE. Diarrhea in day care centers. Pediatr Infect Dis 1982;1:47-52.

December 23, 1983 / 32(50);662-4

Outbreak of Diarrheal Illness Associated with a Natural Disaster -- Utah

On August 8, 1983, the Utah Department of Health was notified by the Tooele County Health Department (TCHD) of an outbreak of diarrheal illness in Tooele, Utah, possibly associated with a contaminated public water supply that resulted from flooding during Utah's spring thaw. By September 30, 1983, 1,272 individuals were identified who met the following case definition: diarrhea lasting more than 5 days or recurrent diarrhea and two or more of the following symptoms: abdominal pain or cramping, bloating, nausea, weight loss, vomiting, or fever over 37.8 C (100 F). A total of 1,230 of the patients resided in Tooele (9.8% of the population of 12,500); the remaining 42 patients resided elsewhere but had visited or worked in Tooele.

Cases were identified from two sources: 1,104 came from Tooele physicians' daily rosters, and 168 responded to announcements by the local news media. Individuals were contacted by telephone and asked standardized questions.

For comparison, individuals living in a city of similar size and sociodemographics 65 miles distant and with its own municipal water system were selected randomly and asked the same questions as the patients. Three (2.9%) of 103 comparison individuals interviewed met the case definition. The difference between the prevalence of diarrheal illness in Tooele and that in the comparison city was statistically significant (p 0.02). Statistical comparison of the patients from Tooele and the individuals from the comparison town failed to incriminate exposure to mountain stream water (a common source of giardiasis in Utah), pet ownership, food, day-care centers, or anal intercourse--all recognized modes of giardiasis exposure.

The age and sex distributions of patients were similar to those of the general population served by the water district. Besides diarrhea, the most common symptoms were abdominal pain or cramping (88%) and bloating (77%). Sixty-seven percent complained of nausea; 32%, of vomiting; and 17%, of fever over 37.8C (100 F). Of 410 individuals submitting stool specimens for bacterial and parasitic examination, 105 (26%) had Giardia lamblia. No other pathogenic parasites were observed, and no Salmonella, Shigella, Yersinia, or Campylobacter were isolated. Approximately 90% of the 1,100 persons receiving medication were treated with metronidazole (Flagyl*); the remainder were treated with quinacrine (Atabrine*) or furazolidone (Furoxone*) or were given symptomatic medications.

Because of complaints about muddy water, the municipal water system in Tooele was inspected during the last week in July, and a pipe damaged by flooding, probably during the week of July 17 when Tooele experienced several days of heavy rain, was identified. During this week, three of five routine bacteriologic samples from this source had unsatisfactory coliform counts. Diarrheal illness peaked on August 1, approximately 2 weeks after the heavy rains and the abnormal coliform counts (Figure 6). The incubation period of waterborne giardiasis has been estimated as 7-14 days (1). On August 1, in response to complaints of murky tap water, the implicated water source was disconnected from the public water system. Ten days later, the system was hyperchlorinated to inactivate G. lamblia cysts.

The number of new cases declined steadily throughout August, and continuing surveillance indicates that no new cases have been epidemiologically linked to the public water system. A detailed cost analysis estimated the direct costs of the giardiasis cases at over $116,000. Reported by DM Perrotta, PhD, CR Nichols, MPA, AP Nelson, MPH, L Scanlon, G Smith, RE Johns, Jr, MD, State Epidemiologist, Utah Dept of Health, D Forster-Burke, M Bateman, G Dalton, MS, Tooele County Health Dept, Utah; Protozoal Diseases Br, Div of Parasitic Diseases, Center for Infectious Diseases, CDC.

Editorial Note: Flooding associated with abnormal weather patterns last year caused extensive damage in many areas in the United States, including Utah. This report illustrates a less obvious consequence of such natural disasters. A similar period of heavy water run-off associated with unseasonably warm weather and ash fall from the Mount St. Helens volcano eruption in 1980 was also linked to an outbreak of diarrhea due to G. lamblia (2).

It is unclear that this present outbreak of diarrheal illness was due solely to giardiasis, although this parasite was the only pathogenic agent identified. Because normal chlorine levels were temporarily unable to control bacterial contamination, some of the diarrhea cases may have been caused by unidentified bacteria or viruses.

Quinacrine (Atabrine) is the drug of choice for adults with giardiasis (3). Although individuals who receive quinacrine often complain of its bitter taste, the drug has not been associated with long-term adverse effects, as has metronidazole. The efficacy of quinacrine is thought to be better than that of metronidazole, and quinacrine costs considerably less (4).

References
1. CDC. Unpublished data.
2. Weniger BG, Blaser MJ, Gedrose J, Lippy EC, Juranek DD. An outbreak of waterborne giardiasis associated with heavy water runoff due to warm weather and volcanic ashfall. Am J Public Health 1983;73:868-72.
3. The Medical Letter. Drugs for parasitic diseases. 1982;24:5-12.
4. Wolfe MS. Giardiasis. New Engl J Med 1978;298:319-21.

March 08, 1985 / 34(9);125-6

Epidemiologic Notes and Reports Pseudo-outbreak of Intestinal Amebiasis -- California

In October 1983, the Los Angeles County (California) Department of Health Services was notified by a local medical laboratory of a large increase in the laboratory's diagnoses of intestinal amebiasis (Entamoeba histolytica infection). Thirty-eight cases were identified from August to October. The laboratory staff estimated that, before August, they had diagnosed approximately one E. histolytica infection per month.

A preliminary investigation failed to identify a common source of the infection. There had been no increase in the number of specimens examined, and although the laboratory served several health facilities, there was no clustering of cases in particular facilities. Finally, most patients did not belong to groups recognized to be at high risk for acquiring amebiasis (such as male homosexuals, tourists to or immigrants from developing countries, or institutionalized persons). The most common complaint of patients was gastrointestinal symptoms, and most improved after treatment with metronidazole. A review of amebiasis diagnoses from other laboratories in Los Angeles County did not reveal other instances of increased reporting.

To evaluate the accuracy of E. histolytica diagnoses, 71 slides from the 38 patients were reexamined by the University of California at Los Angeles Clinical Laboratory or the Los Angeles County Public Health Laboratories. Only four slides from two (5.3%) patients were found to contain E. histolytica. Of specimens from the 36 patients found not to have E. histolytica, 34 contained polymorphonuclear neutrophils and/or macrophages, and two contained nonpathogenic protozoa.

The laboratory reporting the increase follows approved procedures for the collection and examination of stools for protozoa. Permanent slides are prepared from fecal material preserved in polyvinyl alcohol and stained by the Gomori-trichrome method (1). One technician was responsible for reading parasitology slides and had performed that job for the preceding 4 years. The technician's supervisor reviewed all positive slides. The only change in procedure that had been recently introduced was the assignment of a different person to the preparation of the initial smears. This person prepared slides that were "less dense," and the slides were "easier to read." Reported by L Garcia, MT, University of California at Los Angeles Medical Laboratory, F Sorvillo, MPH, M Epstein, MD, K Mori, B Agee, MD, R Barnes, PhD, Los Angeles County Dept of Health Svcs, J Chin, MD, State Epidemiologist, California Dept of Health Svcs; Protozoal Diseases Br, Div of Parasitic Diseases, Center for Infectious Diseases, Laboratory Program Office, CDC.

Editorial Note: This pseudo-outbreak of intestinal amebiasis serves as a reminder that identification of E. histolytica is difficult. Although E. histolytica can be confused with other intestinal protozoa, a more common problem is that leukocytes or macrophages in stool specimens are identified as E. histolytica (2). In 1981, the College of American Pathologists (CAP) conducted a proficiency survey using a stool specimen, which contained many leukocytes, from a patient with inflammatory bowel disease (3). None of 15 referee laboratories but 100 (16.7%) of 599 participating laboratories reported one or more intestinal protozoa, most commonly E. histolytica. Similarly, as shown in a report of seven suspected outbreaks of amebiasis in the United States between 1971 and 1974, three laboratories might have mistakenly diagnosed amebiasis in as many as 1,200 patients a year for 20 years (2).

A summary of proficiency surveys for parasites conducted by the CAP from 1973 to 1977 showed that E. histolytica infections are also often overlooked (4). Twenty-seven percent of participating laboratories overlooked trophozoites, and 37% overlooked cysts of E. histolytica in stool specimens.

Results of CDC's Proficiency Testing Program in Parasitology closely paralleled those reported by the CAP. In 1982, CDC conducted a parasitology proficiency testing survey using a stool specimen that contained no parasites and numerous leukocytes. None of the 17 reference or referee laboratories reported the presence of intestinal parasites; however, 74 (14.0%) of the 528 participant laboratories incorrectly reported one or more intestinal parasites, most commonly E. histolytica cysts. A summary of CDC proficiency testing surveys in parasitology from 1973-1977 also demonstrated that E. histolytica is often overlooked. Twenty-nine percent of participating laboratories overlooked E. histolytica trophozites, and 33% overlooked E. histolytica cysts in stool specimens.

To avoid errors when attempting to diagnose parasitic diseases, physicians should identify laboratories in their areas whose staffs are experienced in diagnostic parasitology and who participate in and score well on proficiency testing for parasitic diseases.

References
1. Garcia LS, Ash LR. Diagnostic parasitology clinical laboratory manual. St. Louis, Missouri: CV Mosby, 1975:16-7.
2. Krogstad DJ, Spencer HC Jr, Healy GR, Gleason NN, Sexton DJ, Herron CA. Amebiasis: epidemiologic studies in the United States, 1971-1974. Ann Intern Med 1978;88:89-97.
3. College of American Pathologists. Special parasitology survey (critique specimen P-12), 1981.

4. Smith JW. Identification of fecal parasites in the special parasitology survey of the College of American Pathologists. Am J Clin Pathol 1979;72:371-3.

October 16, 1998 / 47(40);856-860

Outbreak of Cryptosporidiosis Associated with a Water Sprinkler Fountain -- Minnesota, 1997

Cryptosporidiosis associated with recreational water exposure is becoming recognized more frequently (1). This report summarizes the investigation of a large outbreak of cryptosporidiosis associated with exposure to a water sprinkler fountain at the Minnesota Zoo. The initial cases were not diagnosed as cryptosporidiosis by the health-care system despite patients seeking care, underscoring the need for increased awareness of cryptosporidiosis and routine laboratory diagnostic practices among health-care providers.

On July 10, 1997, the Minnesota Department of Health (MDH) was notified by a parent about four cases of gastroenteritis among a group of 10 children whose only common exposure was a birthday party at the Minnesota Zoo on June 29. The zoo provided MDH with a list of registered groups that had visited the zoo during June 28-30; group members were contacted and interviewed about illness and zoo exposures. Initially, cases were defined as vomiting or diarrhea (defined as three or more loose stools during a 24-hour period) in persons who visited the zoo. Of 120 zoo visitors identified through the registered groups, 11 (9%) had illnesses that met the case definition. All had played in a water sprinkler fountain at the zoo, compared with seven (6%) of 109 controls (relative risk=undefined; p less than 0.001). Cryptosporidium oocysts were identified in nine of 10 stool specimens of case-patients tested at MDH. Two of the laboratory-confirmed case-patients had submitted stool samples previously for ova and parasite examination to their health-care providers; both samples were reported as negative for parasites.

The fountain was closed on July 11, and MDH issued a public statement advising persons who had visited the zoo and subsequently developed diarrheal illness to contact their physician and MDH. The public statement also stated that children who developed diarrhea after exposure to the fountain should not visit swimming beaches, swimming and wading pools, and other recreational water facilities until at least 2 weeks after recovery from diarrheal symptoms. MDH requested that all clinical laboratories in Minnesota specifically test all stools submitted for ova and parasite examination for Cryptosporidium, particularly during the outbreak.

A standard questionnaire was used to document illness history and zoo exposures in persons responding to the public statement. A revised case definition included persons with vomiting or diarrhea persisting at least 3 days, with onset 3-15 days after exposure to the zoo fountain. A total of 369 cases were identified, including the initial 11 cases; 73 (20%) were laboratory confirmed. Petting zoo exposure was reported by 191 (58%) of 332 case-patients, including 37 (55%) of 67 laboratory-confirmed cases. Age data were available for 351 case-patients; the median age was 6 years (range: 0-65 years), and 333 (95%) case-patients were aged less than or equal to 10 years. All but one of the 369 patients reported diarrhea; 317 (86%), abdominal cramps; 287 (78%), vomiting; 233 (63%), fever; and 11 (3%), bloody stools. The median duration of illness was 7 days. Six (2%) patients were hospitalized.

Reported dates of fountain exposure for case-patients were from June 24 through July 11 (Figure_1). Exposure dates for confirmed case-patients were from June 28 through July 1, with 68 (93%) exposures occurring from June 29 through July 1 (Figure_1). The median incubation period after fountain exposure was 6 days. In addition to case-patients with fountain exposure, nine laboratory-confirmed cases of cryptosporidiosis were identified among household contacts of case-patients with fountain exposure.

The implicated water sprinkler fountain was designed and built as a decorative display in 1994. The fountain is comprised of 14 nozzles arranged in five rows and submerged beneath metal grates. The nozzles sprayed jets of water vertically approximately one to six feet. The water drained through the grates, collected in trenches, passed through a sand filter, was chlorinated, and then recirculated. The zoo routinely replaced the water every Monday, Wednesday, and Friday, but the filter was not flushed. Environmental health inspectors from MDH recommended the fountain not be used as an interactive play area. The zoo subsequently erected a fence around the fountain plaza and reopened it as a decorative display only. Water samples collected on July 14 were negative for Cryptosporidium oocysts.

The source of contamination of the fountain was not established, but contamination by a child wearing a diaper and playing in the fountain was suspected. Animals (including ruminants) in a petting zoo approximately 50 yards from the fountain tested negative for Cryptosporidium before being placed in the petting area and again during the outbreak investigation.

A 1997 survey of all clinical laboratories serving Minnesota residents indicated that 13 (22%) of 59 laboratories that perform ova and parasite examinations on site routinely test for Cryptosporidium as part of ova and parasite examinations (i.e., without a specific request from a physician). In a 1997 survey of physicians in Minnesota, 44 (79%) of 56 physicians who thought that their laboratory always tested for Cryptosporidium as part of an ova and parasite examination were incorrect.

Reported by: VC Deneen, MS, PA Belle-Isle, CM Taylor, LL Gabriel, JB Bender, DVM, JH Wicklund, MPH, CW Hedberg, PhD, MT Osterholm, PhD, State Epidemiologist, Minnesota Dept of Health. Div of Parasitic Diseases, National Center for Infectious Diseases; Div of Applied Public Health Training, Epidemiology Program Office; and an EIS Officer, CDC.

Editorial Note: The findings in this report document a novel recreational water source for cryptosporidiosis. Outbreaks of cryptosporidiosis have been documented in a variety of other recreational water settings in the United States since 1988, including a lake, community and hotel pools, a large recreational water park, a wave pool, and a water slide (1). As in several other outbreaks, there was no evidence in this outbreak that inadequate chlorination or filter malfunction contributed to transmission of Cryptosporidium. However, Cryptosporidium oocysts are resistant to disinfection by chlorine at levels generally used in recreational water, and recreational water filtration units that use sand filter media are not effective in removing the 4-6-micron oocysts (1). The zoo fountain in this outbreak was designed as a decorative display and not an interactive play area. However, the fountain was a popular attraction for children on hot summer days. Children would commonly stand directly over the jets and soak their entire bodies, a practice which could explain contamination of the fountain and subsequent transmission associated with ingestion of water. Consumption of foods while walking in the fountain plaza was also a common practice.

Measures that might have reduced the risk for Cryptosporidium contamination of the fountain (e.g., showering before entering the fountain, excluding persons with diarrhea or incontinence, excluding children wearing diapers, and restricting food consumption in the fountain area) were not required or encouraged. Exclusion of persons from decorative water displays not designed for interactive use should be instituted and enforced. For recreational water facilities designed for human use, improved filtration may reduce risk.

Waterborne cryptosporidiosis is probably underrecognized and underreported (1). Laboratory and physician surveys conducted in Minnesota indicate that most laboratories do not routinely test specifically for Cryptosporidium as part of ova and parasite examinations, even though many physicians assumed that they did. Even though cryptosporidiosis is reportable in Minnesota, this large outbreak probably would have remained undetected if not for the parent reporting the cases to the health department. Two of the original ill children had seen physicians, who ordered ova and parasite examinations; however, cryptosporidiosis remained undiagnosed until stool samples were examined specifically for Cryptosporidium at MDH. Because of their small size, Cryptosporidium oocysts can be difficult to detect by routine ova and parasite examination. The magnitude of this outbreak was probably determined only because of the public statement and the request that laboratories test all stools submitted for ova and parasite examination specifically for Cryptosporidium.

To better understand the magnitude of cryptosporidiosis, health-care providers should specifically request testing for suspected cryptosporidiosis. Laboratories should consider routinely testing for Cryptosporidium as part of their ova and parasite examination protocol. Alternatively, when reporting test results back to health-care providers, laboratories should specifically indicate when Cryptosporidium is not tested for as part of a requested ova and parasite examination. Cryptosporidiosis is reportable in 41 states; interpretation of national data would be facilitated by mandatory reporting in all states.

Reference
1. Anonymous. Cryptosporidium and water: a public health handbook. Atlanta, Georgia: Working Group on Waterborne Cryptosporidiosis, 1997.

July 17, 1998 / 47(27);565-567

Foodborne Outbreak of Cryptosporidiosis -- Spokane, Washington, 1997

On December 29, 1997, the Spokane Regional Health District received reports of acute gastroenteritis among members of a group attending a dinner banquet catered by a Spokane restaurant on December 18. The illness was characterized by a prolonged (3-9 days) incubation period and diarrhea, which led public health officials to suspect a parasitic cause of the illness. Eight of 10 stool specimens obtained from ill banquet attendees were positive for Cryptosporidium using both modified acid-fast and auramine-rhodamine staining of concentrated specimens. This report summarizes the epidemiologic investigation of the outbreak, which suggests that foodborne transmission occurred through a contaminated ingredient in multiple menu items.

In a retrospective cohort study, a case was defined as diarrhea or abdominal cramping in a banquet attendee with onset within 10 days after the banquet. Of the 62 attendees, 54 (87%) had illnesses meeting the case definition; they became ill a median of 6 days (range: 3-9 days) after the banquet. Symptoms included diarrhea (98%), fever/chills (61%), headache (59%), body ache (54%), abdominal cramps (50%), nausea (28%), and vomiting (11%). Based on information from initial interviews, the median length of illness was 5 days (range: 1-13 days), but subsequently several persons reported that they had symptoms intermittently for a month or longer. Two persons were hospitalized, and six others sought health care for their illness.

The banquet buffet included 18 separate food and beverage items; seven items contained uncooked produce. No single food was significantly associated with illness. When menu items that contained green onions were combined, foods containing uncooked green onions (au gratin potatoes, romaine salad, and pasta salad) were reportedly eaten by all 51 case-patients who could recall and by three of four persons who were not ill and could recall (undefined relative risk, p=0.07).

The banquet food items were prepared or served by 15 food workers. Stool specimens were available from 14 food workers within 3-4 weeks of the banquet; specimens from two tested positive for Cryptosporidium. One of the two food workers was symptomatic at the same time as banquet attendees; the other was asymptomatic. A stool specimen from another food worker was not available for testing until 5 weeks after the outbreak and was negative; he reported that he worked for 2 days in December while experiencing diarrhea but he could not remember the dates of his illness. All three of these food workers reportedly ate food items served at the banquet associated with the outbreak.

The green onions were not washed before delivery at the restaurant. Food workers at the restaurant reported they did not consistently wash green onions before using them to prepare food or serving them to patrons.

To determine the extent of the outbreak, the health district requested by fax that Spokane area physicians report any patients with symptoms typical of cryptosporidiosis. No other cryptosporidiosis-like illnesses were identified at the time of the outbreak. Two other banquets catered by the restaurant on December 18 and 19 had menus similar to the banquet where the outbreak occurred; no illness was reported in either of these groups.

Reported by: K Quinn, MPA, G Baldwin, P Stepak, MD, K Thorburn, MD, Spokane Regional Health District; C Bartleson, MPH, M Goldoft, MD, J Kobayashi, MD, P Stehr-Green, DrPH, State Epidemiologist, Washington Dept of Health. Div of Parasitic Diseases, National Center for Infectious Diseases, CDC.

Editorial Note: Since 1993, three foodborne outbreaks of cryptosporidiosis have been reported in the United States. In 1993, an outbreak was associated with drinking unpasteurized, fresh-pressed apple cider (1); the apples used for the cider probably were contaminated when they fell to the ground in a cow pasture. In 1995, an outbreak was associated with eating chicken salad that may have been contaminated by a food worker who operated a day care facility in her home (2). In 1996, an outbreak was associated with drinking commercially produced, unpasteurized apple cider (3); the apples used for the cider may have become contaminated when they were washed with well water that had fecal contamination.

The outbreak described in this report had characteristics similar to others in the United States caused by enteric coccidian parasites (Cryptosporidium parvum and Cyclospora cayetanensis) in that case-patients had prolonged diarrhea; the incubation period averaged 6 days; and the attack rates were high (4,5). Physicians and public health officials should have a high index of suspicion for infection with coccidian parasites in patients with severe or prolonged watery diarrhea. Because most laboratories do not routinely test stool for either Cryptosporidium or Cyclospora (6), specific testing for these organisms generally must be ordered by a physician.

The high attack rate among banquet attendees made finding a statistically significant association with a particular menu item difficult. The strongest association between illness and eating a menu item was observed for food items containing uncooked green onions. This suggests that the onions were a possible source, but the data are inadequate to conclusively implicate them as the vehicle of infection. Available data do not exclude the possibility that multiple menu items may have been contaminated before arriving at the restaurant, contaminated by a food worker, or by cross-contamination during preparation.

This outbreak highlights several key issues for food workers. Uncooked produce should be throughly washed before being placed on kitchen work surfaces to prevent contamination of these surfaces. The FDA Food Code prohibits further bare-handed contact with fruits and vegetables after washing when they are intended for use in "ready-to-eat" foods except where approved by the regulating authority (7). Food preparation surfaces should be washed between preparation of different produce to prevent cross-contamination. Food workers should not work when experiencing a gastrointestinal illness. Persons infected with Cryptosporidium may intermittently shed oocysts in stool and remain infectious for up to 60 days after diarrhea has resolved; however, most persons will cease shedding within 2 weeks after resolution of their diarrhea (8). Therefore, food workers should be particularly meticulous about handwashing. Asymptomatic shedding probably occurs in persons exposed to the parasite who have developed some immunity, but the frequency of asymptomatic shedding is unknown.

References
1. Millard PS, Gensheimer KF, Addiss DG, et al. An outbreak of cryptosporidiosis from fresh-pressed apple cider. JAMA 1994;272:1592-6.
2. CDC. Foodborne outbreak of diarrheal illness associated with Cryptosporidium parvum -- Minnesota, 1995. MMWR 1996;45:783-4.
3. CDC. Outbreaks of Escherichia coli O157:H7 infection and cryptosporidiosis associated with drinking unpasteurized apple cider -- Connecticut and New York, October 1996. MMWR 1997;46:4-8.
4. Mac Kenzie WR, Schell WL, Blair KA, et al. Massive outbreak of waterborne Cryptosporidium infection in Milwaukee, Wisconsin: recurrence of illness and risk of secondary transmission. Clin Infect Dis 1995;21:57-62.
5. Herwaldt BL, Ackers ML, The Cyclospora Working Group. An outbreak in 1996 of cyclosporiasis associated with imported raspberries. N Engl J Med 1997;336:1548-56.
6. Boyce TG, Pemberton AG, Addiss DG. Cryptosporidium testing practices among clinical laboratories in the United States. Pediatr Infect Dis J 1996;15:87-8.
7. Food and Drug Administration. Food code, 1997. Rockville, Maryland: US Department of Health and Human Services, Public Health Service, Food and Drug Administration, 1997.
8. Stehr-Green JK, McCaig L, Remsen HM, Rains CS, Fox M, Juranek DD. Shedding of oocysts in immunocompetent individuals infected with Cryptosporidium. Am J Trop Med Hyg 1987;36:338-42.

MORBIDITY AND MORTALITY WEEKLY REPORTS

April 25, 1997 / 46(16);354-358

Outbreaks of Pseudo-Infection with Cyclospora and Cryptosporidium --
Florida and New York City, 1995

Efforts to expand the scope of surveillance and diagnostic testing for emerging infectious diseases
(1) also may increase the potential for identifying pseudo-outbreaks (2,3) (i.e., increases in incidence
that may result from enhanced surveillance) and outbreaks of pseudo-infection (i.e., clusters of false-
positives for infection). This report describes the investigations of outbreaks of pseudo-infection with
Cyclospora in Florida and Cryptosporidium in New York City in 1995 after health departments in
those jurisdictions had initiated surveillance for these emerging organisms. These investigations
emphasize 1) the need for laboratory training in the identification of emerging pathogens and 2) the
importance of confirmation by reference laboratories as an early step in the investigation of any
apparent outbreak caused by an emerging pathogen.

CYCLOSPORIASIS IN FLORIDA

Cyclosporiasis is caused by infection with Cyclospora cayetanensis, a recently identified coccidian
parasite (4) that can cause prolonged, relapsing diarrhea; treatment with trimethoprim-sulfamethoxazole relieves symptoms and accelerates clearance of the parasite (5). Until 1996, most cases of
cyclosporiasis in the United States occurred among international travelers (6), and information about
modes of transmission of C. cayetanensis was limited. Waterborne transmission had been documented, but direct person-to-person transmission was considered unlikely (4).

During the summer of 1995, in response to an outbreak of Cyclospora infection among Florida residents with no history of international travel (7), the state health department initiated surveillance for
the organism. All state laboratories began routine testing with a modified acid-fast stain for C.
cayetanensis in stool specimens submitted for parasitologic examination (8). On July 25, 1995, the
Florida Department of Health (FDH) designated cyclosporiasis a reportable disease.

On August 11, a 3-year-old boy at a children's shelter had onset of diarrhea and abdominal pain;
Giardia cysts were identified in a stool specimen obtained from the child. Because of previous giardiasis outbreaks at the shelter, county public health officials recommended testing the 13 shelter residents who were preschool classmates or roommates of the index patient. State branch laboratory A
reported that stool specimens from six children tested positive for Giardia, and six tested positive for
Cyclospora. The high proportion of specimens positive for Cyclospora prompted testing of 81 persons, including all children residing at the shelter and the shelter's staff and volunteers. Overall,
branch laboratory A identified Cyclospora oocysts in specimens from 31 (86%) of 36 staff, 16 (64%)
of 25 children, and nine (45%) of 20 volunteers. In response to this apparent outbreak, the residence
was closed to new admissions, the children's outside activities were restricted, and trimethoprim-sulfamethoxazole was prescribed for all 25 children.

On September 17, FDH was notified about the apparent outbreak and joined the investigation. The
local community hospital, which had begun testing for Cyclospora in 1995, was contacted for information about laboratory-identified infections in the community during July-September.
Questionnaires were administered to shelter staff, volunteers, the older children, and the infants'
caretakers, and medical records for the children were reviewed. A case was defined as onset of nausea, vomiting, or diarrhea in a resident, employee, or volunteer at the shelter during August-
September. Branch laboratory A sent portions of stool specimens from 23 shelter residents and staff
to the reference laboratories at CDC and the University of Arizona for Cyclospora testing.

Questionnaires were completed for 79 of the 81 children and adults. Symptoms among the 56 persons whose stool specimens were positive for Cyclospora at branch laboratory A included abdominal
pain (30%), nausea (26%), fatigue (22%), diarrhea (three or more loose or watery stools in 24 hours)
with median duration of 2 days (20%), vomiting (19%), anorexia (15%), fever (10%), and weight
loss (8%). The 23 persons with negative stool specimens had symptoms and onset dates similar to
those of persons with positive specimens (Figure_1). In addition, the likelihood of being asymptomatic was similar among persons who were test-positive (35%) and test-negative (46%) (p=0.4).
Potential risk factors (e.g., consumption of food or water at the shelter or participation in field trips)
were not associated with the likelihood of being ill or testing positive. During the time of the appar-

ent outbreak at the shelter, the local community hospital examined 357 stool specimens for ova and parasites and identified Cyclospora oocysts in specimens from two patients; neither person was associated with the shelter.

Of the 23 stool specimens submitted by branch laboratory A to the two reference laboratories, branch laboratory A reported that 17 (74%) specimens were positive for Cyclospora; in comparison, the reference laboratories at both CDC and the University of Arizona reported that all the specimens were negative for Cyclospora. The state central laboratory and the University of Arizona laboratory reviewed slides from branch laboratory A and identified pollen grains and other artifacts similar to Cyclospora oocysts in size and staining characteristics but lacking the appropriate internal morphology. CDC examined stool specimens obtained from 19 other persons at the shelter; rare Cyclospora oocysts were identified in specimens from two children: one child who had been asymptomatic, and one who had vomited.

Based on these findings, FDH asked all laboratories in the state that had reported detecting Cyclospora oocysts in specimens from symptomatic patients in 1995 to forward their positive slides to the state central laboratory or CDC for confirmation. Branch laboratory A submitted slides from 130 patients not associated with the shelter; of these, 38 (29%) were confirmed as Cyclospora, and 92 (71%) were considered to have been false positives.

In response to the investigation, FDH revised the case definition for cyclosporiasis to include confirmation of Cyclospora infection by a reference laboratory, and the state central laboratory initiated a proficiency training program at all state laboratories to teach laboratorians how to identify Cyclospora and Cryptosporidium spp. In 1996, in a subsequent outbreak of cyclosporiasis (9), Florida laboratories initially identified 188 specimens from patients as positive for Cyclospora; 32 (17%) were not confirmed by the state central laboratory.

CRYPTOSPORIDIOSIS IN NEW YORK CITY

To improve disease reporting and identify exposures associated with infection, New York City designated cryptosporidiosis a reportable disease in January 1994, and the New York City Department of Health (NYCDOH) initiated active surveillance in November 1994. Each of the clinical laboratories are routinely contacted (usually monthly) for reports of new cases, and each case is investigated by telephone interview and/or chart review. Of the 289 cases of cryptosporidiosis reported in New York City during 1994, most (72%) occurred among men and among persons aged 20-44 years (63%).

Laboratory B, a commercial laboratory in New York City, examines approximately 400 stool specimens per month for ova and parasites. Although these examinations do not routinely include testing for Cryptosporidium parvum, requests for this test increased fourfold from 1993 (143 {3%} of 4344) to 1995 (587 {11%} of 5333). Before April 1995, laboratory B used a modified acid-fast technique to test for Cryptosporidium oocysts. From January 1994 through March 1995, laboratory B reported four cases of cryptosporidiosis.

In April 1995, after switching to an enzyme-linked immunosorbent assay (ELISA) method to test for Cryptosporidium antigen (ProSpecT Cryptosporidium Microplate Assay 21/96, Alexon Incorporated, Sunnyvale, California *), laboratory B began reporting an increased number of positive tests: 24 in April and a mean of 52 per month from May through September, for a total of 281 in 6 months. Demographic characteristics of these 281 patients differed from those of patients reported to have been positive by other New York City laboratories; specifically, patients who were test-positive by laboratory B were more likely to be aged greater than or equal to 60 years (36% versus 5%, pless than 0.01) and female (59% versus 25%, pless than 0.01).

Because of these findings, in August 1995 the NYCDOH initiated a validation study at laboratory B. Stool specimens submitted to laboratory B for Cryptosporidium testing were split and sent for parallel testing either to the New York City Bureau of Laboratories, which performed ELISA and acid-fast testing, or to the New York State Wadsworth Center, David Axelrod Institute for Public Health, which performed ELISA, direct immunofluorescence testing (MERIFLUOR Cryptosporidium/Giardia Direct Immunofluorescent Detection Procedure, Meridian Diagnostics Incorporated, Cincinnati, Ohio), and modified acid-fast testing. ELISA testing was performed with the same kit used by laboratory B. Of 84 split specimens, laboratory B reported 57 (68%) positive test results, and the two reference laboratories each reported one positive result. Based on these findings, all 280 unconfirmed positive ELISA results for Cryptosporidium identified at laboratory B from April through September were considered to have been false positives. Physicians for these patients were notified that previously reported positive results may have been the result of laboratory error.

Reported by: CR Sterling, PhD, YR Ortega, MS, Dept of Veterinary Science, Univ of Arizona, Tucson. EC Hartwig, Jr, ScD, MB Pawlowicz, MS, MT Cook, Bur of Laboratories, RS Hopkins, MD, State Epidemiologist, Florida Dept of Health. JR Miller, MD, M Layton, MD, Bur of Communicable Disease, A Ebrahimzadeh, PhD, Bur of Laboratories, New York City Dept of Health; J Ennis, J Keithly, PhD, Wadsworth Center, David Axelrod Institute for Public Health, New York State Dept of Health. Div of Parasitic Diseases, National Center for Infectious Diseases; Div of Applied Public Health Training (proposed), Epidemiology Program Office, CDC.

Editorial Note: Pseudo-outbreaks associated with increased surveillance for disease and outbreaks of pseudo-infection resulting from false-positive laboratory results may occur with increasing frequency because of changes in surveillance and rapid developments in the technologies and tests available to identify organisms -- particularly new and emerging pathogens. Cyclosporiasis and cryptosporidiosis are emerging infectious diseases in the United States, and many health departments and laboratories are unfamiliar with the identification and diagnosis of these parasitic infections. The outbreaks described in this report resulted from a combination of laboratory error and enhanced surveillance -- factors identified in previous pseudo-outbreaks (2,10). Consequences of such events include misdirection of resources; unnecessary treatment, anxiety, and disruption of patients' lives; and loss of confidence in laboratories and public health agencies (3).

To prevent the occurrence or minimize the impact of pseudo-outbreaks, the first steps in most outbreak investigations should be to confirm the diagnosis and the occurrence of the outbreak. Confirmation of the diagnosis entails validation of the laboratory findings and assessment of the concordance between the clinical features and test results. In the Florida investigation, an outbreak of pseudo-infection was suspected initially because of an inability to document an association between patients with specific clinical manifestations and positive findings for Cyclospora oocysts, while in New York City, the investigation was prompted, in part, by the atypical demographic characteristics of patients with cryptosporidiosis reported by one laboratory.

Although such patterns were important in the investigations described in this report, they typically are more reliable for well-characterized pathogens than for emerging pathogens, for which critical epidemiologic and clinical information may be limited. Because local laboratories may lack experience and optimal techniques for identifying emerging pathogens, these organisms may be more likely to be associated with outbreaks of pseudo-infection; therefore, confirmation of the diagnosis by an experienced reference laboratory may be critical in confirming outbreaks associated with these pathogens.

The outbreaks of pseudo-infection in Florida and New York City began after laboratory personnel implemented new testing procedures -- in one instance, for a newly-identified pathogen and, in the other, with a different technique. The investigations of these incidents emphasize the potential for the occurrence of such outbreaks when efforts are made to enhance laboratory surveillance. In addition, these incidents indicate the needs for training and proficiency testing in conjunction with the introduction of new laboratory techniques and for reporting laboratories to submit a proportion of their positive and negative specimens for confirmation by a reference laboratory following the initiation of surveillance or testing for new pathogens.

References
1. CDC. Addressing emerging infectious disease threats: a prevention strategy for the United States. Atlanta, Georgia: US Department of Health and Human Services, Public Health Service, 1994.
2. CDC. Enhanced detection of sporadic Escherichia coli O157:H7 infections -- New Jersey, July 1994. MMWR 1995;44:417-8.
3. Shears P. Pseudo-outbreaks. Lancet 1996;347:138.
4. Soave R. Cyclospora: an overview. Clin Infect Dis 1996;23:429-35.
5. Hoge CW, Shlim DR, Ghimire M, et al. Placebo-controlled trial of co-trimoxazole for Cyclospora infections among travelers and foreign residents in Nepal. Lancet 1995;345:691-3.

6. Wurtz R. Cyclospora: a newly identified intestinal pathogen of humans. Clin Infect Dis 1994;18:620-3.

7. Koumans EH, Katz D, Malecki J, et al. Novel parasite and mode of transmission: Cyclospora infection -- Florida {Abstract}. In: Program and abstracts of the 45th Annual Epidemic Intelligence Service (EIS) Conference. Atlanta, Georgia: US Department of Health and Human Services, Public Health Service, CDC, 1996.

8. Garcia LS, Bruckner DA. Diagnostic medical parasitology. 2nd ed. Washington, DC: American Society for Microbiology, 1993:528-32.

9. CDC. Update: outbreaks of Cyclospora cayetanensis infection -- United States and Canada, 1996. MMWR 1996;45:611-2.

10. Casemore DP. A pseudo-outbreak of cryptosporidiosis. Commun Dis Rep Rev 1992;2:R66-R67.

Use of trade names and commercial sources is for identification only and does not imply endorsement by the Public Health Service or the U.S. Department of Health and Human Services.

January 10, 1997 / 46(01);4-8

Outbreaks of Escherichia coli O157:H7 Infection and Cryptosporidiosis Associated with Drinking Unpasteurized Apple Cider -- Connecticut and New York, October 1996

In October 1996, unpasteurized apple cider or juice was associated with three outbreaks of gastrointestinal illness. In the Western United States, an outbreak of Escherichia coli O157:H7 infections associated with unpasteurized commercial apple juice caused illness in 66 persons and one death (1). In addition, one outbreak of apple cider-related E. coli O157:H7 infections and another of cider-related Cryptosporidium parvum infections occurred in the Northeast. Apple cider is a traditional beverage produced and consumed in the fall. Cider often is manufactured locally at small cider mills where apples are crushed in presses, and the cider frequently is not pasteurized before sale. This report summarizes the clinical and epidemiologic features of the two apple cider-related outbreaks, which suggest that current practices for producing apple cider may not be adequate to prevent microbial contamination. Connecticut

On October 11, the Connecticut Department of Public Health (DPH) was notified by staff of the Connecticut site of CDC's Foodborne Diseases Active Surveillance Network of four reported cases of E. coli O157:H7 infection in residents of New Haven County (1995 population: 794,785). An investigation of this cluster was initiated by DPH. A case was defined as onset of diarrhea (i.e., three or more loose stools per day) during October 1-11 in a Connecticut resident and laboratory-confirmed infection with E. coli O157:H7. Additional case-finding was conducted by notifying all Connecticut clinical laboratories of a possible outbreak of E. coli O157:H7 infection and requesting that cases be reported immediately to DPH. As a result of active case-finding, DPH initially identified eight cases with onset during October 3-11.

Of the eight case-patients, six were female, and ages of all eight ranged from 2 to 73 years (mean: 25 years). Case-patients resided in six towns within New Haven County. Manifestations included bloody diarrhea and abdominal pain (eight patients), vomiting (five), and fever (four). Duration of illness ranged from 3 to 11 days (median: 7 days). Five patients were hospitalized, including one with hemolytic uremic syndrome (HUS) and one with thrombotic thrombocytopenic purpura.

On October 17, DPH conducted a matched case-control study to determine probable sources for the outbreak. Controls were selected from telephone-exchange lists and were matched to cases by sex, town of residence, and age group. Controls reported no diarrhea during the 20-day period beginning 10 days before illness onset in their matched cases. Case-patients were asked about food consumption during the 7 days preceding illness, and controls were asked about consumption during the same 7 days as their matched cases. Based on interviews with the first eight case-patients and 21 controls, increased risk for illness was associated with drinking fresh apple cider during the 7 days preceding onset of illness (matched odds ratio {OR}=12.0; 95% confidence interval {CI}=1.3-111.3; pless than 0.01). Specifically, illness was associated with drinking brand A cider (matched OR=undefined; 95% CI=3.5-infinity; pless than 0.01). No other food item (including ground beef, unpasteurized milk, or lettuce) or common event was significantly associated with increased risk for

259

illness. Of the eight patients, seven reported drinking brand A cider during the 7 days preceding illness.

After completion of the case-control study, six additional patients were identified; of these, four had culture-confirmed infection, and two had been hospitalized with HUS but did not have culture-confirmed E. coli O157:H7 infection. All six had a history of drinking brand A cider. Ten of the 12 outbreak-associated isolates of E. coli O157:H7 were sent to CDC for pulse-field gel electrophoresis typing; all 10 were determined to be closely related.

On October 18, DPH and the Connecticut Department of Consumer Protection (DCP) advised Connecticut residents to discard or boil before drinking all brand A cider purchased since the beginning of the cider season in September. DCP coordinated a recall of brand A cider from all retail outlets. Approximately 9000 gallons of the cider had been distributed throughout Connecticut and three neighboring states. DCP and the regional office of the Food and Drug Administration (FDA) notified regulatory agency and state health department personnel in the three neighboring states of the recall.

Brand A cider was pressed at a mill in a residential area from apples purchased from multiple sources. Some of the apples used were "drop" apples (i.e., apples picked up from the ground). All apples were brushed and washed in potable municipal water in a flow-through wash system before pressing in a wooden press. Potassium sorbate 0.1% was added as a preservative; the cider was not pasteurized. New York

During October 10-15, a local hospital laboratory notified the Cortland County Health Department (CCHD) about 10 cases of laboratory-confirmed cryptosporidiosis with recent onset among county residents (1990 population: 48,963). During the same period in 1995, one case of cryptosporidiosis was reported to CCHD. All case-patients had onset of symptoms during September 28-October 10 and reported drinking apple cider produced at a local cider mill (mill A). CCHD, the New York State Department of Health (NYSDOH), and the New York State Department of Agriculture and Markets (NYS A&M) initiated an investigation of this cluster.

A confirmed case was defined as onset of diarrhea during September 28-October 19 in a Cortland County resident and laboratory evidence of Cryptosporidium in a stool specimen. A suspected case was defined as onset of diarrhea during the outbreak period in a household member of a person with confirmed cryptosporidiosis. CCHD conducted active surveillance for additional cases by contacting area clinicians, hospitals, and laboratories.

A total of 20 confirmed and 11 suspected cases were identified from 19 households. The median age was 27 years (range: 1-62 years), and 17 were female. Symptoms included diarrhea (100%), abdominal cramping (55%), vomiting (39%), fever (36%), and bloody diarrhea (10%). The median duration of symptoms was 6 days (range: 1-21 days).

CCHD and NYSDOH conducted a matched case-control study to assess probable sources of the outbreak. One neighborhood-matched control-household was contacted for each household with a laboratory-confirmed case. In each control-household, an adult (age greater than or equal to 18 years) member was asked about history of illness, whether anyone in the household had drank apple cider since September 28, which brand of cider was consumed, and the date the cider was purchased.

Eighteen case-households were included in the matched case-control study. A history of drinking cider from mill A was reported for at least one member of the 18 households, compared with only one of the 18 control-households (matched OR=undefined, pless than 0.01). Specifically, cider pressed during September 28-29 (i.e., opening weekend) was associated with illness: 15 of 17 case-households in which the purchase date was known compared with none of the control-households reported drinking cider pressed on opening weekend (matched OR=undefined, pless than 0.01).

Mill A purchased all apples for cider pressing from one New York orchard. Local and state health departments and NYS A&M inspected the cider mill and apple orchard. The owner of the orchard reported that only picked apples were sold to the cider mill, and drop apples were sold for use in processed or pasteurized foods. Before pressing, the mill washed and brushed the apples using water from a 45-foot drilled well; preservatives were not added to the cider. Although dairy livestock were not maintained by the orchard, the cider mill was located across the road from a dairy farm. Testing of remaining cider samples from opening weekend, swabs of equipment surfaces, and water obtained on October 21 from the drilled well did not yield Cryptosporidium. However, coliform bacteria were detected in four water samples obtained from the well, and E. coli was detected in one sample.

Reported by: PA Mshar, ZF Dembek, PhD, ML Cartter, MD, JL Hadler, MD, State Epidemiologist, Connecticut Dept of Public Health; TR Fiorentino, MPH, RA Marcus, MPH, School of Medicine, Yale Univ, New Haven; J McGuire, MA Shiffrin, Connecticut Dept of Consumer Protection. A Lewis, J Feuss, J Van Dyke, Cortland County Dept of Health; M Toly, M Cambridge, J Guzewich, J Keithly, PhD, D Dziewulski, PhD, E Braun-Howland, PhD, D Ackman, MD, P Smith, MD, State Epidemiologist, New York State Dept of Health; J Coates, J Ferrara, New York State Dept of Agriculture and Markets. Foodborne and Diarrheal Diseases Br, Div of Bacterial and Mycotic Diseases, and Div of Parasitic Diseases, National Center for Infectious Diseases; State Br, Div of Applied Public Health and Training (proposed), Epidemiology Program Office, CDC.

Editorial Note: Unpasteurized apple cider and juice have been associated with outbreaks of E. coli O157:H7 infection, cryptosporidiosis, and salmonellosis (1-4). Animals are the primary reservoir for the pathogenic organisms associated with these outbreaks. In particular, cattle, deer, and sheep can asymptomatically carry E. coli O157:H7 and Cryptosporidium, and many animals, including cattle, chickens, and pigs, can asymptomatically carry Salmonella. Although the exact mechanisms of contamination for these previous outbreaks were not clearly determined, in three of the outbreaks, manure was suspected to have contaminated the apples. For example, in an outbreak of cryptosporidiosis in 1993, drop apples were collected from trees adjacent to an area grazed by cattle whose stool contained Cryptosporidium (3), and in a salmonellosis outbreak in 1974, drop apples had been collected from an orchard fertilized with manure (4). The practice of using drop apples for making apple cider is common (2), and apples can become contaminated by resting on ground contaminated with manure. In an outbreak of E. coli O157:H7 infections in 1991 (2), the cider press operator also raised cattle, and cattle grazed in a field adjacent to the mill. The presence of animals near a cider mill can result in manure inadvertently contacting apples, equipment, or workers' hands. In addition, apples can become contaminated if transported or stored in areas that contain manure, or if rinsed with contaminated water.

These previous outbreaks of illness prompted recommendations to reduce the risk for producing contaminated cider, including 1) preventing the introduction of animal manure into orchards, 2) avoiding use of apples that have fallen to the ground, 3) washing and brushing apples before pressing, 4) using a preservative such as sodium benzoate, and 5) routine pasteurization (3,5). In the outbreaks in Connecticut and New York, some of these recommended production practices had been followed. For example, in Connecticut, apples were washed and brushed before pressing; however, drop apples were used. In New York, the mill reportedly did not use drop apples, and apples were washed and brushed before pressing; however, cattle were present near the farm, and the apples were washed with water from a source later determined to contain E. coli -- an indicator of contamination with animal or human feces.

At least two factors complicate efforts to reduce the risk for transmission of enteric pathogens through unpasteurized apple cider and juice. First, a small number of pathogenic organisms can result in infection -- ingestion of as few as 30 Cryptosporidium (6) and less than 1000 E. coli O157:H7 (7) have caused symptomatic infection in humans. Second, although apple cider and juice usually are acidic (pH of 3-4) (5,8), both Cryptosporidium and E. coli O157:H7 are acid-tolerant, and both organisms can survive in apple cider for up to 4 weeks (3,5). The addition of preservatives to apple cider containing E. coli O157:H7 does not consistently kill the organism (5,8), and Cryptosporidium oocysts are resistant to most common disinfectants (e.g., bleach, iodine, and sodium hydroxide) (9). Pasteurization and boiling kill E. coli O157:H7 and Cryptosporidium, and other methods that might increase the safety of cider are under investigation (10). FDA is evaluating information received at a public meeting held December 16-17, 1996, to determine methods to reduce the risk for illness associated with fresh juices. Until alternative effective methods are developed, consumers can reduce their risk for enteric infections by drinking pasteurized or boiled apple cider and juice.

References
1. CDC. Outbreak of Escherichia coli O157:H7 infections associated with drinking unpasteurized commercial apple juice -- British Columbia, California, Colorado, and Washington, October 1996. MMWR 1996;45:975.

2. Besser RE, Lett SM, Weber JT, et al. An outbreak of diarrhea and hemolytic uremic syndrome from Escherischia coli O157:H7 in fresh-pressed apple cider. JAMA 1993;269:2217-20.

3. Millard PS, Gensheimer KF, Addiss DG, et al. An outbreak of cryptosporidiosis from fresh-pressed apple cider. JAMA 1994;272:1592-6.

4. CDC. Epidemiologic notes and reports: Salmonella typhimurium outbreak traced to a commercial apple cider -- New Jersey. MMWR 1975;24:87-8.

5. Zhao T, Doyle MP, Besser RE. Fate of enterohemorrhagic Escherichia coli O157:H7 in apple cider with and without preservatives. Appl Environ Microbiol 1993;59:2526-30.

6. DuPont HL, Chappell CL, Sterling CR, Okhuysen PC, Rose JB, Jakubowski W. The infectivity of Cryptosporidium parvum in healthy volunteers. N Engl J Med 1995;332:855-9.

7. Griffin PM, Bell BP, Cieslak PR, et al. Large outbreak of Escherichia coli O157:H7 infections in the western United States: the big picture. In: Karmali MA, Goglio AG, eds. Recent advances in verocytotoxin-producing Escherichia coli infections. Amsterdam: Elsevier Science B.V., 1994;7-12.

8. Miller LG, Kaspar CW. Escherichia coli O157:H7 acid tolerance and survival in apple cider. J Food Prot 1994;57:460-4.

9. Campbell I, Tzipori AS, Hutchison G, Angus KW. Effect of disinfectants on survival of Cryptosporidium oocysts. Vet Rec 1982;111:414-5.

10. Anderson BC. Moist heat inactivation of Cryptosporidium sp. Am J Public Health 1985:75:1433-4.

September 13, 1996 / 45(36);783-784

Foodborne Outbreak of Diarrheal Illness Associated with Cryptosporidium parvum -- Minnesota, 1995

On September 29, 1995, the Minnesota Department of Health (MDH) received reports of acute gastroenteritis among an estimated 50 attendees of a social event in Blue Earth County on September 16. This report summarizes the epidemiologic and laboratory investigations of the outbreak, which indicate the probable cause for this foodborne outbreak was Cryptosporidium parvum.

Of the 26 persons who attended the function and who completed telephone interviews with MDH, 15 (58%) reported onset of diarrhea (three or more stools during a 24-hour period) within 14 days after attending the event (range: 1-9 days; median: 6 days). Symptoms included watery diarrhea (100%), abdominal cramps (93%), and chills (79%). The median length of illness was 4 days (range: 1/2 day-14 days). Three persons who sought medical care received outpatient treatment for acute gastroenteritis. Stool specimens obtained from two of these persons were negative for bacterial pathogens and for ova and parasites but were not tested for C. parvum. There were no other reports of cryptosporidiosis in the community at the time of this outbreak.

To identify risk factors for illness, MDH conducted a case-control study using the 15 ill and 11 well attendees. In addition, MDH collected stools from three ill persons, and these were cultured for Salmonella, Shigella, Campylobacter, and Escherichia coli O157:H7; examined for ova and parasites; and tested for C. parvum using acid-fast staining and direct-fluorescent antibody (DFA) methods.

Based on the case-control study, only consumption of chicken salad was associated with increased risk for illness (15 of 15 cases versus two of 11 controls; odds ratio= undefined). Water consumption at the event was not associated with illness.

The chicken salad was prepared by the hostess on September 15 and was refrigerated until served. The ingredients were cooked chopped chicken, pasta, peeled and chopped hard-boiled eggs, chopped celery, and chopped grapes in a seasoned mayonnaise dressing. The hostess operated a licensed day-care home (DCH) and prepared the salad while attendees were in her home. She denied having recent diarrheal illness and refused to submit a stool specimen. In addition, she denied knowledge of diarrheal illnesses among children in her DCH during the week before preparation of the salad. She reported changing diapers on September 15 before preparing the salad and reported routinely following handwashing practices.

Stool specimens from two of the persons whose illnesses met the case definition were obtained by MDH 7 days after resolution of their symptoms; one sample was positive for oocysts and

Cryptosporidium sporozoites on acid-fast staining, but the DFA test was negative. The presence of oocysts containing sporozoites was confirmed by acid-fast tests at two other reference laboratories. Stool specimens obtained from a third person -- the spouse of a case-patient -- who did not attend the event but had onset of diarrhea 8 days after onset of diarrhea in his spouse was positive for C. parvum by acid-fast staining and DFA. All stools obtained by MDH were negative for bacteria and for parasites. No chicken salad was available for testing.

Reported by: JW Besser-Wiek, MS, J Forfang, MPH, CW Hedberg, PhD, JA Korlath, MPH, MT Osterholm, PhD, State Epidemiologist, Minnesota Dept of Health. CR Sterling, PhD, Univ of Arizona, Tucson. L Garcia, PhD, Univ of California at Los Angeles Medical Center. Div of Parasitic Diseases, National Center for Infectious Diseases; Div of Applied Public Health Training (proposed), Epidemiology Program Office, CDC.

Editorial Note: Known modes of transmission of C. parvum include consumption of contaminated surface or ground water (1,2), exposure to contaminated recreational water (3), animal-to-person contact (2), and person-to-person contact (2). Because outbreaks of cryptosporidiosis and asymptomatic carriage of Cryptosporidium have been documented in child-care settings (4), the food preparer in this outbreak may have contaminated the implicated salad after contact with an asymptomatically infected child in the DCH. The salad required extensive handling in preparation, was moist, and was served cold -- conditions conducive to initial contamination and preservation of infectious oocysts.

The outbreak of gastroenteritis described in this report was associated with eating chicken salad at a social function. Despite the small number of stools submitted for testing by ill persons who attended the event, the symptoms, incubation period, and the presence of C. parvum in the stool of an ill attendee all indicate that this was a foodborne outbreak of cryptosporidiosis.

Although foodborne transmission of C. parvum has been suspected previously, evidence supporting this mode has been limited to one report of a point source outbreak associated with raw apple cider (5) and reports of sporadic cases attributed to contaminated foods (6). The reported low infectious dose of C. parvum (ID50=132 organisms) suggests that transmission in food is possible (7). Cryptosporidiosis should be considered in the differential diagnosis of suspected foodborne gastroenteritis.

References
1. Mac Kenzie WR, Hoxie NJ, Proctor ME, et al. A massive outbreak in Milwaukee of Cryptosporidium infection transmitted through the public water supply. N Engl J Med 1994; 331:161-7.
2. Current WL, Garcia LS. Cryptosporidiosis. Clin Microbiol Rev 1991;4:325-58.
3. McAnulty JM, Fleming DW, Gonzalez AH. A community-wide outbreak of cryptosporidiosis associated with swimming at a wave pool. JAMA 1994;272:1597-600.
4. Cordell RL, Addiss DG. Cryptosporidiosis in child care settings: a review of the literature and recommendations for prevention and control. Pediatr Infect Dis J 1994;13:310-7.
5. Millard PS, Gensheimer KF, Addiss DG, et al. An outbreak of cryptosporidiosis from fresh-pressed apple cider. JAMA 1994;272:1592-6.
6. Smith JL. Cryptosporidium and Giardia as agents of foodborne disease. Journal of Food Protection 1993;56:451-61.
7. DuPont HL, Chappell CL, Sterling CR, Okhuysen PC, Rose JB, Jakubowski W. The infectivity of Cryptosporidium parvum in healthy volunteers. N Engl J Med 1995;332:855-9.

May 31, 1996 / 45(21);442-4

Outbreak of Cryptosporidiosis at a Day Camp -- Florida, July-August 1995

On July 27, 1995, the Alachua County Public Health Unit (ACPHU) in central Florida was notified of an outbreak of gastroenteritis among children and counselors at a day camp on the grounds of a public elementary school. This report summarizes the outbreak investigation, which implicated Cryptosporidium parvum as the causative agent and underscores the role of contaminated water as a vehicle for transmission of this organism.

The camp operated from June 12 through August 4 and enrolled 98 children (age range: 4-12 years) and six counselors during the 3 weeks before the outbreak. A confirmed case of cryptosporidiosis was defined as gastrointestinal symptoms (i.e., abdominal pain, nausea, vomiting, and three or more watery stools each day) in a camp attendee during July 20-August 23 with C. parvum isolated in stool. A probable case was defined as gastrointestinal symptoms during July 20-August 23 in a camp attendee who did not submit a stool sample for testing. A questionnaire was administered to each of the 104 persons attending the camp; for some children, information was obtained from parents and camp records.

Of the 104 persons attending the camp, 77 (74%) had symptoms (abdominal pain {74%}, nausea {73%}, diarrhea {71%}, vomiting {57%}, and fever {43%}) with onset during July 20-August 15, including 72 of 98 children and five of six counselors (Figure_1). Follow-up phone calls to 67 of 79 households of those who attended the camp indicated that 24 household members had onset of gastrointestinal symptoms during July 20-August 23.

Stool specimens for bacterial enteric pathogen testing were obtained from 44 camp attendees within 10 days of onset of symptoms; all were negative. Sixteen stool specimens were obtained for testing for ova and parasites; all 16 yielded C. parvum.

Risk for illness was not associated with participating in a particular camp activity or eating a lunch or snack provided by the camp. Water sources for the camp included an outdoor drinking fountain, a sink inside the trailer that served as camp headquarters, and portable coolers. The coolers were filled at either a kitchen sink inside the school or an outdoor faucet with an attached hose and spray nozzle used for washing garbage cans. Although water consumption from any source could not be quantified, virtually all persons at the camp reported drinking water from one of the camp sources during the 3 weeks before the outbreak. Water samples were tested (1) from the city's water treatment plant, all school sources used by campers, and three sinks inside the school. The water treatment plant samples were repeatedly negative. Outdoor faucet samples were positive for total coliforms and C. parvum; other tests from school sites were negative or below detectable limits for total coliform, Escherichia coli, and ova and parasites. The area around the outdoor faucet was not fenced, and feces of unknown origin were observed on several occasions near the faucet and attached hose.

Based on these findings, ACPHU recommended discontinuing use of coolers for water and the outdoor faucet, and enclosing the faucet area by fence. In addition, parents and staff were taught proper handwashing technique and given information about C. parvum. Staff returning to school used alternate water sources until the system was superchlorinated, flushed, and cleared.

Reported by: J Regan, R McVay, M McEvoy, J Gilbert, Water and Wastewater Systems, Gainesville Regional Utilities. R Hughes, T Tougaw, E Parker, PhD, W Crawford, J Johnson, School Board of Alachua County, Gainesville, Florida. J Rose, PhD, Univ of South Florida, St. Petersburg, Florida. S Boutros, PhD, Environmental Associates Ltd, Bradford, Pennsylvania. S Roush, MPH, T Belcuore, MS, C Rains, MD, J Munden, MPH, Alachua County Public Health Unit; L Stark, PhD, E Hartwig, ScD, M Pawlowicz, Florida Dept of Health and Rehabilitative Svcs State Laboratory; R Hammond, PhD, D Windham, R Hopkins, MD, State Epidemiologist, Florida Dept of Health and Rehabilitative Svcs. Div of Field Epidemiology, Epidemiology Program Office, CDC.

Editorial Note: The protozoan parasite C. parvum was first identified as a human pathogen in 1976; since then, the organism has been increasingly recognized as an agent of gastrointestinal illness. In immunocompetent persons, cryptosporidiosis can cause moderately severe watery diarrhea that usually lasts 1-20 days (average: 10 days) (2). In immunocompromised persons (e.g., those with acquired immunodeficiency syndrome {AIDS} or those taking certain chemotherapeutic regimens), the infection can cause severe, unrelenting diarrhea. The antibiotic paromomycin can improve symptoms and decrease parasite excretion in the feces of some persons with AIDS and is the treatment of choice for immunosupppressed patients (3,4).

Cryptosporidiosis is transmitted by the fecal-oral route, most commonly by direct person-to-person transmission or by drinking water that has been contaminated with human or animal feces. In 1993, cryptosporidiosis caused the largest waterborne disease outbreak ever recorded, when an estimated 400,000 persons in Milwaukee became ill after drinking contaminated municipal water (5). The outbreak described in this report most likely was related to drinking contaminated water. Contamination probably occurred at the nozzle of the hose used to fill the water coolers rather than at or near the water treatment plant. Sources of drinking water should be protected from possible fecal contamination, and hoses, which are particularly susceptible to back-syphonage, should not be used to provide drinking water. Public water sources that cannot be protected should be posted as nonpotable.

C. parvum was promptly identified as the source of this outbreak, in part because the Florida State Public Health Laboratory examines all fecal specimens submitted for ova and parasite analysis for C. parvum. The diagnosis of cryptosporidiosis can be delayed or missed when physicians assume incorrectly that diagnostic laboratories routinely perform specific tests for C. parvum when a fecal examination for parasites is requested. A recent national survey of clinical laboratories found that only 5% did so (6). If cryptosporidiosis is suspected in the differential diagnosis, physicians should specifically request testing for C. parvum. In addition, when reporting the results of fecal examinations, clinical laboratories should specify what tests were performed rather than only indicating that no enteric pathogens were identified.

References
1. Messer JW, Fout GS, Schafer FW, Dahling DR, Stetler RE. Information Collection Rule (ICR): Microbiology Laboratory Manual {Draft}. Cincinnati, Ohio: US Environmental Protection Agency, February 1995.
2. American Academy of Pediatrics. Cryptosporidiosis. In: Peter G, ed. 1994 Red book: report of the Committe on Infectious Diseases. 23rd ed. Elk Grove Village, Illinois: American Academy of Pediatrics;1994:171-2.
3. White AC Jr, Chappell CL, Hayat CS, Kimball KT, Flanigan TP, Goodgame RW. Paromomycin for cryptosporidiosis in AIDS: a prospective, double-blind trial. J Infect Dis 1994;170:419-24.
4. Anonymous. Drugs for parasitic infections. Med Lett Drugs Ther 1995;37:99-108.
5. Mac Kenzie WR, Hoxie NJ, Proctor ME, et al. A massive outbreak in Milwaukee of Cryptosporidium infection transmitted through the public water supply. N Engl J Med 1994;331:161-7.
6. Boyce TG, Pemberton AG, Addiss DG. Cryptosporidium testing practices among clinical laboratories in the United States. Ped Infect Dis J 1996;15:87-8.

August 12, 1994 / 43(31);561-563

Cryptosporidium Infections Associated with Swimming Pools -- Dane County, Wisconsin, 1993

In March and April 1993, an outbreak of cryptosporidiosis in Milwaukee resulted in diarrheal illness in an estimated 403,000 persons (1). Following that outbreak, testing for Cryptosporidium in persons with diarrhea increased substantially in some areas of Wisconsin; by August 1, 1993, three of six clinical laboratories in Dane County were testing routinely for Cryptosporidium as part of ova and parasite examinations. In late August 1993, the Madison Department of Public Health and the Dane County Public Health Division identified two clusters of persons with laboratory-confirmed Cryptosporidium infection in Dane County (approximately 80 miles west of Milwaukee). This report summarizes the outbreak investigations.

On August 23, a parent reported to the Madison Department of Public Health that her daughter was ill with laboratory-confirmed Cryptosporidium infection and that other members of her daughter's swim team had had severe diarrhea. On August 26, public health officials inspected the pool where the team practiced (pool A) and interviewed a convenience sample of patrons at the pool. Seventeen (55%) of 31 pool patrons interviewed reported having had watery diarrhea for 2 or more days with onset during July or August. Eight (47%) of the 17 had had watery diarrhea longer than 5 days. Four persons who reported seeking medical care had stool specimens positive for Cryptosporidium.

On August 31, public health nurses at the Dane County Public Health Division identified a second cluster of nine persons with laboratory-confirmed Cryptosporidium infection while following up case-reports voluntarily submitted by physicians. Seven of the nine ill persons reported swimming at one large outdoor pool (pool B). Because of the potential for disease transmission in multiple settings, a community-based matched case-control study was initiated on September 3 to identify risk factors for Cryptosporidium infection among Dane County residents.

Laboratory-based surveillance was used for case finding. A case was defined as Cryptosporidium infection that was laboratory-confirmed during August 1-September 11, 1993, in a Dane County resident who was also the first person in a household to have signs or symptoms (i.e., watery diarrhea of 2 or more days' duration). During the study interval, 85 Dane County residents with stool specimens positive for Cryptosporidium were identified. Sixty-five (77%) persons were interviewed; 36 (55%) had illnesses meeting the case definition. Systematic digit-dialing was used to select 45 controls, who were matched with 34 case-patients by age group and telephone exchange. All study participants were interviewed by telephone using a standardized questionnaire to obtain information on demographics, signs and symptoms, recreational water use, child-care attendance, drinking water sources, and presence of diarrheal illness in household members.

The median age of ill persons was 4 years (range: 1-40 years). Reported signs and symptoms included watery diarrhea (94%), stomach cramps (93%), and vomiting (53%). Median duration of diarrhea was 14 days (range: 1-30 days). Swimming in a pool or lake during the 2 weeks preceding onset of illness was reported by 82% of case-patients and 50% of controls (matched odds ratio {MOR}=6.0; 95% confidence interval {CI}=1.4-25.3). Twenty-one percent of case-patients and 2% of controls (MOR=7.3; 95% CI=0.9-59.3) reported swimming in pool A. Fifteen percent of case-patients and 2% of controls (MOR=undefined {6/0}; p=0.02, paired sample sign test) reported swimming in pool B. When persons reporting pool A or B use were excluded from the analysis, the association with recreational water use was not statistically significant (MOR=3.4, 95% CI=0.8-15.7). Child-care attendance was reported for 74% of case-patients aged less than 6 years and 44% of controls (MOR=2.9; 95% CI=0.8-10.7). Two case-patients reported child-care attendance and use of pool A or pool B. No case-patients reported travel to the Milwaukee area during the March-April outbreak, and no associations were found between illness and drinking water sources.

To limit transmission of Cryptosporidium in Dane County pools, state and local public health officials implemented the following recommendations: 1) closing the pools that were epidemiologically linked to infection and hyperchlorinating those pools to achieve a disinfection (CT *) value of 9600; 2) advising all area pool managers of the increased potential for waterborne transmission of Cryptosporidium; 3) posting signs at all area pools stating that persons who have diarrhea or have had diarrhea during the previous 14 days should not enter the pool; 4) notifying area physicians of the increased potential for cryptosporidiosis in the community and requesting that patients with watery diarrhea be tested for Cryptosporidium; and 5) maintaining laboratory-based surveillance in the community to determine whether transmission was occurring at other sites (e.g., child-care centers and other pools).

On August 27, pool A was closed and hyperchlorinated for 18 hours; on September 3, pool B closed early for the season. Because many control measures were initiated less than 1 week before many pools closed for the season (after September 5), their impact on transmission could not be evaluated adequately.

Reported by: J Bongard, MS, Dane County Public Health Div, Madison; R Savage, MS, Madison Dept of Public Health; R Dern, MS, St. Mary's Medical Center, Madison; H Bostrum, J Kazmierczak, DVM, S Keifer, H Anderson, MD, State Epidemiologist for Occupation and Environmental Health, JP Davis, MD, State Epidemiologist for Communicable Diseases, Bur of Public Health, Wisconsin Div of Health. Div of Parasitic Diseases, National Center for Infectious Diseases; Div of Field Epidemiology, Epidemiology Program Office, CDC.

Editorial Note: Person-to-person, waterborne, and zoonotic transmission of Cryptosporidium has been well documented (2). A marked seasonality has been reported, with peaks occurring in North America during late summer and early fall (3,4). Cryptosporidiosis associated with use of swimming pools has been reported previously (5-7) but is probably underrecognized. Infection with Cryptosporidium resulting from recreational water use may contribute to the observed seasonal distribution.

The March-April 1993 Milwaukee waterborne outbreak stimulated increased testing for Cryptosporidium in Dane County, increasing the likelihood of outbreak detection. However, the number of cases described in this report was not sufficient to conduct a stratified matched analysis. Confounding of the associations found for child-care attendance and pool use is possible, although child-care attendance was reported in only one case for each implicated pool.

Cryptosporidium oocysts are small (4-6 u), are resistant to chlorine, and have a high infectivity. The chlorine CT of 9600 needed to kill Cryptosporidium oocysts is approximately 640 times greater than required for Giardia cysts (8). The ability of pool sand-filtration systems to remove oocysts under field conditions has not been well documented, but would not be expected to be effective. Results of an infectivity study suggest that the infective dose among humans for Cryptosporidium is low (H. DuPont, University of Texas Medical School at Houston, personal communication, 1994). Because of the large number of oocysts probably shed by symptomatic persons, even limited fecal contamination could result in sufficient oocyst concentrations in localized areas of a pool to cause additional human infections.

This investigation underscores the potential for transmission of Cryptosporidium in swimming pools. Health-care providers should consider requesting Cryptosporidium testing of stool specimens from persons with watery diarrhea, and public health departments should consider establishing surveillance for Cryptosporidium to facilitate prompt recognition of outbreaks. Maintaining the high levels of chlorine necessary to kill Cryptosporidium in swimming pools is not feasible; therefore, such recreational water use should be recognized as a potential increased risk for cryptosporidiosis in immumocompromised persons, including those with human immunodeficiency virus infection, in whom this infection may cause lifelong, debilitating illness (9).

References
1. Mac Kenzie WR, Hoxie NJ, Proctor ME, et al. A massive outbreak in Milwaukee of Cryptosporidium infection transmitted through the public water supply. N Engl J Med 1994;331:161-7.
2. Casemore DP. Epidemiologic aspects of human cryptosporidiosis. Epidemiol Infect 1990;104:1-28.
3. Wolfson JS, Richter JM, Waldron WA, Weber DJ, McCarthy DM, Hopkins CC. Cryptosporidiosis in immunocompetent patients. N Engl J Med 1985;312:1278-82.
4. Skeels MR, Sokolow R, Hubbard CV, Andrus JK, Baisch J. Cryptosporidium infection in Oregon public health clinic patients, 1985-1988: the value of statewide laboratory surveillance. Am J Public Health 1990;80:305-8.
5. Sorvillo FJ, Fujioka K, Nahlen B, et al. Swimming-associated cryptosporidiosis. Am J Public Health 1992;82:742-4.
6. Bell A, Guasparini R, Meeds D, et al. A swimming pool-associated outbreak of cryptosporidiosis in British Columbia. Can J Public Health 1993;84:334-7.
7. CDC. Surveillance for waterborne disease outbreaks -- United States, 1991-1992. MMWR 1993;42(no. SS-5):1-22.
8. Current WL, Garcia LS. Cryptosporidiosis. Clin Microbiol Rev 1991;4:305-8.
9. Navin TR, Juranek DD. Cryptosporidiosis: clinical, epidemiologic, and parasitologic review. Rev Infect Dis 1984;6:313-10.

* CT=pool chlorine concentration (in parts per million) multiplied by time (in minutes).

September 16, 1994 / 43(36);661-663,669

Assessment of Inadequately Filtered Public Drinking Water -- Washington, D.C., December 1993

The risk for waterborne infectious diseases increases when filtration and other standard water-treatment measures fail. On December 6, 1993, water-treatment plant operators in the District of Columbia (DC) began to have difficulty maintaining optimal filter effectiveness. On December 7, filter performance worsened, and levels of turbidity (i.e., small suspended particles) exceeded those permitted by U.S. Environmental Protection Agency (EPA) standards. On December 8, DC residents were advised to boil water intended for drinking because of high municipal water turbidity that may have included microbial contaminants. Although adequate chlorination of the DC municipal water was maintained throughout the period of increased turbidity, the parasite Cryptosporidium parvum is highly resistant to chlorination. Because of the increased risk for infection with this organism and other enteric pathogens, the DC Commission of Public Health and CDC conducted four investigations to determine whether excess cases of diarrheal illness occurred because residents drank inadequately filtered water. This report describes the results of these investigations.

The investigations included a random-digit-dialed telephone survey of DC residents and retrospective reviews of records from two emergency departments, two nursing homes, and seven hospital microbiology laboratories. The occurrence of diarrheal illness or presence of organisms in stool during the 2 weeks before the turbidity violation (period 1: November 22-December 5) was compared with that during the 2-3 weeks after the violation was first noted (period 2: December 6-December 21 or 26). The incubation period for cryptosporidiosis typically ranges from 2 to 14 days.

Telephone survey. The telephone survey sampled 1197 household members (0.2% of DC's 600,000 residents) from 462 households in all 22 DC residential ZIP code areas. The percentage of persons who reported having diarrhea (i.e., three or more loose or watery stools in a 24-hour period) were similar for period 1 (the reference period) and period 2 (2.8% versus 3.5%, respectively; relative risk {RR}=1.2; 95% confidence interval {CI}=0.8-1.9). A total of 37% of persons reported that bottled water was their principal source of drinking water at home, and 30% reported that bottled water was their primary source of drinking water both at home and at work. For both periods, reported use of bottled water was similar for persons with and without diarrhea.

Hospital emergency department survey. During the two periods, totals of 2140 (period 1) and 3315 (period 2) persons were evaluated at two DC hospital emergency departments. Medical records were reviewed for all persons with diagnoses suggestive of gastrointestinal illness * (104 and 211 persons for periods 1 and 2, respectively). The percentage of all persons who had diarrhea recorded in their emergency department charts was similar for periods 1 and 2 (1.5% versus 2.0%; RR=1.3; 95% CI=0.9-2.0). For both periods, approximately 70% of patients with diarrheal illness were DC residents. The percentages of stool specimens that were positive for enteric pathogens (i.e., bacteria, parasites, or rotavirus antigen) were similar for the two periods. During each period, two stool specimens were examined for Cryptosporidium: none were positive during period 1, and one was positive during period 2.

Nursing home survey. Medical records were reviewed for all 443 residents from two selected nursing homes (14% of the 3156 nursing home beds in DC). During both periods, the mean numbers of bowel movements per person per day were 1.3. In addition, the daily mean number of residents with loose or large-volume bowel movements were similar (27.1 and 27.8 persons for periods 1 and 2), and antidiarrheal medications were given at the same rate (0.002 doses per person per day) during both periods.

Microbiology laboratory survey. Data were obtained from microbiology laboratories of seven (64%) of the 11 DC hospitals. Although the total number of stool specimens examined for Cryptosporidium increased from period 1 (32 specimens) to period 2 (54 specimens), the percentage positive was lower -- but not statistically different -- for period 2 (12.5% versus 7.4%; RR=0.6; 95% CI=0.2-2.2). The percentages of stools positive for other pathogens (i.e., bacteria, Giardia lamblia, and rotavirus antigen) were similar for both periods.

Reported by: MN Akhter, MD, Commissioner, ME Levy, MD, District Epidemiologist, C Mitchell, R Boddie, District of Columbia Commission of Public Health. N Donegan, B Griffith, M Jones, Washington Hospital Center; TO Stair, MD, Georgetown Univ Medical Center, Washington, DC. Epidemiology Br, Div of Parasitic Diseases, National Center for Infectious Diseases, CDC.

Editorial Note: To ensure safe municipal drinking water supplies, water-treatment programs employ multiple barriers to prevent contaminants from reaching the consumer. These barriers include protection of the watershed, chemical disinfection, and filtration of surface water supplies such as lakes and rivers. When one of these barriers is absent or fails, the risk for waterborne disease may increase. The failure of the filtration process in DC prompted particular concerns about contamination with and exposure to Cryptosporidium.

Outbreaks of cryptosporidiosis resulting from surface water contamination have occurred when turbidity was 0.9-2.0 nephelometric turbidity units (NTU) **. For example, in a waterborne outbreak in Milwaukee in 1993, a peak turbidity of 1.7 NTU was associated with illness in approximately 400,000 persons (1). In DC, the turbidity levels reached 9.0 NTU.

Because Cryptosporidium is highly resistant to chlorination, disinfection of water is not a reliable method for preventing exposure to it. The failure to detect increased rates of illness among residents of DC probably reflects the absence of, or presence of only a small number of, oocysts in the water that supplied the municipal water-treatment plant at the time the filtration failure occurred. In addition, the investigations in DC did not detect any increase in diarrheal illness associated with the elevated water turbidity; however, the sample sizes in these investigations were too small to rule out low-level transmission of waterborne agents. For example, the telephone survey probably would not have detected an outbreak affecting fewer than 12,000 persons.

Cryptosporidium is present in 65%-87% of surface water samples tested throughout the United States (2,3). However, because current techniques to detect Cryptosporidium in water are cumbersome, costly, and insensitive, tests to detect it are not routinely performed by water utilities. During 1995, EPA plans to collect additional information about Cryptosporidium and other microorganisms in surface water used by municipal water-treatment facilities in the United States and to assess the effectiveness of water-treatment methods for removing them. ***

The early detection of waterborne outbreaks of cryptosporidiosis is difficult for at least four reasons: 1) many physicians are unaware that Cryptosporidium can cause watery diarrhea; 2) the symptom complex often resembles a viral syndrome; 3) clinical laboratories often do not routinely test for Cryptosporidium when a physician requests a stool examination for ova and parasites; and 4) few states include cryptosporidiosis as a reportable disease.

Variations in recommendations regarding the duration of boiling during boil-water advisories have reflected uncertainty about how long some organisms can survive. On the basis of a recent literature review, CDC and EPA recommend that water be rendered microbiologically safe for drinking by bringing it to a rolling boil for 1 minute; this will inactivate all major waterborne bacterial pathogens (i.e., Vibrio cholerae, enterotoxigenic Escherichia coli, Salmonella, Shigella sonnei, Campylobacter jejuni, Yersinia enterocolitica, and Legionella pneumophila) and waterborne protozoa (e.g., Cryptosporidium parvum, Giardia lamblia, and Entamoeba histolytica {4-7}). Although information about thermal inactivation is incomplete for waterborne viral pathogens, hepatitis A virus -- considered one of the more heat-resistant waterborne viruses (8) -- also is rendered noninfectious by boiling for 1 minute (9). If viral pathogens are suspected in drinking water in communities at elevations above 6562 ft (2 km), the boiling time should be extended to 3 minutes.

References
1. Mac Kenzie WR, Hoxie NJ, Proctor ME, et al. A massive outbreak in Milwaukee of Cryptosporidium infection transmitted through the public water supply. N Engl J Med 1994;331:161-7.
2. Rose JB, Gerba CP, Jakubowski W. Survey of potable water supplies for Cryptosporidium and Giardia. Environmental Science and Technology 1991;25:1393-400.
3. LeChevallier MW, Norton WD, Lee RG. Occurrence of Giardia and Cryptosporidium spp. in surface water supplies. Appl Environ Microbiol 1991;57:2610-6.
4. Bandres JC, Mathewson JJ, Dupont HL. Heat susceptibility of bacterial enteropathogens. Arch Intern Med 1988;148:2261-3.
5. Anderson BC. Moist heat inactivation of Cryptosporidium sp. Am J Public Health 1985;75: 1433-4.
6. Bingham AK, Jarroll EL, Meyer EA. Giardia sp.: physical factors of excystation in vitro, and excystation vs eosin exclusion as determinants of viability. Exp Parasitol 1979;47:284-91.
7. Boeck WC. The thermal-death point of the human intestinal protozoan cysts. Am J Hygiene 1921;1:365-87.
8. Larkin EP. Viruses of vertebrates: thermal resistance. In: Rechcigl M Jr, ed. CRC handbook of foodborne diseases of biological origin. Boca Raton, Florida: CRC Press, Inc, 1983:3-24.
9. Krugman S, Giles JP, Hammond J. Hepatitis virus: effect of heat on the infectivity and anti-genicity of the MS-1 and MS-2 strains. J Infect Dis 1970;122:432-6.

* Gastroenteritis, diarrhea, nausea, vomiting, gastritis, viral syndrome, dehydration, and hyperemesis gravidarum. ** The American Waterworks Association encourages water utilities to maintain turbidity measurements of water as it leaves the treatment plant at or below 0.1 NTU. *** 59 FR 6332.

MORBIDITY AND MORTALITY WEEKLY REPORTS

May 25, 1990 / 39(20);343-345

Epidemiologic Notes and Reports Swimming-Associated Cryptosporidiosis --
Los Angeles County

From July 13 through August 14, 1988, 44 persons in five separate swimming groups developed a gastrointestinal illness after using a swimming pool in Los Angeles County. The outbreak began several days after an unintentional human defecation in the pool during the first week of July. When the outbreak was reported to the Los Angeles County Department of Health Services (LACDHS) in early August, LACDHS initiated an epidemiologic investigation.

The affected groups had repeated pool contact in July and included a high school water polo team, a SCUBA class, a "masters" group, an elementary school group, and the pool lifeguards. Sixty (73%) of 82 persons from the five groups were interviewed. A case was defined as any person with watery diarrhea or diarrhea plus cramping and/or fever during July or August.

The overall attack rate was 73% (44/60) and ranged from 47% to 100% by group (Table 1). Illness was characterized by watery diarrhea (88%), abdominal cramps (86%), and fever (60%) and was often protracted (median duration: 5 days; range: 1-30 days). Two persons, both from the SCUBA class, were hospitalized. Cryptosporidium was identified in stool specimens by modified acid-fast staining from seven of 11 patients tested. Results of other laboratory examinations, including bacterial culturing for Salmonella, Shigella, and Campylobacter and testing for ova and parasites, were negative. Assessment for viral agents was not performed.

For all persons with pool contact during the outbreak period, the attack rate was highest among those with extensive (greater than 3 total hours) water exposure (p less than 0.01, Fisher's exact test; relative risk=2.2; 95% confidence interval=1.1-4.4). No other common exposures or risk factors were identified. Review of surveillance data revealed no increase of cryptosporidiosis or diarrheal illness during July or August in Los Angeles County or the community affected by the outbreak. Pool water was not tested for Cryptosporidium, and the person who fecally contaminated the pool was not examined for Cryptosporidium infection.

The pool implicated in this outbreak is a 100,000-gallon pool at a school in Los Angeles County. Inspection of the pool during the outbreak period confirmed adequate chlorine levels (2 ppm) but detected a 30% diminished filtration flow rate and established that one of three diatomaceous earth (DE) filters was inoperative. The filtration system was repaired on August 3, and no additional cases of diarrhea were subsequently identified among newly exposed swimmers. Reported by: FJ Sorvillo, MPH, K Fujioka, PhD, M Tormey, MPH, R Kebabjian, RS, W Tokushige, L Mascola, MD, S Schweid, M Hillario, SH Waterman, MD, Los Angeles County Dept of Health Svcs. Parasitic Diseases Br, Div of Parasitic Diseases, Center for Infectious Diseases, CDC.

Editorial Note: Outbreaks of giardiasis, Norwalk gastroenteritis, and adenovirus types 3 and 4 associated with swimming pool contact have been reported (1-4). In each outbreak, inadequate pool maintenance was an important contributing factor.

The clinical features and laboratory findings in this investigation are consistent with an outbreak of cryptosporidiosis. Moreover, the investigation suggests that Cryptosporidium may be acquired through recreational water contact. Resistance of Cryptosporidium to chlorination (5), an inadequately maintained filtration system, and repeated and prolonged exposure may have contributed to the size and extent of this outbreak. Continued pool use and possible ongoing contamination by infected persons, many of whom continued to swim despite their illness, could also have sustained transmission.

Cryptosporidium oocysts are resistant to chlorine. Because Cryptosporidium oocysts are small (4-6 u), rapid sand filters commonly used in swimming pools may not be effective in filtering oocysts. However, evidence suggests that a well-maintained, fine-grade DE filtration system may remove Cryptosporidium (6). Further study is needed to assess the capability of different filtration devices to remove Cryptosporidium oocysts from swimming pool water.

Recommendations for managing swimming pools that have been fecally contaminated include prohibition of swimming until the chlorine level and contact time are sufficient to kill Giardia cysts (1). Given the ineffectiveness of chlorine against Cryptosporidium, greater consideration should be given to control strategies that use effective filtration (e.g., DE filters) or to draining the pool and replacing contaminated filter media in filters not considered effective against Cryptosporidium. In systems that use DE filters, one option may be to close contaminated pools until relatively complete filtration has occurred (typically three turnovers or approximately 1 day).

References
1. Porter JD, Ragazzoni HP, Buchanon JD, Waskin HA, Juranek DD, Parkin WE. Giardia trans mission in a swimming pool. Am J Public Health 1988;78:659-62.
2. Kappus KF, Marks JS, Holman RC, et al. An outbreak of Norwalk gastroenteritis associated with swimming in a pool and secondary person-to-person transmission. Am J Epidemiol 1982;116:834-9.
3. Martone WJ, Hierholzer JC, Keenlyside RA, Fraser DA, D'Angelo LJ, Winkler WG. An outbreak of adenovirus type 3 disease at a private recreation center swimming pool. Am J Epidemiol 1980;111:229-37.
4. D'Angelo LJ, Hierholzer JC, Keenlyside RA, Anderson LJ, Martone WJ. Pharyngoconjunctival fever caused by adenovirus type 4: report of a swimming pool-related outbreak with recovery of virus from pool water. J Infect Dis 1979;140:42-7.
5. Campbell I, Tzipori AS, Hutchison G, Angus KW. Effect of disinfectants on survival of Cryptosporidium oocysts. Vet Rec 1982;11:414-5.
6. Lange KP, Bellamy WD, Hendricks DW, Logsdon GS. Diatomaceous earth filtration of Giardia cysts and other substances. Journal of the American Water Works Association 1986:76-84.

August 28, 1987 / 36(33);561-3

Cryptosporidiosis -- New Mexico, 1986

Between July 1 and October 1, 1986, 78 laboratory-confirmed cases of cryptosporidiosis were reported to the Office of Epidemiology at the New Mexico Health and Environment Department. Because the source of infection in these cases was unclear, investigators conducted a case-control study to establish risk factors for infection.

For study purposes, a patient was defined as a Bernalillo County resident with laboratory-confirmed cryptosporidiosis reported to the Office of Epidemiology from July 1 through October 1, 1986. If more than one laboratory-confirmed case occurred in a household or day-care group, only the person with the earliest onset of symptoms was included in the study.

Fifty-eight (74%) of the 78 patients with cryptosporidiosis lived in Bernalillo County, which includes the city of Albuquerque. Twenty-four of these patients were included in the study. Thirty-two of the remaining patients were household or day-care contacts of these patients, and two were lost to follow-up.

The 24 patients included in the study were matched with 46 controls by age, sex, and neighborhood of residence. Using a questionnaire administered by telephone to both patients and controls, investigators gathered information on household size; day-care-center attendance, employment, or other principal sources of contact; travel; surface-water exposure; pet and domestic animal exposure; and the source of water to the home.

Patients' dates of onset of symptoms ranged from May 28 through September 2, 1986. Symptoms lasted from 5 to 60 days, with a median of 21 days. Ninety-six percent of the patients reported watery, non-bloody diarrhea; 79% reported flatulence; 67%, abdominal pain; 58%, nausea; and 54%, low-grade fever.

Patients ranged in age from 4 months to 44 years, with a median age of 3 years. Seventeen (71%) were less than 10 years of age. Seventeen (71%) of the patients were female, and seven (29%) were male. Thirteen (77%) of the patients less than 10 years of age and four (57%) of those greater than 10 were female.

Univariate analysis suggested that drinking untreated surface water and attending a day-care center where other children were ill with diarrhea were possible risk factors for this infection. There was a strong statistical association between drinking surface water and illness (odds ratio (OR) incalculable, p = 0.0016). None of the five patients who drank surface water had treated it in any way. One of these five patients attended a day-care center, the others had no other risk factors for cryptosporidiosis. None of the 46 controls had drunk surface water.

There may have been an increased risk of illness among those who had swum in surface water (OR = 3.7; 95% confidence interval (CI), 0.71 to 12.6). Exposure to surface water (either through drinking or swimming) had occurred in New Mexico, southern Colorado, and Mexico. If the two patients exposed to surface water in Mexico and their controls are eliminated from the analysis, drinking surface water is still significantly associated with illness (OR incalculable, p = 0.014). The time between exposure to surface water and illness ranged from 4 to 21 days, with a median of 7 days. The average incubation period of cryptosporidiosis is 2 to 10 days. Fourteen (82%) of the 17 household members with exposures to surface water similar to the patients' became ill with diarrhea within 2 to 7 days.

There was no statistically significant difference between patients and controls in attendance at day-care centers or in employment. However, patients were more likely than controls to attend a day-care center reported by a parent as having other children ill with diarrhea (OR = 5; 95% CI, 1.4 to 26.3). A patient was also more likely to be a household contact of a day-care-center attendee or employee, but this did not reach statistical significance (OR = 3.7, 95% CI, 0.95 to 14.2). Reported by: DJ Grabowski, MS, Albuquerque Environmental Health; KM Powers, JA Knott, MV Tanuz, LJ Nims, MS, MI Savitt-Kring, CM Lauren, BI Stevenson, HF Hull, MD, State Epidemiologist, New Mexico Health and Environment Dept. Div of Field Svcs, Epidemiology Program Office; Div of Parasitic Diseases, Center for Infectious Diseases, CDC.

Editorial Note: Cryptosporidium sp. was recognized as a human pathogen in 1976. The illness is associated with significant morbidity, including diarrhea, which is often prolonged and which can be accompanied by severe weight loss. In immunodeficient persons, cryptosporidiosis can cause life-threatening dehydration. There is no known effective therapy.

Previous outbreaks of cryptosporidiosis have occurred among animal handlers, through direct contact with animal feces (1), and in day-care centers, through person-to-person contact (2,3). An outbreak has also been reported from a Texas community where a common water well became contaminated (4).

Although surface water has not been previously recognized as a source of infection with Cryptosporidium, this study demonstrates that it may be. Further evidence was provided in January 1987 when a major waterborne outbreak of cryptosporidiosis in Georgia was traced to a river serving as the municipal water supply (CDC, unpublished data). Cryptosporidium sp. has been isolated from a broad variety of animals, including cattle, sheep, dogs, cats, deer, mice, rabbits, and snakes. Cryptosporidium sp. found in cattle have been shown to be transmitted to humans (1). Surface water might become contaminated through direct deposit of feces into water or by surface runoff that washes feces into water. The seasonal distribution of cryptosporidiosis, which occurs primarily in the summer and early fall (1,5), could be partially explained by the increased outdoor activity during that time of year.

Cryptosporidium species are known to be resistant to most chemical disinfectants, such as chlorine and iodine. Physicians should consider cryptosporidiosis in the differential diagnosis of persons with diarrhea who have a history of drinking surface water that is untreated or treated by chemical means alone.

References
1. Current WL. Cryptosporidiosis. J Am Vet Med Assoc 1985;187:1334-8.
2. CDC. Cryptosporidiosis among children attending day-care centers--Georgia, Pennsylvania, Michigan, California, New Mexico. MMWR 1984;33:599-601.
3. Stehr-Green JK, McCaig L, Remsen HM, Rains CS, Fox M, Juranek DD. Shedding of oocysts in immunocompetent individuals infected with Cryptosporidium. Am J Trop Med Hyg 1987; 36:338-42.
4. D'Antonio RG, Winn RE, Taylor JP, et al. A waterborne outbreak of cryptosporidiosis in normal hosts. Ann Int Med 1985;103:886-8.
5. Mata L. Cryptosporidium and other protozoa in diarrheal disease in less developed countries. Pediatr Infect Dis 1986;5:S117-30.

October 26, 1984 / 33(42);599-601

Epidemiologic Notes and Reports Cryptosporidiosis among Children Attending Day-Care
Centers -- Georgia, Pennsylvania, Michigan, California, New Mexico

During 1984, CDC has received several reports of cryptosporidiosis among children attending
day-care centers. Seven investigations conducted in five states are summarized below.

Georgia: Investigation 1: Two sisters, aged 2 years and 4 years, who attended an Atlanta day-care
center, developed watery diarrhea in late February, and stool specimens showed Cryptosporidium
oocysts. An investigation in April found that 27 (51%) of 53 persons had recent histories of diarrhea.
Stool examinations of 50 children and 11 adult staff members revealed three other children with
Cryptosporidium; all had recent histories of afebrile, diarrheal illness without nausea or vomiting.
No asymptomatic children had cryptosporidiosis. One infected child also had Giardia lamblia cysts.
Eight of 27 symptomatic and six of 26 asymptomatic persons had Giardia. Symptomatic persons had
mild-to-moderate diarrhea, and most sought medical attention. No one was hospitalized.

Investigation 2: On August 27, a 2-year-old day-care center attendee in Atlanta developed severe,
watery diarrhea. Stool examination on September 6 showed Cryptosporidium. Thus far, four (17%)
of 23 children from the same room who were examined have had Cryptosporidium. Two of 12 children tested in other rooms at the day-care center had Cryptosporidium; both children were siblings
of infected children in the original room. Two of the six infected children had no histories of diarrhea and were asypmtomatic at the time of the investigation; the others had mild-to-moderate diarrhea without fever. None required hospitalization, and two children were seen by physicians.

Pennsylvania: Beginning in June, the rate of diarrheal illness increased at a day-care center in
Philadelphia, where 20 (34%) of 59 children were symptomatic. Stool specimens obtained from 45
children were examined for enteropathogenic bacteria, viruses, and parasites. Eleven (65%) of 17
symptomatic children and three (11%) of 28 asymptomatic children had Cryptosporidium.
Enteropathogenic bacteria and viruses were not implicated in the outbreak (1).

Michigan: In September, an investigation of day-care-center-associated diarrhea in Ann Arbor found a
2-year-old with cryptosporidiosis. Review of the day-care center's records showed an increase of diarrhea among children from three rooms--two for toddler-aged children, and one for infants. Stool specimens were obtained from 38 (70%) of the 54 children in the three affected rooms and examined for
parasites, Salmonella, Shigella, Campylobacter, and rotavirus; 21 (55%) had Cryptosporidium. One of
these children also had Salmonella; another also had Giardia. Infected children generally had mild-to-
moderate diarrhea without fever; none required hospitalization, and three children saw physicians.

California: On September 14, a 2-year-old child with a diarrheal illness who regularly attends a day-
care center in San Carlos was found to have cryptosporidiosis. A survey showed that children with
recent histories of diarrhea were limited to the classroom with the index child, where 10 of 11 classmates had been symptomatic. Stool specimens from all 11 children were examined for Salmonella,
Shigella, Campylobacter, Yersinia, Vibrio, Aeromonas, Edwardsiella, Plesiomonas, and parasites. Six
of 10 specimens from symptomatic children were positive for Cryptosporidium. Yersinia enterocolitica serotype 5,27 was recovered from one currently asymptomatic child who had symptoms earlier.
No other bacterial pathogens were isolated. The asymptomatic child had a negative stool examination. Three parents (including both parents of the index patient), who later developed diarrhea, were
positive for Cryptosporidium. Parents of children reported mild-to-moderate diarrhea, and most persons required medical care. No one was hospitalized.

New Mexico: During September, investigation of giardiasis in two children led to the discovery of
widespread diarrheal disease in two day-care centers in Albuquerque.

INVESTIGATION 1: Eighteen (47%) of 38 children attending a day-care center had recently had
diarrhea. Stool specimens from 17 symptomatic and one asymptomatic child were examined for parasites. Cryptosporidium alone was found in specimens from four symptomatic children. Five children had Giardia only; one child was infected with both parasites. Only two of six specimens with
Giardia were examined for Cryptosporidium. Stool specimens were submitted by 11 household
members of symptomatic children. Of seven household members reporting recent diarrheal illness,

273

one had Cryptosporidium, and two had Giardia; one asymptomatic adult had Giardia. Children and adults reported mild but sometimes prolonged diarrhea, and no one was hospitalized.

INVESTIGATION 2: In this day-care center, diarrheal illness was limited to the classroom for toddler-aged children. Thirteen (81%) of 16 children and one of three adults reported recent diarrhea. Of stool specimens from 13 children examined so far, five have shown Cryptosporidium only, and four, Giardia only. Two additional children had both parasites. Two of the specimens with Giardia were not examined for Cryptosporidium. Reported by G Bohan, MD, DeKalb County Health Dept, RK Sikes, DVM, State Epidemiologist, Georgia Dept of Human Resources; G Alpert, MD, L Bell, MD, CE Kirkpatrick, MD, JM Campos, PhD, HM Friedman, MD, SA Plotkin, MD, Children's Hospital of Philadelphia, LD Budnick, MD, RG Sharrar, MD, Philadelphia Dept of Health; ML Collinge, PhD, CL Combee, PhD, JA Gardner, MS, EM Britt, PhD, St Joseph Mercy Hospital, Ann Arbor, KR Wilcox, MD, State Epidemiologist, Michigan Dept of Health; J Bodie, MD, San Mateo County Health Dept, K Hadley, MD, San Francisco General Hospital, C Taclindo, MS, RR Roberto, MD, J Chin, MD, State Epidemiologist, California State Dept of Health Svcs; L Nims, MS, A Salas, HF Hull, MD, State Epidemiologist, New Mexico Health and Environment Dept; Protozoal Diseases Br, Div of Parasitic Diseases, Center for Infectious Diseases, Div of Field Svcs, Epidemiology Program Office, CDC.

Editorial Note: Outbreaks caused by a number of important infectious agents (including Giardia, Shigella, Haemophilus influenza, hepatitis A, rotavirus, and respiratory-tract viruses) have been documented in day-care centers (2). The investigations reported here suggest that the intestinal parasite Cryptosporidium should be added to this list. Although a few children had moderately severe diarrhea, none required hospitalization.

Cryptosporidium is a well-known cause of diarrhea in animals but has been recognized only recently as a cause of human disease. The first case of human cryptosporidiosis was reported in 1976; before 1982, literature exists on only seven human cases of cryptosporidiosis. Since 1982, the number of reported cases increased markedly (3). Initially, this increase was noted in patients with acquired immunodeficiency syndrome (AIDS), but recent reports indicate that cryptosporidiosis is common in immunologically normal persons (4-6). Patients with AIDS and cryptosporidiosis usually have severe, irreversible diarrhea, but persons with normal immunologic function have self-limited, although at times severe, diarrhea. The spectrum of illness caused by Cryptosporodium has yet to be clearly defined, and no satisfactory treatment is currently available.

Public health workers, physicians, parents, and day-care providers need to be alert to cryptosporodiosis as a potential cause of outbreaks of diarrhea in day-care centers. Special concentration and staining techniques for the recovery and isolation of Cryptosporodium are required (7,8), and investigators should notify laboratory personnel that Cryptosporodium is considered a possible pathogen in outbreaks. Knowledge of how Cryptosporidium is transmitted in the day-care setting is presently lacking, and only general guildelines for the prevention and control of enteric infections are available. Cryptosporidiosis outbreaks in day-care centers should be reported to state and local health departments. CDC would also like to be notified so that the spectrum of illness of this organism in this setting can be further defined.

References
1. Alpert G, Bell LM, Kirkpatrick CE, et al. Cryptosporidiosis in a day-care center. N Engl J Med 1984;311:860-1.
2. Pickering LK, Woodward WE. Diarrhea in day care centers. Peditr Infect Dis 1982;1:47-52.
3. Navin TR, Juranek DD. Cryptosporidiosis: clinical, epidemiologic, and parasitologic review. Rev Infect Dis 1984;6:313-27.
4. Tzipori S, Smith M, Birch C, et al. Cryptosporidiosis in hospitalized patients with gastroenteritis. Am J Trop Med Hyg 1983;32:931-4.
5. Jokipii L, Pohjola S, Jokipii AM. Cryptosporidium: a frequent finding in patients with gastrointestinal symptoms. Lancet 1983;II:358-61.
6. Current WL, Reese NC, Ernst JV, Bailey WS, Heyman MB, Weinstein WM. Human cryptosporidiosis in immunocompetent and immunodeficient persons. Studies of an outbreak and experimental transmission. N Engl J Med 1983; 308:1252-7.
7. Ma P, Soave R. Three-step stool examination for cryptosporidiosis

---

October 02, 1998 / 47(38);806-9

Outbreak of Cyclosporiasis -- Ontario, Canada, May 1998

During May-June 1998, the Ontario Ministry of Health and local health departments in Ontario received reports of clusters of cases of cyclosporiasis associated with events held during May. This report describes the preliminary findings of the investigation of a cluster in Toronto, Ontario, and summarizes the findings from investigations of 12 other clusters. These investigations indicated that fresh raspberries imported from Guatemala were linked to the multicluster outbreak.

TORONTO, ONTARIO

On June 2, Toronto Public Health was notified of a laboratory-confirmed case of cyclosporiasis in a person who attended a dinner at a hotel in Toronto on May 8. Six other persons who attended the dinner were reported to have diarrheal illness. A case of cyclosporiasis was defined as onset of any gastrointestinal (e.g., nausea or vomiting) or constitutional (e.g., fever or fatigue) symptom 1-14 days after the dinner and either 1) laboratory confirmation of Cyclospora oocysts in a stool specimen; 2) diarrhea (i.e., three or more loose or watery stools during a 24-hour period); or 3) at least four gastrointestinal symptoms. Of the 174 persons who attended the dinner, 128 (74%) were interviewed. Of these 128 persons, 29 (23%) had illness that met the case definition; three of the 29 persons had laboratory-confirmed cyclosporiasis. The median incubation period was 8 days (range: 1-12 days). All 29 case-patients had diarrhea; the median duration of diarrheal illness was 7 days (range: 1-34 days).

Eating the berry garnish (which included raspberries, blackberries, strawberries, and possibly blueberries) for the dessert was significantly associated with risk for illness. Of the 108 persons who ate or probably ate the berry garnish, 28 (26%) became ill, compared with one (5%) of the 20 persons who did not or probably did not eat the berry garnish (relative risk {RR}=5.2; p=0.04, Fisher's exact test). Among the berries in the garnish, raspberries were the only berries significantly associated with risk for illness. Of the 94 persons who ate or probably ate the raspberries, 27 (29%) became ill, compared with two (6%) of the 32 persons who did not or probably did not eat the raspberries (RR=4.6; 95% confidence interval=1.2-18.3).

OTHER INVESTIGATIONS

Twelve other clusters of cases of cyclosporiasis in addition to the Toronto cluster described above have been investigated; each of the 13 clusters had two or more cases, at least one of which was laboratory confirmed. Based on preliminary data, the 13 clusters comprise 192 cases; 46 (24%) of the 192 were laboratory confirmed. The dates of the events associated with the clusters ranged from May 2 through May 23, 1998.

Fresh raspberries were the only food in common to all 13 events. Raspberries were included in mixtures of various types of berries at 12 events and were the only type of berry served at one event. The median of the event-specific attack rates for the 13 events, irrespective of exposures, was 89% (range: 23%-100%). The median of the event-specific attack rates for persons who ate or probably ate the food items that included raspberries was 100% (range: 26%-100%); the median attack rate for persons who did not or probably did not eat these food items was 0% (range: 0%-67%). Eating the food items that included raspberries was significantly associated with risk for illness for five events; for the other eight events, eating the raspberry-containing food items could account for 60 (92%) of 65 cases. Traceback investigations to identify the source(s) of the raspberries have been completed for eight events, including the event described above; Guatemala was the only source of the raspberries served at the events. Mesclun lettuce and fresh basil, which were implicated in outbreaks of cyclosporiasis in the United States in 1997 (1,2), each were served at two events but were not significantly associated with risk for illness.

Reported by: Toronto Public Health, Toronto; Haliburton-Kawartha-Pine Ridge District Health Unit, Port Hope; Simcoe County District Health Unit, Barrie; York Regional Health Unit, Newmarket; Disease Control Svc, Public Health Br, Ontario Ministry of Health, Toronto; Central Public Health Laboratory, Laboratory Services Br, Ontario Ministry of Health, Toronto. Canadian Food Inspection Agency, Fresh and Processed Plant Products Div, Ottawa, and Food Inspection, Ontario Region, Toronto and Guelph; Bur of Infectious Diseases and Field Epidemiology Training Program,

Laboratory Center for Disease Control, and Food Directorate, Health Canada, Ottawa. Parasitic Disease Surveillance Unit, New York City Dept of Health, New York. Div of Parasitic Diseases, National Center for Infectious Diseases; and an EIS Officer, CDC.

Editorial Note: The findings in this report indicate that fresh raspberries imported from Guatemala were linked to the outbreak of cyclosporiasis in Ontario in May 1998. Outbreaks of cyclosporiasis in North America in the spring of 1996 and 1997 also were linked to Guatemalan raspberries; the mode of contamination of the raspberries was not identified for any of these outbreaks (1,3). No outbreaks were recognized in association with Guatemalan raspberries during Guatemala's fall and winter export seasons in 1996 and 1997.

After the outbreak in 1996, berry growers and exporters in Guatemala, in consultation with the Food and Drug Administration (FDA) and CDC, voluntarily introduced control measures that focused on improving water quality and sanitary conditions on individual farms (1). In the spring of 1997, another outbreak of cyclosporiasis occurred despite the implementation of control measures and the restriction (beginning April 22, 1997) that, during that spring, only farms classified by the Guatemalans as low risk could export to North America (1). In the spring of 1998, FDA did not allow importation of fresh raspberries from Guatemala into the United States. The Canadian Food Inspection Agency reported that fresh raspberries from farms that the Guatemalans had classified as low risk continued to be imported into Canada until June 9, 1998. The occurrence of outbreaks in 1997 and 1998 despite the implementation of control measures on Guatemalan farms suggests either that the control measures may not have been fully implemented by some farms, were not effective, or were not directed against the true source of contamination of the raspberries (1). The Guatemalan Berry Commission and the government of Guatemala are developing a more comprehensive plan for growing and handling raspberries that includes additional control measures and inspection criteria; the plan is being reviewed by U.S. and Canadian officials.

This is at least the third, and possibly the fourth (4), consecutive year in which outbreaks of cyclosporiasis linked to consumption of raw produce have occurred in North America. In addition to Guatemalan raspberries, fresh mesclun lettuce and fresh basil that were not from Guatemala have been implicated in outbreaks in the United States (1,2). The mode of contamination of the produce was not determined for any of the outbreaks, in part because the methods for detecting Cyclospora on produce and in other environmental samples are insensitive for detecting low levels of the parasite. Produce should be washed thoroughly before it is eaten; however, this practice does not eliminate the risk for transmission of Cyclospora (3,5,6).

Health-care providers should consider the diagnosis of Cyclospora infection in persons with prolonged diarrheal illness and specifically request testing of stool specimens for this parasite. The average incubation period for cyclosporiasis is 1 week; in patients who are not treated with trimethoprim-sulfamethoxazole (7), illness can be protracted, with remitting and relapsing symptoms.

Cases of Cyclospora infection unrelated to travel outside of Canada or the United States may be associated with a new outbreak. Newly identified clusters should be investigated to identify the vehicles of infection and to identify the sources and modes of contamination of the implicated vehicles. Although cyclosporiasis is not a reportable disease in any Canadian province or territory, as of June 1998, five states and one municipality in the United States had mandated reporting. In June 1998, the Council of State and Territorial Epidemiologists passed a resolution recommending that cyclosporiasis be made a nationally notifiable disease in the United States. In jurisdictions where formal reporting mechanisms are not yet established, clinicians and laboratorians who identify cases of cyclosporiasis unrelated to travel outside North America are encouraged to inform the appropriate local, provincial, territorial, or state health departments, which in turn are encouraged to contact, in Canada, the Division of Disease Surveillance, Bureau of Infectious Diseases, Laboratory Center for Disease Control, telephone (613) 941-1288; and, in the United States, CDC's Division of Parasitic Diseases, National Center for Infectious Diseases, telephone (770) 488-7760.

References
1. CDC. Update: outbreaks of cyclosporiasis -- United States and Canada, 1997. MMWR 1997;46:521-3.
2. CDC. Outbreak of cyclosporiasis -- Northern Virginia-Washington, D.C.-Baltimore, Maryland, Metropolitan Area, 1997. MMWR 1997;46:689-91.

3. Herwaldt BL, Ackers M-L, Cyclospora Working Group. An outbreak in 1996 of cyclosporiasis associated with imported raspberries. N Engl J Med 1997;336:1548-56.

4. Koumans EH, Katz DJ, Malecki JM, et al. An outbreak of cyclosporiasis in Florida in 1995: a harbinger of multistate outbreaks in 1996 and 1997. Am J Trop Hyg 1998;59:235-42.

5. Robbins JA, Sjulin TM. Scanning electron microscope analysis of drupelet morphology of red raspberry and related Rubus genotypes. Journal of the American Society of Horticultural Science 1988;113: 474-80.

6. Ortega YR, Roxas CR, Gilman RH, et al. Isolation of Cryptosporidium parvum and Cyclospora cayetanensis from vegetables collected in markets of an endemic region in Peru. Am J Trop Med Hyg 1997;57:683-6.

7. Hoge CW, Shlim DR, Ghimire M, et al. Placebo-controlled trial of co-trimoxazole for Cyclospora infections among travellers and foreign residents in Nepal. Lancet 1995;345:691-3.

August 01, 1997 / 46(30);689-691

Outbreak of Cyclosporiasis -- Northern Virginia-Washington, D.C.-Baltimore, Maryland, Metropolitan Area, 1997

During July 1997, state and local health departments in Virginia, the District of Columbia (DC), and Maryland received reports of clusters of cases of cyclosporiasis associated with events (e.g., lunch-eons) held in their jurisdictions during June and July. This report describes the preliminary findings of the investigation of a cluster in Virginia and summarizes the findings from ongoing investigations of the other clusters. Fresh basil has been implicated as the probable vehicle of infection. Alexandria, Virginia

On July 7, a company physician reported to the Alexandria Department of Health (ADOH) that most of the employees who attended a corporate luncheon on June 26 at the company's branch in Fairfax, Virginia, had developed gastrointestinal illness. The luncheon was catered by the Alexandria branch of company A. Company A operates nine stores in the northern Virginia-DC-Baltimore, Maryland, metropolitan area: a central production kitchen and retail food store in Bethesda, Maryland; and eight branch stores, each with a kitchen and retail store.

On July 11, the health department was notified that a stool specimen from one of the employees who attended the luncheon was positive for Cyclospora oocysts. A clinical case of cyclosporiasis was defined as onset of at least four gastrointestinal symptoms, such as diarrhea, nausea, vomiting, or abdominal cramps, 1-14 days after the luncheon. All 54 persons who attended the luncheon on June 26 or who ate leftover food on June 27 were interviewed. Of the 54 persons, 48 (89%) had illness that met the clinical case definition, including 17 whose infections were laboratory confirmed by examination of stool specimens. The median incubation period was 8 days (range: 3-12 days). Of the 48 case-patients, 45 had diarrhea (three or more loose stools during a 24-hour period), with a median number of stools per day of seven (range: three to 35 stools) and a median duration of diarrheal ill-ness of 5 days (range: 1-10 days).

Eating the basil-pesto pasta salad, which was served cold, was the only exposure significantly asso-ciated with risk for illness in univariate analysis; 43 (98%) of the 44 persons who ate this food item became ill, compared with one (17%) of six persons who did not eat it (relative risk=5.9; p less than 0.001, Fisher's exact test; four ill persons did not recall whether they had eaten the salad). The one ill person who did not eat the salad used the spoon from the salad to serve himself leftovers of another food item that he ate on June 27. The salad had been prepared in the Alexandria store with basil-pesto sauce made in the production kitchen in Bethesda. No raspberries or mesclun lettuce, which caused outbreaks of cyclosporiasis in the United States this spring (1), were served at the luncheon. Other Investigations

Twenty-five clusters of cases of cyclosporiasis with at least one laboratory-confirmed case per clus-ter (i.e., confirmed clusters) have been reported in association with events held in the northern Virginia-DC-Baltimore metropolitan area during June and July. In addition, at least 20 possible clus-ters for which laboratory confirmation has not yet been obtained have been reported. The dates of the events associated with confirmed and possible clusters ranged from June 16 to July 8 and from

June 15 to July 12, respectively. Based on preliminary interview data, the 25 confirmed clusters comprise approximately 185 cases (approximately 60 laboratory-confirmed and 125 clinically defined cases), and the 20 possible clusters, approximately 75 clinically defined cases.

All 25 confirmed clusters were associated with events at which at least one food item that contained fresh basil from company A was served (i.e., fresh basil or a prepared food item that contained fresh basil was either purchased at one of its retail stores or served at a meal prepared in one of its kitchens). Six of the nine company A stores have been linked to clusters. For 23 of the 25 events, a basil-containing item that included basil-pesto sauce (e.g., in a pasta salad or on a sandwich) made at the Bethesda store was served. Company A reported that its practice was to wash basil that it used to make pesto sauce. Eating the food item that contained basil was significantly associated (p less than 0.05) or associated (i.e., all ill persons had eaten the item but the p value was greater than or equal to 0.05) with risk for illness for all six events for which preliminary epidemiologic data are available.

At the direction of the ADOH, on July 12, company A terminated production and sales of pesto sauce made with fresh basil and of food items that contained this sauce and terminated sales of fresh basil. On July 18, health departments in Virginia and Maryland issued press releases to inform the public not to consume fresh basil or fresh basil-containing food items previously purchased from company A. State and local health departments, CDC, and the Food and Drug Administration (FDA) are continuing investigations to determine the sources and distribution of the basil; to determine how basil is handled, processed, and distributed by company A; and to identify modes of contamination. FDA and CDC are testing for the presence of Cyclospora oocysts in samples of fresh basil and basil-pesto sauce obtained in mid-July from company A and in leftover pesto sauce obtained from several ill persons.

Reported by: R Pritchett, MPH, C Gossman, V Radke, MPH, J Moore, MHSA, E Busenlehner, K Fischer, K Doerr, C Winkler, M Franklin-Thomsen, J Fiander, J Crowley, E Peoples, L Brembly, J Southard, MSN, L Appleton, D Bowers, MSN, J Lipsman, MD, Alexandria Dept of Health, Alexandria; H Callaway, D Lawrence, R Gardner, Fairfax Dept of Health, Fairfax; B Cunanan, R Snaman, Arlington Dept of Health, Arlington; J Rullan, MD, G Miller, Jr, MD, State Epidemiologist, Virginia Dept of Health; S Henderson, M Mismas, T York, PhD, J Pearson, PhD, Div of Consolidated Svcs, Commonwealth of Virginia. C Lacey, J Purvis, N Curtis, K Mallet, Montgomery County Health Dept, Rockville; R Thompson, Baltimore County Health Dept, Towson; D Portesi, MPH, DM Dwyer, MD, State Epidemiologist, Maryland Dept of Health and Mental Hygiene. M Fletcher, PhD, M Levy, MD, District Epidemiologist, District of Columbia Dept of Health. T Lawford, MD, Fairfax, Virginia. M Sabat, MS, Chicago, Illinois. M Kahn, Atlanta, Georgia. Office of Regulatory Affairs, and Center for Food Safety and Applied Nutrition, Food and Drug Administration. Div of Parasitic Diseases, National Center for Infectious Diseases, CDC.

Editorial Note: The preliminary findings of the investigations described in this report implicate fresh basil from company A as the probable vehicle of infection for the clusters of cases of cyclosporiasis recently identified in the northern Virginia-DC-Baltimore metropolitan area. To date, all of these clusters have been associated with company A, even though the produce distributor that was the sole supplier for company A during the relevant period provided a large (as yet undetermined) proportion of its inventory of fresh basil to other local establishments. Some of the implicated food items from company A did not contain basil-pesto sauce; therefore, basil, rather than the other ingredients of the pesto sauce, is the probable vehicle. The mode of contamination of the basil is being investigated. Cyclospora oocysts are not infectious (i.e., are unsporulated) at the time of excretion. However, the minimum time required for sporulation is unknown, and the conditions in the environment and in foods that expedite sporulation are poorly understood.

In addition to the cases of cyclosporiasis associated with consumption of basil, approximately 1450 other cases of cyclosporiasis, approximately 550 of which have been laboratory confirmed, have been reported in the United States and Canada in 1997. Fresh raspberries imported from Guatemala and mesclun lettuce (specific source not yet determined) have both been implicated as vehicles of infection in outbreak investigations in 1997 (1). The implication of three different vehicles of infection during 1997 highlights the need for strengthened prevention and control measures to ensure the safety of produce that is eaten raw and the need for improved understanding of the epidemiology of Cyclospora.

The average incubation period for cyclosporiasis is 1 week; in patients who are not treated with trimethoprim-sulfamethoxazole (2), illness can be protracted, with remitting and relapsing symptoms. Health-care providers should consider Cyclospora infection in persons with prolonged diarrheal illness and specifically request laboratory testing for this parasite. Cases should be reported to local and state health departments; health departments that identify cases of cyclosporiasis should contact CDC's Division of Parasitic Diseases, National Center for Infectious Diseases, telephone (770) 488-7760. Newly identified clusters should be investigated to identify the vehicles of infection and to identify the sources and modes of contamination of implicated foods.

References
1. CDC. Update: outbreaks of cyclosporiasis -- United States and Canada, 1997. MMWR 1997;46:521-3.
2. Hoge CW, Shlim DR, Ghimire M, et al. Placebo-controlled trial of co-trimoxazole for Cyclospora infections among travellers and foreign residents in Nepal. Lancet 1995;345:691-3.

June 13, 1997 / 46(23);521-523

Update: Outbreaks of Cyclosporiasis -- United States and Canada, 1997

Since April 1997, CDC has received reports of outbreaks of cyclosporiasis in the United States and Canada (1,2). As of June 11, there have been 21 clusters of cases of cyclosporiasis reported from eight states (California, Florida, Maryland, Nebraska, Nevada, New York, Rhode Island, and Texas) and one province in Canada (Ontario). These clusters were associated with events (e.g., receptions, banquets, or time-place-related exposures {meals in the same restaurant on the same day}) that occurred during March 19-May 25 and comprise approximately 140 laboratory-confirmed and 370 clinically defined cases of cyclosporiasis. In addition, four laboratory-confirmed and approximately 220 clinically defined cases have been reported among persons who, during March 29-April 5, were on a cruise ship that departed from Florida. Approximately 70 laboratory-confirmed sporadic cases (i.e., cases not associated with events, the cruise, or recent overseas travel) have been reported in the United States and Canada. The most recent laboratory-confirmed sporadic case occurred in a person who had onset of symptoms on June 3.

Fresh raspberries were served at 19 of the 21 events and were the only food in common to all 19 events, which occurred in April and May. At six of the 19 events, raspberries were the only type of berry served or were served separately from other berries; at 13 events, raspberries were included in mixtures of various types of berries. Eating the food item that included raspberries was significantly associated with risk for illness for seven of the 15 events for which epidemiologic data are currently available (including for three of the events at which raspberries were not served with other types of berries) and was associated with illness but not significantly for six events (i.e., all or nearly all ill persons ate the berry item that was served). The raspberries reportedly had been rinsed in water at 10 (71%) of the 14 events for which such information is available. Guatemala has been identified as one of the possible sources of raspberries for all eight events for which traceback data are currently available (i.e., Guatemala was the source of at least one of the shipments of raspberries that could have been used) and as the only possible source for at least one of these events and perhaps for two others for which the traceback investigations are ongoing.

Fresh raspberries were not served at two events in restaurants in Florida that have been associated with clusters of cases of cyclosporiasis (persons were exposed on March 19 and April 10, respectively, in two different cities). The first cluster was associated with eating mesclun (also known as spring mix, field greens, or baby greens -- a mixture of various types of baby leaves of lettuce); the specific source of the implicated mesclun has not been determined. Mesclun also is suspected as the vehicle for the second cluster.

Reported by: E DeGraw, Leon County Health Dept, Tallahassee; S Heber, MPH, A Rowan, Florida Dept of Health. Other state, provincial, and local health depts. Health Canada. Office of Regulatory Affairs, and Center for Food Safety and Applied Nutrition, Food and Drug Administration. Div of Applied Public Health Training (proposed), Epidemiology Program Office; Div of Parasitic Diseases, National Center for Infectious Diseases, CDC.

Editorial Note: The investigations described in this report indicate that fresh raspberries imported from Guatemala are the probable vehicle of infection for most of the outbreaks of cyclosporiasis identified in 1997. There is no evidence of ongoing transmission of Cyclospora in association with mesclun, which was the vehicle for one, and possibly two, early outbreaks in March and April. In the spring and summer of 1996, an outbreak of cyclosporiasis in the United States and Canada was linked to eating raspberries imported from Guatemala (3). However, the mode of contamination of the raspberries implicated in that outbreak was not determined -- in part because the methods for testing produce and other environmental samples for this emerging pathogen are insensitive and non-standardized. No outbreaks of cyclosporiasis were reported in the United States in association with importation of raspberries from Guatemala during the fall and winter of 1996; however, cyclosporiasis is highly seasonal in some countries.

After the outbreak in 1996, the berry industry in Guatemala, in consultation with the Food and Drug Administration (FDA) and CDC, voluntarily implemented a Hazard Analysis and Critical Control Point system and improved water quality and sanitary conditions on individual farms (3). The occurrence of outbreaks in 1997 suggests either that some farms did not fully implement the control measures or that the contamination is associated with a source against which these measures were not directed.

At FDA's request, on May 30, 1997, the government of Guatemala and the Guatemalan Berries Commission announced their decision to voluntarily suspend exports of fresh raspberries to the United States (the last shipment was May 28). FDA is working with CDC, the government of Guatemala, and the Guatemalan Berries Commission to determine when exports can resume (4). Because of the relatively short shelf life, few, if any, fresh raspberries grown in Guatemala are available now for purchase and consumption in the United States. Cyclospora oocysts, like the oocysts of other coccidian parasites, are expected to be inactivated by temperature extremes (e.g., pasteurization or commercial freezing processes). The minimum time and temperature conditions required to inactivate Cyclospora oocysts by heating or freezing have not yet been determined.

Although exports of fresh raspberries from Guatemala to the United States have been suspended until further notice, cases of cyclosporiasis that are attributable to consumption of raspberries may continue to be identified by health-care providers and health departments. The average incubation period for cyclosporiasis is 1 week; if not treated with trimethoprim-sulfamethoxazole (5), illness can be protracted, with remitting and relapsing symptoms. Health-care providers should consider Cyclospora infection in persons with prolonged diarrheal illness and specifically request laboratory testing for this parasite. Cases should be reported to local and state health departments; health departments that identify cases of cyclosporiasis should contact CDC's Division of Parasitic Diseases, National Center for Infectious Diseases, telephone (770) 488-7760. Newly identified clusters of cases should be investigated to identify the vehicles of infection and to trace the sources of implicated foods.

References
1. CDC. Outbreaks of cyclosporiasis -- United States, 1997. MMWR 1997;46:451-2.
2. CDC. Update: outbreaks of cyclosporiasis -- United States, 1997. MMWR 1997;46:461-2.
3. Herwaldt BL, Ackers M-L, Cyclospora Working Group. An outbreak in 1996 of cyclosporiasis associated with imported raspberries. N Engl J Med 1997;336:1548-56.
4. Food and Drug Administration. Outbreak of cyclosporiasis and Guatemalan raspberries. Rockville, Maryland: US Department of Health and Human Services, Public Health Service, Food and Drug Administration, June 10, 1997. (Talk Paper T97-22).
5. Hoge CW, Shlim DR, Ghimire M, et al. Placebo-controlled trial of co-trimoxazole for Cyclospora infections among travellers and foreign residents in Nepal. Lancet 1995;345:691-3.

May 30, 1997 / 46(21);461-462

Update: Outbreaks of Cyclosporiasis -- United States, 1997

During April and May 1997, CDC received reports of clusters of cases of cyclosporiasis in the United States (1). This report describes the preliminary findings of an investigation of an outbreak in New York and summarizes the findings from on-going investigations in other states. New York

On May 15, the Westchester County Health Department was notified of two laboratory-confirmed cases of cyclosporiasis and other cases of diarrheal illness among persons who attended a wedding reception on April 20 at a private residence in the county. A case of cyclosporiasis was defined as onset of diarrhea (three or more loose stools during a 24-hour period) 1-14 days after the reception. Of the 183 persons who attended the reception, 154 (84%) were interviewed, and 140 were included in this analysis (persons who had loose stools that did not meet the case definition were excluded). Of the 140 persons, 20 (14%) had illness that met the case definition; four cases were laboratory confirmed. The median incubation period was 8 days (range: 3-11 days), and for 19 persons, the duration of diarrheal illness was greater than or equal to 3 days.

Eating raspberries was the exposure most strongly associated with risk for illness in univariate analysis and was the only exposure significantly associated with risk for illness in multivariate logistic regression analysis. Sixteen (36%) of the 45 persons who ate raspberries became ill, compared with three (4%) of the 85 persons who did not eat raspberries (univariate relative risk=10.1; 95% confidence interval=3.1-32.8). The raspberries had not been washed.

OTHER INVESTIGATIONS
CDC has received reports of eight event-associated (e.g., reception) clusters of cases of cyclosporiasis from five states (California, Florida, Nevada, New York {includes Westchester County}, and Texas) and a report of cases among persons who, during March 29-April 5, had been on a cruise ship that left from Florida. The most recent of the eight events occurred on May 8. Approximately 90 event-associated cases of infection have been laboratory confirmed.

Fresh berries were served at six of the eight events. Raspberries were included in mixtures of various types of berries at four events, were served separately from other berries at one event (the event in Westchester County), and were the only type of berry served at one event (in Nevada). Eating the food items that included raspberries was significantly associated with risk for illness for four events, including the two events at which raspberries could be distinguished from other berries (Westchester County and Nevada events); for one of the other two events, all 10 persons ate the berry mixture that was served and became ill. At one event where the implicated food item included a mixture of berries, the source of the raspberries was Guatemala; preliminary traceback data for the other events at which raspberries were served indicate that both Guatemala and Chile may be sources (i.e., each country was the source of at least one of the shipments of raspberries that could have been used).

State and local health departments, CDC, and the Food and Drug Administration (FDA) are continuing the investigations to identify the vehicles of infection, to trace the sources of implicated foods, and to determine whether transmission is ongoing.

Reported by: G Jacquette, MD, F Guido, MPA, J Jacobs, Westchester County Dept of Health, Hawthorne; P Smith, MD, State Epidemiologist, New York State Dept of Health. Other state and local health depts. D Adler, San Francisco, California. Office of Regulatory Affairs, and Center for Food Safety and Applied Nutrition, Food and Drug Administration. Foodborne and Diarrheal Diseases Br and Childhood and Respiratory Diseases Br, Div of Bacterial and Mycotic Diseases, and Div of Parasitic Diseases, National Center for Infectious Diseases, CDC.

Editorial Note: The preliminary findings of the investigations described in this report suggest that raspberries imported from Guatemala and possibly from Chile were the likely vehicle of infection for some of the outbreaks of cyclosporiasis during April and May. In the spring and summer of 1996, an outbreak of cyclosporiasis in the United States and Canada was linked to eating raspberries imported from Guatemala (2). However, the mode of contamination of the implicated raspberries in that outbreak was not determined -- in part because the methods for testing produce and other environmental samples for Cyclospora are insensitive.

Produce should always be thoroughly washed before it is eaten. This practice should decrease, but may not eliminate, the risk for transmission of Cyclospora. Because raspberries are fragile and replete with crevices (3), even thorough washing may not eliminate contamination of the fruit. State and local health departments, CDC, and FDA are evaluating the findings from the investigations to determine the need for additional public health measures.

Health-care providers should consider Cyclospora infection in persons with prolonged diarrheal illness and specifically request laboratory testing for this parasite. Cases should be reported to local and state health departments; health departments that identify cases of cyclosporiasis should contact CDC's Division of Parasitic Diseases, National Center for Infectious Diseases, telephone (770) 488-7760.

References
1. CDC. Outbreaks of cyclosporiasis -- United States, 1997. MMWR 1997;46:451-2.
2. Herwaldt BL, Ackers M-L, Cyclospora Working Group. An outbreak in 1996 of cyclosporiasis associated with imported raspberries. N Engl J Med 1997;336:1548-56.
3. Robbins JA, Sjulin TM. Scanning electron microscope analysis of drupelet morphology of red raspberry and related Rubus genotypes. J Am Soc Horticult Sci 1988;113:474-80.

May 23, 1997 / 46(20);451-452

Outbreaks of Cyclosporiasis -- United States, 1997

In April and May 1997, CDC received reports of seven event-associated clusters of cases of cyclosporiasis from five states (California, Florida, Nevada, New York, and Texas). Approximately 80 cases of infection with human-associated Cyclospora, a recently characterized coccidian parasite (1), have been laboratory-confirmed. State and local health departments, CDC, and the Food and Drug Administration are conducting investigations to identify the vehicles of infection.

Both foodborne and waterborne outbreaks of cyclosporiasis have previously been reported in the United States during spring and summer months (2-4). In 1996, a total of 978 laboratory-confirmed cases of cyclosporiasis in the United States and Canada were reported in association with a wide-spread foodborne outbreak (3). The average incubation period of cyclosporiasis is 1 week. Illness can be protracted (from days to weeks) with frequent, watery stools and other gastrointestinal symp-toms; symptoms may remit and relapse. Health-care providers should consider Cyclospora infection in persons with prolonged diarrheal illness and specifically request laboratory testing for this parasite (5,6), which is not routinely performed by most laboratories.

Cyclosporiasis can be treated with a 7-day course of oral trimethoprim (TMP)-sulfamethoxazole (SMX) (for adults, 160 mg TMP plus 800 mg SMX twice daily; for children, 5 mg/kg TMP plus 25 mg/kg SMX twice daily) (7). Treatment regimens for patients who cannot tolerate sulfa drugs have not yet been identified.

Health departments that identify cases of cyclosporiasis should contact CDC's Division of Parasitic Diseases, National Center for Infectious Diseases, telephone (770) 488-7760.

Reported by: State and local health departments. Office of Regulatory Affairs, and Center for Food Safety and Applied Nutrition, Food and Drug Administration. Div of Parasitic Diseases, National Center for Infectious Diseases, CDC.

References
1. Ortega YR, Sterling CR, Gilman RH, Cama VA, Dæaz F. Cyclospora species -- a new protozoan pathogen of humans. N Engl J Med 1993;328:1308-12.
2. CDC. Update: outbreaks of Cyclospora cayetanensis infection -- United States and Canada, 1996. MMWR 1996;45:611-2.
3. Herwaldt BL, Ackers M-L, Cyclospora Working Group. An outbreak in 1996 of cyclosporiasis associated with imported raspberries. N Engl J Med 1997;336:1548-56 (in press).
4. Huang P, Weber JT, Sosin DM, et al. The first reported outbreak of diarrheal illness associated with Cyclospora in the United States. Ann Intern Med 1995;123:409-14.
5. Soave R. Cyclospora: an overview. Clin Infect Dis 1996;23:429-37.
6. Garcia LS, Bruckner DA. Diagnostic medical parasitology. 3rd ed. Washington, DC: American Society for Microbiology, 1997:66-9.
7. Hoge CW, Shlim DR, Ghimire M, et al. Placebo-controlled trial of co-trimoxazole for Cyclospora infections among travellers and foreign residents in Nepal. Lancet 1995;345:691-3.

April 25, 1997 / 46(16);354-358

Outbreaks of Pseudo-Infection with Cyclospora and Cryptosporidium --
Florida and New York City, 1995

Efforts to expand the scope of surveillance and diagnostic testing for emerging infectious diseases
(1) also may increase the potential for identifying pseudo-outbreaks (2,3) (i.e., increases in incidence
that may result from enhanced surveillance) and outbreaks of pseudo-infection (i.e., clusters of false-
positives for infection). This report describes the investigations of outbreaks of pseudo-infection with
Cyclospora in Florida and Cryptosporidium in New York City in 1995 after health departments in
those jurisdictions had initiated surveillance for these emerging organisms. These investigations
emphasize 1) the need for laboratory training in the identification of emerging pathogens and 2) the
importance of confirmation by reference laboratories as an early step in the investigation of any
apparent outbreak caused by an emerging pathogen.

CYCLOSPORIASIS IN FLORIDA

Cyclosporiasis is caused by infection with Cyclospora cayetanensis, a recently identified coccidian
parasite (4) that can cause prolonged, relapsing diarrhea; treatment with trimethoprim-sulfamethoxazole relieves symptoms and accelerates clearance of the parasite (5). Until 1996, most cases of
cyclosporiasis in the United States occurred among international travelers (6), and information about
modes of transmission of C. cayetanensis was limited. Waterborne transmission had been documented, but direct person-to-person transmission was considered unlikely (4).

During the summer of 1995, in response to an outbreak of Cyclospora infection among Florida residents with no history of international travel (7), the state health department initiated surveillance for
the organism. All state laboratories began routine testing with a modified acid-fast stain for C.
cayetanensis in stool specimens submitted for parasitologic examination (8). On July 25, 1995, the
Florida Department of Health (FDH) designated cyclosporiasis a reportable disease.

On August 11, a 3-year-old boy at a children's shelter had onset of diarrhea and abdominal pain;
Giardia cysts were identified in a stool specimen obtained from the child. Because of previous giardiasis outbreaks at the shelter, county public health officials recommended testing the 13 shelter residents who were preschool classmates or roommates of the index patient. State branch laboratory A
reported that stool specimens from six children tested positive for Giardia, and six tested positive for
Cyclospora. The high proportion of specimens positive for Cyclospora prompted testing of 81 persons, including all children residing at the shelter and the shelter's staff and volunteers. Overall,
branch laboratory A identified Cyclospora oocysts in specimens from 31 (86%) of 36 staff, 16 (64%)
of 25 children, and nine (45%) of 20 volunteers. In response to this apparent outbreak, the residence
was closed to new admissions, the children's outside activities were restricted, and trimethoprim-sulfamethoxazole was prescribed for all 25 children.

On September 17, FDH was notified about the apparent outbreak and joined the investigation. The
local community hospital, which had begun testing for Cyclospora in 1995, was contacted for information about laboratory-identified infections in the community during July-September.
Questionnaires were administered to shelter staff, volunteers, the older children, and the infants'
caretakers, and medical records for the children were reviewed. A case was defined as onset of nausea, vomiting, or diarrhea in a resident, employee, or volunteer at the shelter during August-
September. Branch laboratory A sent portions of stool specimens from 23 shelter residents and staff
to the reference laboratories at CDC and the University of Arizona for Cyclospora testing.

Questionnaires were completed for 79 of the 81 children and adults. Symptoms among the 56 persons whose stool specimens were positive for Cyclospora at branch laboratory A included abdominal
pain (30%), nausea (26%), fatigue (22%), diarrhea (three or more loose or watery stools in 24 hours)
with median duration of 2 days (20%), vomiting (19%), anorexia (15%), fever (10%), and weight
loss (8%). The 23 persons with negative stool specimens had symptoms and onset dates similar to
those of persons with positive specimens (Figure_1). In addition, the likelihood of being asymptomatic was similar among persons who were test-positive (35%) and test-negative (46%) (p=0.4).
Potential risk factors (e.g., consumption of food or water at the shelter or participation in field trips)
were not associated with the likelihood of being ill or testing positive. During the time of the appar-

283

ent outbreak at the shelter, the local community hospital examined 357 stool specimens for ova and parasites and identified Cyclospora oocysts in specimens from two patients; neither person was associated with the shelter.

Of the 23 stool specimens submitted by branch laboratory A to the two reference laboratories, branch laboratory A reported that 17 (74%) specimens were positive for Cyclospora; in comparison, the reference laboratories at both CDC and the University of Arizona reported that all the specimens were negative for Cyclospora. The state central laboratory and the University of Arizona laboratory reviewed slides from branch laboratory A and identified pollen grains and other artifacts similar to Cyclospora oocysts in size and staining characteristics but lacking the appropriate internal morphology. CDC examined stool specimens obtained from 19 other persons at the shelter; rare Cyclospora oocysts were identified in specimens from two children: one child who had been asymptomatic, and one who had vomited.

Based on these findings, FDH asked all laboratories in the state that had reported detecting Cyclospora oocysts in specimens from symptomatic patients in 1995 to forward their positive slides to the state central laboratory or CDC for confirmation. Branch laboratory A submitted slides from 130 patients not associated with the shelter; of these, 38 (29%) were confirmed as Cyclospora, and 92 (71%) were considered to have been false positives.

In response to the investigation, FDH revised the case definition for cyclosporiasis to include confirmation of Cyclospora infection by a reference laboratory, and the state central laboratory initiated a proficiency training program at all state laboratories to teach laboratorians how to identify Cyclospora and Cryptosporidium spp. In 1996, in a subsequent outbreak of cyclosporiasis (9), Florida laboratories initially identified 188 specimens from patients as positive for Cyclospora; 32 (17%) were not confirmed by the state central laboratory.

CRYPTOSPORIDIOSIS IN NEW YORK CITY

To improve disease reporting and identify exposures associated with infection, New York City designated cryptosporidiosis a reportable disease in January 1994, and the New York City Department of Health (NYCDOH) initiated active surveillance in November 1994. Each of the clinical laboratories are routinely contacted (usually monthly) for reports of new cases, and each case is investigated by telephone interview and/or chart review. Of the 289 cases of cryptosporidiosis reported in New York City during 1994, most (72%) occurred among men and among persons aged 20-44 years (63%).

Laboratory B, a commercial laboratory in New York City, examines approximately 400 stool specimens per month for ova and parasites. Although these examinations do not routinely include testing for Cryptosporidium parvum, requests for this test increased fourfold from 1993 (143 {3%} of 4344) to 1995 (587 {11%} of 5333). Before April 1995, laboratory B used a modified acid-fast technique to test for Cryptosporidium oocysts. From January 1994 through March 1995, laboratory B reported four cases of cryptosporidiosis.

In April 1995, after switching to an enzyme-linked immunosorbent assay (ELISA) method to test for Cryptosporidium antigen (ProSpecT Cryptosporidium Microplate Assay 21/96, Alexon Incorporated, Sunnyvale, California *), laboratory B began reporting an increased number of positive tests: 24 in April and a mean of 52 per month from May through September, for a total of 281 in 6 months. Demographic characteristics of these 281 patients differed from those of patients reported to have been positive by other New York City laboratories; specifically, patients who were test-positive by laboratory B were more likely to be aged greater than or equal to 60 years (36% versus 5%, pless than 0.01) and female (59% versus 25%, pless than 0.01).

Because of these findings, in August 1995 the NYCDOH initiated a validation study at laboratory B. Stool specimens submitted to laboratory B for Cryptosporidium testing were split and sent for parallel testing either to the New York City Bureau of Laboratories, which performed ELISA and acid-fast testing, or to the New York State Wadsworth Center, David Axelrod Institute for Public Health, which performed ELISA, direct immunofluorescence testing (MERIFLUOR Cryptosporidium/Giardia Direct Immunofluorescent Detection Procedure, Meridian Diagnostics Incorporated, Cincinnati, Ohio), and modified acid-fast testing. ELISA testing was performed with the same kit used by laboratory B. Of 84 split specimens, laboratory B reported 57 (68%) positive test results, and the two reference laboratories each reported one positive result. Based on these findings, all 280 unconfirmed positive ELISA results for Cryptosporidium identified at laboratory B from April through September were considered to have been false positives. Physicians for these patients were notified that previously reported positive results may have been the result of laboratory error.

Reported by: CR Sterling, PhD, YR Ortega, MS, Dept of Veterinary Science, Univ of Arizona, Tucson. EC Hartwig, Jr, ScD, MB Pawlowicz, MS, MT Cook, Bur of Laboratories, RS Hopkins, MD, State Epidemiologist, Florida Dept of Health. JR Miller, MD, M Layton, MD, Bur of Communicable Disease, A Ebrahimzadeh, PhD, Bur of Laboratories, New York City Dept of Health; J Ennis, J Keithly, PhD, Wadsworth Center, David Axelrod Institute for Public Health, New York State Dept of Health. Div of Parasitic Diseases, National Center for Infectious Diseases; Div of Applied Public Health Training (proposed), Epidemiology Program Office, CDC.

Editorial Note: Pseudo-outbreaks associated with increased surveillance for disease and outbreaks of pseudo-infection resulting from false-positive laboratory results may occur with increasing frequency because of changes in surveillance and rapid developments in the technologies and tests available to identify organisms -- particularly new and emerging pathogens. Cyclosporiasis and cryptosporidiosis are emerging infectious diseases in the United States, and many health departments and laboratories are unfamiliar with the identification and diagnosis of these parasitic infections. The outbreaks described in this report resulted from a combination of laboratory error and enhanced surveillance -- factors identified in previous pseudo-outbreaks (2,10). Consequences of such events include misdirection of resources; unnecessary treatment, anxiety, and disruption of patients' lives; and loss of confidence in laboratories and public health agencies (3).

To prevent the occurrence or minimize the impact of pseudo-outbreaks, the first steps in most outbreak investigations should be to confirm the diagnosis and the occurrence of the outbreak. Confirmation of the diagnosis entails validation of the laboratory findings and assessment of the concordance between the clinical features and test results. In the Florida investigation, an outbreak of pseudo-infection was suspected initially because of an inability to document an association between patients with specific clinical manifestations and positive findings for Cyclospora oocysts, while in New York City, the investigation was prompted, in part, by the atypical demographic characteristics of patients with cryptosporidiosis reported by one laboratory.

Although such patterns were important in the investigations described in this report, they typically are more reliable for well-characterized pathogens than for emerging pathogens, for which critical epidemiologic and clinical information may be limited. Because local laboratories may lack experience and optimal techniques for identifying emerging pathogens, these organisms may be more likely to be associated with outbreaks of pseudo-infection; therefore, confirmation of the diagnosis by an experienced reference laboratory may be critical in confirming outbreaks associated with these pathogens.

The outbreaks of pseudo-infection in Florida and New York City began after laboratory personnel implemented new testing procedures -- in one instance, for a newly-identified pathogen and, in the other, with a different technique. The investigations of these incidents emphasize the potential for the occurrence of such outbreaks when efforts are made to enhance laboratory surveillance. In addition, these incidents indicate the needs for training and proficiency testing in conjunction with the introduction of new laboratory techniques and for reporting laboratories to submit a proportion of their positive and negative specimens for confirmation by a reference laboratory following the initiation of surveillance or testing for new pathogens.

References
1. CDC. Addressing emerging infectious disease threats: a prevention strategy for the United States. Atlanta, Georgia: US Department of Health and Human Services, Public Health Service, 1994.
2. CDC. Enhanced detection of sporadic Escherichia coli O157:H7 infections -- New Jersey, July 1994. MMWR 1995;44:417-8.
3. Shears P. Pseudo-outbreaks. Lancet 1996;347:138.
4. Soave R. Cyclospora: an overview. Clin Infect Dis 1996;23:429-35.
5. Hoge CW, Shlim DR, Ghimire M, et al. Placebo-controlled trial of co-trimoxazole for Cyclospora infections among travelers and foreign residents in Nepal. Lancet 1995;345:691-3.
6. Wurtz R. Cyclospora: a newly identified intestinal pathogen of humans. Clin Infect Dis 1994;18:620-3.

7. Koumans EH, Katz D, Malecki J, et al. Novel parasite and mode of transmission: Cyclospora infection -- Florida {Abstract}. In: Program and abstracts of the 45th Annual Epidemic Intelligence Service (EIS) Conference. Atlanta, Georgia: US Department of Health and Human Services, Public Health Service, CDC, 1996.

8. Garcia LS, Bruckner DA. Diagnostic medical parasitology. 2nd ed. Washington, DC: American Society for Microbiology, 1993:528-32.

9. CDC. Update: outbreaks of Cyclospora cayetanensis infection -- United States and Canada, 1996. MMWR 1996;45:611-2.

10. Casemore DP. A pseudo-outbreak of cryptosporidiosis. Commun Dis Rep Rev 1992;2:R66-R67.

July 19, 1996 / 45(28);611-612

Update: Outbreaks of Cyclospora cayetanensis Infection -- United States and Canada, 1996

Since May 1996, CDC has received reports of clusters and sporadic cases of infection with the parasite Cyclospora cayetanensis that occurred in May and June in the United States and Canada (1). This report describes preliminary findings of an investigation by the New Jersey Department of Health and Senior Services (NJDHSS) and updates the findings of other ongoing investigations.

NEW JERSEY

During June 17-26, 1996, NJDHSS received reports of 42 sporadic cases of laboratory-confirmed Cyclospora infection (by light microscopic examination of a stool specimen) among New Jersey residents. To assess possible risk factors for infection among persons with sporadic cases, NJDHSS conducted a case-control study. A case was defined as laboratory-confirmed Cyclospora infection and symptoms of gastroenteritis (e.g., diarrhea) with onset during May 1-June 20, 1996, in a New Jersey resident aged greater than or equal to 18 years. Two age-matched (plus or minus 10 years) controls (aged greater than or equal to 18 years) were selected by random-digit dialing; to be eligible, controls could not have had loose stools during the 2-week period before onset of symptoms for the referent case-patient (i.e., the period of interest). In addition, case-patients and matched controls must have been in New Jersey during the period of interest and not have traveled outside the United States or Canada during the month before symptom onset. Investigators interviewed 30 case-patients and 60 controls by telephone and used a standardized questionnaire that asked about possible exposures (including consumption of 17 fruits and 15 vegetables, water and soil exposures, and animal contact) during the period of interest.

Case-patients and controls were similar by age (median age of case-patients: 47.5 years {range: 20-81 years}), sex, and educational level. Twenty (69%) of 29 case-patients and four (7%) of 60 controls had eaten raspberries. In multivariate conditional logistic regression analysis, only consumption of raspberries was significantly associated with illness (odds ratio and 95% confidence interval were undefined because of a denominator of 0, p less than 0.001 {computed using the score test}). Consumption of strawberries was not significantly associated with illness. Other Investigations

Approximately 850 cases of laboratory-confirmed Cyclospora infection in persons residing in the United States and Canada whose onset of illness was in May and June 1996 have been reported to CDC and Health Canada. Approximately 14% of all cases have been reported from Ontario, Canada; nearly all (approximately 99%) of the other cases have been reported from states east of the Rocky Mountains. Fourteen states, the District of Columbia, and Ontario are each investigating clusters of cases related to specific events (e.g., a luncheon) and/or at least 30 sporadic cases (i.e., not related to any identified event). Six other states have each reported less than or equal to 10 sporadic cases. Most sporadic and event-related cases have occurred in immunocompetent adults. Fifteen case-patients have been hospitalized, but no deaths have been reported. The most recent event associated with cases occurred on June 8 (i.e., exposure date), and the most recent laboratory-confirmed sporadic case occurred in a person with onset of symptoms on June 27.

With the possible exception of a few events for which limited information is available, raspberries were served at the 42 events under investigation. For 12 (29%) of the events, raspberries were either the only berry served or were served separately from other berries. Initial investigations of three events that occurred in May had attributed risk for Cyclospora infection to consumption of strawberries; however, further investigation indicated that raspberries and other berries also were served (one event) or may have been served (two events). Preliminary findings of case-control studies by health departments in Florida and New York City also indicate an association between consumption of raspberries and risk for Cyclospora infection.

The Food and Drug Administration (FDA), CDC, and other health and food-safety agencies in the United States and Canada are tracing the sources of the raspberries that were served at the events. Findings from the first 21 tracebacks completed by CDC and state agencies indicate that raspberries grown in some regions of Guatemala either definitely were or could have been served at each of these events; for 17 of these 21 events, the only source of raspberries was Guatemala. Efforts are ongoing to identify the specific source(s) of the raspberries and possible modes of contamination.

Reported by: Health Protection Br, Health Canada. J Hofmann, MD, Z Liu, MD, C Genese, MBA, G Wolf, MBA, W Manley, MA, K Pilot, E Dalley, MA, L Finelli, DrPH, Acting State Epidemiologist, New Jersey Dept of Health and Senior Svcs. Prevention Effectiveness Activity, Office of the Director, and Div of Field Epidemiology, Epidemiology Program Office; Foodborne and Diarrheal Diseases Br, Div of Bacterial and Mycotic Diseases, and Div of Parasitic Diseases, National Center for Infectious Diseases, CDC.

Editorial Note: The multistate outbreak of infection with the emerging pathogen Cyclospora has been investigated by state and local health departments, CDC, health officials in Canada, and other organizations. Although the findings of these investigations have demonstrated consistent associations between risk for Cyclospora infection and antecedent consumption of raspberries, some case-patients have not reported raspberry consumption; this finding may reflect poor recall and, for some persons with cases not related to events, different sources of infection.

The preliminary investigations indicate that some regions of Guatemala were the most likely sources of the epidemiologically implicated raspberries. The growing season in Guatemala is ending, and recent imports of raspberries from that country have markedly decreased. The specific mode of contamination of the raspberries and whether contamination occurred in Guatemala or after the raspberries had been shipped from the country have not yet been determined. CDC, FDA, the government of Guatemala, growers, exporters, and trade associations are collaborating in ongoing investigations to evaluate these issues. Since the latter half of June, FDA has begun to examine shipments of raspberries from Guatemala for Cyclospora. Cyclospora oocysts have not been found on any of the raspberries that have been tested to date. FDA, CDC, and others are developing standardized methods for such testing and are evaluating their sensitivity.

As always, produce should be thoroughly washed before it is eaten. This practice should decrease but may not eliminate the risk for transmission of Cyclospora. Health departments that identify cases of Cyclospora infection should contact CDC's Division of Parasitic Diseases, National Center for Infectious Disease, telephone (770) 488-7760.

Reference
1. CDC. Outbreaks of Cyclospora cayetanensis infection -- United States, 1996. MMWR 1996; 45:549-51.

June 28, 1996 / 45(25);549-551

Outbreaks of Cyclospora cayetanensis Infection -- United States, 1996

Cyclospora cayetanensis (previously termed cyanobacterium-like body) is a recently characterized coccidian parasite (1); the first known cases of infection in humans were diagnosed in 1977 (2). Before 1996, only three outbreaks of Cyclospora infection had been reported in the United States (3-5). This report describes the preliminary findings of an ongoing outbreak investigation by the South Carolina Department of Health and Environmental Control (SCDHEC) and summarizes the findings from investigations in other states. South Carolina

On June 14, the SCDHEC was notified of diarrheal illness among persons who attended a luncheon near Charleston on May 23. A case of Cyclospora infection was defined as diarrhea (three or more loose stools per day or two or more stools per day if using antimotility drugs) after attending the luncheon. All 64 attendees were interviewed. Of the 64 persons, 37 (58%) had Cyclospora infection, including seven with laboratory-confirmed infection. The median incubation period was 7.5 days (range: 1-23 days).

Based on univariate analysis by the SCDHEC, food items associated with illness included raspberries (RR=5.6; 95% CI=2.3-13.7), strawberries (RR=2.2; 95% CI=1.0-5.1), and potato salad (RR=1.9; 95% CI=1.3-2.7). On May 23, a total of 95 persons attended a luncheon in an adjacent room and were served strawberries obtained from the same source but were not served raspberries; no cases were identified among these persons. One person who ate raspberries at the establishment that evening developed laboratory-confirmed infection; she had not attended either luncheon or eaten strawberries. Other investigations

In May and June 1996, social event-related clusters of cases and/or sporadic cases of Cyclospora infection were reported in at least 10 states and in Ontario, Canada. Several hundred laboratory-confirmed cases have been reported to CDC. Most cases have occurred in immunocompetent adults.

Preliminary evidence suggests that, in these outbreaks, consumption of fresh fruit -- raspberries and mixtures of berries and other fruits (precluding determination of which fruit in the mixture was associated with illness) -- may be associated with Cyclospora infection. CDC, the Food and Drug Administration (FDA), and health officials in state and local health departments and Canada are collaborating to determine the extent and causes of the outbreaks, the sources of contamination, and whether transmission is ongoing. Additional efforts include the use of the five-site CDC/U.S. Department of Agriculture/FDA active foodborne diseases surveillance network (established in 1995; collaborating sites include Atlanta and portions of California, Connecticut, Minnesota, and Oregon). Although standardized methods are not yet available, FDA, CDC, and others are testing samples of produce for Cyclospora.

Reported by: J Chambers, MD, S Somerfeldt, MS, L Mackey, S Nichols, MS, Trident Health District; R Ball, MD, D Roberts, MPH, N Dufford, MS, A Reddick, PhD, J Gibson, MD, State Epidemiologist, South Carolina Dept of Health and Environmental Control. Center for Food Safety and Applied Nutrition, and Office of Regulatory Affairs, Food and Drug Administration. Div of Field Epidemiology, Epidemiology Program Office; Foodborne and Diarrheal Diseases Br, Div of Bacterial and Mycotic Diseases, and Epidemiology Br, Div of Parasitic Diseases, National Center for Infectious Diseases, CDC.

Editorial Note: Although Cyclospora is transmitted by the fecal-oral route, direct person-to-person transmission is unlikely because excreted oocysts require days to weeks under favorable environmental conditions to become infectious (i.e., sporulate). Whether animals serve as sources of infection for humans is unknown. Most reported cases have occurred during spring and summer. The average incubation period is 1 week, and illness may be protracted (from days to weeks) with frequent, watery stools and other gastrointestinal symptoms; symptoms may remit and relapse.

The diameter of Cyclospora oocysts is 8-10 um, approximately twice that of Cryptosporidium parvum. Oocysts can be identified in stool by examination of wet mounts under phase microscopy, use of modified acid-fast stains (oocysts are variably acid-fast), or demonstration of autofluorescence with ultraviolet epifluorescence microscopy. However, these procedures are not routine for most clinical laboratories, and confirmation of the diagnosis by an experienced reference laboratory is recommended. Demonstration of sporulation provides definitive evidence for the diagnosis (1). Infection with Cyclospora can be treated with a 7-day course of oral trimethoprim (TMP)-sulfamethoxazole (SMX) (for adults, TMP 160 mg plus SMX 800 mg twice daily; for children, TMP 5 mg/kg plus SMX 25 mg/kg twice daily) (6). Treatment regimens for patients who cannot tolerate sulfa drugs have not yet been identified.

The preliminary findings of these investigations suggest that consumption of some fresh fruits has been associated with increased risk for illness. However, the investigations have not yet determined specific sources or modes of contamination. Potential sources of infection include seasonal produce that orginates from different domestic and international locations at different times of the year; the complex distribution routes and handling of these foods complicate tracebacks and other key aspects of the investgations. As always, produce to be eaten raw should be thoroughly washed. This practice may not entirely eliminate the risk of transmission of Cyclospora. Health-care providers should consider Cyclospora infection in persons with prolonged diarrheal illness and specifically request laboratory testing for this parasite; cases should be reported to local and state health departments. Health departments that identify cases of Cyclospora infection should contact CDC's Division of Parasitic Diseases, National Center for Infectious Diseases, telephone (770) 488-7760.

References
1. Ortega YR, Sterling CR, Gilman RH, Cama VA, Daz F. Cyclospora species -- a new protozoan pathogen of humans. N Engl J Med 1993;328:1308-12.
2. Ashford RW. Occurrence of an undescribed coccidian in man in Papua New Guinea. Ann Trop Med Parasitol 1979;73:497-500.
3. Huang P, Weber JT, Sosin DM, et al. The first reported outbreak of diarrheal illness associated with Cyclospora in the United States. Ann Intern Med 1995;123:409-14.
4. Carter RJ, Guido F, Jacquette G, Rapoport M. Outbreak of cyclosporiasis at a country club -- New York, 1995 {Abstract}. In: 45th Annual Epidemic Intelligence Service (EIS) Conference. Atlanta, Georgia: US Department of Health and Human Services, Public Health Service, April 1996:58.
5. Koumans EH, Katz D, Malecki J, et al. Novel parasite and mode of transmission: Cyclospora infection -- Florida {Abstract}. In: 45th Annual Epidemic Intelligence Service (EIS) Conference. Atlanta, Georgia: US Department of Health and Human Services, Public Health Service, April 1996:60.
6. Hoge CW, Shlim DR, Ghimire M, et al. Placebo-controlled trial of co-trimoxazole for cyclospora infections among travellers and foreign residents in Nepal. Lancet 1995;345:691-3.

July 23, 1982 / 31(28);383-4,389

Epidemiologic Notes and Reports Intestinal Perforation Caused by Larval Eustrongylides -- Maryland

CDC recently received reports that three fishermen in Baltimore, Maryland, swallowed live minnows and developed severe abdominal pain within 24 hours.

Patient 1, a 23-year-old male, was seen at a community hospital on March 21, 1982, 2 days after swallowing two live minnows, because of progressive abdominal cramping pain of 24-hours' duration. During surgery, two roundworms were found, one penetrating the cecum, the other in the abdominal cavity. The transverse colon was found to be ecchymotic with punctate hemorrhage and exudates. On April 7, patient 2, a 25-year-old fisherman, was brought to the emergency room of the same hospital with similar symptoms 24 hours after swallowing one minnow. At laparotomy on April 9, two roundworms were found near a perforated cecum. Patient 3, a fisherman who swallowed minnows from the same source, later developed similar symptoms, which resolved 4 days later without surgery. Twelve other persons who also ingested live minnows reported no symptoms during 4 weeks of follow-up.

Sixty-seven minnows, collected in East Baltimore waters and secured from the same store at which the patients obtained their fish, were examined; 32 (48%) were infected with roundworms identical to those recovered from the two patients described above. Of the infected fish, six had two worms, one had three worms, and 26 had one worm each. The worms, 1-2 mm in diameter and 80-120 mm long, were identified as 4th-stage larval nematodes of the genus Eustrongylides.* Reported by PF Guerin, MD, S Marapudi, MD, L McGrail, RN, CL Moravec, MD, E Schiller, DSc, Baltimore, EW Hopf, MD, R Thompson, Baltimore County Health Dept, FYC Lin, MD, E Israel, MD, State Epidemiologist, Maryland State Health Dept; JW Bier, PhD, GJ Jackson, PhD, Bureau of Foods, Div of Microbiology, US Food and Drug Adminstration; Parasitic Diseases Div, Center for Infectious Diseases, CDC.

Editorial Note: Nematodes of the genus Eustrongylides (Family Dioctophymidae Railliet, 1915) are parasitic as adults in the gastrointestinal tract of fish-eating birds and as larvae in the connective tissue or body cavity of freshwater fish (1). Amphibians, reptiles, and mammals (rarely) may become infected with larval Eustrongylides spp. and may play an ecological role as paratenic or transport hosts. Moreover, extensive larval migration in accidentally and experimentally infected reptilian, amphibian, and avian hosts has been observed and has sometimes been associated with high mortality (1-3), suggesting a possible pathologic role for Eustrongylides spp. However, no human infections have been reported to CDC.** Although data are incomplete, infection by larval Eustrongylides spp. is widespread and common in numerous species of freshwater fish. The high rates of infection for minnows (Fundulus spp.) reported here and earlier (3) may indicate a high degree of risk for persons who choose to eat these fish without cooking them first.

References
1. Lichtenfels JR, Lavies B. Mortality in red-sided garter snakes, Thamnophis sirtalis parietalis, due to larval nematode, Eustrongylides sp. Lab Anim Sci 1976;26:465-7.
2. Abram JB, Lichtenfels JR. Larval Eustrongylides sp. (Nematoda: Dioctophymatoidea) from otter, Lutra canadensis, in Maryland. Proceeding of the Helminthological Society of Washington 1974;41:253.
3. Von Brand T, Cullinan RP. Physiological observations upon a larval Eustrongylides. V. The behavior in abnormal warmblooded hosts. Proceeding of the Helminthological Society of Washington 1943; 10:29-33. *Larval specimens have been deposited with the U.S. Department of Agriculture (USDA) in the U.S. National Museum, Helminthological Collection. **The USDA National Helminthological Collection contains a single larval specimen obtained from a human (2).

June 27, 1986 / 35(25);405-8

Acanthamoeba Keratitis Associated with Contact Lenses -- United States

Twenty-four patients with Acanthamoeba keratitis have been reported to CDC from 14 states in the last 9 months (Table 1). Although onset of illness for some patients dates to as early as 1982, most had onset of illness in 1985 or 1986. In two patients, the infected eye was enucleated; 12 patients underwent corneal transplantation.

Twenty (83%) of the patients wore contact lenses. Of these, two wore hard lenses (one hard, the other rigid gas-permeable); four wore extended-wear soft lenses; and 14 wore daily-wear soft lenses. Ten of these 20 patients cleaned their lenses with home-made saline solution prepared by mixing salt tablets with bottled, distilled, nonsterile water; four used commercially available lens-cleaning solutions followed by a tap water rinse; one used commercial bottled saline; and one cleaned lenses with tap water pumped from a private well. No lens-care information was available for four patients.

Twenty-two (90%) of the 24 patients were initially diagnosed as having corneal herpes simplex virus (HSV) infections; in the other two patients, corneal lesions were attributed to autoimmune disease. Acanthamoeba keratitis was diagnosed by examination of stained corneal scrapings or tissues (67%) and/or tissue indirect fluorescent antibody (IFA) test (52%) using species-specific antisera. Acanthamoebae were isolated from the corneal scrapings/biopsies of 17 (71%) of the patients. Three of the 17 patients' lens cases containing home-made saline solution were also cultured; all were positive for Acanthamoeba. Contact lens cases from other patients were not cultured. Patients' ages ranged from 17 years to 55 years; half were females. The right eye was affected in 13 (54%) patients and the left eye, in 11. A. castellanii was identified from nine (38%); A. polyphaga, from eight (33%); A. rhysodes, from four (17%); A. culbertsoni, from three (13%); and A. hatchetti, from one (4%). The species of Acanthamoeba was not determined for six (25%) patients. More than one species of Acanthamoeba was cultured from samples from four patients. Reported by C Newton, MD, Louisville, Kentucky; WT Driebe, Jr, MD, University of Florida, Gainesville, LR Groden, MD, G Genvert, MD, JH Brensen, PhD, University of South Florida, Tampa; AD Proia, MD, GK Clintworth, MD, M Cobo, MD, D Klein, PhD, Duke University Medical Center, Durham, P Morton, MD, Raleigh, North Carolina Dept of Human Resources; T Wolf, MD, University of Oklahoma, Oklahoma City; DB Jones, MD, RL Font, MD, M Osata, PhD, Baylor College of Medicine, Houston, MC Kincaid, University Health Science Center at San Antonio, MB Moore, MD, R Silvany, University of Texas Health Science Center at Dallas, Texas; RJ Epstein, MD, LA Wilson,

MD, Emory University, Atlanta, Georgia; RA Miller, MD, P Gardner, MD, RC Tripathi, MD, DF Sahm, PhD, University of Chicago, Illinois; JS Wolfson, MD, S Foster, MD, MA Waldrom, Massachusetts General Hospital and Harvard University, Boston; CF Bahn, MD, Naval Hospital, Dept of the Navy, Bethesda, Maryland; G Rao, MD, FS Nolte, PhD, University of Rochester Medical Center, Rochester, New York; C Parlato, MD, JC Davis, PhD, Mountainside Hospital, Montclair, New Jersey; E Cohen, MD, Wills Eye Hospital and Thomas Jefferson University, Philadelphia, Pennsylvania; MJ Mannis, MD, CE Thirkill, PhD, University of California, Davis; Protozoal Disease Br, Div of Parasitic Diseases, Center for Infectious Diseases, CDC.

Editorial Note: Members of the genus Acanthamoeba are the most common free-living amoebae in fresh water and soil. They have been isolated from brackish and sea water, airborne dust, and hot tubs. Acanthamoebae have also been recovered from the nose and throat of humans with impaired respiratory function and from apparently healthy persons, suggesting that these organisms are commonly inhaled (1). It is, therefore, not surprising that acanthamoebae may contaminate contact lenses or lens-cleaning/soaking fluids.

The first case of Acanthamoeba keratitis in the United States was reported in 1973 in a South Texas rancher with a history of trauma to his right eye (1). A. polyphaga was repeatedly cultured from his cornea, and both trophozoite and cyst forms of the organism were demonstrated in the corneal sections. Since then, 31 patients have been diagnosed in the United States (excluding those reported here). Nineteen of these 31 cases have been published (2-12); seven occurred before 1981; four occurred in 1981; one, in 1982; five, in 1983; and two, in 1984. The 24 Acanthamoeba keratitis cases described here represent a striking increase over those reported in previous years. A similar increase has been observed in the use of contact lenses during the past 5 years, from 14.5 million in 1980 to 23.1 million in 1985 (13).

Review of the 19 published cases indicates that nearly all infections were preceded by some degree of ocular trauma and/or exposure to contaminated water. Only recently has it been suggested that wearing contact lenses or using contaminated lens-cleaning/soaking solution may predipose the wearer to developing Acanthamoeba keratitis (10). Although information on contact lens use was not specified in all the published reports, at least 13 of the 19 patients were known users, and in the present report, 20 (83%) of 24 patients wore contact lenses.

Acanthamoebae are resistant to killing by freezing, dessication, a variety of antimicrobial agents, and levels of chlorine that are routinely used to disinfect municipal drinking water, swimming pools, and hot tubs (14). Recent studies indicate that thermal disinfection systems for contact lenses are superior to cold chemical disinfection in preventing the growth of Acanthamoeba (15). Although 10 of the 20 patients who wore contact lenses used home-made saline cleaning solutions, it is not known how many of them heat-sterilized the solutions before use.

Since the clinical characteristics of Acanthamoeba keratitis, especially the irregular epithelial lesions, the stromal infiltrative keratitis, and edema seen in most patients may resemble HSV keratitis, many patients are initially diagnosed and treated for this infection. Until recently, the correct diagnosis was made only after detailed histologic examination of corneal tissue removed at the time of transplantation. The following clinical features are suggestive of Acanthamoeba keratitis: (1) severe ocular pain; (2) a characteristic 360-degree or partial paracentral stromal ring infiltrate; (3) recurrent corneal epithelial breakdown; and (4) a corneal lesion refractory to the usual medications. The diagnosis can be confirmed by vigorously scraping the cornea with a swab or platinum-tipped spatula, staining the material obtained with Giemsa or trichrome stain, and examining it at 400X with a standard light microscope. In addition, some of the corneal scrapings should be cultured on non-nutrient agar seeded with Escherichia coli (1).

Medical management of Acanthamoeba keratitis is complicated by the resistance of these organisms to most of the commonly used antibacterial, antifungal, antiprotozoal, and antiviral agents. Although some patients have recently been treated successfully using ketoconazole, miconazole, and propamidine isethionate (Brolene*), penetrating keratoplasty usually has been necessary to recover useful vision (5,7-11). Further studies are needed to better estimate the true risk of infection, to improve diagnostic and treatment methods, and to evaluate the ability of different lens cleaning/soaking solutions to prevent growth of Acanthamoeba.

References
1. Visvesvara GS. Free-living pathogenic amoebae. In: Lennette EH, Balows A, Hausler, WJ Jr, Truant JP, eds. Manual of Clinical Microbiology, 3rd edition. 1980:704-8.
2. Jones DB, Visvesvara GS, Robinson NM. Acanthamoeba polyphaga keratitis and Acanthamoeba uveitis associated with fatal meningoencephalitis. Trans Ophthalmol Soc UK 1975;95:221-32.
3. Key SN, III, Green WR, Willaert E, Stevens AR, Key SN, Jr. Keratitis due to Acanthamoeba castellanii: a clinicopathologic case report. Arch Ophthalmol 1980;98:475-9.
4. Ma P, Willaert E, Juechter KB, Stevens AR. A case of keratitis due to Acanthameoba in New York, New York, and features of 10 cases. J Infect Dis 1981;143:662-7.
5. Hirst LW, Green WR, Merz W, et al. Management of Acanthamoeba keratitis. A case report and review of the literature. Ophthalmology 1984;91:1105-11.
6. Blackman HJ, Rao NA, Lemp MA, Visvesvara GS. Acanthamoeba keratitis successfully treated with penetrating keratoplasty: suggested immunogenic mechanisms of action. Cornea 1984:3:125-30.
7. Samples JR, Binder PS, Luibel FJ, Font RL, Visvesvara GS, Peter CR. Acanthamoeba keratitis possibly acquired from a hot tub. Arch Ophthalmol 1984;102:707-10.
8. Scully RE, Mark EJ, McNealy BN, et al. Case 10-1985. N Engl J Med 1985;312:634-41.
9. Cohen EJ, Buchanan HW, Laughrea P, et al. Diagnosis and management of Acanthamoeba keratitis. Am J Ophthalmol 1985;100:389-95.
10. Moore MB, McCulley JP, Luckenbach M, et al. Acanthamoeba keratitis associated with soft contact lenses. Am J Ophthalmol 1985;100:396-403.
11. Theodore FH, Jakobiec FA, Juechter KB, et al. The diagnostic.

VIRUSES

July 16, 1993 / 42(27);526-529

Foodborne Hepatitis A -- Missouri, Wisconsin, and Alaska, 1990-1992

Person-to-person spread is the predominant mode of transmission of hepatitis A virus (HAV) infection. However, based on findings for national surveillance for viral hepatitis, since 1983, 3%-8% of reported hepatitis A cases have been associated with suspected or confirmed foodborne or waterborne outbreaks (1). This report summarizes three recent foodborne outbreaks of hepatitis A and addresses the prevention of this problem.

MISSOURI

On November 26, 1990, hepatitis A was diagnosed in an employee of a restaurant in Cass County, Missouri. The employee's duties involved washing pots and pans in the restaurant. From December 7, 1990, through January 9, 1991, hepatitis A was diagnosed in 110 persons, including four waitresses, who had eaten at the restaurant; two persons died as a result of fulminant hepatitis.

To identify risk factors for hepatitis A in restaurant patrons, CDC, in collaboration with the Missouri Department of Health (MDH), conducted a case-control study. A case was defined as an anti-HAV immunoglobulin M (IgM)-positive diagnosis in a person who had eaten at the restaurant three or more times during the 6-week period before onset of illness. Eating companions of case-patients were selected as controls. Twenty-three case-patients and 31 controls were included. Case-patients were asked about risk factors for hepatitis A (including contact {i.e., sexual, household, or other} with a person with hepatitis A, employment as a food handler, injecting-drug use, recent international travel, association with child care centers, consumption of raw shellfish, and eating at other restaurants in town) during the 2-6 weeks before onset of illness. Foods at the restaurant that were either uncooked or were handled after cooking were included in a food-history questionnaire.

Case-patients were more likely than controls to have consumed a salad (odds ratio {OR}=8.6; 95% confidence interval {CI}=2.0- 40.6). In addition, case-patients (100%) were more likely than controls (48%) to have eaten lettuce, either in a salad or as a garnish for a sandwich (OR=undefined; lower 95% confidence limit=6.2). On follow-up interview, the index case-patient reported that he occasionally helped unpack fresh produce and prepare lettuce for salads. From December 1990 through January 1991, immune globulin (IG) was administered to 22 restaurant employees and approximately 3000 potentially exposed restaurant patrons. No cases of hepatitis A were reported among restaurant patrons after January 9, 1991.

WISCONSIN

On April 10, 1991, a food handler employed at sandwich shops in downtown Milwaukee and at a university campus sought medical attention following onset of fatigue, loss of appetite, diarrhea, and fever. He was jaundiced and excluded from work. Acute hepatitis A was diagnosed serologically, and the case was reported to the Milwaukee Health Department (MHD).

Inspection by the MHD of the downtown shop found no health-code violations, and medical histories and serologies obtained from other employees were negative for evidence of hepatitis A. The case-patient reported his hygiene to be good, although this report could not be confirmed by his supervisor. His coworkers received prophylaxis with IG. Because of the report of good hygiene and a good report following inspection of the facility, the risk to patrons was considered minimal. Because 2 weeks had elapsed since the employee had last worked in the campus sandwich shop, this shop was not inspected, and IG was not administered to other employees.

On April 27, eight students presented to the student health service of a university in Milwaukee with symptoms of hepatitis. On April 28, 60 additional persons with hepatitis A were reported to local public health agencies. Review of food histories from these patients suggested both the downtown and university sandwich shops as probable sources. Because no new cases were identified among food handlers, and because a 2-week period had passed between the food handler's last working at the campus sandwich shop and recognition of the outbreak, IG was not offered to restaurant patrons.

The two sandwich shops were owned by the same person and received some produce from the same commercial suppliers; no other common links were identified. Although the infected food handler reported his personal hygiene to be good, one coworker and several customers reported his hygiene was poor. To prevent secondary transmission of hepatitis from shop customers who might be food handlers, more than 350 centrally located restaurants were visited by MHD inspectors and advised on proper precautions.

Overall, outbreak-related hepatitis A was diagnosed in 230 persons: 50 reported eating at the university sandwich shop and 180 reported eating at the downtown sandwich shop during April 17-May 29, 1992. The 2-week peak period for onset of jaundice (in 85% of cases) occurred approximately 1 month after the 2-week period in which the infected food handler staffed both shops. Because 228 of the 230 case-patients ate exclusively at one of the two shops and because no prepared food was shared between them, food was considered to have been contaminated independently at each site. Through July 15, one second generation case (in a household contact of a sandwich shop patron) was documented.

ALASKA
On May 4, 1992, a food handler who routinely prepared uncooked sandwiches at a fast-food restaurant in Juneau, Alaska, had onset of nausea, vomiting, and diarrhea. Although his employer instructed him not to handle food, he was allowed to continue work. On May 8, he sought medical attention and was jaundiced; IgM anti-HAV was negative. On May 18, repeat testing was positive for IgM anti-HAV. The case-patient reported his hygiene to be good, and this was confirmed by his supervisor and coworkers.

From June 1 through June 11, 11 cases of acute hepatitis A were diagnosed in residents of or visitors to Juneau. To identify risk factors for infection, the Alaska Department of Health and Social Services conducted a case-control study. A case was defined as an anti-HAV IgM-positive diagnosis in a Juneau resident or visitor with onset of illness during June 1-11. Twenty-four controls were selected from among coworkers of case-patients. Case-patients were asked about risk factors for hepatitis A, including contact (i.e., sexual, household, or other) with a person with hepatitis A, employment as a food handler, injecting-drug use, recent international travel, association with child care centers, consumption of raw shellfish, and eating at restaurants in town. All case-patients, compared with six (25%) controls, ate at least once during May 4-8 at the fast-food restaurant where the index case-patient worked (OR= undefined; lower 95% confidence limit=5.1). Because 2 weeks had elapsed between the index case-patient's onset of illness and serologic confirmation of HAV infection, IG was not administered to coworkers or restaurant patrons.

Reported by: M Skala, C Collier, CJ Hinkle, HD Donnell, Jr, MD, State Epidemiologist, Missouri Dept of Health. T Schlenker, MD, K Fessler, M Hotelling, Milwaukee Health Dept; D Hopfensperger, Div of Health, Wisconsin Dept of Health and Social Svcs. M Schloss, JP Middaugh, MD, State Epidemiologist, Alaska Dept of Health and Social Svcs. Div of Field Epidemiology, Epidemiology Program Office; Hepatitis Br, Div of Viral and Rickettsial Diseases, National Center for Infectious Diseases, CDC.

Editorial Note: Foodborne hepatitis A outbreaks are most often caused by contamination of food during preparation by an infected food handler. An important method of prevention is attention to personal hygiene, including frequent handwashing during all phases of food preparation. In addition, when hepatitis A is diagnosed in a food handler, IG should be administered to all other food handlers at the establishment. Administration of IG to patrons should be considered if 1) the infected person is directly involved in handling, without gloves, foods that will not be cooked before they are eaten; 2) the hygienic practices of the food handler are deficient or the food handler has had diarrhea; and 3) patrons can be identified and treated within 2 weeks of exposure (2,3).

The outbreaks in this report highlight several important aspects concerning recognition and reporting of persons with hepatitis A and decisions on the use of IG. Restaurant employees other than food handlers may handle food and, if infected with hepatitis A virus, pose a risk for foodborne transmission. Therefore, regardless of their job description and duties, restaurant employees with hepatitis A should be asked about any handling of uncooked food during the period that they may have been infectious.

In the Milwaukee outbreak, despite the self-reported good hygienic practices of the food handler, criteria were sufficient to recommend IG to restaurant patrons. Without the presence of diarrhea in a food handler with hepatitis A, a self-report of good hygienic practice may be inadequate to assess the level of risk to patrons. Evaluation of the hygienic practices of an infected food handler should include interviews with supervisors and coworkers.

In the outbreak in Alaska, all criteria were met for the consideration of administration of IG to restaurant customers. However, because the food handler was initially IgM anti-HAV negative at the time of jaundice, diagnosis was delayed beyond the 2-week interval for recommended use of IG. Even though specific antibody is almost always present at the time of the onset of symptoms (4-8), in food handlers with acute onset of jaundice and no identified cause, retesting for IgM anti-HAV is recommended.

Factors that are essential in the prevention and control of foodborne hepatitis A include accurate assessment of the hygienic status of food handlers; identification of food handlers and other restaurant employees with hepatitis A; and rapid diagnosis and reporting of cases in food handlers. Because IG must be administered within 2 weeks of exposure to HAV to be effective, health-care providers should promptly evaluate food handlers with symptoms of hepatitis and report food handlers with hepatitis A to appropriate public health agencies.

References
1. CDC. Hepatitis surveillance report no. 54. Atlanta: US Department of Health and Human Services, Public Health Service, 1992:16-17.
2. CDC. Protection against viral hepatitis: recommendations of the Immunization Practices Advisory Committee (ACIP). MMWR 1990;39(no. RR-2):2-5.
3. Carl M, Francis DP, Maynard JE. Food-borne hepatitis A: recommendations for control. J Infect Dis 1983;148:1133-5.
4. Lemon SM. Type A viral hepatitis: new developments in an old disease. N Engl J Med 1985;313:1059-67.
5. Decker RH, Overby LR, Ling CM, Frosner C, Deinhardt F, Boggs J. Serologic studies of transmission of hepatitis A in humans. J Infect Dis 1979;139:74-82.
6. Bradley DW, Fields HA, McCaustland KA, et al. Serodiagnosis of viral hepatitis A by a modified competitive binding radioimmunoassay for immunoglobulin M anti-hepatitis A virus. J Clin Microbiol 1979;9:120-7.
7. Lemon SM, Brown CD, Brooks DS, Simms TE, Bancroft WH. Specific immunoglobulin M response to hepatitis A virus determined by solid-phase radioimmunoassay. Infect Immun 1980;28:927-36.
8. Locarnini SA, Ferris AA, Lehmann NI, Gust ID. The antibody response following hepatitis A infection. Intervirology 1977;8:309-

----

April 13, 1990 / 39(14);228-232

Epidemiologic Notes and Reports Foodborne Hepatitis A -- Alaska, Florida, North Carolina, Washington

From 1983 through 1989, the incidence of hepatitis A in the United States increased 58% (from 9.2 to 14.5 cases per 100,000 population). Based on analysis of hepatitis A cases reported to CDC's national Viral Hepatitis Surveillance Program in 1988, 7.3% of hepatitis A cases were associated with foodborne or waterborne outbreaks (1). This report summarizes recent foodborne-related outbreaks of hepatitis A in Alaska, Florida, North Carolina, and Washington.

ALASKA
Between June 18 and July 20, 1988, 32 serologically confirmed hepatitis A cases among persons who resided in or had visited Peters Creek, Alaska (population 4000), were reported to the Alaska Department of Health and Social Services (Figure 1). Patients ranged in age from 1 to 54 years (median: 13 years). Between July 8 and August 14, 23 additional (secondary) cases occurred among household contacts of the original patients.

To examine potential sources of infection, the Alaska Department of Health and Social Services conducted a case-control study of the first 14 reported patients and 22 asymptomatic household members. All 14 patients and seven (32%) household members had consumed an ice-slush beverage purchased from a local convenience market between May 23 and June 10 (odds ratio (OR) cannot be calculated; 95% confidence interval (CI)=3.4-infinity). No other food-consumption or exposure category (including social events, restaurants, grocery stores, or international travel) was statistically associated with illness. The 18 other patients had also consumed the ice-slush beverage.

The ice-slush beverage mixture was prepared daily with tap water from a bathroom sink using utensils stored beside a toilet. All five employees of the market denied having hepatitis symptoms; four of these were tested and were negative for IgM antibody to hepatitis A virus (IgM anti-HAV). The fifth employee, who was one of the two persons who prepared the ice-slush beverage, refused to be tested. However, a household contact of this employee had had serologically confirmed hepatitis A in early June and reported that the employee had been jaundiced concurrently with her illness.

FLORIDA
In August 1988, the Alabama Department of Public Health noted an increase in cases of serologically confirmed hepatitis A in persons living in several areas of the state. Within 6 weeks before onset of illness, most affected persons had eaten raw oysters harvested from coastal waters of Bay County, Florida. The Florida Department of Health and Rehabilitative Services (FDHRS) contacted state health departments in neighboring and other states about hepatitis A cases in July or August 1988 in persons who had attended events serving seafood within 10-50 days of becoming ill. The 61 persons who were identified resided in five states: Alabama (23 persons), Florida (18), Georgia (18), Hawaii (one), and Tennessee (one). Patients ranged in age from 8 to 60 years (median: 31 years); all were white, and 49 (80%) were male. Fifty-nine (97%) had eaten raw oysters; one, raw scallops; and one, baked oysters. All the oysters and scallops were traced to the same growing area of Bay County coastal waters. The median incubation period between consumption of raw oysters and onset of illness was 29 days (range: 16-48 days).

To further study oyster consumption as a potential risk factor for hepatitis A, the FDHRS conducted a case-control study using uninfected eating companions of the patients as controls. Fifty-three patients who had serologically confirmed hepatitis A and 64 controls were interviewed by telephone; 51 (96%) of the patients and 33 (52%) of the controls had eaten raw oysters (OR=24; 95% CI=5.4-252.6). Consumption of other seafoods (i.e., clams, mussels, and shrimp) was not statistically associated with illness.

The implicated oysters apparently had been illegally harvested from outside approved coastal waters of Bay County. Sources of human fecal contamination were identified near oyster beds unapproved for harvesting and included boats with inappropriate sewage disposal systems and a local sewage treatment plant with discharges containing high levels of fecal coliforms.

NORTH CAROLINA
Beginning September 30, 1988, hepatitis A cases among employees of businesses located in east Greensboro were reported to county health departments in central North Carolina. Only day-shift employees became ill. Preliminary investigation suggested a common exposure to one nearby restaurant (restaurant A), which served as many as 400 meals per day to regular clientele. A total of 32 outbreak-associated cases was eventually reported.

The North Carolina Department of Human Resources conducted a case-control study to assess a possible association between illness and exposure to restaurant A. Twenty-seven patients and 50 controls (randomly selected from co-workers) were interviewed about exposures to different restaurants since August 15. Patients were more likely than controls to have eaten at restaurant A (OR=4.1; 95% CI=1.3-14.4). No other restaurant was statistically associated with illness.

Based on additional information obtained from 16 patients and 20 controls who reported eating lunch at restaurant A 2-6 weeks before the outbreak, only consumption of iced tea (OR=8.1; 95% CI=0.8-387.8) or hamburgers (OR=11.4; 95% CI=1.1-551.3) was associated with illness. However, 15 (94%) of the ill persons drank iced tea, whereas only six (38%) of the ill persons reported eating hamburgers.

All foodhandlers at the restaurant were tested for IgM anti-HAV; one employee, who was IgM anti-HAV-positive, denied symptoms of and risk factors for hepatitis A. However, this employee was a suspected intravenous (IV)-drug user and had job tasks that included preparation of fountain drinks and sandwiches.

Immune globulin (IG) was given to all foodhandlers at the restaurant. Because primary/secondary-case status and infectiousness of the IgM anti-HAV-positive foodhandler were unknown and because her hygiene and foodhandling practices were questionable, the local health department recommended administration of IG to all patrons who had eaten at the restaurant within 2 weeks before the association between hepatitis A and the restaurant had been determined. More than 1000 IG doses were given. The restaurant voluntarily closed for 24 days, and no persons with hepatitis A were identified with onset after November 8.

WASHINGTON
In May 1989, the Seattle-King County Department of Public Health (SKCDPH) received reports of and investigated 213 cases of hepatitis A--a threefold increase over the average of 68 cases reported in each of the first 4 months of 1989. Onsets of illness clustered during April 28-May 5. One hundred seventeen (55%) of the patients had eaten at one outlet of a Seattle-area restaurant chain (chain A). One of the patients was a recent employee and three were current employees of three of the chain's restaurants. Interviews with past and present chain A employees did not identify any worker with illness during the period of likely exposure for most patients (2-6 weeks before onset of illness). All other current workers in the three restaurants were tested for IgM anti-HAV. None were positive, and all were given IG. Because two of the ill employees had poor hygiene and had worked while ill with diarrhea, the SKCDPH recommended IG for patrons who had eaten at two of the restaurants from May 3 through May 6.

The SKCDPH conducted a case-control study to further examine the potential role of chain A restaurants in the outbreak. Sixteen patients were randomly selected and re-interviewed by telephone; 16 age-group- and sex-matched controls were obtained by increasing each patient's telephone number by one. Exposure to 11 multi-outlet restaurant chains (including chain A) was ascertained for patients during the 2-6 weeks before onset and for controls during April 14-May 12. Mean total of any restaurant visits was higher among patients (7.7) than among controls (4.3). In addition, patients (89%) were more likely than controls (25%) to have eaten at restaurants from chain A (OR=11.0; 95% CI=2.2-56.0); differences in exposure to the 10 other multi-outlet restaurants were not statistically significant.

Follow-up investigation did not detect deficiencies in sanitation practices or history of recent hepatitis among employees of chain A's distributors of foodstuffs, paper goods, and related supplies. The cause of the outbreak remains undetermined. Reported by: ME Jones, MD, SA Jenkerson, MSN, JP Middaugh, MD, State Epidemiologist, Alaska Dept of Health and Social Svcs. J Benton, MD, P Sylvester, MD, Bay County Health and Rehabilitative Public Health Unit, Panama City; KC Klontz, MD, MH Wilder, MD, RA Calder, MD, State Epidemiologist, Florida Dept of Health and Rehabilitative Svcs. CH Woernle, MD, State Epidemiologist, Alabama Dept of Public Health. RK Sikes, DVM, State Epidemiologist, Georgia Dept of Human Resources. E Veuthy, SW Wyrick, MPH, BD Weant, E Tysinger, C Rocco, MS, J Holliday, MD, Guilford County Health Dept, Greensboro; CJ Staes, MPH, RA Meriwether, MD, JN MacCormack, MD, State Epidemiologist, North Carolina Dept of Human Resources. JF Hogan, MPH, S Cummings, N Harris, DVM, CM Nolan, MD, Seattle-King County Dept of Public Health; JM Kobayashi, MD, State Epidemiologist, Washington Dept of Health. J Black, Food and Dairy Div, Oregon Dept of Agriculture; D Fleming, MD, LR Foster, MD, State Epidemiologist, State Health Div, Oregon Dept of Human Resources. Div of Field Svcs, Epidemiology Program Office; Hepatitis Br, Div of Viral and Rickettsial Diseases, Center for Infectious Diseases, CDC.

Editorial Note: The outbreaks reported here illustrate two principal modes of transmission associated with foodborne hepatitis A outbreaks: 1) contamination of food during preparation by a foodhandler infected with hepatitis A virus and 2) contamination of food, such as shellfish, before it reaches the food service establishment.

Contamination of food during preparation by a hepatitis A-infected foodhandler is the most common mode of transmission in foodborne outbreaks. The Alaska and North Carolina outbreaks are atypical in that ice or drinks as vehicles are rare; usually the vehicles are sandwiches or green salads that are not cooked or are improperly handled after cooking. The outbreak in North Carolina is also consistent with a nationwide phenomenon of increased reports of hepatitis A among IV-drug users (2), who can become sources of foodborne outbreaks if they are also foodhandlers.

Contamination of food with virus before the food reaches the service establishment is less common. Shellfish filter large quantities of water during feeding and in the process can concentrate microorganisms, including enterically transmitted viruses such as hepatitis A (3). Transmission to humans occurs when contaminated shellfish are consumed raw or undercooked. Hepatitis A outbreaks attributed to consumption of contaminated shellfish have been reported intermittently in the United States and abroad (4-8); in 1988, an outbreak associated with clams involved more than 250,000 cases in Shanghai, People's Republic of China (7). The Florida outbreak reported here is the largest attributed to shellfish in the United States since 1973 (4) and the largest ever reported in Florida. Outbreaks due to pre-retail contamination of products other than shellfish have rarely been reported. In 1988, a multifocal outbreak linked to lettuce possibly contaminated before local distribution occurred in Louisville, Kentucky (9).

Measures to prevent foodborne hepatitis A outbreaks include training of food handlers regarding proper hygiene and foodhandling practices, investigation of food handlers who have symptoms of hepatitis or are otherwise ill, prompt reporting by health-care providers to local health departments of patients with suspected foodborne hepatitis A, and prompt investigation by health departments of possible sources of infection. Consistent maintenance of good handwashing and other personal hygiene measures by foodhandlers is important because the source patient in foodborne outbreaks is often asymptomatic (as apparently occurred in North Carolina and Alaska). Prevention of hepatitis A outbreaks associated with shellfish relies on surveillance of water beds where shellfish are harvested to ensure that there is no evidence of fecal contamination. Transmission and infection from shellfish also can be prevented by thorough cooking and proper storage and handling before and after cooking.

When a foodhandler is diagnosed with hepatitis A, IG is usually recommended for other foodhandlers at the same establishment (10). IG is generally not recommended for patrons because commonsource transmission is infrequent; however, it may be considered if the infected person handles highrisk foods, has poor hygiene, or has diarrhea during the early stages of illness and if patrons can be identified and treated within 2 weeks after exposure (10). Once a foodborne hepatitis outbreak has occurred, it is usually too late to prevent further cases because the 2-week period after exposure during which IG is effective has already passed. The increasing number of hepatitis A cases nationwide underscores the importance of focusing on food handlers with hepatitis A and decisions regarding IG administration to food service patrons.

References
1. CDC. Hepatitis surveillance report no. 52. Atlanta: US Department of Health and Human Services, Public Health Service, 1989:19-21.
2. CDC. Hepatitis A among drug abusers. MMWR 1988;37:297-300,305.
3. Gerba CP, Goyal SM. Detection and occurrence of enteric viruses in shellfish: a review. J Food Protection 1978;41:743-54.
4. Portnoy BL, Mackowiak PA, Caraway CT, Walker JA, McKinley TW, Klein CA. Oyster-associated hepatitis: failure of shellfish certification programs to prevent outbreaks. JAMA 1975;233:1065-8.
5. Dienstag JL, Lucas CR, Gust ID, Wong DC, Purcell RH. Mussel-associated viral hepatitis, type A: serological confirmation. Lancet 1976;1:561-4.
6. Ohara H, Naruto H, Watanabe W, Ebisawa I. An outbreak of hepatitis A caused by consumption of raw oysters. J Hyg 1983;91:163-5.
7. Xie H, Cai Y, Davis LE. Guillain-Barre syndrome and hepatitis A: lack of association during a major epidemic. Ann Neurol 1988;24:697-8.
8. Mele A, Rastelli MG, Gill ON, et al. Recurrent epidemic hepatitis A associated with consumption of raw shellfish, probably controlled through public health measures. Am J Epidemiol 1989;130:540-6.
9. Rosenblum LS, Mirkin I, Allen D, Safford S, Hadler S. Multifocal outbreak of hepatitis A, Louisville, Kentucky (Abstract). In: Program of the Epidemic Intelligence Service 38th Annual Conference. Atlanta: US Department of Health and Human Services, Public Health Service, CDC, 1989:72. 10. Carl M, Francis DP, Maynard JE. Food-borne hepatitis A: recommendations for control. J Infect Dis 1983;148:1133-5.

December 23, 1983 / 32(50);652-4,659

Food-borne Hepatitis A -- Oklahoma, Texas

Two unrelated outbreaks of hepatitis A, involving a total of 326 people, occurred in Oklahoma and Texas during September and October 1983. Both were associated with restaurant food.

OKLAHOMA: The first outbreak occurred in Marietta in Love County (county population approximately 7,800), where 203 persons became ill from August 15 to October 10 (Figure 3). Hepatitis A was defined as: (1) jaundice or (2) serum glutamic oxalacetic transaminase enzyme (SGOT) greater than 100 mIU/ml plus nausea, vomiting, or fever or (3) a positive serum anti-hepatitis A virus (HAV) immunoglobulin (IgM). Twelve outbreak-related cases were reported elsewhere--10 in Texas and two in California. Patients ranged in age from 2 to 66 years (median 22 years); 52% were male.

Of 175 patients interviewed about exposures, 161 (92%) had eaten at a drive-in restaurant 2-6 weeks before onset of illness. Twenty-nine patients were employed as foodhandlers at eight other restaurants in town. Two worked on icing and cream-filling machines at a local bakery that distributed cookies nationwide.

The index patient, a 22-year-old foodhandler at the drive-in restaurant, developed jaundice on August 19. Investigation into his personal hygiene suggested that his handwashing practices were good, although he developed diarrhea on August 15 and continued to work up to the onset of his jaundice.

To identify risk factors of the outbreak, a survey was conducted of local high-school students. Twenty-two (13%) of 169 students who completed questionnaires had hepatitis A. The only exposure associated with illness was eating at the same drive-in restaurant during August. Twenty-one (19%) of 110 students who had eaten there became ill, compared with one (2%) of 59 who had not eaten at the restaurant (p 0.01). Attack rates increased with the number of meals eaten. No single food or drink could be implicated as a vehicle for transmission.

Most of the town's foodhandlers either had been exposed at the drive-in restaurant or were coworkers of infected foodhandlers; therefore, on September 16, the Oklahoma State Health Department recommended that immune globulin (IG) be given to patrons of five restaurants in Marietta where ill foodhandlers had prepared uncooked foods and to all foodhandlers who worked in the town. A total of 5,500 doses were given. The drive-in restaurant voluntarily closed for a month; in addition, following a U.S. Food and Drug Administration investigation, the bakery, at which two hepatitis A patients worked, voluntarily recalled selected products. No additional cases have been reported.

TEXAS: The second outbreak occurred in Lubbock, a city of 180,000 people. From October 5, through October 28, 1983, 123 physician-diagnosed cases of hepatitis A were reported to the Lubbock City Health Department. One hundred of these patients had eaten at a salad bar-type restaurant in the city 14-60 days before illness (Figure 4). Eight of the patients, including three cooks, were employed at the restaurant. Patients with restaurant-associated hepatitis A ranged in age from 7 to 64 years (mean 31 years); 65% were male; and 92% became jaundiced.

A case-control study was performed using 50 patients and 59 controls who had eaten at the restaurant only once between August 24 and September 17; controls had eaten with the patients and had sera negative for anti-HAV. Eating lettuce, tomatoes, or pickles on sandwiches was strongly associated with illness (p 0.001); eating these vegetables at the salad bar, which was prepared by different foodhandlers, was not.

Eighty-seven of the restaurant's 96 employees, including all the cooks, completed questionnaires and underwent screening for anti-HAV immunoglobulin G (IgG) and IgM. One sandwich-maker experienced nausea and vomiting in mid-September but was never jaundiced. Two of his household members contracted hepatitis A during the outbreak, despite never having eaten at the restaurant, and only he made the implicated sandwiches during periods when patients were known to have been exposed. An anti-HAV IgM drawn on November 2 was negative; however, an anti-HAV IgG was positive.

On October 8, the Lubbock City Health Department advised that the following persons receive

immune globulin (IG) as prophylaxis against hepatitis A: (1) all employees of the restaurant, (2) anyone who had eaten at the restaurant during the previous 2 weeks, and (3) all household contacts of persons with hepatitis A. Patrons were included because of the possibility of continuing food contamination by frequent sewage backups in the restaurant's kitchen. During October 1983, an estimated 15,000-20,000 doses of IG were given in the Lubbock area, mostly by private physicians.
Reported by M Gaither, JP Lofgren, MD, State Epidemiologist, Arkansas State Dept of Health; G Empey, Kern County Health Dept, AF Taylor, MPH, TG Stephenson, MPH, GA Pettersen, MD, San Bernadino Dept of Public Health, J Chin, MD, State Epidemiologist, California Dept of Health Svcs; B Baylor, MD, G Gwin, P Hunt, JT O'Connor, DO, V Smith, DO, Love County Health Center, W Baber, S Butler, R Campbell, M Claborn, P Claborn, L Douglas, MT, LL Jones, Y McGinnis, B Smith, Love County Health Dept, S Makintubee, J Mallonee, MPH, G Istre, MD, Acting State Epidemiologist, Oklahoma State Dept of Health; AB Way, MD, Lubbock City Health Dept, C Reed, MPH, L Sehulster, PhD, TL Gustafson, MD, CE Alexander, MD, Acting State Epidemiologist, Texas Dept of Health; Hepatitis Br, Div of Viral Diseases, Center for Infectious Diseases, Div of Field Svcs, Epidemiology Program Office, CDC.

Editorial Note: Hepatitis A outbreaks remain a highly visible health problem in the United States, although only a small proportion of hepatitis cases are traceable to such outbreaks. In 1982, less than 7% of hepatitis cases reported to the Viral Hepatitis Surveillance Program were associated with food-borne or waterborne outbreaks (1).

Despite substantial numbers of hepatitis A infections reported each year among foodhandlers, only a few food-borne outbreaks result from such infections. In 1982, 691 infected foodhandlers were reported to CDC, but only eight food-borne or waterborne epidemics were reported (1). This suggests that contamination of food by infected foodhandlers is uncommon. Since cooking inactivates the virus, food-borne outbreaks of hepatitis A almost always involve only foods that remain uncooked between contamination and consumption. Most authorities accept handwashing as the single, most important environmental barrier preventing transfer of virus from feces to food. As demonstrated in the first outbreak, the presence of diarrhea in the index patient may increase risk of disease transmission in spite of a history of good handwashing.

Since the 1940s, immune globulin (IG) has been used successfully in the prophylaxis of hepatitis A if given within 2 weeks of exposure (2). In established food-borne outbreaks, which are usually recognized about 4 weeks (one incubation period) after exposure has occurred, IG is generally not useful in preventing illness.

Health departments are often asked to evaluate situations in which a lone foodhandler at a restaurant has contracted hepatitis A. If the diagnosis has been confirmed by a positive serum anti-HAV IgM, IG should be administered to all other foodhandlers at the restaurant. Because of the low risk of hepatitis transmission by a foodhandler, only rarely is IG prophylaxis recommended for patrons of the restaurant. CDC has recommended that such a program not be undertaken unless the following conditions exist: (1) the foodhandler has a positive anti-HAV IgM; (2) the foodhandler handles, without gloves, cold foods that will not be cooked before consumption; (3) the foodhandler has inadequate personal hygiene, especially failure to wash hands after defecation; (4) the patrons have had repeated exposures to these foods; (5) IG can be administered within 2 weeks of the last possible exposure (3).

References
1. CDC. Unpublished data.
2. Seeff LB, Hoofnagle JH. Immunoprophylaxis of viral hepatitis. Gastroenterology 1979;77:161-82.
3. CDC. Hepatitis Surveillance Report No. 45. May 1980.

April 02, 1982 / 31(12);150-2

Epidemiologic Notes and Reports Outbreak of Food-borne Hepatitis A -- New Jersey

An increase in the number of hepatitis cases in Monmouth County, New Jersey, was reported to the New Jersey Department of Health on June 15, 1981. Investigation by state and local area health departments revealed that 56 cases of hepatitis had occurred during the first 3 weeks of June in an area of Monmouth County where the usual average is 3-4 cases/ month. Patients for whom appropriate laboratory tests had been done were confirmed to have hepatitis A.

Detailed food histories revealed that, within the appropriate incubation period for hepatitis A, 55 of the 56 patients had eaten at a Mexican style restaurant (Figure 1). Interviews of a control group matched for age, sex, and neighborhood of residence, showed that 10% of the controls ate food from this restaurant over a time period comparable with that for 98% of the patients. The restaurant agreed to close voluntarily pending further investigation.

Of the patients whose illness was related to the Mexican restaurant, 71% were male, 68% were between the ages of 15 and 29 years, and 4 were children under 15 years. A case-control study using 46 non-ill patrons revealed that patients were more likely to have eaten nachos, beans, and jalapeno peppers. Both beans and jalapeno peppers were used in preparing nachos.

Ten individuals including the 2 owners worked in the restaurant; all handled food at one time or another. Interviews on June 18 revealed that 1 employee who frequently ate food from the restaurant was ill with hepatitis at the time of the interview. Another employee had symptoms compatible with hepatitis on May 9. He had worked all day May 9, but felt too ill to work thereafter; the diagnosis of hepatitis A was confirmed for him on May 16. This employee prepared food--including grating cheese, shredding lettuce, and occasionally cutting meat--measured portions of meat, beans, jalapeno peppers, onions, cheeses, and lettuce into shells, and served the customers.

Because a food handler was recently ill with hepatitis and because the restaurant was implicated in the spread of hepatitis, immune globulin was offered to all individuals who ate in the restaurant from June 5 until it closed. A total of 1,430 people were immunized at a 2-day clinic held June 19 and 20. Reported by R Hary, Matawan Borough, S McKee, Middletown Township, S Scapricio, Hazlet Township, L. Jargowski, Monmouth County Health Dept, F Richart, Red Bank, R Altman, MD, P Marzinsky, B Mojica, MD, WE Parkin, DVM, State Epidemiologist, New Jersey State Dept of Health; Field Svcs Br, Hepatitis Laboratory Div, Center for Infectious Diseases, Field Svcs Div, Epidemiology Program Office, CDC.

Editorial Note: Hepatitis A virus (HAV) can be transmitted by food contaminated with feces from an infected food handler. If acute hepatitis A has been confirmed in a food handler by testing for IgM-specific HAV antibody, immunoglobulin prophylaxis (IG, gamma globulin) may be considered for patrons, depending on the probability of transmission of infectious virus and the probability of successful intervention in transmission by using IG. However, few food handlers actually appear to transmit disease via food, and IG prophylaxis of patrons is seldom warranted. Although for the past few years approximately 1,000 food handlers with non-B hepatitis have been reported annually to CDC, an average of 4 outbreaks of food-borne hepatitis A have been reported each year.

Transmission of HAV is affected by the amount of virus excreted by the food handler, the type of food handled, and the food handler's hygiene practices. Because the amount of virus excreted peaks 7-10 days before onset of symptoms and declines rapidly thereafter (1), food-borne outbreaks of hepatitis commonly originate from foods prepared before the food handler has clinical symptoms (2,3). As in this outbreak, most reported outbreaks have been traced to symptomatic rather than asymptomatic excreters. Uncooked foods have most frequently been associated with food-borne hepatitis because normal cooking temperatures inactivate HAV (4). However, cooked foods that were handled after cooling and foods that were contaminated and then cooked with insufficiently high internal temperature to inactivate HAV have also been implicated (2,5). Although poor hygiene practices among food handlers increase the chance of transmission of virus, outbreaks have occurred even when food handlers' personal hygiene practices were described as "acceptable" and "generally good" (3). Hygiene practices should be assessed by interviewing the ill food handler, coworkers, and employer. If deficiencies occurred and the ill food handler did not wear gloves, prophylaxis may be considered for patrons who ate implicated food items during the appropriate time period. Successful intervention in disease transmission depends on identifying persons at risk and administering IG within 2 weeks after exposure (6).

Other employees who have been regularly exposed to the index case are at risk of acquiring infection. If they do become infected, they may serve as additional sources of infection for future food consumers. These employees should be extremely conscientious in their hygiene practices, and those who handle high-risk foods should be given IG. Screening of coworkers for elevated liver enzymes or antibodies to HAV does not appear justified because the enzymes are not specific for hepatitis A, and both enzymes and antibodies appear after most virus excretion has occurred.

References
1. Bradley DW, Gravelle CR, Cook EM, Fields RM, Maynard JE. Cyclic excretion of hepatitis A virus in experimentally infected chimpanzees: biophysical characterization of the associated HAV particles. J Med Virol 1977;1:133-8.
2. Leger RT, Boyer KM, Pattison CP, Maynard JE. Hepatitis A: report of a common-source outbreak with recovery of a possible etiologic agent. I. Epidemiologic studies. J Infect Dis 1975;131:163-6.
3. Denes AE, Smith JL, Hindman SM, et al. Foodborne hepatitis A infection: a report of two urban restaurant-associated outbreaks. Am J Epidemiol 1977;105:156-62.
4. Krugman S, Gocke DJ. Viral hepatitis. Philadelphia: WB Saunders Company, 1978.
5. Peterson DA, Wolfe LG, Larkin EP, Deinhardt FW. Thermal treatment and infectivity of hepatitis A virus in human feces. J Med Virol 1978;2:201-6.
6. Brachott D, Lifschitz I, Mosley JW, Kendrick MA, Sgouris JT. Potency of fragmented IgG: two studies of postexposure prophylaxis in Type A hepatitis. J Lab Clin Med 1975;85:281-6.

September 18, 1987 / 36(36);597-602

Epidemiologic Notes and Reports Enterically Transmitted Non- A, Non-B Hepatitis -- Mexico

Two outbreaks of enterically transmitted non-A, non-B (ET-NANB) hepatitis occurred during the late summer and fall of 1986 in rural villages in the State of Morelos, Mexico. This is the first reported instance of epidemic transmission of this disease in the Americas.

HUITZILILLA, MORELOS. In September 1986, an outbreak of hepatitis among adult residents of Huitzililla, Morelos, was reported to the Mexican Secretariat of Health. A census of the 1,757 inhabitants of this rural town identified 94 persons who had developed an illness with jaundice since June 1. Onsets were between June 5 and October 16, and the overall attack rate was 5%. The outbreak lasted 20 weeks, with the peak incidence in the second week of August (Figure 1). The first case occurred about 1 month after the seasonal rains began. Ninety-eight percent of the patients had anorexia and discolored urine; 97% had malaise; 87%, abdominal pain; 78%, arthralgias; and 53%, fever. Five of the six patients for whom sera were tested had abnormal liver-function tests (alanine aminotransferase (ALT)). Two patients, both nonpregnant adult women, died. One patient was a woman in the third trimester of pregnancy; neither she nor her infant suffered any detectable complications.

The attack rate was significantly higher for persons over 15 years of age (10%) than for younger persons (1%) (pless than 0.01) but did not vary significantly by sex. Attack rates by block of residence in the town varied widely and ranged from 0% to 29%, with the highest rates being in blocks that bordered on two small streams.

The town has no system for disposal of human feces, and, at the time of investigation, human fecal material was present on the banks of both streams. The wells of families living next to the streams were very shallow (3 to 6 feet). Nineteen (56%) of the 34 well-water samples tested exceeded 2 fecal coliforms per 100 ml.

A case-control study was carried out to determine risk factors associated with illness. Thirty-two patients who had the initial case of hepatitis in their families were compared with 19 persons from families without illness. Illness was highly associated with water-related factors: families with illness were more likely than families without illness to have well-water with visible turbidity or particulate matter (91% compared with 21%; odds ratio (OR) = 36.3; 95% confidence interval (CI), 5.9 to 278.1); families with illness were less likely to have wells with protective walls (38% compared with 84%; OR = 0.11; 95% CI, 0.03 to 0.54) and were less likely to boil water for drinking (23% compared with 56%; OR = 0.23; 95% CI, 0.05 to 0.96). Contact with an ill member of a different household was also a significant risk factor. Other factors, such as consumption of specific foods and receipt of injections, were not associated with the risk of hepatitis.

Sera were collected from 62 patients and stools from 8 patients with recent onset of disease. Sixty (97%) of the serum samples were positive for antibody to hepatitis A virus (anti-HAV), but none had measurable IgM anti-HAV. None were positive for hepatitis B surface antigen (HBsAg); five (9%) were positive for antibody to hepatitis B core antigen (anti-HBc), but none were positive for IgM anti-HBc. Stool specimens were examined by immune electron microscopy (IEM) using sera from Asian patients with known ET-NANB hepatitis and sera from patients in this outbreak as an antibody source. Three of the 8 stools were positive for 28- to 34-nm viruslike particles similar to those seen by IEM in cases of ET-NANB hepatitis from Central Asia, Nepal, and Burma. In addition, a pool of the first four serum samples tested aggregated non-A, non-B hepatitis viruslike particles obtained from a patient during a recent outbreak in the Soviet Union.

TELIXTAC, MORELOS. In October, while the outbreak in Huitzililla was being investigated, a cluster of hepatitis cases among young adults was reported from Telixtac, Morelos. Telixtac is a small rural community of 2,194 inhabitants about 30 miles from Huitzililla. A census identified 129 persons who had developed jaundice since June 1. Onsets were between August 20 and January 9, and the overall attack rate was 6%. This outbreak lasted 21 weeks, with the peak incidence occurring during the third week of September (Figure 2). The first case occurred about 3 months after the beginning of the seasonal rains in May. Ninety-three percent of the patients had discolored urine; 91% had malaise; 90%, anorexia; 85%, abdominal pain; 81%, arthralgias; and 62%, fever. All three patients for whom sera were tested had abnormal liver-function tests (ALT). One person, an adult nonpregnant woman, died. One pregnant woman was ill; she made an uneventful recovery, but, 15 days after onset of icterus, she delivered a premature infant of 32 weeks gestation. The infant weighed 2.2 kg and died at 3 months of age of unknown causes.

The attack rate was significantly higher for persons over 15 years of age (10%) than for younger persons (2%) (pless than 0.01) and did not vary by sex. Attack rates varied from 0% to 40% by block of residence.

At the time of the investigation, most families were getting their drinking water from deep irrigation wells located at the edge of the community. However, during the May through September rainy season, many families had gotten water from two small streams that run through the center of the village but that are dry most of the year. Like Huitzililla, Telixtac has no system for disposal of human feces.

A case-control study similar to the one in Huitzililla was performed. Fifty-four patients who had the initial case of hepatitis in their families were compared with 67 persons from families without illness. Families with illness were more likely than families without illness to obtain drinking water from the local stream (20% compared with 2%; OR = 16.9; 95% CI, 2.1 to 98.1); families with illness were also more likely to use stream water for cooking and washing dishes (16% compared with 2%; OR = 12.6, 95% CI, 1.5 to 84.6). No risk could be demonstrated for obtaining drinking water from any of the deep wells. Contact with an ill person outside the household was also a risk factor for illness, but other factors, such as consumption of specific foods, attendance at social events, and receipt of injections, were not significantly different between patients and controls.

Sera were collected from 53 patients and stools from 8 patients with recent onset of disease. All serum samples were positive for anti-HAV antibody; only two (4%) had detectable IgM anti-HAV. None were positive for HBsAg; only one was positive for anti-HBc, but this serum was negative for IgM anti-HBc. The same IEM technique used to evaluate the stool samples from the Huitzililla patients was used in studying this outbreak. Numerous 28- to 34-nm viruslike particles similar to those detected for the Huitzililla and Asian patients with ET-NANB hepatitis were identified in one stool. These viruslike particles were aggregated by sera from the Huitzililla and Asian patients. Reported by: C Tavera, Secretariat of Health, State of Morelos; O Velazquez, MD, C Avila, MD, G Ornelas, MD, C Alvarez, MD, Field Epidemiology Training Program; J Sepulveda, MD, Director, Div of Epidemiology, Secretariat of Health, Mexico. Field Epidemiology Training Program, International Health Program Office; Hepatitis Br, Div of Viral Diseases, Center for Infectious Diseases, CDC.

Editorial Note: Non-A, non-B hepatitis is caused by at least two distinct viral agents with different modes of spread. The first, post-transfusion non-A, non-B hepatitis, is epidemiologically similar to hepatitis B and is believed to be the most common type of non-A, non-B hepatitis in North America and Europe. The second, ET-NANB hepatitis, is transmitted by the fecal-oral route and has caused large outbreaks in India (1,2), Nepal (3), Burma (4), Pakistan (5), and the Soviet Union (6). More recently, ET-NANB outbreaks have been reported from Africa (7,8). Although person-to-person transmission takes place, most of this epidemic transmission has occurred following heavy rains in populations with inadequate sewage disposal. Mortality rates for pregnant women have been as high as 20% in many of the large outbreaks (3,8).

As in other large outbreaks, disease transmission via contaminated water was important in both Mexican outbreaks. It was most apparent in Huitzililla, where the outbreak coincided with seasonal rains, and the shallow, poorly protected wells were easily contaminated with inadequately disposed human feces. The outbreak in Telixtac differed in that, even though it began during the seasonal rains when impure stream water was available, only a minority (20%) of patients used this water source. Thus, the majority of cases in Telixtac may have resulted from person-to-person transmission.

In almost all reported outbreaks of ET-NANB hepatitis, clinical illness is much more common among adults than among children. In most outbreaks, it is likely that children and adults have been exposed at comparable frequencies and that the observed differences in rates of clinical illness are due to differential expression of disease by age similar to that seen for hepatitis A.

May 01, 1987 / 36(16);241-4

Epidemiologic Notes and Reports Enterically Transmitted Non-A, Non-B Hepatitis -- East Africa

Outbreaks of enterically transmitted non-A, non-B hepatitis occurred in 1985 and 1986 at refugee camps for Ethiopians in Somalia and the Sudan.

Somalia. From January 1985 to September 1986, more than 2,000 cases and 87 deaths occurred at four refugee camps in Somalia; 40 (46%) of the persons who died were pregnant women. The first outbreak among refugees occurred in Bixin Dhule, a holding camp in northwestern Somalia. During the period January-March 1985, there were 699 cases of acute hepatitis and 13 deaths. Adults accounted for 81% of the cases and 92% of the deaths. From April-June 1985, Gannet refugee camp had more than 400 cases and 16 deaths, including nine (56%) among pregnant women.

After an outbreak was recognized at the Tug Wajale B refugee camp in northwestern Somalia, intensive epidemiologic investigation and serologic testing of cases were begun. In January 1986, there had been 2,500 refugees in this camp; an influx of new refugees had increased the population to approximately 32,000 by August 1986. Starting in April 1986, medical personnel at Tug Wajale B noticed a sharp increase in the number of hepatitis cases among adult Ethiopian refugees. In addition, a number of staff members had contracted hepatitis. Cases of hepatitis (diagnosed by the presence of scleral icterus) were identified by reviewing camp medical records. The peak number of cases occurred from mid-May to mid-June (Figure 1), about 6 to 7 weeks after the beginning of a rainy season. The majority (89%) of these persons with clinical cases were young adults; an equal number of males and females were affected. Symptoms associated with hepatitis were nausea, vomiting, dark urine, fever, abdominal pain, itching, fatigue, and headache.

During this period, there were 30 deaths due to hepatitis. Sixteen of those who died were pregnant women; four were non-pregnant women; nine were men; and one was a child. Only four maternal deaths from other causes were recorded in these months. The fatality rate for second- and third-trimester women with hepatitis was 17%.

A tent-to-tent survey involving 2,000 refugees revealed a 3% point prevalence of jaundice in adults and an overall attack rate (April to mid-June) of 8%. Among children 15 years of age, the point prevalence of jaundice was 0.2%, and the overall attack rate was 1.8%. Estimates indicated that over 2,000 cases of clinical hepatitis occurred during the study period. Among the Somali national staff the attack rate was 17%, whereas in expatriate medical personnel, the attack rate was 42%.

Serum samples were obtained from 84 patients and 50 age- and sex-matched controls, and stool specimens were obtained from 21 patients who had been jaundiced for less than or equal to 1 week. Nine patients (10%) and two controls (4%) were positive for hepatitis B surface antigen. Of these, only one patient was positive for IgM anti-core antibody, which is indicative of recent hepatitis B infection. None of the patients or controls were positive for IgM class antibody to hepatitis A virus. Stool specimens were examined by immune electron microscopy (IEM) using serum from a Pakistani patient with known enterically transmitted non-A, non-B hepatitis (1); 27-nm virus-like particles, similar to those seen by IEM in cases from Central Asia, Nepal, and Burma, were found in 13 of 21 samples. These particles cross reacted with sera from patients of enterically transmitted non-A, non-B hepatitis from Central Asia.

Sudan. In mid-1985, when outbreaks of hepatitis were occurring at the refugee camps in Somalia, there were reports of an increase in cases of acute jaundice in Eritrean and Tigrean refugees from Ethiopia residing in refugee camps in eastern Sudan. The investigation of this occurrence included intensified surveillance in four large reception centers (Wad Sherife, Shagarab East 1, Shagarab East 2, and Wad Kowli) and a case-control study in one camp (Wad Kowli).

Active case detection by expatriate health staffs, refugee health workers, and refugee organizations revealed an increase in cases of acute illness with scleral icterus among refugees from June-October (Figure 2), beginning approximately 6 weeks after the onset of heavy rains in eastern Sudan. The majority of patients were adults 15 years of age (66%); only 6.3% were children 5 years of age. There were almost twice as many cases reported among males as among females. Reported fatality rates ranged from 1.3%-4.7% and averaged 3.1% in the four camps. Eleven of the 63 persons who died were pregnant women.

Serum samples were obtained from 175 acutely jaundiced refugees. Seven patients (4%) were positive for hepatitis B surface antigen, and one of these was positive for IgM anti-core antibody. Three other patients (2%) had only IgM anti-core antibody, also indicative of recent hepatitis B infection. Eleven patients (6%) were positive for IgM-class antibody to hepatitis A virus and were considered to have acute cases of hepatitis A. The remaining 154 patients were considered to have non-A, non-B hepatitis. A pool of serum collected from non-A, non-B hepatitis patients cross reacted with stool samples from a Pakistani patient with known enterically transmitted non-A, non-B hepatitis (1).

A questionnaire regarding the onset of acute jaundice among expatriate staff while working in eastern Sudan refugee camps during 1985 has been distributed to 17 agencies involved. In addition, epidemiologic and clinical data are still being collected. Reported by: S Gove, MD, MPH, A Ali-Salad, MD, MA Farah, MD, D Delaney, MJ Roble, J Walter, Somalia Ministry of Health. N Aziz, MBBS, Sudan Commission on Refugees Health Unit. International Health Program Office; Hepatitis Br, Div of Viral Diseases, Center for Infectious Diseases, CDC.

Editorial Note: Non-A, non-B hepatitis, which continues to be a diagnosis of exclusion, is considered to have two distinct forms, which are transmitted by different routes and presumably caused by different viruses. The first, initially recognized as post-transfusion non-A, non-B hepatitis, is seen commonly in North America and Europe, is epidemiologically similar to hepatitis B, and is recognized most commonly after blood transfusions and parenteral drug abuse. The second, enterically transmitted non-A, non-B hepatitis, is transmitted by the fecal-oral route. This disease is known to cause large outbreaks of viral hepatitis and has been reported in the Indian subcontinent (2-7), Burma (8), and Algeria (9). Frequently, large outbreaks have been linked to a fecally contaminated water source or have occurred after heavy rains in areas without systems for adequate sewage disposal. Person-to-person transmission can occur.

Enterically transmitted non-A, non-B hepatitis has several characteristic epidemiologic features. Its incubation period is approximately 40 days (as opposed to 30 days for hepatitis A and 60-180 days for hepatitis B). Clinical disease is common among adults, but infrequent among children. Pregnant women have a dramatically high mortality rate. Large outbreaks of acute viral hepatitis among adults in areas where the population is immune to hepatitis A should alert public health authorities to the presence of enterically transmitted non-A, non-B hepatitis.

Signs and symptoms of enterically transmitted non-A, non-B hepatitis are similar to those of other forms of viral hepatitis, although generalized pruritus may be more common. The majority of patients who are not pregnant recover completely, and there is no evidence of chronic liver disease as a long-term sequela. Outbreaks of disease may be identified by the suggestive epidemiologic pattern (especially the high mortality rate among pregnant women) and the exclusion, through serologic testing, of other forms of viral hepatitis. Post-transfusion non-A, non-B hepatitis has not been documented in communitywide outbreaks.

Currently, no serologic test is available for diagnosis; however, 27- to 30-nm virus-like particles have been found by IEM in stool samples of patients in the early acute phase of infection (1,7,10), and hepatitis can be induced in two different species of primates with this agent. Acute-phase antibody in sera may also be demonstrated by IEM.

In an outbreak situation, emphasis must be placed on preventing transmission. Water sources should be examined for fecal contamination. If the water supply is contaminated, all water should be boiled or chlorinated before consumption. Efforts to reduce person-to-person transmission by improving sanitation should be stressed. Immune globulin (IG) manufactured in the West does not appear to be effective in preventing disease. The efficacy of IG from endemic areas is unknown.

These reports mark the first time that this disease has been described as a problem in refugee camps and the first time that the characteristic virus-like particles have been identified in Africa. Refugee camps represent a fertile setting for the transmission of enterically transmitted non-A, non-B hepatitis. These camps usually have inadequate sanitation and are overcrowded. While contaminated drinking water was not a factor in this outbreak, this problem may exist in other refugee camps.

July 16, 1999 / 48(27);577-581

Intussusception Among Recipients of Rotavirus Vaccine -- United States, 1998-1999

On August 31, 1998, a tetravalent rhesus-based rotavirus vaccine (RotaShield[Registered]*, Wyeth Laboratories, Inc., Marietta, Pennsylvania) (RRV-TV) was licensed in the United States for vaccination of infants. The Advisory Committee on Immunization Practices (ACIP), the American Academy of Pediatrics, and the American Academy of Family Physicians have recommended routine use of RRV-TV for vaccination of healthy infants (1,2). During September 1, 1998-July 7, 1999, 15 cases of intussusception (a bowel obstruction in which one segment of bowel becomes enfolded within another segment) among infants who had received RRV-TV were reported to the Vaccine Adverse Event Reporting System (VAERS). This report summarizes the clinical and epidemiologic features of these cases and preliminary data from ongoing studies of intussusception and rotavirus vaccine.

VAERS
VAERS is a passive surveillance system operated by the Food and Drug Administration (FDA) and CDC (3,4). Vaccine manufacturers are required to report to VAERS any adverse event reported to them, and health-care providers are encouraged to report any adverse event possibly attributable to vaccine. Vaccine recipients and their families also can report adverse events to VAERS. For this report, VAERS case reports of intussusception following rotavirus vaccination were reviewed, and health-care providers, parents, or guardians of patients were contacted by telephone for additional clinical and demographic information. Data on RRV-TV distribution were obtained from the manufacturer. To estimate the expected rate of intussusception among infants aged less than 12 months, hospital discharge data from New York for 1991-1997 were reviewed.

Of the 15 infants with intussusception reported to VAERS, 13 (87%) developed intussusception following the first dose of the three-dose RRV-TV series, and 12 (80%) of 15 developed symptoms within 1 week of receiving any dose of RRV-TV (Table 1). Thirteen of the 15 patients received concurrently other vaccines with RRV-TV. Intussusception was confirmed radiographically in all 15 patients. Eight infants required surgical reduction, and one required resection of 7 inches (18 cm) of distal ileum and proximal colon. Histopathologic examination of the distal ileum indicated lymphoid hyperplasia and ischemic necrosis. All infants recovered. Onset dates of reported illness occurred from November 21, 1998, to June 24, 1999 (Figure 1). The median age of patients was 3 months (range: 2-11 months). Ten were boys. Intussusception among RRV-TV recipients was reported from seven states (Table 1). Of the 15 cases reported to VAERS, 14 were spontaneous reports and one was identified through active postlicensure surveillance.

The rate of hospitalization for intussusception among infants aged less than 12 months during 1991-

1997 (before RRV-TV licensure) was 51 per 100,000 infant-years** in New York (95% confidence interval [CI]=48-54 per 100,000). The manufacturer had distributed approximately 1.8 million doses of RRV-TV as of June 1, 1999, and estimated that 1.5 million doses (83%) had been administered. Given this information, 14-16 intussusception cases among infants would be expected by chance alone during the week following receipt of any dose of RRV-TV. Fourteen of the 15 case-patients were vaccinated before June 1, 1999, and of those, 11 developed intussusception within 1 week of receiving RRV-TV.

POSTLICENSURE STUDIES OF ADVERSE EVENTS FOLLOWING RRV-TV
As part of a preliminary analysis of ongoing postlicensure surveillance of adverse events following vaccination with RRV-TV, cases of intussusception during December 1, 1998-June 10, 1999, were identified among infants aged 2-11 months at Northern California Kaiser Permanente (NCKP) by review of hospital discharge diagnoses, admitting diagnoses for the records for which discharge summaries were not yet complete, and computerized records of all barium enemas performed on children aged less than 1 year. Relative risks were age-adjusted because of differences in the ages of vaccinated and unvaccinated infants, and p values were calculated by Poisson regression.

At NCKP, 16,627 doses of RRV-TV were administered to 9802 infants during December 1, 1998-June 10, 1999. Nine cases of intussusception among infants were identified with onset during that same period, all of which were radiographically or surgically confirmed. Three were among vaccinated children, with intervals of 3, 15, and 58 days following vaccination. The rate of intussusception among never-vaccinated children was 45 per 100,000 infant-years, and among children who had received RRV-TV was 125 per 100,000 infant-years (age-adjusted relative risk [RR]=1.9, 95% CI=0.5-7.7, p=0.39). The rate among children who had received RRV-TV during the preceding 3 weeks was 219 per 100,000 infant-years (age-adjusted RR=3.7, 95% CI=0.7-19, p=0.12). Among children who had received RRV-TV during the previous week, the rate was 314 per 100,000 infant-years (age-adjusted RR=5.7, 95% CI= 0.7-50, p=0.11).

MINNESOTA
In Minnesota, intussusception cases were identified among infants aged 30 days-11 months who were born after April 1, 1998, and were hospitalized with radiographically or surgically confirmed intussusception with onset during November 1, 1998- June 30, 1999. During October 1, 1998-June 1, 1999, 62,916 doses of vaccine were distributed. Eighteen cases of intussusception were identified, five of which were among infants who had received RRV-TV. Vaccinated children had a median age of 4 months (range: 3-5 months), and unvaccinated children had a median age of 7 months (range: 5-9 months). Four of the five RRV-TV recipients with intussusception required surgical reduction, and five of 13 unvaccinated children required surgical reduction. Intussusception occurred after receipt of dose one (two children), dose two (two children), and dose three (one child). The five RRV-TV recipients developed intussusception within 2 weeks of receipt of vaccine; intervals were 6 days (two children), 7 days, 10 days, and 14 days after receipt of vaccine. Assuming 85% of RRV-TV doses distributed in Minnesota were administered, the observed rate of intussusception within 1 week of receipt of RRV-TV was 292 per 100,000 infant-years.

Reported by: K Ehresman, MPH, R Lynfield, MD, R Danila, PhD, Acting State Epidemiologist, Minnesota Dept of Health. S Black, MD, H Shinefield, MD, B Fireman, MS, S Cordova, MS, Kaiser Permanente Vaccine Study Center, Oakland, California. Div of Biostatistics and Epidemiology, Food and Drug Administration. Viral Gastroenteritis Section, Respiratory and Enteric Viruses Br, and Office of the Director, Div of Viral and Rickettsial Diseases, National Center for Infectious Diseases; Vaccine Safety Datalink Team; Statistical Analysis Br, Data Management Div; Vaccine Safety and Development Activity; Child Vaccine Preventable Diseases Br, Epidemiology and Surveillance Div, National Immunization Program; and EIS officers, CDC.

Editorial Note: Rotavirus is the most common cause of severe gastroenteritis in infants and young children aged less than 5 years in the United States, resulting in approximately 500,000 physician visits, 50,000 hospitalizations, and 20 deaths each year. Worldwide, rotavirus is a major cause of childhood death, accounting for an estimated 600,000 deaths annually among children aged less than 5 years. Rotavirus vaccines offer the opportunity to reduce substantially the occurrence of this disease (1).

In prelicensure studies, five cases of intussusception occurred among 10,054 vaccine recipients and one of 4633 controls, a difference that was not statistically significant (5). Three of the five cases among vaccinated children occurred within 6-7 days of receiving rotavirus vaccine. On the basis of these data, intussusception was included as a potential adverse reaction on the package insert, and the ACIP recommended postlicensure surveillance for this adverse event following vaccination (1).

Because of concerns about intussusception identified in prelicensure trials, VAERS data were analyzed early in the postlicensure period. The number of reported intussusception case-patients with illness onset within 1 week of receiving any dose of vaccine is in the expected range; however, because reporting to VAERS of adverse events following vaccination is incomplete (6), the actual number of intussusception cases among RRV-TV recipients may be substantially greater than that reported.

In response to the VAERS reports, a preliminary analysis of data from an ongoing postlicensure study at NCKP was performed, and a multistate investigation was initiated to determine whether an association exists between administration of RRV-TV and intussusception in infants. Preliminary data from Minnesota and from NCKP also suggest an increased risk for intussusception following receipt of RRV-TV. Observed rates of intussusception among recently vaccinated children were similar in both studies. However, the number of cases of intussusception among vaccinated children is small at both NCKP and in Minnesota, and neither study has adequate power to establish a statistically significant difference in incidence of intussusception among vaccinated and unvaccinated children. Available data suggest but do not establish a causal association between receipt of rotavirus vaccine and intussusception, and additional studies are ongoing.

Although neither these studies nor the VAERS reports is conclusive, the consistency of findings from these three data sources raises strong concerns. Because more data are anticipated within several months and rotavirus season is still 4-6 months away in most areas of the United States, CDC recommends postponing administration of RRV-TV to children scheduled to receive the vaccine before November 1999, including those who already have begun the RRV-TV series. Parents or caregivers of children who have recently received rotavirus vaccine should promptly contact their health-care provider if the infant develops symptoms consistent with intussusception (e.g., persistent vomiting, bloody stools, black stools, abdominal distention, and/or severe colic pain). Health-care providers should consider intussusception in infants who have recently received RRV-TV and present with a consistent clinical syndrome; early diagnosis may increase the probability that the intussusception can be treated successfully without surgery. Vaccine providers, parents, and caregivers should report to VAERS intussusception and other adverse events following vaccination.

Information on reporting to VAERS and case report forms can be requested 24 hours a day by telephone, (800) 822-7967, or the World-Wide Web, http://www.nip.gov/nip/vaers.htm.

References
1. CDC. Rotavirus vaccine for the prevention of rotavirus gastroenteritis among children-- recommendations of the Advisory Committee on Immunization Practices. MMWR 1999;48(no. RR-2).
2. Committee on Infectious Diseases, American Academy of Pediatrics. Prevention of rotavirus disease: guidelines for use of rotavirus vaccine. Pediatrics 1998;102:1483-91.
3. Chen RT, Rastogi SC, Mullen JR, et al. The Vaccine Adverse Event Reporting System (VAERS). Vaccine 1994;542-50.
4. Niu MT, Salive ME, Ellenberg SS. Post-marketing surveillance for adverse events after vaccination: the national Vaccine Adverse Event Reporting System (VAERS). Food and Drug Administration Medwatch Continuing Education Article, November 1998. Available at http://www.fda.gov/medwatch/articles/vaers/vaersce.pdf. Accessed July 1, 1999.
5. Rennels MB, Parashar UD, Holman RC, Le CT, Chang H-C, Glass RI. Lack of an apparent association between intussusception and wild or vaccine rotavirus infection. Pediatr Infect Dis J 1998;17:924-5.
6. Rosenthal S, Chen R. The reporting sensitivities of two passive surveillance systems for vaccine adverse events. Am J Public Health 1995;85:1706-9.

* Use of trade names and commercial sources is for identification only and does not imply endorsement by CDC or the U.S. Department of Health and Human Services.
** An infant-year is a unit of measurement combining infants and time used as a denominator in calculating incidence. In this report, it is the sum of the individual units of time (days, weeks, or months) converted to years that the infants in the study population have been followed.

November 20, 1998 / 47(45);978-980

Laboratory-Based Surveillance for Rotavirus -- United States, July 1997-June 1998

Rotavirus infections are the leading cause of severe gastroenteritis among infants and young children worldwide (1,2). Each year in the United States, rotavirus causes an estimated 2.7 million cases of gastroenteritis among children aged less than 5 years, resulting in approximately 500,000 outpatient clinic and emergency department visits and 49,000 hospitalizations (3,4). In addition, rotavirus accounts for an estimated $264 million in health-care costs and approximately $1 billion in total medical and nonmedical costs (3). The large disease burden and cost associated with rotavirus have led to the development of rotavirus vaccines. In August 1998, the first live attenuated rotavirus vaccine (Rotashield{registered} {Wyeth Lederle Vaccines and Pediatrics}) * was approved for use in infants by the Food and Drug Administration. The Advisory Committee on Immunization Practices has recommended that this vaccine be given as a three-dose schedule to infants aged 2, 4, and 6 months. Since 1991, rotavirus activity in the United States has been prospectively monitored by the National Respiratory and Enteric Virus Surveillance System (NREVSS), a voluntary, laboratory-based system (5). This report summarizes surveillance data from NREVSS during the 1997-1998 rotavirus season and reviews issues related to rotavirus surveillance that are important for a national rotavirus vaccine program.

From July 1997 through June 1998, 66 laboratories in 41 states participated in NREVSS. Each laboratory reported weekly to CDC the number of stool specimens tested and the number positive for rotavirus by antigen-detection and electron microscopy methods. Of 22,912 fecal specimens examined, 5343 (23%) were positive for rotavirus. Seasonal increases in rotavirus detection were noted throughout the United States, and the timing of peak rotavirus activity varied by geographic location. Activity peaked first in the Southwest during November-December 1997 and last in the North and Northeast during April-May. Temporal and geographic trends during the July 1997-June 1998 reporting period varied slightly from trends during previous years (5), with late-season peaks in some laboratories in the western United States. Laboratories in Montana, Nevada, and Washington reported peak rotavirus activity during April, and an additional laboratory in Nevada reported peak activity during May, substantially later than the usual December-January peak for these sites. Data from Alaska and Hawaii were not available.

Reported by: National Respiratory and Enteric Virus Surveillance System collaborating laboratories, National Rotavirus Strain Surveillance System collaborating laboratories; Viral Gastroenteritis Section, Respiratory and Enteric Viruses Br, Div of Viral and Rickettsial Diseases, National Center for Infectious Diseases, CDC.

Editorial Note: Rotavirus causes seasonal peaks of gastroenteritis each year in the United States, and the temporal and geographic patterns observed during the July 1997-June 1998 reporting period were generally characteristic of trends noted during previous years (5). The late-season (April-May 1998) peaks reported by laboratories in the southwestern United States is unusual and, so far, unexplained. The annual seasonal peaks of activity and the proportion of total specimens positive for rotavirus noted in this surveillance system are consistent with data collected from other temperate countries (6).

Surveillance systems for rotavirus are particularly important as a means to measure the impact of the licensure of the first rotavirus vaccine for use in U.S. infants. NREVSS is the largest, nationally representative system for surveillance of rotavirus infections in the United States (5). Participating laboratories transmit reports to CDC weekly by using an automated telephone reporting system, which allows for timely analysis of rotavirus activity. NREVSS has been an important tool for characterizing the geographic and temporal trends of rotavirus infections in the United States, and its findings have been validated by disease-based surveillance studies (4,7,8). Initiation of a new vaccine program against rotavirus gastroenteritis will generate additional surveillance needs, such as the capability for monitoring rotavirus strain prevalences and assessing disease burden over time. Although the implications for the potential effectiveness of the vaccine are unclear, these findings highlight the importance of laboratory-based surveillance to monitor for the emergence of novel or unusual rotavirus strains following the introduction of the new vaccine. In addition, disease-based rotavirus surveillance systems will be initiated during the 1998-99 rotavirus season to monitor the effectiveness of rotavirus vaccine programs.

To monitor rotavirus strain circulation in the United States, CDC, in collaboration with state and local public health laboratories, established the National Rotavirus Strain Surveillance in 1996 (9). During November 1996-May 1997, 10 laboratories submitted rotavirus-positive stool specimens to CDC for strain characterization. During this period, the four rotavirus strains that predominate worldwide and that are represented in the licensed vaccine accounted for 83% of isolates tested. However, 9% of strains characterized had not been detected previously in the United States and are not represented in the current vaccine. The implications for the potential effectiveness of the vaccine are unclear.

NREVSS will continue to monitor for changes in the epidemiology of rotavirus following implementation of a vaccine program, and will provide a foundation for expansion of U.S. strain surveillance. The combination of NREVSS, strain surveillance, and disease-based surveillance will make it possible to monitor the impact of the new vaccine program.

References
1. De Zoysa I, Feachem RG. Interventions for the control of diarrhoeal disease among young children: rotavirus and cholera immunization. Bull World Health Organ 1985;63:569-83.
2. Glass RI, Kilgore PE, Holman RC, et al. The epidemiology of rotavirus diarrhea in the United States: surveillance and estimates of disease burden. J Infect Dis 1996;174:S5-S11.
3. Tucker AW, Haddix AC, Bresee JS, Holman RC, Parashar UD, Glass RI. Cost-effectiveness analysis of a rotavirus immunization program for the United States. JAMA 1998;279:1371-6.
4. Parashar UD, Holman RC, Clarke MJ, Bresee JS, Glass RI. Hospitalizations associated with rotavirus diarrhea in the United States, 1993 through 1995: surveillance based on the new ICD-9-CM rotavirus-specific diagnostic code. J Infect Dis 1997;177:13-7.
5. Torok TJ, Kilgore PE, Clarke MJ, Holman RC, Bresee JS, Glass RI. Visualizing geographic and temporal trends in rotavirus activity in the United States, 1991 to 1996. Pediatr Infect Dis J 1997;16:941-6.
6. Cook SM, Glass RI, LeBaron CW, Ho M-S. Global seasonality of rotavirus infections. Bull World Health Organ 1990;68:171-7.
7. Jin S, Kilgore PK, Holman RC, Clarke MJ, Gangarosa EJ, Glass RI. Trends in hospitalizations for diarrhea in United States children from 1979-1992: estimates of the morbidity associated with rotavirus. Ped Infect Dis J 1996;15:397-404.
8. Ho M-S, Glass RI, Pinsky PF, Anderson LJ. Rotavirus as a cause of diarrheal morbidity and mortality in the United States. J Infect Dis 1988;158:1112-6.
9. Ramachandran M, Gentsch JR, Parashar UD, et al. Detection and characterization of novel rotavirus strains in the United States. J Clin Microbiol 1998;36:3223-9.

Use of trade names and commercial sources is for identification only and does not imply endorsement by the U.S. Department of Health and Human Services or CDC.

May 22, 1998 / 47(19);394-396

Plesiomonas shigelloides and Salmonella serotype Hartford Infections Associated with a Contaminated Water Supply -- Livingston County, New York, 1996

On June 24, 1996, the Livingston County (New York) Department of Health (LCDOH) was notified of a cluster of diarrheal illness following a party on June 22, at which approximately 30 persons had become ill. This report summarizes the findings of the investigation, which implicated water contaminated with Plesiomonas shigelloides and Salmonella serotype Hartford as the cause of the outbreak.

The party was held at a private residence on June 22 and was attended by 189 persons. Food was provided by a local convenience store that sells gasoline, packaged goods, sandwiches, and pizza and prepares food for catered events. The convenience store had not catered any parties during the preceding 5 days but catered two parties on June 23. LCDOH contacted the organizers of these events and found no other reports of illness.

To determine the source and extent of the outbreak and mechanism of contamination, LCDOH conducted a cohort study, an environmental investigation, and micro-biologic examinations of stool specimens, leftover food items, and water samples. A menu and guest list were obtained and guests were interviewed by telephone. A probable case was defined as diarrhea (greater than 3 loose stools during a 24-hour period) in a person who attended the party and became ill within 72 hours. Persons with a confirmed case had either Plesiomonas shigelloides or Salmonella serotype Hartford or both isolated from stool. The caterer and facility employees were interviewed to obtain information on food preparation, and the water source was inspected.

Of the 189 attendees, 98 (52%) were interviewed. Sixty persons reported illness; 56 (57%) of 98 respondents had illnesses meeting the case definition. The mean age for case-patients was 41 years (range: 2-85 years), and 32 (57%) were male. Stool specimens were obtained from 14 ill attendees: nine yielded only P. shigelloides, three only Salmonella serotype Hartford, and two had both organisms. One person with culture-confirmed Salmonella serotype Hartford was hospitalized. The clinical profiles of the culture-confirmed (n=14) and probable (n=42) cases were similar.

Twenty food and beverage items were served at the party. Three food items were associated with illness: macaroni salad, potato salad, and baked ziti. Of 56 attendees who ate macaroni salad, 43 (77%) became ill, compared with 17 (40%) of 42 who did not eat macaroni salad (relative risk {RR}=2.6; 95% confidence interval {CI}=1.5-4.4). Of 49 guests who ate potato salad, 36 (73%) became ill, compared with 20 (44%) of 45 who did not eat potato salad (RR=2.1; 95% CI=1.2-3.6). Of 46 attendees who ate baked ziti, 36 (78%) became ill, compared with 20 (42%) of 48 that did not eat baked ziti (RR=2.7; 95% CI=1.5-4.9).

Leftover food samples of these three items were collected on June 25 and sent for microbiologic examination. Salmonella serotype Hartford was isolated from the macaroni salad and baked ziti. Both Salmonella serotype Hartford and P. shigelloides were isolated from the potato salad. Escherichia coli was isolated from a water sample collected on June 27 from the tap in the store. Water samples collected on July 8 from the well that supplied water to the store contained both Salmonella serotype Hartford and P. shigelloides.

Preparation of the salads and the baked ziti began on June 21, and prepared food items were stored in a walk-in cooler overnight. On June 22, the ziti was prepared by heating the tomato sauce, pouring it over the meat and pasta, and heating in an oven for 50 minutes at an unknown temperature. The ziti remained in the oven with the heat off until it and the salads were transported to the party.

All foodhandlers denied gastrointestinal illness with onset before June 22. However, three foodhandlers reported illness beginning after June 22; all three reported having eaten foods prepared for the party. P. shigelloides was recovered from stool specimens from these three workers only.

The New York State Department of Agriculture and Markets found nine sanitary violations at the caterer's facilities. The water source, an unprotected dug well approximately 10 feet deep, served only the store. The well was fed by shallow ground water and may have received surface runoff from surrounding tilled and manured farm land and water from adjacent streams. A small poultry farm was located approximately 1600 feet upstream of the well. Farm field drainage systems discharged into the source water stream just above the well. A water sample collected at the store on June 27 showed no chlorine residual, indicating that the pellet chlorinator was off-line at the time of the event. The pellet chamber was empty and the system did not contain any filtration mechanism. Well water used for food preparation (i.e., rinsing pasta used in salads, mixing ingredients, cooking food items, and cleaning equipment) was probably contaminated as a result of rainfall on June 19 and June 20 that transported pathogens from the surrounding farmland. The improperly maintained chlorinator allowed these pathogens to reach the food preparation area. After the outbreak, the store was prohibited from preparing food until an adequate water-treatment system that met drinking water standards could be provided. Store employees and the public were instructed not to drink the water.

Reported by: R Van Houten, D Farberman, J Norton, J Ellison, Livingston County Dept of Health, Mt. Morris; J Kiehlbauch, PhD, T Morris, MD, P Smith, MD, State Epidemiologist, New York State Dept of Health. Foodborne and Diarrheal Diseases Br, Div of Bacterial and Mycotic Diseases, National Center for Infectious Diseases, CDC.

Editorial Note: The findings in this report implicated a deficient water supply system as the cause of an outbreak of diarrheal illness caused by Salmonella serotype Hartford and P. shigelloides.

Unfiltered, untreated surface water led to contamination of food during its preparation.

Most infections with P. shigelloides have been associated with drinking untreated water, eating uncooked shellfish, or with travel to developing countries (1-3). P. shigelloides (previously Aeromonas shigelloides) are ubiquitous, facultatively anaerobic, flagellated, gram-negative rods (3). Although they are widespread in the environment, few waterborne or foodborne outbreaks have been reported (4). P. shigelloides have been isolated from a variety of sources, including wild and domestic animals (2). Infection is characterized by self-limited diarrhea with blood or mucus, abdominal cramps, and vomiting or fever (5). Symptoms usually occur within 48 hours of exposure. Fecal leukocytes and erythrocytes have been found on stool smears (1); however, the exact mechanism of the diarrhea (secretory versus inflammatory) is unknown.

Salmonella serotype Hartford is a rare serotype that has been isolated from porcine and bovine sources. In May 1995, freshly squeezed, unpasteurized commercial orange juice was implicated as the cause of an outbreak (6). Contamination was thought to have originated from inadequate sanitization of the exterior surfaces of oranges.

In this outbreak, the well water most likely became contaminated with both P. shigelloides and Salmonella serotype Hartford through runoff from nearby farms. The outbreak could have been prevented if effective public health measures had been in place. Routine testing of well water for total fecal coliform bacteria, turbidity, and chlorine residual may enable early detection of fecal contamination and rapid decontamination. Filtration and chlorination of potable water systems have substantially reduced waterborne outbreaks and subsequent morbidity and mortality. Where possible, water sources subject to contamination from agricultural runoff should not be used for drinking or food preparation. Disinfection and filtration of water from any source can further reduce the risk for waterborne illness.

References
1. Soweid AM, Clarkston WK. Plesiomonas shigelloides: An unusual cause of diarrhea. Am J Gastroenterol 1995;90:2235-6.
2. Jeppesen C. Media for Aeromonas spp., Plesiomonas shigelloides and Pseudomonas spp. food and environment. Int J Food Microbiol 1995;26:25-41.
3. San Joaquin VH. Aeromonas, Yersinia, and miscellaneous bacterial enteropathogens. Pediatr Ann 1994;23:544-8.
4. Schofield GM. Emerging foodborne pathogens and their significance in chilled foods. J Appl Bacteriol 1992;72:267-73.
5. Holmberg SD, Wachsmuth IK, Hickman-Brenner FW, Blake PA, Farmer JJ. Pleisiomonas enteric infections in the United States. Ann Intern Med 1986;105:690-4.
6. Cook KA, Swerdlow D, Dobbs T, et al. Fresh-squeezed Salmonella: an outbreak of Salmonella Hartford associated with unpasteurized orange juice -- Florida {Abstract}. EIS Conference Abstract 1996;38-9.

February 08, 1991 / 40(5);80,87

Current Trends Rotavirus Surveillance -- United States, 1989-1990

Rotavirus infection is the most common cause of dehydrating diarrhea in children in the United States (1). In January 1989, CDC established a National Rotavirus Surveillance System (NRSS) to monitor national patterns in the epidemiology of rotavirus. This report summarizes findings from the NRSS from January 1989 through November 1990.

In January 1989, 99 laboratories began submitting monthly reports of positive detections, numbers of specimens tested, and laboratory methods used to detect rotavirus. Of those laboratories, 72 in 48 states also provided retrospective data for 1984-1988; these data indicate a temporal and geographic sequence of peaks in reported positive detections that begins in the southwest in November and ends in the northeast in March (2).

From January 1989 through November 1990, 56 laboratories submitted reports every month; they included 12 pediatric, 17 community, and 23 university hospital laboratories; two public health laboratories; and two commercial laboratories. To detect rotavirus, most (46 (82%)) of these laboratories used enzyme immunoassay techniques, four used a latex agglutination test, and six used electron microscopy.

For the 23-month period, 48,035 specimens were tested for rotavirus; 9639 (20%) were positive. The total number of specimens tested each month varied from 1410 in September 1990 to 3275 in January 1990. For all centers combined, the percentage of positive specimens was highest in February 1990 (1056 (36%) of 2925) and lowest in October 1990 (103 (6%) of 1817) (Figure 1).

October 1989 through May 1990 was the first full rotavirus season for prospective surveillance in the United States. During that period, peaks in the positive detection rate varied by region, beginning in December in the West (36% positive detections), January-February in the South (32%-33%), February in the North Central (49%), and March in the Northeast (47%). By June, no region had more than 16% positive detections, and three of the four regions had less than 10% positive detections. For the 1990-91 rotavirus season, an increase in positive detections was reported in the West during November 1990 (positive rate of 21%) when compared with August-October (1%-4%). Reported by: National Rotavirus Surveillance System laboratories. Viral Gastroenteritis Section, Respiratory and Enteric Virus Br, Div of Viral and Rickettsial Diseases, Center for Infectious Diseases, CDC.

Editorial Note: Rotavirus, the most important cause of pediatric gastroenteritis in the United States, is responsible for an estimated one third of all hospitalizations for diarrhea in children less than 5 years of age (3). These hospitalizations occur predominantly in the winter, and in one large children's hospital, rotavirus accounted for 3% of all hospital days (4). Rotavirus disease-associated hospitalization rates are highest for children less than 2 years of age (3,4).

From 1979 through 1985, an average of 500 children died annually from diarrheal disease in the United States (5); an estimated 20% of these deaths were caused by rotavirus infection (3). Death rates for diarrheal disease were highest in the South and among black children less than 6 months of age (5). Patterns of childhood mortality related to diarrheal disease reflect the winter seasonality of rotavirus (3).

Because national rotavirus surveillance data suggest an increase in the risk for rotavirus infections from October through May, health-care providers should consider rotavirus as a cause of diarrhea in groups at risk and be familiar with approaches for management of this disease. Many deaths and hospitalizations may be prevented by the aggressive use of oral rehydration therapy, which is underused (6-8). Vaccines for prevention or modification of rotavirus diarrhea are under development but are unlikely to be available for 3-5 years.

For most children hospitalized with rotavirus gastroenteritis, no laboratory diagnosis is made (4), and only a small number of deaths from rotavirus infection have been virologically confirmed (9). Because the ninth revision of the International Classification of Diseases (ICD) did not include a rubric for rotavirus enteritis, proxy codes (3-5) were used to reflect this cause of death; however, the 10th revision will introduce a specific rubric (National Center for Health Statistics, unpublished data). The wider use of rapid diagnostic tests for rotavirus, combined with the use of a specific ICD rubric, will permit improved surveillance of rotavirus hospitalizations and deaths.

References
1. Kapikian AZ, Chanock RM. Rotaviruses. In: Fields BN, Knipe DM, Chanock RM, Hirsch MS, Melnick JL, Monath TP, eds. Virology. Vol 2. 2nd ed. New York: Raven Press, 1990:1353-404.
2. LeBaron CW, Lew J, Glass RI, et al. Annual rotavirus epidemic patterns in North America: results of a 5-year prospective survey of 88 centers in Canada, Mexico, and the United States. JAMA 1990;264:983-8.
3. Ho MS, Glass RI, Pinsky PF, Anderson LJ. Rotavirus as a cause of diarrheal morbidity and mortality in the United States. J Infect Dis 1988;158:1112-6.
4. Matson DO, Estes MK. Impact of rotavirus infection at a large pediatric hospital. J Infect Dis 1990;162:598-604.
5. Ho MS, Glass RI, Pinsky PF, et al. Diarrheal deaths in American children: are they preventable? JAMA 1988;260:3281-5.

6. Santosham M, Daum RS, Dillman L, et al. Oral rehydration therapy of infantile diarrhea: a controlled study of well-nourished children hospitalized in the United States and Panama. N Engl J Med 1982;306:1070-6.

7. Avery ME, Snyder JD. Oral therapy for acute diarrhea: the underused simple solution. N Engl J Med 1990;323:891-4.

8. Mauer AM, Dweck HS, Finberg L, et al. American Academy of Pediatrics Committee on Nutrition: use of oral fluid therapy and posttreatment feeding following enteritis in children in a developed country. Pediatrics 1985;75:358-61.

9. Carlson JAK, Middleton PJ, Szymanski MT, Huber J, Petric M. Fatal rotavirus gastroenteritis: an analysis of 21 cases. Am J Dis Child 1978;132:477-9.

March 26, 1999 / 48(11);225-227

Norwalk-Like Viral Gastroenteritis in U.S. Army Trainees -- Texas, 1998

During August 27-September 1, 1998, 99 (12%) of 835 soldiers in one unit at a U.S. Army training center in El Paso, Texas, were hospitalized for acute gastroenteritis (AGE). Their symptoms included acute onset of vomiting, abdominal pain, diarrhea, and fever. Review of medical center admission records for AGE during the previous year indicated that fewer than five cases occurred each month. This report describes the outbreak investigation initiated on August 30 by a U.S. Army Epidemiologic Consultation Service (EPICON) team; the findings indicated the outbreak was caused by a Norwalk-like virus (NLV).

The EPICON team reviewed data from the inpatient records of 90 ill soldiers. AGE was defined as three or more loose stools and/or vomiting within a 24-hour period in a soldier or employee at the training center during August 26-September 1. Illness was accompanied by a minimally elevated leukocyte count, mild thrombocytopenia, and low-grade fever. The median duration of hospitalization was 24 hours (range: 12-72 hours). Stool samples collected from persons with AGE on hospital admission were negative for bacterial and parasitic pathogens. Of 24 stool specimens sent to CDC for viral agent identification, 17 were positive by reverse transcriptase poly-merase chain reaction assays for NLVs (genogroup 2).

Interviews with foodhandlers in the base's two dining facilities (DF1 and DF2) revealed illness in a confection baker, who had become ill in DF1 while baking crumb cake, pie, and rolls on August 26. One other DF1 employee who was not a foodhandler also reported self-limited gastrointestinal illness during August 27-29. No worker in DF2 reported illness.

Cultures of food specimens from the ice cream dispenser in DF1 grew nonpathogenic coliform bacteria (Citrobacter diversus and Serratia liquefaciens); however, the sample was at room temperature before culture. Enterobacter cloacae coliform bacteria were cultured from the soda fountain in DF2. Water samples taken from multiple sites in the training compound and from elsewhere on post were all negative for coliform contamination.

A questionnaire about food preferences, based on the previous week's menu, was administered to 86 hospitalized soldiers (84 of whom had eaten in DF1 during the 10 days before answering the questionnaire) and to 237 randomly selected soldiers from the training unit. Of the 237 nonhospitalized soldiers, 41 (17%) did not eat at DF1 during the 10 days before answering the questionnaire; 40 (17%) had illnesses that met the case definition. Thus, cases of AGE were characterized in 126 soldiers.

To determine the point source of the outbreak, cases with onset during August 27-28 (n=98) were analyzed separately for odds ratios (ORs) of selected exposures (Table_1). The univariate OR for illness associated with dining at DF1 during the week before the outbreak was 9.8 (95% confidence interval=2.8-40.2). Two soldiers who ate exclusively at DF2 became ill, and one ill soldier reported not eating at either facility. Food items (crumb cake, pie, cinnamon rolls, and ice cream) and soda fountain dispensers were associated with illness by univariate analysis. Using multivariate analysis, only DF1 and the carbonated beverage dispensers remained strongly associated with illness.

Reported by: M Arness, MD, M Canham, MPH, B Feighner, MD, E Hoedebecke, DVM, J Cuthie, PhD, C Polyak, US Army Center for Health Promotion and Preventive Medicine, Edgewood,

Maryland. DR Skillman, MD, J English, C Jenkins, T Barker, MD, William Beaumont Army Medical Center, El Paso, Texas. T Cieslak, MD, US Army Medical Research Institute of Infectious Diseases, Frederick, Maryland. DN Taylor, MD, Walter Reed Army Institute of Research, Washington, DC. Viral Gastroenterology Section and Infectious Disease Pathology Activity, Div of Viral and Rickettsial Diseases, National Center for Infectious Diseases, CDC.

Editorial Note: NLVs, previously known as small round-structured viruses, are the most common cause of nonbacterial gastroenteritis outbreaks in adults (1,2). Classified in the family Caliciviridae (1,2), NLVs are transmitted by the fecal-oral route and have been implicated in 42%-71% of viral outbreaks associated with contaminated water and food since the Norwalk virus was identified (1,3,4). NLV outbreaks have been caused by eating contaminated raw shellfish and by unsanitary food preparation practices by foodhandlers (1,3-6). NLVs are hardy, ubiquitous, and extremely persistent in the environment, resisting disinfection and chlorination, and have caused serial gastroenteritis outbreaks (1,3,4).

The epidemiologic evidence described in this report indicates that the outbreak was a point-source, propagated, foodborne viral illness. Although cases occurred before the onset of acute illness in the confection baker, he could have been the point source because he probably shed virus before the onset of clinical symptoms. The strong association with drinking carbonated beverages is not easily explained and may represent increased thirst among ill persons. The use of the Army hospital as a quarantine bay probably decreased secondary propagation of the illness.

Prevention of future outbreaks of NLVs in U.S. military dining facilities or any food service establishment depends on vigilance and rigorous enforcement of simple measures to prevent food contamination. These measures include handwashing, exclusion of ill foodhandlers from the workplace, and basic hygiene and sanitation measures.

References
1. Kapikian AZ, Estes MK, Chanock RM. Norwalk group of viruses. In: Fields BN, Knipe DM, Howley PM, et al, eds. Fields virology. 3rd ed. Philadelphia, Pennsylvania: Lippincott-Raven Publishers, 1996:783-810.
2. Levett PN, Gu M, Luan B, et al. Longitudinal study of molecular epidemiology of small round-structured viruses in a pediatric population. J Clin Microbiol 1996;34:1497-501.
3. Hedberg CW, Osterholm MT. Outbreaks of food-borne and waterborne viral gastroenteritis. Clin Microbiol Rev 1993;6:199-210.
4. CDC. Viral agents of gastroenteritis: public health importance and outbreak management. MMWR 1990;39(no. RR-5).
5. Kuritsky JN, Osterholm MT, Greenberg HB, et al. Norwalk gastroenteritis: a community outbreak associated with bakery product consumption. Ann Intern Med 1984;100:519-21.
6. Parashar UD, Dow L, Fankhauser FL, et al. An outbreak of viral gastroenteritis associated with consumption of sandwiches: implications for the control of transmission by food handlers. Epidemiol Infect 1998;121:615-21.

January 20, 1995 / 44(02);37-39

Epidemiologic Notes and Reports Multistate Outbreak of Viral Gastroenteritis Associated with Consumption of Oysters -- Apalachicola Bay, Florida, December 1994- January 1995

On January 3, 1995, the Florida Department of Health and Rehabilitative Services (HRS) was notified of an outbreak of acute gastroenteritis associated with eating oysters. The subsequent investigation by HRS has identified 34 separate clusters of cases, many of which were associated with oysters harvested during December 29-31 from 13 Mile Area and Cat Point in Apalachicola Bay. Oysters were shipped to other states, but additional clusters of illness associated with these oysters have been reported only in Georgia. Most of these oysters were served steamed or roasted. This report summarizes the preliminary findings of the ongoing investigation of this outbreak.

On January 4, Apalachicola Bay was closed to harvesting even though levels of fecal coliforms in the water and in the oyster meat were within acceptable limits. The preliminary investigation identified no gross breaches of sanitation; however, during the holiday season, the bay was used heavily by recreational boaters and commercial fishermen. Clusters of cases identified since the bay was closed prompted concern regarding the continued marketing of these oysters as unshelled and as shucked product both in Florida and other states.

Following the detection of cases associated with oysters from Apalachicola Bay, enhanced surveillance detected three additional clusters of cases in Florida and two in Texas initially linked to oysters harvested in Galveston Bay. As a result, on January 13, Galveston Bay was closed to harvesting.

Norwalk-like viruses have been detected by electronmicroscopy in stool specimens from seven of 11 persons who ate oysters from Apalachicola Bay. Reported by: C Aristeguieta, MD, Dept of Family Medicine, Univ of Miami; I Koenders, Districts 1 and 2 Health Office, Tallahassee; D Windham, Districts 3 and 13 Health Office, Ocala; K Ward, MSEH, Districts 4 and 12 Health Office, Daytona Beach; E Gregos, Districts 5 and 6 Health Office, Tampa; L Gorospe, E Ngo-Seidel, MD, Nassau County Public Health Unit, Fernandina Beach; J Walker, MD, District 4 Health Office, Jacksonville; WG Hlady, MD, R Hammond, PhD, RS Hopkins, MD, State Epidemiologist, Florida Dept of Health and Rehabilitative Svcs. DM Simpson, MD, State Epidemiologist, Texas Dept of Health. Viral Gastroenteritis Section, Respiratory and Enteric Viruses Br, Div of Viral and Rickettsial Diseases, National Center for Infectious Diseases; Div of Field Epidemiology, Epidemiology Program Office, CDC.

Editorial Note: Outbreaks of oyster-associated gastroenteritis affect substantially more persons than those identified in the few documented sentinel clusters (1-3). An important feature of these outbreaks is the inherent delays in removing contaminated oysters from the market. Although oyster tags permit traceback to the general harvest areas, they are not sufficiently detailed to allow recall of oysters from a specific site, and they can be lost when oysters are shucked. In this outbreak, the continued occurrence of cases 1 week after the bay was closed and the product was recalled suggests that the contaminated product was still available to consumers. Cooking (i.e., steaming and roasting) did not always render the oysters noninfectious. In addition, enhanced surveillance in Florida prompted by the investigation led to the closing of an oyster bed in Texas. The observation that both the quality of water in the Florida beds and the meat in the implicated oysters met national standards underscores the inherent limitations of the existing methods and the urgent need for improved indicators of viral contamination. In the absence of such indicators, it is difficult to determine when a bed can be safely reopened.

The findings in this investigation indicate the outbreak resulted from consumption of oysters contaminated with Norwalk-like virus. In a previous oyster-associated Norwalk virus outbreak, identification of the identical sequence of the virus genome in specimens from patients in five states established a clear link between those cases and the oysters from one harvest site (1,2). For the outbreaks described in this report, molecular analysis of fecal specimens will be required to determine the number of linked outbreaks and help assess the usefulness of the specific control measures (4,5). However, the preliminary findings suggest that Apalachicola Bay oysters may have become contaminated by sewage dumped overboard by recreational and commercial boaters. Long-term solutions to eliminate fecal contamination of oyster beds will require either that boaters not be permitted to dump sewage overboard or that beds used for harvesting be limited to those in pristine waters. Improved methods to detect virus in these oysters are needed to understand the extent of the contamination and to strengthen prevention efforts and enforcement.

References
1. Kohn MA, Farley TA, Ando T, et al. A large outbreak of Norwalk virus gastroenteritis associated with eating raw oysters: implications for maintaining safe oyster beds. JAMA 1995 (in press).
2. Dowell SF, Groves C, Kirkland KB, et al. A multistate outbreak of oyster-associated gastroenteritis: implications for interstate tracing of contaminated shellfish. J Infect Dis 1994 (in press).
3. CDC. Viral gastroenteritis associated with consumption of raw oysters -- Florida, 1993. MMWR 1994;43:446-9.
4. Ando T, Monroe SS, Gentsch JR, Jin Q, Lewis DC, Glass RI. Detection and differentiation of antigenically distinct small round-structured viruses (Norwalk-like viruses) by reverse transcription-PCR and Southern hybridization. J Clin Microbiol 1995;33:64-71.
5. Lew JF, LeBaron CW, Glass RI, et al. Recommendations for collection of laboratory specimens associated with outbreaks of gastroenteritis. MMWR 1990;39(no. RR-14).

June 24, 1994 / 43(24);446-449

Viral Gastroenteritis Associated with Consumption of Raw Oysters -- Florida, 1993

During November 20-30, 1993, four county public health units (CPHUs) of the Florida Department of Health and Rehabilitative Services (HRS) in northwestern Florida conducted preliminary investigations of seven separate outbreaks of foodborne illness following consumption of raw oysters. On December 1, the HRS State Health Office initiated an investigation to characterize the illness, examine risk factors for oyster-associated gastroenteritis, and quantify the dose-response relation. This report presents the findings of these two investigations. Preliminary Investigations by the HRS CPHUs

In November 1993, private physicians notified the CPHUs of 20 persons with possible foodborne illness. These 20 ill persons identified seven well meal companions. Raw oysters were the only common food item eaten by all ill persons; no well meal companions had eaten oysters. At the request of the HRS State Health Office, CPHUs initiated active surveillance for cases of raw oyster-associated gastroenteritis among patients of hospital emergency departments, urgent-care centers, and private physicians in northwestern Florida. A case was defined as sudden onset of nausea, vomiting, diarrhea, or abdominal cramps within 72 hours of eating raw oysters. Twenty-five additional cases of gastroenteritis associated with eating raw oysters were detected.

Traceback of implicated oysters by the CPHUs and the Florida Department of Environmental Quality indicated the oysters had been harvested from Apalachicola Bay in northwestern Florida during November 15-23. Epidemiologic Investigation by the HRS State Health Office

The 45 persons with raw oyster-associated gastroenteritis reported by the CPHUs identified 26 well meal companions who had eaten oysters during the same meal as ill persons, but did not become ill. Of 44 ill persons for whom data were available, 36 (82%) had developed diarrhea; 34 (77%), nausea; 33 (75%), abdominal cramps; 25 (57%), vomiting; 17 (39%), fever; 15 (34%), headache; and 14 (32%), myalgia. The attack rate was 63%. Of the 45 ill persons, 10 were hospitalized for 24 hours or longer. For 30 persons for whom data were available, the median incubation period was 31 hours (range: 2-69 hours). For 26 persons for whom data were available, the median duration of illness was 48 hours (range: 10 hours-7 days); for 13 persons, duration of illness was more than 3 days. No household contacts of ill persons developed gastroenteritis.

No differences were identified between persons who became ill and well meal companions in preexisting medical conditions or medications. Consumption of alcohol or food (e.g., crackers and hot sauce) with the oysters was not associated with risk for illness. Based on the 33 cases for which data were available, a dose-response relation was observed between illness and number of raw oysters eaten (chi square for trend=3.98; p=0.05). The attack rate was highest among raw-oyster eaters who had consumed more than 5 dozen oysters (91%) and lowest among those who had consumed less than 1 dozen oysters (46%).

Paired serum specimens from 10 patients were tested for antibody to Norwalk-like virus by enzyme immunoassay (1); three pairs demonstrated a fourfold or greater rise in titer. Seven stool specimens were examined by electron microscopy (EM) and reverse transcription-polymerase chain reaction (RT-PCR). In four specimens, small round-structured viruses were detected by EM; in one specimen, a Norwalk-like genome was confirmed by RT-PCR (2,3). This Norwalk-like virus strain had a nucleotide sequence distinct from similar viruses in nearly simultaneous outbreaks associated with consumption of oysters harvested along the Louisiana coast (4).

No confirmed evidence of improper handling (e.g., inadequate refrigeration time or temperature) of the implicated oysters was detected. However, three ill persons had purchased oysters from retail establishments that were not licensed seafood dealers.

The National Shellfish Sanitation Program (NSSP) requires fecal coliform testing at least once each month. Fecal coliform testing of water drawn from 39 monitoring sites in Apalachicola Bay on October 3, November 21, and November 24 indicated that water quality in the bay met the criteria of the NSSP (5). No environmental source of pollution was identified. Sanitation procedures at the oyster-processing facilities where seafood dealers purchased oysters met standards set by the Florida Department of Environmental Protection (FDEP). However, based on the epidemiologic evidence of

317

illness associated with oysters harvested from those waters, FDEP temporarily closed the shellfish-harvesting area of Apalachicola Bay during December 1-7. No cases of gastroenteritis related to consumption of oysters harvested after December 7 have been reported.

Reported by: C Davis, A Smith, MD, R Walden, Bay County Public Health Unit, Panama City; G Bower, K Cummings, B Dean, J Rigsby, Jackson County Public Health Unit, Marianna; P Justice, C Anderson, N Brown, J Minor, Washington County Public Health Unit, Chipley; EF Geiger, MD, V Laxton, District 1 Health Office, Pensacola; L Crockett, MD, W McDougal, District 2 Health Office, Tallahassee; WG Hlady, MD, RS Hopkins, MD, State Epidemiologist, State Health Office, Florida Dept of Health and Rehabilitative Svcs. Food and Drug Administration. Viral Gastroenteritis Section, Respiratory and Enterovirus Br, Div of Viral and Rickettsial Diseases, National Center for Infectious Diseases; Div of Field Epidemiology, Epidemiology Program Office, CDC.

Editorial Note: This report documents outbreaks of viral gastroenteritis in Florida linked to consumption of raw oysters from waters that apparently met the standards for shellfish sanitation. Clinical and epidemiologic features of the outbreaks are similar to recently reported multistate outbreaks of viral gastroenteritis associated with eating oysters harvested in Louisiana (4). RT-PCR with sequencing identified different strains of the virus in the multistate outbreak and the Florida outbreak, suggesting independent sources of oyster contamination.

Although infection with the oysterborne Norwalk-like virus caused no fatalities in this outbreak, raw oyster consumption has been linked in Florida to 30 fatal cases of infection with Vibrio vulnificus during 1981-1992 among persons with preexisting liver disease (6). V. vulnificus is a ubiquitous organism found in seawater. In Florida, consumer information statements (required as labels on bags of oysters and in restaurants) emphasize the risk for Vibrio infection among persons with underlying liver disease and other preexisting illnesses (6). In addition, these statements suggest that such persons eat oysters fully cooked and consult with their physician if uncertain about whether they are at risk.

States conduct monitoring programs to assure clean oyster beds, legal harvesting, and proper handling of oysters. However, at both the Louisiana and Florida oyster harvest sites, routine fecal coliform water-quality monitoring conducted once each month did not detect oyster-bed contamination. Furthermore, the outbreak reported in Florida was identified in part because of publicity about the larger outbreaks associated with oysters harvested in Louisiana. These findings suggest that monitoring waters for fecal coliforms may be insufficient to indicate the presence of viruses (e.g., Norwalk-like virus). Continued surveillance for outbreaks of gastroenteritis associated with consumption of raw oysters is needed to assess efficacy of the NSSP in preventing human illness. Public health officials should consider raw oyster consumption as a possible source of infection during the evaluation of gastroenteritis outbreaks.

References
1. Monroe SS, Stine SE, Jiang XI, Estes MK, Glass RI. Detection of antibody to recombinant Norwalk virus antigen in specimens from outbreaks of gastroenteritis. J Clin Microbiol 1993; 31:2866-72.
2. Moe CL, Gentsch J, Ando T, et al. Application of PCR to detect Norwalk virus in fecal specimens from outbreaks of gastroenteritis. J Clin Microbiol 1994;32:642-8.
3. Ando T, Mulders MN, Lewis DC, Estes MK, Monroe SS, Glass RI. Comparison of the polymerase region of small round structured virus strains previously classified in three antigenic types by solid-phase immune electron microscopy. Arch Virol 1994;135:217-26.
4. CDC. Multistate outbreak of viral gastroenteritis related to consumption of oysters -- Louisiana, Maryland, Mississippi, and North Carolina, 1993. MMWR 1993;42:945-8.
5. Office of Seafood, Shellfish Sanitation Branch, Food and Drug Administration. Sanitation of shellfish growing areas, part 1. {Section C.3.c}. In: National Shellfish Sanitation Program manual of operations. Washington, DC: US Department of Health and Human Services, Public Health Service, 1992:C8-C9.
6. Hlady WG, Mullen RC, Hopkins RS. Vibrio vulnificus from raw oysters: leading cause of reported deaths from foodborne illness in Florida. J Fla Med Assoc 1993;80:536-8.

December 17, 1993 / 42(49)

Multistate Outbreak of Viral Gastroenteritis Related to Consumption of Oysters -- Louisiana, Maryland, Mississippi, and North Carolina, 1993

On November 17, 1993, the state health departments of Louisiana, Maryland, and Mississippi notified CDC of several outbreaks of gastroenteritis occurring in their states since November 12. Preliminary epidemiologic investigations identified consumption of oysters as the primary risk factor for illness. On November 16, the Louisiana Department of Health and Hospitals (LDHH) had identified the Grand Pass and Cabbage Reef harvesting areas off the Louisiana coast as the source of oysters associated with outbreaks in Louisiana and Mississippi. Tagged oysters associated with outbreaks in Maryland were traced to the same oyster beds. The oysters harvested from these areas had been distributed throughout the United States. On November 18 and 19, the LDHH and CDC notified state epidemiologists of the potential for oyster-associated illness; outbreaks of oyster-associated gastroenteritis subsequently were identified in Florida and North Carolina. Collaborative investigations by state health officials, the Food and Drug Administration (FDA), and CDC were initiated to determine the magnitude and characteristics of the multistate outbreak, identify the etiologic agent, and trace the oysters. This report summarizes the preliminary findings of the ongoing investigation.*

As of December 2, the investigation had identified 23 separate clusters of ill persons in four states. These clusters have accounted for acute gastroenteritis in at least 180 persons who consumed oysters in a variety of settings, ranging from an individual family meal to a 3-day festival attended by 19,000 persons. Similar clinical features of gastroenteritis predominated in all clusters. In Maryland, where 90 ill persons were identified, clinical features included diarrhea (83 {92%}), vomiting (64 {71%}), nausea (60 {67%}), abdominal cramps (55 {61%}), and fever (40 {44%}). For ill persons from Louisiana, Maryland, and Mississippi, the median incubation period was 34 hours (n=146 persons), and median duration of illness was 37 hours (n=137).

Raw or steamed oysters were the only food associated with illness; attack rates among the 23 groups ranged from 43% to 100%. Oysters from 20 of 23 outbreaks were traced to the implicated harvest area; oysters or their tags were not available from the other three clusters. Three persons were hospitalized, and at least four cases of secondary transmission have been reported. In one Maryland cluster, associated with a 3-day event beginning on November 12, primary cases first occurred on November 13; secondary cases first occurred on November 18.

Stool specimens were examined by electron microscopy (EM) and reverse transcription-polymerase chain reaction (RT-PCR) methods. Small round structured viruses or Norwalk-like viruses were detected by EM and confirmed by RT-PCR in 13 of 26 stool specimens from ill persons in Louisiana, Maryland, Mississippi, and North Carolina. Oysters associated with several of the outbreaks are being analyzed for the presence of Norwalk-like viruses by RT-PCR.

In addition to the notification of state and territorial epidemiologists by LDHH and CDC on November 18 and 19, four public health measures were implemented to prevent further outbreaks associated with the contaminated oysters. First, on November 16, LDHH implemented National Shellfish Sanitation Program (NSSP) procedures for shellfish harvesting closures and recall procedures for oysters from the implicated harvest area (1). Second, on November 18, public health officials in Maryland, North Carolina, and Virginia initiated investigations to identify, detain, and recall all Grand Pass and Cabbage Reef oysters harvested during November 9-11 that had reached the retail markets in their states. Third, on November 23, FDA issued a statement advising consumers that all oysters harvested before November 16 from the Grand Pass and Cabbage Reef areas should not be consumed. Fourth, on November 24, CDC issued a follow-up memorandum to all state and territorial epidemiologists and public health laboratory directors alerting them to the outbreaks and instructing appropriate handling of laboratory specimens if additional outbreaks are suspected.

The continuing investigation in Louisiana, Maryland, Mississippi, and North Carolina includes efforts to trace contaminated oysters from the implicated harvest area through large distributors to retailers and consumers. Reported by: C Conrad, Seafood Sanitation Program; K Hemphill, Molluscan Shellfish Program; S Wilson, L McFarland, DrPh, State Epidemiologist, Office of Public Health, Louisiana Dept of Health and Hospitals. K Coulbourne, Talbot County Health Dept, Easton; S Qarni, MD, Baltimore County Health Dept, Baltimore; S Poster, Harford County Health Dept, Bel

Air; C Groves, MS, C Slemp, MD, E Butler, D Matuszak, MD, D Dwyer, MD, E Israel, MD, State Epidemiologist, Maryland State Dept of Health and Mental Hygiene. J Cirino, Bureau of Marine Resources; D Cumberland, L Pollack, MD, B Brackin, MPH, M Currier, MD, State Epidemiologist, Mississippi State Dept of Health. H Morris, Beaufort County Health Dept, Washington; M Bissett, S Evans, Craven County Health Dept, New Bern; B Respess, Pitt County Health Dept, Greenville; B Jenkins, J Maillard, MD, R Meriwether, MD, JN MacCormack, MD, State Epidemiologist, North Carolina Dept of Environment, Health, and Natural Resources. B Creasy, J Veazey, K Calci, S Rippey, PhD, G Hoskin, Food and Drug Administration. Div of Field Epidemiology, Epidemiology Program Office; Viral Gastroenteritis Section, Respiratory and Enteric Viruses Br, Div of Viral and Rickettsial Diseases, National Center for Infectious Diseases, CDC.

Editorial Note: Because oysters from the beds implicated in this outbreak were shipped to at least 14 states, ** public health officials, health-care providers, and the public should be informed of the possibility that consumption of oysters from these beds may be associated with clusters and isolated cases of acute gastroenteritis in their states. The cases of gastrointestinal illnesses identified by this investigation were recognized because they occurred as part of discrete clusters; however, it is likely that many isolated cases occurred but were not recognized or reported. For example, a previous study of persons who attended a national convention in Louisiana determined that the risk for acute gastroenteritis was higher among persons who consumed raw shellfish than among those who did not, even though no "outbreaks" were identified (2).

Oysters can be traced to their harvest beds because of the regulation requiring sacks of oysters to carry a tag identifying their harvest date and the bed from which they were harvested (1). In this multistate outbreak, these tags facilitated the rapid identification and closing of contaminated beds, provided the link for illness occurring simultaneously in several states, and enabled a product recall.

Investigations of shellfish-associated outbreaks of gastroenteritis have implicated a variety of pathogens, including Vibrio species, Salmonella typhi, Campylobacter species, hepatitis A, and Norwalk-like viruses. For most reported outbreaks, however, an etiologic agent is not identified; these outbreaks may be of viral origin (3). Gastrointestinal illness associated with the consumption of virally contaminated oysters characteristically is self-limited and not life-threatening. However, the likelihood of more severe disease may be increased for persons who are immunocompromised or have other chronic problems (e.g., alcoholism; hepatic, gastrointestinal, or hematologic disorders; cancer; diabetes; or kidney disease).

The etiology of this multistate outbreak was determined rapidly because specimens were collected and handled appropriately and new PCR-based assays were available (4,5). To enable examination of specimens for viral agents in such outbreaks, the following methods are recommended: 1) collection of large-volume stool specimens in clean, dry containers during the first 48 hours of illness and storage at 39 F (4 C) and 2) collection of acute- (within 1 week of onset of illness) and convalescent-phase (3-4 weeks after onset) serum specimens.

FDA, NSSP, and the Interstate Shellfish Sanitation Conference have developed guidelines to protect consumers by controlling the harvesting, handling, and processing of shellfish products (6). Additional efforts are required to develop new assays for screening for viral pathogens in these products before distribution to consumers and to evaluate the effectiveness of various food-preparation practices in decreasing the risk for infection associated with the consumption of molluscan shellfish.

References
1. Office of Seafood, Shellfish Sanitation Branch, Food and Drug Administration. National Shellfish Sanitation Program manual of operations: part II, 1992 revision. Washington, DC: US Department of Health and Human Services, Public Health Service, Food and Drug Administration, 1992.
2. Lowry PW, McFarland LM, Peltier BH, et al. Vibrio gastroenteritis in Louisiana: a prospective study among attendees of a scientific congress in New Orleans. J Infect Dis 1989;160:978-3.
4. Morse DL, Guzewich JJ, Hanrahan JP, et al. Widespread outbreaks of clam- and oyster-associated gastroenteritis: role of Norwalk virus. N Engl J Med 1986;314:678-81.
5. CDC. Recommendations for collection of laboratory specimens associated with outbreaks of gastroenteritis. MMWR 1990;39(no. RR-14).
6. Jiang X, Wang J, Graham DY, Estes MK. Detection of Norwalk virus in stool by polymerase

chain reaction. J Clin Microbiol 1992;30:2529-34.

7. Ahmed FE, ed. Seafood safety. Washington, DC: National Academy Press, 1991.

* Because the outbreaks in Florida have been linked to consumption of oysters from harvest areas other than the Louisiana coast, those outbreaks are not included in this report. ** Alabama, California, Florida, Illinois, Louisiana, Maryland, Mississippi, Missouri, New Jersey, North Carolina, South Carolina, Tennessee, Texas, and Virginia.

February 12, 1988 / 37(5);69-71

Epidemiologic Notes and Reports Viral Gastroenteritis -- South Dakota and New Mexico

The following reports describe two outbreaks of viral gastroenteritis associated with contaminated water.

South Dakota. An outbreak of diarrhea occurred among the 331 participants in an outing held at a South Dakota campground on August 30 and 31, 1986. During the event, in which participants hiked 10 or 20 km, water and a reconstituted soft drink were available at rest stands. The State Department of Health conducted a survey of 181 participants: 135 (75%) of these persons reported a gastrointestinal illness. Symptoms most frequently reported were diarrhea (69%), explosive vomiting (55%), nausea (49%), headache (47%), abdominal cramping (46%), and fever (36%). None of the participants required hospitalization. Attack rates by sex and age of patients were virtually equal. Onset of illness occurred 35 hours (mean) after arrival at the campground, and duration of illness was about 33 hours.

A biotin-avidin immunoassay performed at CDC yielded a fourfold rise in antibody titer to Norwalk virus in seven of 11 paired human serum specimens. No pathogenic bacterial or parasitic agents were identified from stool samples. Illness was strongly associated with the consumption of water or the reconstituted powdered soft drink made with water. No other foodstuffs were implicated. The implicated water came from a well at the campground. A yard hydrant was located next to a septic dump station, where sewage from self-contained septic tanks and portable toilets in the park was collected. Water from this hydrant had been used to fill water coolers and to prepare the powdered soft drink. Laboratory analyses of remaining water and reconstituted soft drink samples showed bacterial contamination (fecal coliforms greater than 1,600 cfu/100 mL). Chlorine was stored in a tank and then drawn directly into the water system by a pump without a monitoring system. Water samples obtained from various locations in the campground had excess coliforms when the chlorination system was not operating. Fluorescent dye injected into a 5,000-gallon septic tank situated uphill from the well confirmed that the well was contaminated with sewage.

This campground was closed immediately and voluntarily by the owner. Corrective measures included relocating the well, installing an alarm system to detect malfunctions in the chlorination system, reconstructing the chlorination system to ensure that chlorine remains in contact with water in a storage tank for 30 minutes before the water is distributed, maintaining a daily log on chlorine residuals and sample collection points, and posting the yard hydrant as a nonpotable source of water.

New Mexico. An outbreak of gastroenteritis occurred among the 92 guests and staff at a cabin lodge in northern New Mexico over the Labor Day weekend in 1986. The guests arrived Friday, August 29, and provided their own food for the weekend. The first persons to become ill developed diarrhea on Saturday morning, within 24 hours after arrival. By Wednesday, 36 of the guests and staff members reported symptoms: 34 had diarrhea; 9, vomiting; 14, fever; 22, abdominal cramps; and 1, bloody stools. There were no deaths or hospitalizations.

A questionnaire was administered to all 92 guests and staff to ascertain risk factors for gastroenteritis. Guests consisted of unrelated groups, and they stayed in 18 separate cabins. All 36 of the patients and 37 of the 56 unaffected attendees had drunk water at their cabin. A dose-response relationship was demonstrated between the amount of water consumed and the attack rate. No illness occurred among the persons who did not drink water; 33% of those drinking 1-2 cups and 59% of those drinking greater than or equal to3 cups became ill. Five of the 18 cabins were unaffected; three of these belonged to families who were residents or frequent visitors at the lodge.

Assuming guests were exposed upon arrival or when they first drank water, the median incubation period was 41 hours (range = 7-110 hours). Symptoms lasted from 2-17 days, with a median of 5 days.

The cabins were supplied with water taken from a stream and processed through a small chlorinator and a storage tank that was periodically iodized. A filter had been removed recently from the pipe because it repeatedly became plugged with debris. A severe rainstorm occurred the evening the guests arrived, resulting in increased water turbidity.

Water samples taken at the cabins and the surface stream that supplied the cabins were positive for total coliforms and fecal coliforms. Stool samples from ill patrons were negative for pathogenic bacteria and parasites, except for one sample, from which Giardia was isolated. Convalescent-phase sera were submitted to CDC for 13 cases and 26 controls (2 per case), matched for age within 5 years, gender, and city of residence. Controls were selected from health department personnel who had not visited the lodge. No difference in Norwalk titers was found between five cases and five controls.

Under the supervision of state environmentalists, the water system was renovated before the lodge reopened, with particular emphasis on filters, the chlorinator, and the storage tank. Reported by: PA Bonrud, MS, AL Volmer, TL Dosch, W Chalcraft, D Johnson, B Hoon, M Baker, KA Senger, State Epidemiologist, South Dakota State Dept of Health. CF Martinez, TO Madrid, MPA, RM Gallegos, MS, SP Castle, MPH, CM Powers, JA Knott, RM Gurule, MS Blanch, LJ Nims, MS, PW Gray, PA Gutierrez, MS, M Eidson, DVM, MV Tanuz, HF Hull, MD, State Epidemiologist, New Mexico Health and Environment Dept. Respiratory and Enteroviral Br, Div of Viral Diseases, Center for Infectious Diseases, CDC.

Editorial Note: The two outbreaks of gastroenteritis described above are representative of those frequently reported to CDC. They demonstrate the need for an improved, specific laboratory approach to identify the agents (many of which are presumed to be viral) responsible for these outbreaks (1,2). Transmission of these viruses is often associated with fecal contamination of water sources used for drinking, swimming, or producing ice (3). Additionally, the contamination of coastal water poses a special problem, since the consumption of seafood is a risk factor for acquiring Norwalk agent infection and other enteric viral agents.

The two best-known enteric viral agents, rotavirus (group A) and Norwalk agent, were first seen in the stools of diarrhea patients by means of electron microscopy in the early 1970s. Both agents have proven to be important causes of gastroenteritis in this country, with rotavirus being the most common agent for diarrhea in young children (4) and Norwalk agent being common in adults (5). In recent years, enteric adenoviruses, non-group A rotavirus, and several 27- to 32-nm enteric viruses, including other Norwalk-like agents, caliciviruses, astroviruses, and other enteric viral pathogens, reportedly have been associated with gastroenteritis (1,6). Recent advances in identifying and diagnosing some of these viruses should make it possible to reduce the number of undiagnosed outbreaks in future investigations. Methods for serologic and antigenic tests are available for some agents, but the examination of stool samples by electron microscopy offers the possibility of identifying agents for which no specific tests are available. The probability of detecting viral particles by electron and diagnosing some of these viruses should make it possible to reduce the number of undiagnosed outbreaks in future investigations. Methods for serologic and antigenic tests are available for some agents, but the examination of stool samples by electron microscopy offers the possibility of identifying agents for which no specific tests are available. The probability of detecting viral particles by electron microscopy is greatest if stool specimens are collected during the early stages of illness, preferably within 12 hours and no later than 48 hours after onset. Some viral particles may be more stable if stool samples are stored at 4 ISDC. The following guidelines are currently recommended for specimen collection specifically for diagnosing outbreaks of viral gastroenteritis.

1. Stool specimens should be collected in bulk volume as soon after the time of disease onset as possible and no later than 48 hours after the onset of symptoms.
2. Stool specimens should be refrigerated, not frozen, and shipped to the laboratory on the same day that the specimen is collected.
3. Paired serum specimens that are collected within 1 week of the disease onset (acute phase) and 3 to 4 weeks after the onset of symptoms (convalescent phase) from both ill patients and controls are

required to establish the causal association between agents seen in the stools and the illness.

References
1. Dolin R, Treanor JJ, Madore HP. Novel agents of viral enteritis in humans. J Infect Dis 1987;155:365-76.
2. Ciba Foundation. Novel diarrhoea viruses. Chichester, United Kingdom: Wiley, 1987. (Symposium no. 128).
3. Centers for Disease Control. Outbreak of viral gastroenteritis--Pennsylvania and Delaware. MMWR 1987;36:709-11.
4. Brandt CD, Kim HW, Rodriguez WJ, et al. Pediatric viral gastroenteritis during eight years of study. J Clin Microbiol 1983;18:71-8.
5. Kapikian AZ, Chanock RM. Norwalk group of viruses. In: Fields BN, ed. Virology. New York: Raven Press, 1985:1495-517.
6. Cubitt WD, Blacklow NR, Herrmann JE, Nowak NA, Nakata S, Chiba S. Antigenic relationships between human caliciviruses and Norwalk virus. J Infect Dis 1987;156:806-14.

June 13, 1986 / 35(23);383-4

Epidemiologic Notes and Reports Gastroenteritis Outbreaks on Two Caribbean Cruise Ships

Three outbreaks of gastroenteritis occurred on two Caribbean cruise ships between April 26, and May 10, 1986. More than 1,200 persons developed gastrointestinal illness; no deaths were reported. At least one of the outbreaks appears to be associated with Norwalk virus.

Two outbreaks occurred on two consecutive 1-week cruises of the Holiday, a Carnival Cruise Line ship. Between April 26 and May 3, a total of 392 (25%) of 1,550 passengers and 30 (4%) of 679 crew who completed questionnaires developed gastroenteritis. Eighty-six percent had diarrhea; 62%, vomiting; 36%, headache; and 26%, subjective symptoms of fever. The outbreak peaked on the fifth and sixth days of the cruise. On the next voyage, from May 3 to May 10, a second outbreak occurred on the Holiday in which 321 (22%) of 1,470 passengers and 48 (7%) of 658 crew developed gastroenteritis. A sanitation inspection initiated by CDC on May 3 revealed deficiencies related to water chlorination record-keeping, food preparation and holding, and potential contamination of food. A detailed account of these deficiencies was provided to the ship's management at the end of the investigation on May 3, and recommendations were made to prohibit food-service personnel from working while ill and to correct the sanitation deficiencies. A week later, when the inspection was completed, several deficiencies similar to those of the previous week were noted. The final vessel sanitation inspection score on May 10 was 18 out of a possible 100 points (passing = 85).

An outbreak of gastroenteritis also occurred on Holland America Cruises' Rotterdam. Between May 3 and May 10, 405 (37%) of 1,108 passengers and 35 (6%) of 554 crew who completed questionnaires had a gastrointestinal illness. Eighty percent of ill passengers had diarrhea; 78%, vomiting; 41%, headache; and 32%, subjective symptoms of fever. Mean duration of illness was 2.4 days, and 76% of ill passengers were confined to their cabins during the illness. A sanitation inspection by CDC on May 9 and May 10 revealed numerous deficiencies related to food and water sanitation; the sanitation inspection score was 16 out of a possible 100 points. A detailed account of the deficiencies was presented to the ship's management on May 10 following the inspection, and recommendations were made to prohibit food service personnel from working while ill and to correct the sanitation deficiencies.

Bacterial cultures of stool specimens from the first Holiday outbreak did not yield any recognized pathogens. However, an eightfold or greater rise in antibodies to Norwalk virus was demonstrated by biotin-avidin immunoassay (1) in paired sera obtained from three ill Holiday crew members who had suffered gastroenteritis during the April 26-May 3 voyage; Norwalk antigen was detected by biotin-avidin immunoassay in two of six ill passengers from the same voyage. Laboratory studies of specimens and epidemiologic analysis of questionnaires from ill passengers and crew from all three outbreaks are continuing. Reported by Div of Quarantine, Center for Prevention Svcs, Enteric Diseases Br, Div of Bacterial Diseases, Respiratory and Enterovirus Br, Div of Viral Diseases, Center for Infectious Diseases, CDC.

Editorial Note: Outbreaks of gastrointestinal illness on cruise vessels have been caused in the past by contaminated water and by food consumed on the ships or on shore visits (2). Person-to-person transmission has also been strongly suspected on some occasions--in one case, in the setting of repeated outbreaks on consecutive cruises (3).

Laboratory findings implicated Norwalk virus as the pathogenic agent on the first of the Holiday outbreaks. However, all three outbreaks had epidemiologic features characteristic of epidemics of Norwalk virus gastroenteritis. These include: (1) a high attack rate in adults; (2) a high frequency of vomiting; (3) short duration of illness; and (4) absence of identified bacterial pathogens (4).

It is not yet clear whether food or water were vehicles of infection in these outbreaks or whether sanitary deficiencies contributed to the risk of outbreaks of viral enteric disease on these cruise ships.

References
1. Gary GW, Kaplan JE, Stine SE, Anderson LJ. Detection of Norwalk virus antibodies and antigen with a biotin-avidin immunoassay. J Clin Microbiol 1985:22:274-8.
2. Merson MH, Hughes JM, Wood BT, Yashuk JC, Wells JG. Gastrointestinal illness on passenger cruise ships. JAMA 1975:231:723-7.
3. Gunn AG, Terranova WA, Greenberg HB, et al. Norwalk virus gastroenteritis aboard a cruise ship: an outbreak on five consecutive cruises. Am J Epidemiol 1986:112:820-7.
4. Kaplan JE, Gary GW, Baron RC, et al. Epidemiology of Norwalk gastroenteritis and the role of Norwalk virus in outbreaks of acute, non-bacterial gastroenteritis. Ann Int Med 1982:96:756-61.

November 28, 1997 / 46(47);1109-1112

Viral Gastroenteritis Associated with Eating Oysters -- Louisiana, December 1996-January 1997

Viral gastroenteritis outbreaks caused by caliciviruses (i.e., Norwalk-like viruses or small round-structured viruses) have been associated with eating contaminated shellfish, particularly oysters (Crassostrea virginica) (1-3). This report describes the findings of the investigation of an outbreak of oyster-associated viral gastroenteritis in Louisiana during the 1996-97 winter season and implicates sewage from oyster harvesting vessels as the probable cause of contaminated oysters.

On December 30, 1996, the Louisiana Office of Public Health (LOPH) was notified about a cluster of six persons who had onset of gastroenteritis after eating raw oysters on December 25. During December 30, 1996-January 3, 1997, three additional clusters were identified. In all four clusters, ill persons had eaten oysters harvested from Louisiana waterways. LOPH notified all state epidemiologists in the United States about the apparent association of gastroenteritis with eating oysters and requested reports of suspected cases.

A case of gastroenteritis was defined as three or more watery stools or vomiting within a 24-hour period, with onset during December 15-January 9. A cluster of oyster-related cases was defined as a group of three or more persons who had shared a common meal, at least one of whom had eaten oysters and at least one of whom developed gastroenteritis. Sixty clusters comprising 493 persons were reported from Alabama, Florida, Georgia, Louisiana, and Mississippi, and all were included in the subsequent traceback investigation. Of the 60 clusters, data were included in the descriptive analysis of the illness only for those 34 clusters for whom all persons in a cluster could be interviewed. The 34 clusters comprised 290 persons who completed interviews and were included in the descriptive analysis; 271 of 290 persons supplied information on oyster consumption.

Onsets of illness occurred during December 21-January 7 (Figure_1). Of the 290 persons interviewed, 179 (62%) had symptoms that met the case definition. The most common symptoms were diarrhea (83%), abdominal cramps (78%), vomiting (58%), headache (50%), and fever (50%). The median incubation period was 38 hours (range: 8-90 hours), and the median duration of illness was 2 days (range: 1-14 days). The median age of case-patients was 42 years (range: 14-83 years), and 111 (62%) were male. The number of reported cases peaked during December 31-January 5 (Figure_1); the harvest dates of subsequently implicated oysters ranged from December 15 to January 1. Of 201 persons who ate raw oysters, 153 (76%) became ill, compared with 13 (19%) of 70 persons who did not eat raw oysters (risk ratio=4.0). Small round-structured viruses were found by direct electron microscopy in fecal specimens from eight of 11 ill persons. Sequence analysis of nucleic acid from eight specimens representing six clusters demonstrated three unique genetic sequences that corresponded with oysters harvested from three separate harvest sites. Small round-structured viruses were detected in oysters, but genetic sequencing could not be conducted.

The LOPH traced oysters eaten by ill persons to retailers, wholesalers, and harvesters. Restaurants and seafood markets were inspected to observe handling and storage of shellfish, and tags that identified the date and site of harvesting and the harvester's identification number were obtained from purchasers and retailers of sacks that were definitely or possibly implicated. Retailer records were cross- referenced with records from wholesalers and harvesters to establish the accuracy of information about harvester and site of harvest. Oysters associated with the 60 clusters were traced to 26 retailers, 11 wholesalers, and 20 harvesters. Records from several wholesalers did not agree with the information on the oyster sack tag.

As of February 15 (6 weeks after notification of the outbreak), LOPH, despite repeated attempts, had been successful in completing interviews with only three of 20 harvesters about the date and specific location of harvesting of potentially contaminated oysters. However, with the assistance of Louisiana Department of Wildlife and Fisheries, 12 additional harvesters were interviewed. Of eight oyster harvesting boats inspected, seven had inadequate sewage collection and disposal systems.

Testing by the LOPH Molluscan Shellfish Program determined that a toxic algal bloom, which causes paralytic shellfish poisoning, was present in Louisiana's northeastern waterways beginning November 13, 1996; these findings prompted LOPH to close these waterways that day and required harvesters to move to southeastern harvest sites. In addition, on November 15, a freshwater diversion was opened to decrease the salinity and eliminate the algal bloom in the northeastern waters; the diversion also decreased the salinity in the southeastern waters.

On January 3, 1997, LOPH mandated an emergency closure of eight waterways with suspected contamination southeast of the Mississippi River, and on January 6, LOPH recalled oysters harvested from these sites after December 22, 1996. On January 23, 1997, harvesting was permitted to resume, and no additional cases of oyster-associated gastroenteritis were reported.

Reported by: TA Farley, MD, L McFarland, PhD, Epidemiology Section, Louisiana Dept of Health and Hospitals. M Estes, K Schwab, Baylor Univ, Dept of Virology, Houston, Texas. Viral Gastroenteritis Section, Respiratory and Enteric Viruses Br, Div of Viral and Rickettsial Diseases, National Center for Infectious Diseases; Div of Applied Public Health Training (proposed), Epidemiology Program Office, CDC.

Editorial Note: Caliciviruses are small single-stranded RNA viruses that cause acute gastroenteritis characterized by vomiting and/or diarrhea (4). The viruses are difficult to detect, requiring relatively sophisticated molecular methods to identify the virus in fecal specimens and in oysters. There is no reliable marker for indicating presence of the virus in oyster harvesting waters.

This report represents the third oyster-related gastroenteritis outbreak attributed to calicivirus in Louisiana since 1993. An outbreak in 1993 accounted for cases of illness in 73 persons in Louisiana and approximately 130 persons in other states (5) who had consumed oysters from Louisiana. In that outbreak, a harvester with a high level of immunoglobulin A to Norwalk virus reported having been ill before the outbreak and admitted to dumping sewage directly into harvest waters. The findings of the investigation of that outbreak suggested that one ill harvester could contaminate large quantities of oysters in a relatively large oyster bed (6). An oyster-associated outbreak in 1996 was attributed to a malfunctioning sewage disposal system on an oil rig on which some workers had been ill with Norwalk-like gastroenteritis (LOPH, unpublished data, 1997). However, harvesters dumping feces overboard could not be excluded as an additional source of oyster contamination. In both outbreaks, recommendations focused on proper sewage disposal and its regulation.

In this outbreak, the link to the large number of wholesalers and retailers suggests that the oyster contamination preceded distribution and probably occurred in the oyster beds. In addition, harvest sites were 12-15 miles from the nearest community sewage outlet, recreational boating was infrequent in December, commercial boating traffic was infrequent because of the shallow depth of the water, and all oil rigs were considered to have had adequate sewage facilities. The only known source of caliciviruses, such as that implicated in this outbreak, is feces from ill persons. Therefore, based on these considerations, the probable source of human sewage found in the implicated waterways was oyster harvesters, who admitted to routinely discharging their sewage overboard, despite recent recommendations in Louisiana for proper sewage collection and disposal (6; LOPH, unpublished data, 1997).

In previous outbreak investigations, molecular tracebacks generally identified a single strain from a single source. A distinguishing feature of this outbreak was its protracted duration and involvement of three geographically separate harvest sites, each associated with a unique strain of calicivirus. These characteristics suggest a contributory role for different oyster harvesters who were concurrently infected with genetically distinct strains of calicivirus, and each of whom dumped their sewage in different waterways, possibly when environmental conditions (e.g., low water temperatures and decreased salinity) facilitated contamination of oysters with calici-viruses.

Findings in this investigation underscore some of the inadequacies in both the current sewage-disposal practices of oyster harvesting vessels and the oyster tagging system designed to reduce the risk for and magnitude of oyster-associated gastroenteritis outbreaks. Oyster-related outbreaks of viral gastroenteritis probably will continue unless seafood regulators and the oyster industry develop, adopt, and enforce standards for the proper disposal of human sewage from oyster harvesting vessels. Traceback investigations of oysters in outbreaks such as this are difficult because of the prevalence of mislabeling in wholesalers' records and on oyster tags and because harvest identification numbers cannot be consistently traced to harvesters. In this investigation, the inability to accurately trace many of the contaminated oysters hampered efforts to contain the outbreak and prevent recurrences and caused a recall of more products than may have been necessary.

Prevention of oyster-related outbreaks of gastroenteritis requires intensified efforts to 1) develop and enforce laws for appropriate sewage containers on oyster harvesting boats with dump-pumpout stations at docks, 2) educate workers in the oyster industry about the consequences of improper sewage disposal, 3) improve record- keeping by oyster harvesters, wholesalers, and retailers to enhance the reliability of traceback investigations, and 4) further assess the relation between environmental conditions and contamination of oysters.

References
1. Murphy AM, Grohmann GS, Christopher PJ, Lopez WA, Davey GR, Millsom RH. An Australia-wide outbreak of gastroenteritis from oysters caused by Norwalk virus. Med J Aust 1979;2:329-33.
2. Gill ON, Cubitt WD, McSwiggan DA, Watney BM, Bartlett CL. Epidemic of gastroenteritis caused by oysters contaminated with small round structured viruses. Br Med J {Clin Res} 1983;287:1532-4.
3.
Gunn RA, Janowski HT, Lieb S, Prather EC, Greenberg HB. Norwalk virus gastroenteritis following raw oyster consumption. Am J Epidemiol 1982;115:348-51.
4. Estes MK, Atmar RL, Hardy ME. Norwalk and related diarrhea viruses in clinical virology. New York, New York: Churchill Livingstone, 1997:1073-95.
5. Dowell SF, Groves C, Kirkland KB, et al. A multistate outbreak of oyster-associated gastroenteritis: implications for interstate tracing of contaminated shellfish. J Infect Dis 1995;171: 1497-503.
6. Kohn MA, Farley TA, Ando T, et al. An outbreak of Norwalk virus gastroenteritis associated with eating raw oysters: implications for maintaining safe oyster beds. JAMA 1995;273:466-71.

NATURAL TOXINS

August 28, 1998 / 47(33);692-694

Ciguatera Fish Poisoning -- Texas, 1997

On October 21, 1997, the Southeast Texas Poison Center was contacted by a local physician requesting information about treatment for crew members of a cargo ship docked in Freeport, Texas, who were ill with nausea, vomiting, diarrhea, and muscle weakness. This report summarizes an investigation of this outbreak by the Texas Department of Health (TDH), which indicated that 17 crew members experienced ciguatera fish poisoning resulting from eating a contaminated barracuda.

On October 12 and 13, gastrointestinal illness developed in crew members aboard a Norwegian cargo ship. After the ship had docked, on October 22 interviews were conducted with 23 (85%) of 27 crew members. A case was defined as ciguatera fish poisoning if there was a combination of gastrointestinal symptoms (i.e., nausea, vomiting, diarrhea, or abdominal cramps) and neurologic symptoms (i.e., muscle pain, weakness, dizziness, numbness or itching of the mouth, hands, or feet) in a crew member after eating fish on October 12. Of the 23 interviewed, 17 (74%) crew members reported the following symptoms: diarrhea (17 {100%}), abdominal cramps (14 {82%}), nausea (13 {76%}), and vomiting (13 {76%}). Symptoms occurred within 2-16 hours (median: 4.5 hours) after eating fish at approximately 7 p.m. on October 12. By October 14, all ill crew members had experienced neurologic symptoms characteristic of ciguatera poisoning: 15 (88%) reported muscle weakness and pain; 13 (76%), numbness or itching of the mouth; 11 (65%), pruritus of the feet and/or hands; 11 (65%), temperature sensation reversal; 10 (59%), dizziness; and eight (47%), aching or loose-feeling teeth.

On October 21, all 17 ill crew members sought medical care at a clinic. None of the crew members were hospitalized; treatment consisted of supportive measures to reduce discomfort from symptoms. All patients were men aged 23-46 years.

Based on food histories from the 23 crew members, TDH suspected consumption of a barracuda caught by crew members while fishing near the Cay Sal Bank of the Bahamas on October 11 as the source of illness. Seventeen crew members ate the barracuda, and all became ill. None of the eight crew members who did not eat barracuda became ill. Although crew members also ate red snapper and grouper at the same meal, neither of these fish were linked epidemiologically with illness.

Results of cultures of stool samples from 16 crew members were negative for Salmonella, Shigella, Campylobacter, Yersinia, and Vibrio. Three samples of leftover raw barracuda and red snapper that were caught simultaneously with the barracuda that was eaten were recovered from cold storage and then tested for ciguatoxin using an experimental membrane immunobead assay at the Department of Pathology, University of Hawaii. The samples from both fish tested positive for ciguatoxin.

Reported by: W Smith, MD, US Health Works, Freeport; B Lieber, Southeast Texas Poison Center, Galveston; DM Perrotta, PhD, Bur of Epidemiology, Texas Dept of Health. Y Hokama, Univ of Hawaii, Manoa. Foodborne and Diarrheal Diseases Br, Div of Bacterial and Mycotic Diseases, National Center for Infectious Diseases; Div of Applied Public Health Training, Epidemiology Program Office; and EIS officers, CDC.

Editorial Note: Ciguatera poisoning occurs throughout the Caribbean and tropical Pacific regions, where outbreaks have been reported among both residents and tourists. From 1983 through 1992 in the United States, 129 outbreaks of ciguatera poisoning involving 508 persons were reported to CDC; no ciguatera-related deaths were reported (1,2). Most outbreaks were reported from Hawaii (111) and Florida (10), although outbreaks and sporadic cases in California (two), Vermont (one), New York (one), and Illinois (one) also have been associated with consumption of fish imported from tropical waters (3,4). The outbreak described in this report was recognized in an area not typically associated with ciguatera intoxication and underscores that ciguatera poisoning can occur among travelers returning from areas where ciguatera is endemic or among persons consuming fish imported from those areas.

Ciguatera toxins are produced by dinoflagellates, which herbivorous fish consume. These fish are then eaten by large, predatory reef fish (e.g., barracuda, grouper, and amberjacks), which appear to be unharmed by the toxin; because the toxins are lipid-soluble, they accumulate through the food chain. The toxin may be most concentrated in the head, viscera, and roe. Ciguatoxin-containing fish may be highly localized; islands may have some reefs where the fish are inedible because of the toxin and other reefs where the fish are unaffected. No deep-sea fish (e.g., tuna, dolphin, or wahoo) have been found to carry ciguatoxin.

As in this outbreak, ciguatera fish poisoning is diagnosed by the characteristic combination of acute gastrointestinal symptoms (developing within 3-6 hours after ingestion of contaminated fish; watery diarrhea, nausea, and abdominal pain occur and typically lasting approximately 12 hours) and neurologic symptoms (circumoral and extremity paresthesia, severe pruritus, and hot-cold temperature reversal) in persons who eat large, predatory reef fish. Neurologic symptoms may be worsened by alcohol consumption, exercise, sexual intercourse, or changes in dietary behavior, such as dieting or high-protein meals (5; R.W. Dickey, Ph.D., Center for Food Safety and Applied Nutrition, Food and Drug Administration, personal communication, 1998). Occasionally, hypotension, respiratory depression, and coma develop in patients. Mean duration of acute illness is typically 8.5 days, although neurologic symptoms may last for months (6). Because there is no approved human assay for ciguatoxin, the diagnosis is based on clinical findings and by the detection of toxin in samples of fish. No known antidote for ciguatoxin poisoning has been proven, and treatment is primarily for relief of symptoms. Intravenous mannitol may be effective early in the course of illness, but the results of a randomized, placebo-controlled trial of mannitol therapy have not been reported (7-9).

Ciguatoxins are odorless, colorless, tasteless, and unaffected by either cooking or freezing; therefore, persons living in or traveling to areas where ciguatera toxin is endemic should follow these general precautions: 1) avoid consuming large, predatory reef fish, especially barracuda; 2) avoid eating the head, viscera, or roe of any reef fish; and 3) avoid eating fish caught at sites with known ciguatera toxins. Persons traveling to areas where ciguatera is endemic should contact local health officials for more specific recommendations pertaining to that area. Fishermen should avoid known ciguatera-contaminated areas, and vendors should not sell fish caught in those areas.

Ill persons with suspected ciguatera poisoning should promptly seek medical care and save any uneaten portions of fish in a freezer. Suspected cases should be reported to state or local public health officials to assist with the investigation and control of a possible outbreak. Additional information is available about ciguatoxin testing of implicated fish from the Gulf Coast Seafood Laboratory of the Food and Drug Administration (FDA) in Dauphin Island, Alabama, telephone (334) 694-4480, or the University of Hawaii, Honolulu, telephone (808) 956-8682. For general information about seafood safety, call FDA's Seafood Hotline, telephone (800) 332-4010.

References
1. CDC. Foodborne disease outbreaks, 5-year summary, 1983-1987. MMWR 1990;39(no. SS-1):15-57.
2. CDC. Surveillance for foodborne disease outbreaks, United States, 1988-1992. MMWR 1996;45(no. SS-5).
3. CDC. Ciguatera fish poisoning -- Vermont. MMWR 1986;35:263-4.
4. Swift AEB, Swift TR. Ciguatera. Clin Toxicol 1993;31:1-29.
5. Lange WR. Ciguatera fish poisoning. Am Fam Physician 1994;50:579-84.
6. Hughes JM, Merson MH. Fish and shellfish poisoning. N Engl J Med 1976;295:1117-20.
7. Palafox NA, Jain LG, Pinano AZ, et al. Successful treatment of ciguatera fish poisoning with intravenous mannitol. JAMA 1988;259:2740-2.
8. Blythe DG, De Sylva DP, Fleming LE, et al. Clinical experience with IV mannitol in the treatment of ciguatera. Bull Soc Path Ex 1992;85:425-6.
9. Bagnis R, Spiegel A, Boutin JP, et al. Evaluation of the mannitol's efficiency in the treatment of ciguatera in French Polynesia. Med Tropicale 1992;52:67-73.

June 04, 1993 / 42(21);417-418

Ciguatera Fish Poisoning -- Florida, 1991

Twenty cases of ciguatera fish poisoning from consumption of amberjack were reported to the Florida Department of Health and Rehabilitative Services (HRS) in August and September 1991.

This report summarizes the investigation of these cases by the Florida HRS.

On August 9, the Florida HRS was notified of eight persons who developed one or more of the following symptoms: cramps, nausea, vomiting, diarrhea, and chills and sweats within 3-9 hours (mean: 5 hours) after eating amberjack at a restaurant on August 7 or August 8; duration of symptoms was 12-24 hours. Three persons were hospitalized. By August 12, patients began to report pruritus of the hands and feet, paresthesia, dysesthesia, and muscle weakness. Based on initial food histories, the Florida HRS suspected consumption of amberjack as the source of illness. On August 14, three additional persons with similar symptoms who also had eaten amberjack at the restaurant on August 8 were reported.

Results of cultures of stool and vomitus samples from the hospitalized persons were negative for Salmonella, Shigella, Campylobacter, and Yersinia. No cooked amberjack was available from the same lot from the restaurant for further testing. Although minor sanitation and safety violations were observed at the restaurant, they did not appear related to the outbreak. Because of the unique symptomology and common denominator of amberjack, investigators suspected either scombroid or ciguatera poisoning.

The shipment of amberjack was traced to a seafood dealer in Key West, Florida, who had distributed the fish through a dealer in north Florida. The second dealer subsequently had sold the fish to the restaurant, another restaurant in Alabama, and a third dealer who sold the fish to two grocery stores in Alabama and north Florida. On August 20 and on September 20, the Florida HRS received reports of additional suspected cases among persons who had bought amberjack at the Alabama grocery store (six persons) and at the grocery store in north Florida (three), respectively.

The Food and Drug Administration evaluated 19 amberjack samples believed to have originated from a single lot from the Key West dealer and obtained from restaurants and grocery stores in Florida and Alabama for ciguatera-related toxin. Forty percent of the specimens tested by mouse bioassay were positive for ciguatera-related biotoxins.

Reported by: RM Hammond, PhD, Office of Restaurant Programs, RS Hopkins, MD, State Epidemiologist, Florida Dept of Health and Rehabilitative Svcs. R Dickey, PhD, Div of Seafood Research, Food and Drug Administration, Dauphin Island, Alabama. Scientific Information and Communications Program, Office of the Director, Epidemiology Program Office, CDC.

Editorial Note: Ciguatera is a naturally, sporadically occurring fish toxin that affects a wide variety of popularly consumed reef fish; ciguatera becomes more bioconcentrated as it moves up the food chain. Ciguatera and related toxins are derived from dinoflagellates, which herbivorous fish consume while foraging through macro-algae (1). Larger predator reef fish (e.g., barracuda, grouper, amberjack, surgeon fish, sea bass, and Spanish mackerel) have been implicated in previous outbreaks (2,3).

Humans ingest the toxin by consuming either herbivorous fish or carnivorous fish that have eaten contaminated herbivorous fish (4,5). The toxin is tasteless, and because it is heat-stable, cooking does not render the fish safe for consumption. As in this outbreak, ciguatera fish poisoning is diagnosed by the characteristic combination of gastrointestinal and neurologic symptoms in a person who eats a suspected fish (6,7). The diagnosis is supported by detection of ciguatoxin in the implicated fish. No specific, effective treatment for ciguatera fish poisoning has been proven; supportive treatment is based on symptoms (4,7).

Further study of seafood toxins is required to develop routine detection tests for the fishing industry, diagnostic tests to evaluate clinical cases, and effective treatment for persons who ingest ciguatera toxins.

References
1. Bagnis R, Chanteau S, Chungue E, Hurtel JM, Yasumoto T, Inoue A. Origins of ciguatera fish poisoning: a new dinoflagellate, Gambierduscis toxicus Adachi and Fukuyo, definitively involved as a causal agent. Toxicon 1980;18:199-208.
2. CDC. Ciguatera fish poisoning -- Bahamas, Miami. MMWR 1982;31:391-2.
3. Craig CP. It's always the big ones that should get away {Editorial}. JAMA 1980;244:272-3.
4. Monis JG Jr, Lewin P, Smith CW, Blake PA, Schneider R. Ciguatera fish poisoning: epidemiology of the disease on St. Thomas, U.S. Virgin Islands. Am J Trop Med Hyg 1982;31:574-8.

5. Halstead BW. Class osteicthyes: poisonous ciguatoxic fishes. In: Halstead BW. Poisonous and venomous marine animals of the world. Princeton, New Jersey: Darwin Press, Inc, 1978: 325-402.
6. Hughes JM, Merson JH. Fish and shellfish poisoning. N Engl J Med 1976;295:1117-20.
7. Baldy L. 1992 Ciguatera. Florida Journal of Environmental Health 1992;137:10-3.

April 25, 1986 / 35(16);263-4

Epidemiologic Notes and Reports Ciguatera Fish Poisoning -- Vermont

On October 29, 1985, the Epidemiology Division, Vermont Department of Health, learned of two persons with symptoms consistent with ciguatera fish poisoning. Both had eaten barracuda at a local restaurant on October 19. One ill person, a 48-year-old woman, had vomiting, diarrhea, myalgia, and chills 4 hours after the meal, followed the next morning by pruritus, flushing, burning of the tongue, and reversal of hot and cold temperature sensation of objects held in her hands. The second ill person, a 30-year-old male bartender at the restaurant, sought medical attention for severe myalgia and gingival and dental dysesthesia several hours after eating barracuda. In both patients, most symptoms subsided; however, some pruritus and temperature reversal persisted 6 weeks later. A third patron reported pruritus to the restaurant after the meal but was lost to follow-up. No additional cases were identified by contacting the two local emergency rooms and requesting case reports in the Vermont Disease Control Bulletin.

The restaurant had served 24 portions of barracuda received fresh by air from a fish distributor in Florida. Two other restaurants in Burlington had received barracuda from the same shipment. One served 44 portions, and the second froze all portions received. The fish distributor reported that the fish was purchased from boats fishing in Florida's coastal waters but could not identify the exact location. The distributor ships to locations throughout the contiguous United States. No information was available about the distribution of other fish from the same catch.

All portions of a single barracuda frozen by one restaurant and tested for ciguatoxin by enzyme immunoassay at the Department of Pathology, University of Hawaii, were positive for ciguatoxin. Reported by RL Vogt, MD, State Epidemiologist, Vermont Dept of Health; AP Liang, MD, State Epidemiologist, Hawaii Dept of Health; Div of Field Svcs, Epidemiology Program Office, Enteric Diseases Br, Div of Bacterial Diseases, Center for Infectious Diseases, CDC.

Editorial Note: Human ciguatera poisoning can occur after consumption of a wide variety of coral reef fish, such as barracuda, grouper, red snapper, amberjack, surgeonfish, and sea bass (1,2). Ciguatoxin and related toxins are derived from dinoflagellates, which herbivorous fish consume while foraging through the macro-algae (3). Humans ingest the toxin by consuming either herbivorous fish or carnivorous fish that have eaten the contaminated herbivores. Larger, more predacious reef fish are generally more likely to be toxic (4,5). Since the toxin is heat-stable, cooking does not make the fish safe to eat.

As the domestic and imported fish industry expands its market, the diagnosis of this "tropical" disease must be considered even in areas to which coral-reef fish are not native. Ciguatera fish poisoning can be diagnosed by the characteristic combination of gastrointestinal and neurologic symptoms in a person who ate a suspect fish (6). The diagnosis can be supported by detection of ciguatoxin in the implicated fish.

Hawaii now uses a "stick test" immunoassay to detect ciguatoxin in fish (7). The test is sensitive, specific, inexpensive, and easy to use in the field. In Hawaii, if an outbreak-related fish tests positive for ciguatoxin, the reef area of catch is posted to discourage further fishing in that area. In Miami, Florida, because barracuda have been frequently associated with ciguatera poisoning, a city ordinance bans the sale of barracuda (8).

References
1. Morris JG Jr, Lewin P, Smith CW, Blake PA, Schneider R. Ciguatera fish poisoning: epidemiology of the disease on St. Thomas, U.S. Virgin Islands. Am J Trop Med Hyg 1982;31:574-8.
2. Halstead BW. Class osteichthyes: poisonous ciguatoxic fishes. In: Halstead BW. Poisonous and

venomous marine animals of the world. Princeton: Darwin Press, Inc., 1978:325-402.

3. Bagnis R, Chanteau S, Chungue E, Hurtel JM, Yasumoto T, Inoue A. Origins of ciguatera fish poisoning: a new dinoflagellate, Gambierdiscus toxicus Adachi and Fukuyo, definitively involved as a causal agent. Toxicon 1980;18:199-208.

4. CDC. Ciguatera fish poisoning--Bahamas, Miami. MMWR 1982;31:391-2.

5. Craig CP. It's always the big ones that should get away . JAMA 1980;244:272-3.

6. Hughes JM, Merson MH. Fish and shellfish poisoning. N Engl J Med 1976;295:1117-20.

7. Hokama Y. A rapid, simplified enzyme immunoassay stick test for the detection of ciguatoxin and related polyethers from fish tissues. Toxicon 1985;23:939-46.

8. Lawrence DN, Enriquez MB, Lumish RM, Maceo A. Ciguatera fish poisoning in Miami. JAMA 1980;244:254-8.

July 23, 1982 / 31(28);391-2

Epidemiologic Notes and Reports Ciguatera Fish Poisoning -- Bahamas, Miami

On March 6, 1982, the U.S. Coast Guard in Miami, Florida, received a request for medical assistance from an Italian freighter located in waters off Freeport, Bahamas. Numerous crew members were ill with nausea, vomiting, and muscle weakness and required medical evacuation for hospitalization and treatment.

A total of 14 ill crew members were airlifted to three Florida hospitals. Three were seen in emergency rooms and later released. Eleven were hospitalized; seven required admission to intensive care units. All patients were Italian males, age 24-40 years; symptoms included diarrhea--12 patients (86%), vomiting--11 (79%), paresthesias--11 (79%), hypotension--10 (71%), peripheral muscular weakness--9 (65%), nausea--8 (57%), abdominal cramping--6 (43%), pruritis--4 (29%), and peripheral numbness--2 (14%). These findings were consistent with ciguatera fish poisoning, and an epidemiologic investigation was initiated.

The ship employed 26 crew members and is permanently based near Freeport, where it ferries petroleum products ashore from large tankers. On March 4, a crew member caught a 25-pound barracuda while fishing from the ship. On March 6, 14 crew members cooked and ate the barracuda; all became ill within 6 hours. None of the 12 crew members who did not eat the barracuda became ill. Six of the ill crew members reported becoming sick 45 minutes to 6 hours after the implicated meal (median: 2.5 hours). All 14 crew members eventually recovered without sequelae and returned to work. Median length of hospital stay was 6 days. Reported by SC Royal, Miami Quarantine Station, MA Poli, TJ Mende, DG Baden, Dept of Biochemistry, University of Miami, School of Medicine, RM Galbraith, TB Higerd, Dept of Clinical Immunology and Microbiology, Medical University of South Carolina, M Enriquez, MD, Dade County Health Dept, HJ Janowski, MPH, Acting State Epidemiologist, Florida Dept of Health and Rehabilitative Svcs; Field Services Div, Epidemiology Program Office, CDC.

Editorial Note: Ciguatera is a human intoxication syndrome associated with the consumption of marine tropical reef fishes. Although recent surveys indicate that poisonings are relatively uncommon in Florida (1,2), one investigator recorded 280 intoxications from January 1978 to June 1980 (2).

The ichthyosarcotoxins are thought to be accumulated through the food chain, the toxins being produced by microalgae known as dinoflagellates (3,4). The toxins are lipid-soluble and appear to accumulate in the flesh, fatty tissue, and viscera of large predatory species of fish, such as barracuda, grouper, and snapper (5,6). The isolation, purification, and characterization of the suspected toxins have been hampered by limited availability of authentic ciguatoxic fish, lack of a specific sensitive assay, and the low concentration and heterogeneity of toxins present in specimens.

The assessment of toxicity most often used is the mouse bio-assay. Based on signs elicited following intraperitoneal (IP) injection, it includes, but is not limited to inactivity, diarrhea, labored breathing, cyanosis, piloerection, tremors, paralysis, and staggering gait. Death occurs when the injection is given in higher doses, with a lethal dose, 50% kill (LD((50))), of 0.45 ug/kg for purified toxin (5). Thus, ciguatoxin is one of the most potent marine toxins known. The barracuda's head was toxic by mouse bio-assay with an LD((50)) (IP) of 2-5 gram equivalents of original fish meat. Thin-layer chromatographic separation of extracts revealed the presence of at least two major toxins. Further purification is under way to define more clearly the toxin(s) implicated in this outbreak.

References
1. Lawrence DN, Enriquez MB, Lumish RM, Maceo A. Ciguatera fish poisoning in Miami. JAMA 1980;244:254-8.
2. Poli MA. A review of ciguatera, with special reference to the Caribbean, and an investigation into its significance and incidence in Florida. Master's thesis, University of Miami, 1982.
3. CDC. Ciguatera fish poisoning--St. Croix, Virgin Islands of the United States. MMWR 1981;30:138-9.
4. Bagnis R, Chanteau S, Chunque E, Hurtel JM, Yasumoto T, Inoue A. Origins of ciguatera fish poisoning: a new dinoflagellate, Gambierdiscus toxicus Adachi and Fukuyo, definitely involved as a causal agent. Toxicon 1980;18:199-208.
5. Withers NW. Ciguatera fish poisoning. Ann Rev Med 1982;3:97-111.
6. Banner AH. Ciguatera: a disease from coral reef fish. In: Jones OA, Endean R, eds. Biology and geology of coral reefs. New York: Academic Press, 1975; III:177-213.

March 15, 1991 / 40(10);157-161

Epidemiologic Notes and Reports Paralytic Shellfish Poisoning -- Massachusetts and Alaska, 1990

Paralytic shellfish poisoning (PSP) is a foodborne illness caused by consumption of shellfish or broth from cooked shellfish that contain either concentrated saxitoxin, an alkaloid neurotoxin, or related compounds. This report summarizes outbreaks of PSP that occurred in Massachusetts and Alaska in June 1990.

MASSACHUSETTS
On June 6, 1990, the Massachusetts Department of Public Health (MDPH) was notified that, on June 5, foodborne illness had occurred in six fishermen aboard a fishing boat in the Georges Bank area off the Nantucket coast. Onset of illness occurred after the men had eaten blue mussels (Mytilus edulis) harvested in deep water about 115 miles from the island of Nantucket.

The six men (age range: 24-47 years) developed symptoms 1-2 hours after consuming the shellfish (Table 1). Symptoms included numbness of mouth (five men), vomiting (four), paresthesia of extremities (four), numbness and tingling of tongue (two), numbness of face (two), numbness of throat (one), and periorbital edema (one). In all six men, lower back pain occurred approximately 24 hours after onset. The median duration of neurologic symptoms was 14 hours, and for lower back pain, 3.3 days. Approximately 10 hours after onset, when the fishermen presented to a local hospital emergency room, four were recovering; however, two, including one who had recovered from loss of consciousness, required hospitalization for 2-3 days.

The six fishermen had boiled the mussels for approximately 90 minutes before consuming them with baked fish, boiled rice, boiled potatoes, green salad, and other food items. They did not consume alcoholic beverages with the implicated meal.

Laboratory examination of the uneaten mussels detected saxitoxin concentrations of 24,400 ug/100 g in the raw mussels and 4280 ug/100 g in the cooked mussels (maximum safe level: 80 ug/100 mg). The difference in the levels of PSP toxin between raw and cooked mussels suggested that much of the saxitoxin had dissipated into the boiling water.

The implicated mussels had been harvested in an area of the Georges Bank where contamination of surf clams and sea scallops with saxitoxin had been detected through a deep-sea sampling survey conducted by the MDPH. The same area had been identified in a warning issued 2 weeks earlier by the MDPH and the National Marine Fisheries Service. The warning had been based on a report of a fisherman and his wife who had developed symptoms compatible with PSP after eating mussels obtained from that area. Because of the sampling survey and the first reported incident, the Georges

Bank had been closed to harvesting of all shellfish except the adductor muscles of sea scallops shucked at sea. The closure notice had been sent to all appropriate Coast Guard stations and fishing vessels in the area; however, the six fishermen involved in the outbreak reported they had not received it.

ALASKA

On June 26, 1990, a physician reported to the Alaska Department of Health and Social Services (ADHSS) that a Native Alaskan man had died after consuming shellfish collected from a beach on the Alaska Peninsula. On the evening of June 25, while aboard a fishing boat, the decedent had consumed 25-30 steamed butter clams and 2 teaspoons of butter clam broth. Within an hour, he complained of numbness and tingling around his mouth, face, and fingers. Two hours later, he suffered a cardiopulmonary arrest; despite cardiopulmonary resuscitation efforts by emergency personnel, the patient died. Based on the symptoms reported, PSP was diagnosed. The patient's gastric contents contained 370 ug/100 g of PSP toxin, and a sample of the butter clam broth from the meal contained 2650 ug/100 g.

Two other crewmembers had also consumed butter clams. One developed numbness and tingling of the face and hands and dizziness approximately 1 hours later and recovered uneventfully; the other had no symptoms. Four crewmembers from two other fishing boats also had shared the butter clams presumed to be the vehicle for illness; all four had symptoms consistent with PSP.

As a consequence of this episode, ADHSS identified three additional episodes in the region of the Alaska Peninsula and Kodiak Island during June 1990. Each episode involved consumption of shellfish collected from a different area (Figure 1). When aggregated, the four episodes constituted a PSP outbreak with 13 cases among 21 persons (attack rate: 62%) who had consumed the implicated shellfish.

The four episodes occurred during June 17-25. Onset of symptoms ranged from 0 to 2 hours (median: 1 hour) after consumption of shellfish. Duration of illness ranged from 1 to 24 hours (median: 7.5 hours). Seven (54%) persons sought medical care. Only the index patient died (case-fatality rate: 8%); the others recovered uneventfully.

Seven (54%) cases resulted from consumption of butter clams, and six (46%), from mussels. Shellfish were consumed raw, boiled, or steamed. Affected persons consumed three to 30 shellfish each (median: four shellfish).

Because shellfish from the four episodes were not available for testing for PSP toxin, samples were collected from the four sites where the shellfish had been harvested. Butter clam samples from Volcano Bay and King Cove contained 7750 ug/ 100 g of PSP toxin, and mussel samples from Sand Point and Kodiak contained 1925-12,960 ug/100 g.

ADHSS, in conjunction with the Alaska Department of Environmental Conservation, issued a statewide press release warning of the risk for PSP for persons consuming shellfish collected from Alaskan beaches. Reported by: K Sharifzadeh, DVM, N Ridley, MS, R Waskiewicz, MS, P Luongo, Div of Food and Drugs, GF Grady, MD, A DeMaria, MD, RJ Timperi, MPH, J Nassif, MS, Massachusetts Dept of Public Health. M Sugita, Kodiak Health Clinic; V Gehrman, King Cove Clinic; P Peterson, Sand Point Clinic; A Alexander, Cold Bay Clinic; R Barrett, K Ballentine, Alaska Dept of Environmental Conservation; JP Middaugh, MD, State Epidemiologist, Alaska Dept of Health and Social Svcs. I Somerset, MS, Food and Drug Administration. Enteric Diseases Br, Div of Bacterial and Mycotic Diseases, Center for Infectious Diseases; Div of Field Epidemiology, Epidemiology Program Office, CDC.

Editorial Note: The neurotoxins that cause PSP are among the most potent toxins known and can impair sensory, cerebellar, and motor functions. Saxitoxin is heat-stable and unaffected by standard cooking or steaming (1), is water-soluble, and can be concentrated in broth. Symptoms usually occur within 2 hours after ingestion of shellfish; high doses can lead to diaphragmatic paralysis, respiratory failure, and death (1,2).

The diagnosis of PSP is based on patient exposure history and clinical manifestations and on epidemiologic information. Predominant manifestations include paresthesia of the mouth and extremities, ataxia, dysphagia, and muscle paralysis (3-5); gastrointestinal symptoms are less common. Coma, total muscular paralysis, and respiratory arrest with death can occur. The prognosis is favorable for patients who survive beyond 12-18 hours (3). Because PSP has no specific treatment or antidote, treatment is supportive. Prompt evacuation of stomach contents may help by removing the remaining toxin-containing shellfish.

During 1973-1987, state health departments reported 19 PSP outbreaks (mean size: eight persons) to CDC's Foodborne Disease Outbreak Surveillance System. Outbreaks were caused by consumption of mussels, clams, oysters, scallops, and cockles. Outbreaks on the west coast have been reported from May through October, and on the east coast from August through October.

Most cases of PSP occur in individuals or small groups who gather shellfish for personal consumption. Although PSP has traditionally been considered a risk only in shellfish harvested from cold water, the incidence in tropical areas may be increasing: outbreaks have been reported recently from Central and South America, Asia, and the Pacific region (2,6).

The PSP-associated death in Alaska was the first in that state in greater than 14 years. From 1976 through 1989, 42 PSP outbreaks (accounting for 94 cases) were documented in Alaska. Butter clams were implicated in 23 (55%) of the outbreaks. Other shellfish implicated in Alaskan outbreaks included mussels, cockles, steamer clams, sea snails, and razor clams. Thirty-one (74%) of the 42 outbreaks occurred during May-July.

Shellfish can become toxic when toxin-producing dinoflagellates create massive algal blooms known as "red tides." However, shellfish can become toxic even in the absence of such blooms; detoxification may require a month or more in clean waters.

To prevent outbreaks of PSP and other shellfish intoxications, samples of susceptible mollusks are periodically collected in the coastal states and tested for toxin by mouse bioassay. When toxin levels exceed 80 ug/100 g, affected growing areas are quarantined, and sale of shellfish is prohibited. Warnings posted in shellfish-growing areas and on beaches and placed in the news media can alert the public to the hazard.

References
1. Hughes JM, Merson MH. Fish and shellfish poisoning. N Engl J Med 1976;295:1117-20.
2. Rodrigue DC, Etzel RA, Hall S, et al. Lethal paralytic shellfish poisoning in Guatemala. Am J Trop Med Hyg 1990;42:267-71.
3. Eastaugh J, Shepherd S. Infectious and toxic syndromes from fish and shellfish consumption. Arch Intern Med 1989;149:1735-40.
4. Sakamoto Y, Lockey RF, Krzanowski JJ. Shellfish and fish poisoning related to the toxic dinoflagellates. South Med J 1987;80:866-72.
5. Wallace J. Disorders caused by venoms, bites, and stings. In: Isselbacher KJ, Adams RD, Braunwald E, Petersdorf RG, Wilson JD, eds. Harrison's principles of internal medicine. 9th ed. New York: McGraw-Hill Book Company, 1980:927.
6. Maclean J, White A. Toxic dinoflagellate blooms in Asia: a growing concern. In: Anderson DM, White AW, Baden DG, eds. Toxic dinoflagellates. New York: Elsevier, 1985:517-30.

April 12, 1991 / 40(14);242

Notices to Readers Paralytic Shellfish Poisoning - Massachusetts and Alaska, 1990 Errata: Vol. 40, No. 10

In the article, "Paralytic Shellfish Poisoning -- Massachusetts and Alaska, 1990," the maximum safe level of saxitoxin concentration given in the third-to-last line of page 157 should be 80 ug/100 g.

Epidemiologic Notes and Reports Scombroid Fish Poisoning -- Illinois, South Carolina

Scombroid fish poisoning is an acute syndrome resulting from consumption of fish containing high levels of histamine. This report summarizes investigations of two outbreaks of scombroid fish poisoning in Illinois and South Carolina in 1988.

ILLINOIS. On February 26, 1988, eight cases of scombroid fish poisoning occurred in Chicago in five patrons and three employees of a private club who had eaten a buffet lunch. Six of the ill persons experienced symptoms that included headache, nausea, flushing, dizziness, and diarrhea 90 minutes after the meal. The median duration of symptoms was 9.5 hours. Investigation by the Illinois Department of Public Health revealed that seven of the ill persons had eaten mahi mahi with dill sauce; the eighth had eaten the dill sauce scraped from the serving pan that held the fish. Three persons noted that the fish tasted "Cajun," and one stated that it had a hot or spicy taste.

The club had purchased 10.5 pounds of frozen mahi mahi from a suburban Chicago distributor the week before it was served. The distributor's records revealed that fish from one of two lots of mahi mahi had been sent to the club. On March 1, the state health department placed both lots under embargo. The lots included boxes with evidence of freezer burn, a sign of thawing and refreezing, but these boxes were held by the distributor as damaged goods and not used. The Food and Drug Administration (FDA) tested fish from 17 boxes in these lots; no fish from boxes with evidence of freezer burn were sampled. Six samples had histamine levels greater than or equal to 50 mg/ 100 g (range: 50-160 mg). The fish was kept in the club's freezer at 0-5 F (œm-15.0 C- œm-17.8 C) until February 26, when it was thawed by placing it under running water for 15 minutes. The fish was then cut into portions, placed flat in pans in the cooler, and baked as needed during lunch until the supply was depleted.

SOUTH CAROLINA. In September 1988, nine cases of scombroid fish poisoning in Charleston were investigated by the South Carolina Department of Health and Environmental Control. Of the nine cases, five occurred after consumption of a midday meal at a restaurant September 9, one case followed an evening meal at a second restaurant September 10, and three cases occurred after an evening meal of fish prepared at home but obtained from the first restaurant.

The median age of the nine ill persons was 55 years (range: 18-64 years); five were women. Illness occurred 5-60 minutes after the meal (median time to onset of symptoms: 38 minutes). Symptoms included flushing, diarrhea, headache, feverishness, nausea, rapid pulse, pruritus, dizziness, vomiting, facial swelling, numbness around the mouth, and stomach pain. Symptoms resolved in all persons within 10 hours (median: 6 hours). Fiv patients required emergency room treatment, and one was admitted for observation because of underlying cardiac disease.

Two persons noted that the fish had a slight peppery taste, and one person noted a metallic taste. All had eaten yellow-fin tuna supplied by the same local distributor. FDA analyses of two samples from the yellow-fin tuna revealed histamine levels of 728 mg/100 g and 583 mg/100 g, respectively.

The yellow-fin tuna were probably caught 1 day before purchase in waters off the coast of New Jersey, Rhode Island, and Virginia, and were cleaned and packed in ice on the boat. They were then obtained from docks in Cape May and Barnegat Light, New Jersey, by a regional supplier in Philadelphia 3 days before the outbreak. After purchase, the fish were repacked in ice and delivered by truck to Philadelphia, where they were divided into two lots and repacked in ice for shipment to wholesalers. They left the Philadelphia supply plant by refrigerated truck 12 hours after arrival. The wholesaler in Charleston received 188 pounds of yellow-fin tuna from the supply truck 1 day before the outbreak, processed the tuna into steaks, and shipped 17 pounds of steaks from the same fish to each of the two restaurants implicated in the outbreak. Tuna steaks from the same shipment were supplied to 12 other Charleston restaurants, all of which reported receiving the fish in ice. Both implicated restaurants kept the fish packed in ice and refrigerated before it was broiled and served to customers.

One day after the outbreak, a telephone survey of emergency rooms in the Charleston area revealed no other cases suggestive of scombroid poisoning. All restaurants that had received yellow-fin tuna supplied by the Charleston wholesaler from this shipment were notified.

Reported by: LR Murray, MD, LC Edwards, MD, City of Chicago Dept of Health; RJ Martin, DVM, D Rogers, DS, CW Langkop, MSPH, CD Cuda, S Redschlag, MS, BJ Francis, MD, State Epidemiologist, Illinois Dept of Health. MD Lawhead, MD,AMI East Cooper Community Hospital, Charleston; FU Davis, RL Mackey, Charleston County Health Dept; JC Chambers, MD, Trident Public Health District; WB Gamble, MD, State Epidemiologist, South Carolina Dept of Health and Environmental Control. Food and Drug Administration. Div of Field Svcs, Epidemiology Program Office; Enteric Diseases Br, Div of Bacterial Diseases, Center for Infectious Diseases, CDC.

Editorial Note: During 1973-1986, 178 outbreaks of scombroid poisoning affecting 1096 persons (median: two cases/outbreak) were reported to CDC's Foodborne Disease Outbreak Surveillance System; no fatal cases were reported. Outbreaks have been reported from 30 states and the District of Columbia, with Hawaii reporting the largest number of outbreaks (51), followed by California (29), New York (24), Washington (19), and Connecticut (nine). The fish species was known in 143 (80%) of the scombroid outbreaks; the most commonly reported types were mahi mahi (66 outbreaks), tuna (42 outbreaks), and bluefish (19 outbreaks).

Scombroid poisoning is named for the family Scombridae, which includes tuna and mackerel, but this illness can occur after ingestion of any dark-fleshed nonscombroid species containing high levels of free histidine (1). When these fish are improperly refrigerated, free histidine is broken down to histamine by surface bacteria. This latter compound is thought to produce the clinical manifestations of illness (2); hence, some investigators have termed this syndrome histamine poisoning (2).

Illness begins minutes to hours after ingestion of the toxic fish. Symptoms resemble a histamine reaction and frequently include dizziness, headache, diarrhea, and a burning sensation or peppery taste in the mouth. Facial flushing, tachycardia, pruritus, and asthma-like symptoms can also occur. Illness is usually mild and duration is short, making treatment unnecessary. For more severe cases or in patients with underlying medical conditions, oral antihistamines may be beneficial (3). Intravenous cimetidine has been anecdotally reported to ameliorate symptoms but its use warrants further study (4).

Scombroid poisoning is diagnosed by history and clinical symptoms combined with the measurement of histamine levels in implicated fish. Fresh fish normally contains less than 1 mg/100 g of histamine; levels of 20 mg/100 g in some species have been reported to produce symptoms (5). The FDA has established 50 mg/100 g of histamine as a hazardous level in tuna (6), a level exceeded in both outbreaks in this report. Investigation failed to reveal evidence of improper storage. Experimental studies indicate that histamine formation is low at refrigerator temperatures and negligible in fish stored at less than or equal to 32 F (less than or equal to 0 C) (2). As these outbreaks demonstrate, cooking toxic fish is not protective. Therefore, the key to prevention of scombroid poisoning is continuous icing or refrigeration of all potentially scombrotoxic fish from the time they are caught until they are cooked.

References
1. Etkind P, Wilson ME, Gallagher K, Cournoyer J, Working Group on Foodborne Illness Control. Bluefish-associated scombroid poisoning: an example of the expanding spectrum of food poisoning from seafood. JAMA 1987;258:3409-10. 2.Taylor SL. Histamine food poisoning: toxicology and clinical aspects. CRC Crit Rev Toxicol 1986;17:91-128. 3.Hughes JM, Merson MH. Fish and shellfish poisoning. N Engl J Med 1976;295:1117-20. 4.Blakesley ML. Scombroid poisoning: prompt resolution of symptoms with cimetidine. Ann Emerg Med 1983;12:104-6. 5.Bartholomew BA, Berry PR, Rodhouse JC, Gilbert RJ, Murray CK. Scombrotoxic fish poisoning in Britain: features of over 250 suspected incidents from 1976 to 1986. Epidemiol Infect 1987;99:775-82. 6.Food and Drug Administration. Defect action levels for histamine in tuna; availability of guide. Federal Register 1982;47:40487.

July 29, 1988 / 37(29);451

Scombroid Fish Poisoning -- New Mexico, 1987

In July 1987, state and local public health officials in New Mexico investigated two cases of scombroid fish poisoning (histamine poisoning) in persons living in Albuquerque. The New Mexico Health and Environment Department was initially consulted by an Albuquerque physician regarding two patients, a husband and wife, who had become ill within 45 minutes after eating dinner. Their symptoms included nausea, vomiting, diarrhea, headache, fever, flushing, and rapid pulse rate. An investigation by the Albuquerque Environmental Health Department found that the couple had shared a meal of grilled mahi mahi, pasta, salad, water, and wine. Their dog had eaten some of the fish and had vomited; however, their daughter, who had eaten no fish, did not become ill. Both of the patients had been treated with Benadryl, activated charcoal, and ipecac in a hospital emergency room. Their symptoms resolved within 36 hours of onset of illness.

Samples of the remaining mahi mahi were sent to the Food and Drug Administration laboratory in Seattle. Histamine was detected in the samples at a ratio of 20 mg/100 g, a level sufficient to cause symptoms (1). Samples from a different shipment of fish were obtained from the store in Albuquerque where the mahi mahi was purchased. These samples yielded histamine levels of 3 mg/100 g of sample and were negative for ciguatera toxin.

The fish had been imported from Taiwan through California and shipped frozen to the Albuquerque distributor, where it was thawed and sold from iced refrigerator cases. The patients had frozen the fish after they bought it. Later, they thawed it for 3 hours at room temperature and then grilled the still icy fish. Reported by: NB Rieder, MD; NI Goertz, RS, JD Hall, DrPH, Albuquerque Environmental Health Dept; M Eidson, DVM, HF Hull, MD, State Epidemiologist, New Mexico Health and Environment Dept. Albuquerque Resident Post, Food and Drug Administration. Enteric Diseases Br, Div of Bacterial Diseases, Center for Infectious Diseases, CDC.

Editorial Note: Of all varieties of fish, the scombroid species (tuna, bonito, and mackerel) and certain other dark-meat fish, such as mahi mahi, are the most likely to develop high levels of histamine. When fresh scombroid fish are not continuously iced or refrigerated, bacteria may convert the amino acid histidine, which occurs naturally in the muscle of the fish, to histamine. Since histamine is resistant to heat, cooking the fish generally will not prevent illness. Histamine levels may not be correlated with any obvious signs of decomposition of the fish. Thus, prompt and proper refrigeration or icing from the time the fish is caught until it is preserved, processed, or cooked is essential to prevent scombroid fish poisoning. Antihistamines may be useful for symptomatic treatment.

Because histamine is metabolized by intestinal flora, even large doses of ingested pure histamine usually do not cause symptoms. Thus, although histamine is a marker for fish that could cause scombroid fish poisoning, the actual mechanism for the poisoning must depend on an additional cofactor. Experimental evidence indicates that other substances produced in fish by putrefactive bacteria inhibit the metabolism of histamine and permit its absorption and circulation (2).

References
1. Bartholomew BA, Berry PR, Rodhouse JC, Gilbert RJ, Murray CK. Scombrotoxic fish poisoning in Britain: features of over 250 suspected incidents from 1976 to 1986. Epidemiol Infect 1987;99:775-82.
2. Taylor SL. Histamine food poisoning: toxicology and clinical aspects. CRC Crit Rev Toxicol 1986;17:91-128.

April 25, 1986 / 35(16);264-5

Epidemiologic Notes and Reports Restaurant-Associated Scombroid Fish Poisoning -- Alabama, Tennessee

Between December 31, 1985, and January 4, 1986, three restaurants in Alabama and Tennessee received complaints of illness from nine customers and one employee who ate Pacific amberjack fish (also called yellowtail or kahala). Detailed information was obtained on four of the 10 persons. Illness onset occurred 10-90 minutes after eating (median 23 minutes). Symptoms included red facial rash (4/4), body rash (2/4), severe headache (2/4), oral paresthesias (1/4), shortness of breath (2/4), vomiting (1/4), and diarrhea (3/4). Of the three persons who sought medical evaluation, one had diastolic hypotension, and one had bronchospasm. All three were diagnosed as having food or fish allergy and were treated with an antihistamine. Rash persisted for 2-5 hours (median 3 hours), and all other symptoms resolved in 3-36 hours (median 14 hours). One restaurant cook, who did not eat the fish, reported a transient red rash on the hands shortly after handling the fish.

Ill persons reported no other menu items in common. The fish meals were prepared by grilling or frying, and none of the restaurants reported using food preservatives or monosodium glutamate (MSG) on the fish.

In November 1985, a Florida seafood company procured 1,100 pounds of fresh amberjack from southern California. A 120-pound portion was resold December 30 to a distributor that in turn supplied the fish to nine restaurants in Alabama, Kentucky, and Tennessee. After receiving complaints from three of the restaurants, the distributor promptly notified all recipient restaurants and collected 20 pounds of amberjack. Analysis of the leftover fish by the U.S. Food and Drug Administration (FDA) showed 19 of 20 subsamples had markedly elevated levels of histamine (257-430 mg%). (Fresh fish normally contains less than 1 mg% of histamine.) The remaining fish, which had not been distributed, was destroyed under FDA supervision. Reported by JFE Shaw, MPA, Jefferson County Health Dept, Birmingham, WE Birch, DVM, State Epidemiologist, Alabama State Dept of Public Health; RH Hutcheson, MD, State Epidemiologist, Tennessee Dept of Health and Environment; Investigations Br, Nashville District Office, W Staruszkiewicz, Div of Food Technology, Center for Food Safety and Applied Nutrition, US Food and Drug Administration; Div of Field Svcs, Epidemiology Program Office, Enteric Diseases Br, Div of Bacterial Diseases, Center for Infectious Diseases, CDC.

Editorial Note: The symptoms of scombroid fish poisoning resemble those of a histamine reaction; the illness is characterized by flushing, headache, dizziness, burning of the mouth and throat, abdominal cramps, nausea, vomiting, and diarrhea. Urticaria and generalized pruritus often occur (1). In severe cases, bronchospasm and respiratory distress may develop (2). Some victims complain that the toxic food has a sharp or peppery taste. Typical incubation periods are less than 1 hour, although wide variations can occur among individuals (1).

Scombroid means mackerel-like; mackerel, tuna, and bonito are related species that are often implicated in outbreaks of scombroid poisoning. However, nonscombroid species, such as the amberjack reported here, have also been implicated in scombroid poisoning (2,3). Of the 73 outbreaks of scombroid poisoning reported to CDC during the 5-year period 1978-1982, 31 (42%) implicated mahi-mahi (dolphin fish), a nonscombroid fish (4).

Poisoning is caused by the ingestion of spoiled fish. Histamine and probably other toxic byproducts are produced by bacterial action on histidine, a normal muscle constituent of dark-meat fishes (5). Scombroid poisoning is a response to toxic by-products--not an allergic reaction to fish. Once formed, the toxins are heat-stable, so the best defense against poisoning is prompt storage of freshly caught fish at 0 C (32 F) or below (6). Laboratory confirmation of scombroid fish poisoning is based on demonstrating elevated histamine levels in incriminated fish (1). Public health authorities should be notified when this or other fish-related illness is suspected so that the distribution of the implicated food can be determined.

References
1. Hughes JM, Merson MH. Fish and shellfish poisoning. N Engl J Med 1976; 295:1117-20.
2. Halstead BW. Class osteichthyes: poisonous scombrotoxic fishes. In: Halstead BW. Poisonous and venomous marine animals of the world. Princeton: Darwin Press, Inc., 1978:417-35.
3. CDC. Foodborne and waterborne disease outbreaks, annual summary 1977.
4. CDC. Foodborne disease surveillance, annual summary reports, 1978 to 1982.
5. Arnold SH, Brown WD. Histamine toxicity from fish products. Adv in Food Res 1978;23:113-54.
6. Behling AR, Taylor SL. Bacterial histamine production as a function of temperature and time of incubation. J Food Sci 1982;47:1311-4.

May 17, 1996 / 45(19);389-391

Tetrodotoxin Poisoning Associated With Eating Puffer Fish Transported from Japan -- California, 1996

On April 29, 1996, three cases of tetrodotoxin poisoning occurred among chefs in California who shared contaminated fugu (puffer fish) brought from Japan by a co-worker as a prepackaged, ready-to-eat product. The quantity eaten by each person was minimal, ranging from approximately 1/4 to 1

1/2 oz. Onset of symptoms began approximately 3-20 minutes after ingestion, and all three persons were transported by ambulance to a local emergency department (ED). This report summarizes the investigation of these cases by the San Diego Department of Environmental Health (SDEH) and the Food and Drug Administration (FDA). Case Reports

Case 1. A 23-year-old man ate a piece of fugu "the size of a quarter" (approximately 1/4 oz). Approximately 10-15 minutes later, he had onset of tingling in his mouth and lips followed by dizziness, fatigue, headache, a constricting feeling in his throat, difficulty speaking, tightness in his upper chest, facial flushing, shaking, nausea, and vomiting. His legs weakened, and he collapsed. On examination in the ED, his blood pressure was 150/90 mmHg; heart rate, 117 beats per minute; respiratory rate, 22 per minute; temperature, 99.3 F (37.4 C); and oxygen saturation, 99% on room air.

Case 2. A 32-year-old man ate three bites of fugu (approximately 1 1/2 oz) over 2-3 minutes. While eating his third bite, he noticed tingling in his tongue and right side of his mouth followed by a "light feeling," anxiety, and "thoughts of dying." He felt weak and collapsed. At the ED, his blood pressure was 167/125 mmHg; heart rate, 112 beats per minute; respiratory rate, 20 per minute; and oxygen saturation, 96% on room air.

Case 3. A 39-year-old man ate approximately 1/4 oz of fugu after eating a full meal. Approximately 20 minutes after eating the fugu, he had onset of dizziness and mild chest tightness. At the ED, his blood pressure was 129/75 mmHg; heart rate, 84 beats per minute; respiratory rate, 22 per minute; temperature, 97.2 F (36.2 C); and oxygen saturation, 97% on room air. Diagnosis and Treatment

A presumptive diagnosis of tetrodotoxin poisoning in all three men was based on clinical presentation in the ED and the history of recent consumption of fugu. All were treated with intravenous hydration, gastric lavage, and activated charcoal. Symptoms gradually resolved, and the men were discharged the following day with no residual symptoms. Follow-Up Investigation

The chef who brought the fugu from Japan failed to declare this item through customs. The remaining fugu was obtained for toxin analysis at FDA. SDEH contacted health authorities in Japan and relayed the product label information for identification of the product manufacturer to assist in their local follow-up investigation.

Reported by: P Tanner, San Diego Dept of Environmental Health; G Przekwas, R Clark, MD, San Diego Regional Poison Center, Univ of California at San Diego Medical Center; M Ginsberg, MD, San Diego County Health Dept; S Waterman, MD, State Epidemiologist, California Dept of Health Svcs. Food and Drug Administration. Div of Environmental Hazards and Health Effects, National Center for Environmental Health; Div of Field Epidemiology, Epidemiology Program Office, CDC.

Editorial Note: The order Tetraodontoidea includes ocean sunfishes, porcupine fishes, and fugu, which are among the most poisonous of all marine life (1). These species inhabit the shallow waters of the temperate and tropical zones and can be exported from China, Japan, Mexico, the Philippines, and Taiwan. The liver, gonads, intestines, and skin of these fish contain tetrodotoxin, a powerful neurotoxin that can cause death in approximately 60% of persons who ingest it (2). Other animals (e.g., California newt and the eastern salamander) also possess tetrodotoxin in lethal quantities (3) (Table_1).

Tetrodotoxin is heat-stable and blocks sodium conductance and neuronal transmission in skeletal muscles. Paresthesias begin 10-45 minutes after ingestion, usually as tingling of the tongue and inner surface of the mouth. Other common symptoms include vomiting, lightheadedness, dizziness, feelings of doom, and weakness. An ascending paralysis develops, and death can occur within 6-24 hours, secondary to respiratory muscle paralysis. Other manifestations include salivation, muscle twitching, diaphoresis, pleuritic chest pain, dysphagia, aphonia, and convulsions. Severe poisoning is indicated by hypotension, bradycardia, depressed corneal reflexes, and fixed dilated pupils. Diagnosis is based on clinical symptoms and a history of ingestion. Treatment is supportive, and there is no specific antitoxin (6). Despite the high death rate associated with tetrodotoxin poisoning, the three persons described in this report survived probably because of the small amount of toxin ingested and rapid stomach evacuation by the ED.

Although personal importation of fugu into the United States is prohibited, FDA has permitted fugu to be imported and served in Japanese restaurants by certified fugu chefs on special occasions. A cooperative agreement with the Japanese Ministry of Health and Welfare ensures fugu is properly processed and certified safe for consumption before export by the government of Japan. If cleaned and dressed properly, the fugu flesh or musculature is edible and considered a delicacy by some persons in Japan, who may pay the equivalent of $400 U.S. for one meal. Despite careful preparation, fugu remains a common cause of fatal food poisoning in Japan, accounting for approximately 50 deaths annually (7).

Although arriving travelers are required to declare all food products brought into the United States, control measures rely primarily on the traveler. Other foodborne outbreaks in the United States have occurred after consumption of illegally imported food products (8). Persons who travel to countries where fugu is served should be aware of the potential risk of eating this fish.

References
1. Halstead BW. Dangerous marine animals: that bite-sting-shock-are non-edible. Cambridge, Maryland: Cornell Maritime Press, 1959.
2. Ellenhorn MJ, Barceloux DG. Medical toxicology: diagnosis and treatment of human poisoning. New York: Elsevier Science Publishing Company, Inc., 1988.
3. Bradley SG, Kilka LJ. A fatal poisoning from the Oregon roughskinned newt (Taricha granulosa). JAMA 1981;246:247.
4. Kim S. Food poisoning: fish and shellfish. In: Olson KR, ed. Poisoning and drug overdose. 2nd ed. Norwalk, Connecticut: Appleton and Lange, 1994.
5. Gellert GA, Ralls J, Brown C, Huston J, Merryman R. Scombroid fish poisoning: underreporting and prevention among noncommercial recreational fishers. West J Med 1992;157:645-7.
6. Anonymous. Poisindex toxicologic managements. Vol 88: tetrodotoxin. Englewood, Colorado: Micromedex, Inc., 1974-1996.
7. Torda TA, Sinclair E, Ulyatt DB. Puffer fish (tetrodotoxin) poisoning: clinical record and suggested management. Med J Aust 1973;1:599-602.
8. CDC. Cholera associated with food transported from El Salvador -- Indiana, 1994. MMWR 1995; 44:385-6.

June 06, 1997 / 46(22);489-492

Amanita phalloides Mushroom Poisoning -- Northern California, January 1997

The popular interest in gathering and eating uncultivated mushrooms has been associated with an increase in incidents of serious mushroom-related poisonings (1). From December 28, 1996, through January 6, 1997, nine persons in northern California required hospitalization after eating Amanita phalloides (i.e., "death cap") mushrooms; two of these persons died. Risks associated with eating these mushrooms result from a potent hepatotoxin. This report describes four cases of A. phalloides poisoning in patients admitted to a regional referral hospital in northern California during January 1997 and underscores that wild mushrooms should not be eaten unless identified as nonpoisonous by a mushroom expert. Case 1. A 32-year-old man gathered and ate wild mushrooms that he believed were similar to other mushrooms he had previously gathered and eaten. Eight hours later, he developed vomiting and profuse diarrhea; he was admitted to a hospital 19 hours after ingestion. On admission, he was dehydrated, and laboratory findings included an aspartate aminotransferase (AST) level of 81 U/L (normal: 0-48 U/L), prothrombin time (PT) of 12.3 seconds (normal: 11.0-12.8 seconds), and bilirubin level of 0.9 mg/dL (normal: 0-0.3 mg/dL). He received intravenous fluids, intravenous penicillin, repeated oral doses of activated charcoal, and oral N-acetylcysteine. Although the diarrhea resolved after 24 hours, his PT and AST and bilirubin levels continued to rise. On the third day after eating the mushrooms, abnormal findings included an AST level of 2400 U/L, alanine aminotransferase (ALT) level of 4100 U/L (normal: 0-53 U/L), PT of greater than 60 seconds, and total bilirubin level of 11 mg/dL. Six days after eating the mushrooms, his bilirubin level was 16 mg/dL, and his AST level had decreased to 355 U/L; he developed metabolic acidosis and hypotension. Seven days after eating the mushrooms, he developed hepatic encephalopathy, oliguric renal failure, and adult respiratory distress syndrome requiring intubation and mechanical ventilation. He died from multiple organ failure 9 days after eating the mushrooms. One mushroom cap remaining after the meal was identified as A. phalloides. Case 2. A 42-year-old man developed vomiting and diarrhea 11 hours after eating wild mushrooms, and he was admitted to a hospital 14 hours after eating the mushrooms. His transaminase levels were elevated 24 hours after ingestion (AST and ALT levels both at 100 U/L); his PT was 12.1 seconds, and his bilirubin level was 0.2 mg/dL. His PT became prolonged the next day and peaked at 35 seconds on the fourth day. His transaminase levels also peaked on the fourth day (AST level of 3000 U/L and ALT level of 6000 U/L); his bilirubin level was 7.8 mg/dL. He was given repeated doses of activated charcoal and oral N-acetylcysteine. His transaminase levels and PT gradually decreased, and he was discharged on the seventh day after eating the mushrooms without sequelae. Case 3. A 30-year-old man used a guidebook to assist in the collection of wild mushrooms. Twelve hours after eating the mushrooms he had gathered, he developed vomiting and severe diarrhea. He was admitted to a hospital 17 hours after ingestion because of orthostatic hypotension and dehydration. Abnormal laboratory findings indicated an AST level of 75 U/L, blood urea nitrogen level of 22 mg/dL (normal: 6-20 mg/dL), and creatinine level of 2.8 mg/dL (normal: 0.6-1.3 mg/dL). He was treated with intravenous fluids. Although renal function indicators were within normal limits 1 day after admission, his liver enzyme and PT levels began to increase; on the fourth day, transaminase levels peaked (AST level of 1900 U/L and ALT level of 2800 U/L), total bilirubin level was 1.6 mg/dL, and PT was 18 seconds. His clinical status continued to improve, and he was discharged 7 days after eating the mushrooms. Case 4. A 68-year-old man ate mushrooms he had collected on a golf course. Two days after eating the mushrooms, he was admitted to a hospital because of diarrhea and weakness. His AST level was 630 U/L, and he had renal failure. On the third day after eating the mushrooms he required hemodialysis, and his transaminase levels and his PT continued to increase; on the fifth day, his AST level was 3500 U/L; ALT level, 4600 U/L; PT, 34 seconds; and bilirubin, 9.7 mg/dL. He developed hepatic encephalopathy and died 6 days after eating the mushrooms.

Reported by: S Zevin, MD, D Dempsey, MD, K Olson MD, California Poison Control System, Div of Clinical Pharmacology and Experimental Therapeutics, Univ of California, San Francisco. Environmental Hazards Epidemiology Section, Health Studies Br, Div of Environmental Hazards and Health Effects, National Center for Environmental Health, CDC.

Editorial Note: Ingestion of A. phalloides may account for approximately 90% of deaths attributable to mushroom ingestion worldwide (1-5); the proportion of cases of mushroom poisoning attributable to A. phalloides in the United States is unknown. In the United States, this species is found primarily in the cool coastal regions of the west coast, but it also grows in several other regions, including the mid-Atlantic coast and in the northeast (1,2). These mushrooms flourish in favorable weather conditions during the fall or the rainy season (2,6). The mature cap usually is metallic green but varies from light yellow to greenish-brown (1-3). A. phalloides, like most mushroom species, is not unique in appearance and can be mistaken for nonpoisonous species; it has no distinct taste or smell, and the toxins are not destroyed by cooking or drying (3,5,6). The principal toxins (amatoxins) are taken up by hepatocytes and interfere with messenger RNA synthesis, suppressing protein synthesis and resulting in severe acute hepatitis and possible liver failure. Radioimmunoassay of amatoxins can be obtained from serum and urine; the tests are performed at referral laboratories (1,2).

Since 1979, A. phalloides has been found in the region from northern California to Washington state, and since 1995, it has appeared in greater numbers because of abundant rainfall during winter months. During the winter of 1995-96, at least 13 persons in northern California were hospitalized for treatment of poisonings after eating A. phalloides; one patient died, and another required a liver transplant. The cluster of mushroom poisoning in northern California described in this report probably occurred because warm, heavy rainfall created optimal conditions for the growth of A. phalloides in unprecedented numbers. In addition, this mushroom grew in places where it had not grown before (e.g., backyards), which increased the likelihood that persons gathering these mushrooms could mistake them for a nonpoisonous species.

Patients may not associate their symptoms with ingestion of wild mushrooms because of the delayed onset. As illustrated by the cases described in this report, symptoms typically occur in a progression through three stages. During the first stage, which occurs 6-24 hours after ingestion, symptoms may include abdominal pain, nausea, vomiting, severe diarrhea, fever, tachycardia, hyperglycemia, hypotension, and electrolyte imbalance. During the second stage, which occurs during the next 24-48 hours, symptoms appear to abate even as hepatic and renal functions deteriorate. During the third stage, which occurs 3-5 days after the ingestion, hepatocellular damage and renal failure may progress, resulting in jaundice and hepatic coma (1-5). Possible sequelae include cardiomyopathy, coagulopathy, and seizures (1,2,5). Death from A. phalloides poisoning usually results from hepatic and/or renal failure and may occur 4-9 days after ingestion. Fatal outcomes are associated with age less than 10 years, a short latency between ingestion and onset of symptoms, and severe coagulopathy (1,4). The fatality rate among persons treated for A. phalloides poisoning is 20%-30% (1,2,4), and the median lethal dose is 0.1 mg to 0.3 mg of the toxin per kg of body weight (1,5).

A. phalloides poisoning has no specific antidote. The main treatment is vigorous intravenous fluid replacement and correction of electrolyte disturbances (1-5); correction of coagulopathy, if present, also may be indicated. Physicians should perform gastric lavage and administer repeated doses of activated charcoal to remove any unabsorbed Amanita and to interrupt the enterohepatic circulation of the toxin (2,4,5). Although some therapeutic regimens have included the administration of penicillin, cimetidine, silibinin, or N-acetylcysteine, these treatments have not been confirmed by clinical trials to be effective. Hemodialysis and hemoperfusion may be effective in removing the toxin if initiated within 24 hours of ingestion (7). The only definitive treatment may be liver transplantation once fulminant liver failure occurs (1,2,4).

Unintentional ingestion of A. phalloides can be prevented by ensuring that wild mushrooms are not eaten unless identified as nonpoisonous by a competent mycologist. Education campaigns should be established in areas where A. phalloides is common to educate the public about the potentially lethal consequences associated with eating uncultivated mushrooms. Field guides do not provide sufficient details to differentiate toxic from nontoxic species. Health-care providers should report cases of mushroom poisoning to poison-control centers; these centers can provide expertise in the clinical management of mushroom poisoning.

References
1. Bryson PD. Mushrooms. In: Bryson PD. Comprehensive review in toxicology for emergency clinicians. 3rd ed. Washington, DC: Taylor and Francis, 1996:685-93.
2. Klein AS, Hart J, Brems JJ, Goldstein L, Lewin K, Busuttil RW. Amanita poisoning: treatment and the role of liver transplantation. Am J Med 1989;86:187-93.
3. Lampe KF, McCann MA. AMA handbook of poisonous and injurious plants. Chicago, Illinois: American Medical Association, 1985.
4. Pinson CW, Daya MR, Benner KG, et al. Liver transplantation for severe Amanita phalloides mushroom poisoning. Am J Surg 1990;159:493-9.
5. Koppel C. Clinical symptomatology and management of mushroom poisoning. Toxicon 1993; 31:1513-40.
6. Nicholls DW, Hyne BE, Buchanan P. Death cap mushroom poisoning {Letter}. N Z Med J 1995; 108:234.
7. Feinfeld DA, Mofenson HC, Caraccio T, Kee M. Poisoning by amatoxin-containing mushrooms in suburban New York -- report of four cases. Clin Toxicol 1994;32;715-21.

June 04, 1982 / 31(21);287-8

Mushroom Poisoning among Laotian Refugees -- 1981

In the period, December 1-3, 1981, 7 Laotian refugees were seen at a Sonoma County California hospital for apparent mushroom poisoning; 6 had nausea, vomiting, diarrhea, dehydration, and elevated liver enzymes. All persons had eaten mushrooms that were gathered and eaten on November 30, although 1 week earlier 20-30 Laotians had eaten mushrooms gathered in the same area without incident. The incubation period was variable, but most patients experienced gastrointestinal distress within 8 hours. Three persons were treated in the intensive care unit, but all recovered and were discharged within 7 days.

Several remaining cooked mushrooms were examined at Sonoma State University; all but 1 were identified as Russula species. The remaining specimen could not be identified, probably because of cooking. The mushrooms examined may not have been representative of those actually consumed.

Laotians customarily gather wild mushrooms in their homeland and attempt to identify poisonous species by boiling the mushrooms with rice; if the rice turns red, the mushrooms are deemed poisonous. Because the Sonoma County mushrooms did not cause a color reaction, it was assumed they were safe to eat. Reported by K Rattanvilay, N Tavares, G Eliaser, MD, G Hands, MD, S Boynton, Community Hospital, Santa Rosa; J Young, PHN, R Holtzer, MD, Sonoma County Health Dept, TG Tong, PharmD, San Francisco Poison Control Center, RR Roberto, MD, SB Werner, MD, Infectious Disease Section, California Dept of Health Svcs; Enteric Bacteriology and Epidemiology Branch, CDC.

Editorial Note: Mushroom poisoning can be produced by about 100 of the 2,000 species known. In the United States, mushrooms of the genera Amanita and Galerina produce amanitins and phallotoxins, which are common causes of mushroom poisoning. The most feared fungi are those that produce amanitin, which include the "deathcap" Amanita phalloides. A. phalloides has become increasingly common in the San Francisco bay region in recent years. The odor of fresh A. phalloides is similar to raw potatoes. Symptoms generally begin 6-24 hours after ingestion and may include the explosive onset of violent abdominal pain, vomiting, diarrhea, hematuria, fever, tachycardia, hypotension, rapid volume depletion, fluid and electrolyte imbalance, and extreme thirst. After a short phase of improvement, hepatic, renal, and central nervous system damage may ensue. The mortality rate is 50%, and those who recover do so slowly. Treatment is supportive, and thioctic acid, charcoal hemoperfusion, and vitamin C may be useful (1,2).

Other genera of mushrooms, including the Russula genus implicated in the California outbreak, produce less lethal toxins. The Russula toxin has not been identified, but it results in a shorter incubation period--1-2 hours--followed by minor gastrointestinal and parasympathetic symptoms and hallucinations (1). Russula emetica can produce additional toxins, including muscarine.

Most cases of mushroom poisoning occur in late summer and early fall. Early abundant rains and mild temperatures in northern California have produced a profusion of mushrooms, some of which are poisonous. Nontoxic mushrooms may grow in the same area with toxic ones, and even trained mycologists may confuse toxic varieties with edible ones because of the extensive variations among species. There are no simple tests to identify poisonous mushrooms and no safe ways to detoxify the poisonous varieties.

Identification of implicated mushrooms may be difficult if specimens have been prepared and cooked. Since a variety of mushrooms may have been ingested in most poisoning situations, reliance cannot be placed on the initial symptoms. Gastric contents, stool, and mushroom samples may be assayed for toxins by radioimmunoassay (3).

The San Francisco Poison Control Center recommends routine administration of ipecac after ingestion of any wild mushrooms of questionable identification. The Infectious Disease Section offers its assistance in suspected mushroom poisoning, and information is available from local mycological societies, colleges, and poison control centers.

In the last 5 years, 16 outbreaks involving 44 cases of mushroom poisoning were reported to CDC; 23 cases were from California. In 1981 in California, 1 death in Santa Cruz County and 2 in Marin County were attributed to mushroom poisoning.

References
1. Becker CE, Tong TE, Boerner U, Roe RL, Scott RAT, MacQuarrie MB. Diagnosis and treatment of Amanita phalloides-type mushroom poisoning. West J Med 1976;125:100-9.
2. Wauters JP, Rossel C, Farquet JJ. Amanita phalloides poisoning treated by early charcoal haemoperfusion. Brit Med J 1978;2:1465.
3. Mushroom poisoning. Lancet 1980;2:351-2.

INDEX

INDEX

Intern Pocket Survival Guides

The Intern Pocket Survival Guide	$7.50
The CCU Intern Pocket Survival Guide	$7.50
The ER Intern Pocket Survival Guide	$7.50
The ICU Intern Pocket Survival Guide	$7.50
The Surgical Intern Pocket Survival Guide	$7.50
The Oncology Intern Pocket Survival Guide	$7.50
The OB/GYN Intern Pocket Survival Guide	$9.95
The Pocket Guide to Eponyms and Subtle Signs of Disease	$7.50
The Intern Pocket Admission Book *contains blank forms*	$4.00

Clinical Practice Handbooks

Pocketful of Prevention	$7.50
Locatelli & Singh's Handbook of Neurology	$19.95
The Ophthalmology Resident Pocket Survival Guide	$9.95
The Rehab Pocket Survival Guide	$9.95
The Gastroenterology Resident Pocket Survival Guide	$9.95
The EKG Pocket Survival Guide	$6.00
The ACLS Pocket Survival Guide	$6.00
The PALS Pocket Survival Guide	$6.00
Medical Management of the Chemical Casualties Handbook	$9.95
Medical Management of the Radiological Casualties Handbook	$9.95
USAMRIID's Medical Management of the Biological Casualties Handbook	$9.95
Antibiotic Prophylaxis	$6.00

U.S. Public Health Service and other federal agencies

Clinician's Handbook of Preventative 2nd ed.	$20.00
Guide to Clinical Preventative Services 2nd and 3rd ed.	$24.00
Healthy People 2010	$49.50
Physical Activity and Health	$17.00

In addition to your address, please include an e-mail or phone number so that we may contact you with any questions about your order.

Name: _____

Address: _____

City, State, Zip: _____

Daytime phone/e-mail: _____

Please send a Personal Check or Credit Card (circle one):

      Visa   Mastercard

Card Number_____ Expiration _____

Signature _____

Title	Qty	Price	Total
Bad Bug Book		$21.95	
		*Tax (see below)	
		**S/H (see below)	
		Total	

*Sales Tax 5% in MD, 4.5% in VA, and 6% in WV.
**Shipping charges are $5.95 for the first book and $0.50 (50 cents) for each additional book. Please call for other rates.
Please send your order to:
      International Medical Publishing, Inc.
      500 Monocacy Blvd.
      Frederick, MD 21701

Information and ordering: (800) 530-4146   Fax: (301) 695-3632
E-mail inquiries to: orders@medicalpublishing.com
online: www.medicalpublishing.com